PROGRESS IN COLLOID & POLYMER SCIENCE

Editors: F. Kremer (Leipzig) and G. Lagaly (Kiel)

Volume 110 (1998)

Trends in Colloid and Interface Science XII

Guest Editors:

G. J. M. Koper (Leiden), D. Bedeaux (Leiden)
W. F. C. Sager (Twente), C. Cavaco (Leiden)

Springer

ISBN 978-3-662-16062-6
ISSN 0340-255 X

Die Deutsche Bibliothek –
CIP-Einheitsaufnahme

Progress in colloid & polymer science. –

Früher Schriftenreihe
Reihe Progress in colloid & polymer
science zu: Colloid & polymer science
ISSN 0340-255X

Trends in colloid and interface science … :
… European Colloid and Interface Society
(ECIS) conference …
(Progress in colloid & polymer science ;
Vol. 110)
ISBN 978-3-662-16062-6
ISBN 978-3-7985-1653-3 (eBook)
DOI 10.1007/978-3-7985-1653-3

© 1998 by Springer-Verlag Berlin Heidelberg
Originally published by Dr. Dietrich Steinkopff Verlag GmbH & Co. KG, Darmstadt in 1998
Softcover reprint of the hardcover 1st edition 1998

Chemistry Editor:
Dr. Maria Magdalene Nabbe;
Production: Holger Frey, Ajit Vaidya.

Typesetting and Copy-Editing:
Macmillan Ltd., Bangalore, India

Prog Colloid Polym Sci (1998) V
© Steinkopff Verlag 1998

PREFACE

The 11th Conference of the European Colloid and Interface Society (ECIS 97) was held September 14–19, 1997 in the Congress Center *De Blije Werelt*, Lunteren, The Netherlands. The conference was attended by 205 scientists from 24 different countries. The scientific program contained 50 oral and 142 poster contributions and covered theoretical, experimental and technical aspects of modern colloid and interface science. This volume contains a selection of the contributions presented at the conference and is divided into the following sections:
– New topics in colloid science,
– Polymer colloids,
– Surfactant colloids,
– Polymers and surfactants at interfaces,
– Rheology.
In the Overview (page XXX) various contributions to the conference are discussed. During the conference three poster contributions were selected that, according to the Organizing Committee, were of outstanding quality. The contributors, H. Behrens (see page XXX), T. Iwanaga (see page XXX), and K. Marinova (see page XXX), received the Poster Prize in the form of a book of choice from the exhibition at the conference site.

The Organizing Committee wishes to thank all participants for their scientific contributions which resulted in a very successful conference. We are especially grateful to the members of the Scientific Committee:

Peter Schurtenberger, ETH Zürich, Switzerland
Otto Glatter, Universität Graz, Austria
Dominique Langevin, Centre de Récherche Paul Pascal, France
Thomas Zemb, CEN Saclay, France
Vittorio Degiorgio, Università di Pavia, Italy
Adrian Rennie, University of Cambridge, United Kingdom
Bob Thomas, University of Oxford, United Kingdom
Hakan Wennerström, University of Lund, Sweden
Martien Cohen Stuart, Agricultural University, The Netherlands
Jaap Leyte, University of Leiden, The Netherlands
Jorrit Mellema, University of Twente, The Netherlands
Grégoire Porte, Université Montpellier II, France

who helped us in the difficult task of selecting the contributions for oral presentations.

We gratefully acknowledge the financial support from the European Colloid and Interface Society, the Royal Academy of Sciences, the Leiden Institute of Chemistry, the Faculty of Mathematics and Science of the Leiden University, Shell Nederland, DSM Research, Akzo Nobel, Unilever, and the Institut für Terrestische Ökologie of the Eidgenössische Technische Hochschule in Zürich. Their support enabled us to give 13 grants covering the conference fee and accommodation to young researchers and scientists from Eastern European countries.

On behalf of the Organizing Committee:
Ger J. M. Koper, Leiden Institute of Chemistry, The Netherlands, chair
Dick Bedeaux, Leiden Institute of Chemistry, The Netherlands
Wiebke F. C. Sager, University of Twente, The Netherlands
Carolina Cavaco, Leiden Institute of Chemistry, The Netherlands

Prog Colloid Polym Sci (1998) VI
© Steinkopff Verlag 1998

Wiebke F. C. Sager
University of Twente

ECIS 97 – an overview

This volume contains contributions presented at the 11th Conference of the European Colloid and Interface Society *(ECIS 97)* held in the Congress Center De Blije Werelt, Lunteren, The Netherlands, on September 14–19, 1997. Topics covered at the conference were: *New topics in colloid science, Rheology, Surfactant colloids, Polymer colloids, Surfactants at interfaces, and Polymers at interfaces.* Each topic formed a separate section of the conference comprising a 45 minute key note lecture (KL), a 35 minute invited lecture (IL), five or six 15 minute contributed talks, and a dedicated part of the poster session. In the evenings, informal scientific lectures were given by leading Dutch scientists related to the field. These proceedings comprise 40 % of the oral and 25 % of poster contributions presented at the conference. A somewhat larger number of papers were submitted for the topics *New topics in colloid science and Polymer colloids.* The contributions are ordered according to the conference topics of which *Surfactants at interfaces and Polymers at interfaces* have been merged into one section. In the following, a brief overview of selected talks which are not published in this volume is given. In order to facilitate the access to recent publications for interested readers we include a reference list of the subjects presented. The conference started with the topic:

New topics in colloid science
Chairman: Otto Glatter, Universität Graz

KL   Mode selective optical receivers: Application to colloidal systems. J. Ricka, University of Berne [1, 2, 3]

IL   Quantitative real-space analysis of colloidal structures and processes. A. van Blaaderen, Utrecht University [4, 5, 6]

Ricka started his lecture by discussing the advantages of replacing the classical pair of pinholes in optical devices by a single mode fiber, thus, selecting a single mode [1, 2, 3]. After introducing the principles of optical single mode receivers, he focused on two techniques whose performance improved due to the superior sensitivity and simplicity of single mode fiber set-ups, namely, dynamic light scattering and confocal scanning microscopy. In the last part of his talk, he presented new results on an interesting coil-globule transition occurring in aqueous solutions of the mixed polymer/surfactant system PNIPAM (poly(N-isopropylacrylamide))/SDS. Phenomena observed upon addition of SDS to a solution of the thermosensitive polymers, which collapse from the coil state into the globular state upon increasing temperature, are more complex than expected from polymer/surfactant couples studied previously. Below the coil-globule transition temperature polymer-bound SDS micelles that incorporate the polymer backbone as well as quasi-free micelles form, depending on the SDS concentration. In the course of the temperature-induced coil-globule transition of the macromolecules, the mixed polymer surfactant aggregates undergo a profound restructuring. The surfactant remains firmly associated with the polymer and above the transition temperature the polymeric globules seem to be coated with a surfactant layer, stabilizing the globules against aggregation observed in SDS free systems.

Van Blaaderen showed how fluorescence confocal scanning light microscopy can be used to obtain 3D coordinates in optically matched dispersions of, e.g., fluorescently labeled core-shell particles [4, 5, 6]. Examples were given for real-space structures of hard-sphere glasses, crystals grown by colloidal epitaxy, electrorheological fluids, and binary structures. At the end of the lecture he demonstrated that so far fast temporal information can only be obtained in a 2D plane albeit in the bulk of a dispersion.

Rheology

Chairman: J. Mellema, University of Twente

KL   The rheology of strongly interacting colloidal suspensions: Experimental elucidation of reversible shear thickening. N. Wagner, University of Delaware, Newark [7, 8, 9]

IL   The role of disorder in emulsion rheology. F. Lequeux, Université Louis Pasteur, Strasbourg [10, 11]

Among a variety of interesting phenomena observed for concentrated suspensions under shear, Wagner discussed the reversible shear thickening transition in the context of inter-particle interaction and the underlying microstructure on a colloidal level [7, 8, 9]. He demonstrated that combining rheological measurements with optical and neutron scattering is a versatile approach to reveal the mechanism of shear thickening and its dependence on the type of colloidal stabilization employed.

In the invited lecture, Lequeux demonstrated that quenched disorder can explain complex phenomena observed in the rheological behavior of concentrated oil-in-water emulsion [10, 11]. The non-linearity and the frequency dependence of the loss modulus cannot be understood in terms of the classical Princen models which, over the last 2 decades, have been successfully applied to describe yield stress and the elastic modulus. The mechanical response of an emulsion can, in fact, be divided in at least three domains: The first regime is characterized by very small strains and a linear response which can be related to thermal fluctuations. In the second regime, for intermediate strain values below the yield stress, the response is non-linear, but the elastic modulus is apparently linear. Above the yield stress, in the third regime, the system flows with fracture. To demonstrate the origin of non-linearities a new technique, the position echo technique, was developed which allows one to measure the motion of the oil droplets in a periodic macroscopic flow. The results demonstrate that in the second and third mechanical regime, only some of the droplets contribute to the irreversibility of the flow.

Surfactant colloids

Chairman: J. C. Earnshaw, The Queens University of Belfast

KL   Structure and phase equilibria of microemulsions. U. Olsson, University of Lund [12, 13, 14]

IL   Recent advantages in cryogenic transmission electron microscopy. D. Danino, Israel Institute of Technology, Haifa [15]

Olsson first reviewed state-of-the-art knowledge on the stability and the phase behavior of microemulsions introducing the Helfrich free energy concept and emphasized the crucial role of the spontaneous curvature, H_o, of the surfactant monolayer [12, 13, 14]. The invariance of the curvature energy with respect to the transformation $H_o \rightarrow -H_o$ and ϕ_o $(1-\phi_o)$, ϕ_o being the volume fraction of oil, implies a symmetric phase diagram, as observed with nonionic surfactants of the C_iE_j type, for which H_o is found to be a strong and approximately linear function of temperature. At constant ϕ_s/ϕ_o, ϕ_s being the surfactant volume fraction, plates, cylinders, spheres, and spheres plus excess oil form when H_o deviates further from zero. He focused in particular on features of spherical o/w droplets and the L_3 or sponge phase. For the droplet phase, characterized over a large volume fraction essentially by hard sphere interactions, new osmotic pressure measurements were presented for the ternary system as well as for droplets charged up by the addition of SDS. While the binodal corresponds to $\Pi = 0$, Π being the osmotic pressure, it might be possible to identify the spinodal from $\partial\Pi/\partial\phi = 0$ inside the coexistence region. When H_o is near zero, the phase diagram reveals also an L_3 phase, which joins up with the bicontinuous microemulsion upon bilayer swelling and may crystallize into a cubic phase at higher surfactant concentrations. The L_3 phase has a finite swelling and exists with excess solvent.

De Vries (Delft University of Technology) gave an interesting presentation on undulations in lamellar charged fluid membranes [16, 17]. He first introduced a theoretical account for the interplay between electrostatic interactions and thermal undulations, both in the presence and absence of added electrolyte. In the second part of his talk he showed that this theoretical description can quantitatively explain backscattering experiments obtained on dilute lamellar $C_{12}E_5$ phases, which were charged up by the addition of small amounts of SDS.

Pedersen (Risø National Laboratory) presented Monte Carlo simulation studies on semidilute solutions of semiflexible polymers and worm-like micelles [18, 19]. The simulations were performed to calculate the scattering functions for the full system as well as the single chains including excluded-volume effects.

Appell (Université Montpellier II) presented results of a small angle neutron scattering study of oil swollen micelles to which hydrophobically endgrafted hydrosoluble polymers were added. Successive addition of polymers grafted at only one end induces an increasing repulsive interaction between the micelles leading to the occurrence of a pronounced peak in the scattering curves. For double sided grafted polymers attenuation of the introduced repulsion was explained by bridging occurring between the micelles.

In the invited lecture Danino showed recent advances in cryogenic transmission electron microscopy [15]. Special attention was given to the vitrification of viscous and oil-continuous phases which allows to follow directly microstructural

Prog Colloid Polym Sci (1998)
© Steinkopff Verlag 1998

changes induced by, e.g., temperature or concentration changes.

Polymer colloids

Chairman: V. Degiorgio, Università di Pavia

KL Charge, structure, and stability of aqueous polymer-colloid suspensions. M. Borkovec, ETH Zürich [20, 21, 22]

IL Light scattering studies of adsorption of surfactant and polymers on colloids. R. Piazza, Università di Pavia (see page 89)

After reviewing the different mechanisms that can lead to the build-up of charges on macromolecules and colloidal particles, Borkovec showed how these concepts can be applied to the understanding of electrostatic forces, which determine the structure and aggregation behavior of colloidal particle suspensions [20, 21, 22]. In the low salt region the properties of the stable suspensions are mainly determined by effective charges derived from the far limit of the electrostatic potentials. The suspensions thus display fluid-like structures, aggregation processes are slow and colloidal crystals may form. At higher salt concentrations the suspensions become unstable, aggregation processes are fast and lead initially to the formation of doublets and at the later stages to larger clusters, and ultimately to gelation.

Surfactants at interfaces

Chairman: U. Olsson, University of Lund

KL Forces between surfactant monolayers adsorbed at oil-water interfaces. P. D. I. Fletcher, University of Hull [23, 24]

IL Light scattering from surfactants at interfaces. J. C. Earnshaw, The Queens University of Belfast [25, 26, 27, 28]

Fletcher reported the first results obtained with a newly developed liquid surface force apparatus, which allows direct determination of the interaction force exerted, the film radius, and thickness, when a micrometer sized oil droplet coated with a surfactant monolayer approaches a surfactant monolayer adsorbed at an oil/water interface [23, 24]. The measurements also yield results for the disjoining pressure as a function of separation between the monolayers interacting across the thin oil-water-oil emulsion film formed when the apex of the oil drop is close to the oil/water interface. Results for a range of ionic and nonionic surfactant monolayers showing repulsive interactions were presented. The films formed have a film thickness greater than approximately 15 nm. The variation of disjoining pressure with film thickness is found to be in accordance with electrostatic theory. In addition SDS monolayers have been studied at various salt concentrations to modify the attractive forces between the monolayers. Such monolayers show a short range adhesive interaction and exhibit a strong hysteresis in their force-distance behavior as the surfaces are pushed together and pulled apart indicating that additional forces are required to unstick the oil droplet from the interface. Measurements of forces between surfaces at small separation is of central importance in attempting to gain a fundamental understanding of many aspects of the complex behavior of colloidal systems, such as factors controlling emulsion stability.

Treiner (Université Pierre et Marie Curie, Paris) presented adsorption isotherms at a silica/water interface of nonionic surfactants of the alkylpolyoxyethylated series for pure compounds as well as polydispersed commercial detergents [29, 30]. At low surface coverage polydispersity effects in the alkyl chain length or the ethoxy group number are negligible and the oligomers behave essentially as pure, single surfactants. At higher surface coverage differences in the tendency to form large structures, disks or patched bilayers, which decreases with increasing length of the apolar part, are shown to be cancelled out in the presence of oligomers with longer and shorter alkyl chain lengths.

The question whether first order phase transitions can occur in soluble monolayers at water/air interfaces has been addressed by Lunkenheimer (Max Planck Institute for Colloid and Interface Science, Berlin) who gave a presentation on transitional behavior and phase transitions in soluble adsorption layers [26, 31, 32]. Adsorption of soluble amphiphiles at water/air interfaces is normally characterized by a continuous transition region connecting at least two distinct surface states attributed to the different surface configurations of the amphiphiles. Recently, strong water evaporation retardation and a constant orientation of the amphiphilic layer above the transition concentration has been observed for a still soluble system indicating that even for soluble monolayers surface interactions might become strong enough to meet the conditions for phase separation following the theory of regular surface behavior.

Riegler (Max Planck Institute for Colloid and Interface Science, Berlin) presented a comparative study on the relation between film topology and molecular ordering for rodlike monomers with amphiphilic and purely hydrophobic character at solid/vapor interfaces [33, 34]. Fatty acids as well as alkanes show, at the solid/vapor interface, similar equilibrium film topologies despite their different characters. Below the bulk melting temperature a monolayer forms on the substrate surface which is covered by mesalike islands or a closed film, depending on the surplus concentration. For fatty acids and alkanes with chain length between 14 and 50 C-atoms, small droplets form out of the surplus material above the melting point. Whereas alkanes with shorter chain lengths or at much higher temperatures show nearly complete wetting.

In the invited lecture Earnshaw demonstrated that the scattering of light by thermally excited capillary waves, which has in the past been successfully applied to study liquid surfaces of pure fluids and spread monolayers, can reveal new phenomena in surfactant solutions [25, 26, 27, 28]. These include demonstrating the existence of transitional effects in adsorbed films associated with the changes in the dynamics of molecular orientation and the subsequent observation of viscoelastic relaxation of the dilatational modulus of such films. Anomalous behavior of the observed capillary waves suggests an electrostatic adsorption barrier.

Polymers at interfaces

Chairman:

M. Cohen Stuart, Agricultural University Wageningen

KL   Layer formation and exchange kinetics of adsorbed polymer chains. A. Chakrabarti, Kansas State University (see page 291)

IL   Properties of bridging polymer chains. T. van de Ven, McGill University, Montréal [35, 36, 37]

The last lecture of the conference was presented by Van de Ven, who gave an overview on possible mechanisms of polymer bridging [35, 36, 37]. A prerequisite for particle flocculation by polymer bridging is that the adsorbing polymer coats the particles only partly, leaving enough room on the particle surface for bridging to occur. Even particles onto which the polymer does not adsorb can be bridged in the presence of particles onto which the polymer adsorbs, leading to heteroflocculation due to asymmetric polymer bridging. In association induced polymer bridging, non-adsorbing polymers can be turned into adsorbing molecules which are able to bridge colloidal particles. After the bridge is formed, polymers continue to rearrange their configuration, similar to regularly

adsorbed polymers, leading usually to an increase in the bond strength with time.

Special lectures

Phase transitions, aggregation, and gelation in colloid polymer mixtures. Henk Lekkerkerker, Utrecht University [38, 39, 40]

Depletion interactions among globular proteins and semidilute polymer. Theo Odijk, Delft University of Technology [41, 42, 43]

Simulating rare events in complex liquids. Daan Frenkel, FOM Institute for Atomic and Molecular Physics, Amsterdam [44, 45]

Poster prizes

During the conference prizes for the three best posters were awarded by a scientific jury and the Organizing Committee. Papers of all three prize winning poster contributions are published in this volume (see pages 66, 245, and 225). The prizes went to:

H. Behrens, M. Schudel, M. Semmler, M. Borkovec, P. Schurtenberger, and H. Sticher
Stability of colloidal suspensions: A DLVO-treatment accounting for surface heterogeneity.

K. G. Marinova, T. D. Gurkov, T. D. Dimitrova, R. G. Alargova, and D. Smith
Oscillatory structural interactions in thin emulsion films containing micelles of ionic surfactant.

T. Iwanaga, Y. Shiogai, and H. Kunieda
Phase behavior of polyoxyethylene modified silicone with water.

Z Kardiol 87: Suppl 2 (1998)
© Steinkopff Verlag 1998

References

1. Ricka J (1993) Appl Opt 32: 4846–4851
2. Walter R, Ricka J, Quellet C, Nyffenegger R, Binkert T (1996) Macromolecules 29: 4019–4028
3. Brown RGW (1987) Appl Opt 26: 4846–4851
4. Van Blaaderen A, Ruel R, Wiltzius P (1997) Nature 385: 321–324
5. Van Blaaderen A, Wiltzius P (1995) Science 270: 1177–1179
6. Van Blaaderen A (1993) Advanced Materials 5: 52–54
7. Bender J, Wagner NJ (1996) J Rheology 40: 899–916
8. Butera RJ, Wolfe MS, Bender J, Wagner NJ (1996) Phys Rev Lett 77: 2117–2120
9. Rastogi SR, Wagner NJ, Lustig SR (1996) J Chem Phys 104: 9249–9258
10. Hébraud P, Lequeux F, Munch JP, Pine DJ (1997) Phys Rev Lett 78: 4657–4660
11. Sollich P, Lequeux F, Hébraud P, Cates ME (1997) Phys Rev Lett 78: 2020–2023
12. Leaver MS, Olsson U, Wennerstrom H, Strey R, Würz U (1995) J Chem Soc Faraday Trans 91: 4269–4274
13. Olsson U, Wennerstrom H (1994) Adv Colloid Interface Sci 49: 113–146
14. Daicic J, Olsson U, Wennerstrom H (1995) Langmuir 11: 2451–2458
15. Talmon Y (1996) Ber Bunsenges Phys Chem 100: 364–372
16. De Vries R (1994) J de Physique II 4: 1541–1555
17. De Vries R (1997) Phys Rev E 56: 1879–1886
18. Pedersen JS, Laso M, Schurtenberger P (1996) Phys Rev E 54: R5917–R5920
19. Pedersen JS, Schurtenberger P (1996) Macomolecules 29: 7602–7612
20. Gisler T, Schulz SF, Borkovec M, Sticher H, Schurtenberger P, Daguanno B, Klein R (1994) J Chem Phys 101: 9924–9936
21. Borkovec M, Daicic J, Koper GJM (1997) PNAS 94: 3499–3503
22. See also H. Behrens et al., page 66 this volume
23. Aveyard R, Binks BP, Cho W-G, Fisher LR, Fletcher PDI, Klinkhamer F (1996) Langmuir 12: 6561–6569
24. Cho W-G, Fletcher PDI (1997) J Chem Soc Faraday Trans 93: 1389–1395
25. Earnshaw JC (1997) Appl Optics 36: 7583–7592
26. Earnshaw JC, Nugent CP, Lunkenheimer K, Hirte R (1996) J Phys Chem 100: 5004–5010
27. Sharpe D, Earnshaw JC (1997) J Chem Phys 107: 7493–7501
28. Earnshaw JC, Sharpe DJ (1996) J Chem Soc Faraday Trans 92: 611–618
29. Portet F, Desbene PL, Treiner C (1997) J Colloid Interface Science 194: 379–391
30. Desbene PL, Portet F, Treiner C (1997) J Colloid Interface Sci 190: 350–356
31. Lunkenheiner K, Zembala M (1997) J Colloid Interface Sci 188: 363–371
32. Bae S, Harke M, Goebel A, Lunkenheimer K, Motschmann H, Prescher D (1997) Langmuir 13: 6274–6278
33. Asmussen A, Riegler H (1996) J Chem Phys 104: 8159–8164
34. Merkl C, Pfohl T, Riegler H (1997) Phys Rev Lett 79: 4625–4628
35. Van de Ven TGM, Alince B (1996) J Colloid Interface Sci 181: 73–78
36. Van de Ven TGM, Alince, B (1996) J Pulp Paper Sci 22: J 257–J 263
37. Takase H, van de Ven TGM (1996) Colloids Surfaces A 118: 115–120
38. Lekkerkerker HNW: Strong, weak and metastable liquids (1997) Physica A 244: 227–237
39. Verhaegh NAM, Asnaghi D, Lekkerkerker HNW, Giglio M, Cipelleti L (1997) Physica A 242: 104–118
40. Verhaegh NAM, Van Duijneveldt JS, Ghont JKG, Lekkerkerker HNW (1996) Physica A 230: 409–436
41. Odijk T (1996) Macromolecules 29: 1842–1843
42. Odijk T (1997) J Chem Phys 106 (1997) 3402–3406
43. Odijk T (1997) Langmuir 13: 3579–3581
44. Bolhuis PG, Frenkel D (1997) Physica A 244: 45–58
45. Ten Wolde PR, Frenkel D (1997) Science 277: 1975–1978

Prog Colloid Polym Sci (1998) XI
© Steinkopff Verlag 1998

CONTENTS

Progr Colloid Polym Sci (1998) 110:1–3
© Steinkopff Verlag 1998

N. Kochurova

Hydrophobic hydration and CMC

N. Kochurova (✉)
St. Petersburg University
Department of Chemistry
199034 St. Petersburg
Russia

Abstract One of the most important problems of the physical chemistry of surfactant – the nature of the hydrophobic effect is examined here. Usually this effect is explained by means of iceberg formation around the hydrocarbon groups of surfactant in water. Therefore, it is interesting to know the type of the hydration of the surfactant ions.

In this work parameters of the hydration are calculated from data containing electroconductivity of the surfactant aqueous solutions. Measuring of electroconductivity is carried out in a broad region of temperatures (15–50) °C and concentrations (0.1–8 mM l^{-1}). Experiments and calculations show that the surfactant ions may condense or brake the structure of water as in the case of the nonorganic ions.

The type of the hydration may influence the value of the CMC.

Key words Hydration – surfactant ions – CMC

Introduction

Various aspects of interactions between surfactants and solvents have been studied for many years. Despite the long history of these studies the investigations are still in progress. The main attention concerns water–surfactant interaction – the so-called hydrophobic effect, which explains such phenomena as adsorption and micellization. Therefore, this fundamental study is important to almost every application of surface-active substances.

The molecular-kinetic parameter of the hydration was obtained by Samoilov [1]. In Samoilov's model the hydration is characterized by the influence of the ion on the transport mobility of the water molecules which are near the ion. This influence is determined by means of the modification of the activation energy ΔE_{tr}^0 of water molecule to jump from one local equilibrium to another. ΔE_{tr}^0 is calculated in [1] as

$$E_{tr}^0 = -R \frac{1}{U_i^0 \eta} \frac{d(U_i^0 \eta)}{d(1/T)}, \tag{1}$$

where $R = 8.31$ J mol^{-1} K^{-1}, η is the viscosity of water and T the temperature.

Samoilov has proved that at $E_{tr}^0 > 0$ there is a positive hydration of ion (the structure-making ion) and at $\Delta E_{tr}^0 < 0$ there is a negative hydration of ion (structure-breaking ion).

The thermodynamic parameter of hydration is ΔS_{II} – the change of the water entropy at the hydration [2], the change in such process, when ion (M) in gas (M_g) and m molecules of water (R) in liquid-state (mR_l) are becoming as ion in solution (mR_{sol}): $M_g + mR_l$ to $M_{sol} + mR_{sol}$. The hydration is positive if $\Delta S_{II} < 0$ (effect of putting in order) and negative if $\Delta S_{II} > 0$ (effect of not putting in order). The hydration is positive if $\Delta S_{II} < 0$ and negative if $\Delta S_{II} > 0$.

This value in the case of the surfactant solutions may be estimated with great difficulty because these effects are small at small concentrations of solutions before CMC (the critical concentration of the micellization). In literature these data are absent.

Gurikov [3] suggested the hydrodynamic parameter of hydration – the relative local viscosity η_i/η_0

$$\frac{\eta_i}{\eta_0} = \frac{82.5}{\eta_0} \frac{z}{\lambda_i^0 R_i}. \tag{2}$$

In formula (2) z is the ion valency, R_i the ion radius (Å). η_0 the water viscosity, λ_i^0 is the local viscosity near ion i, λ_i^0 is measured in $cm^2\,\Omega^{-1}\,mol^{-1}$, viscosity – in cPois. If the $[(\eta_i\,\eta_0) - 1]$ is larger than zero, then the hydration is positive, in the opposite case it is negative.

The aim of this paper is to discuss the type of hydration of the surfactant ions and the influence of it on the CMC.

Experimental

The experimental method of investigation was to measure the electroconductivity of the surfactant aqueous solutions. Measurements are described in [4]. From data of the specific electroconductivity the values of the equivalent electroconductivity at infinite dilution λ_i^0 and the maximum mobility of the surfactant ions U_i^0 were obtained (i is the kind of the ion). In these calculations the contribution of the counterion was taken into account. By means of λ_i^0 and U_i^0 we can estimate the parameters of the hydration. Here we present the results of the investigation of the following surfactants:

$$\bigcirc\!\!-N^+\!-RBr^-(Cl^-), \quad R = C_{10}H_{21}, C_{12}H_{25}, C_{16}H_{33}$$

$$(CH_3)_3N^+C_{16}H_{33}Br^-$$

in a broad region of temperatures (15–50) °C and concentrations (0.1–8 mM l^{-1}). We also calculated the radii of the surfactant ions from their molal volumes (method by Le Ba [5]). We have obtained that the ion $C_{16}H_{33}N(CH_3)_3$ (I) has radius 5.59 Å, the ion $C_{10}H_{21}N^+C_5H_5$ (II) – 5.02 Å, the ion $C_{12}H_{25}N^+C_5H_5$ (III) – 5.24 Å, the ion $C_{16}H_{33}N^+C_5H_5$ (IV) – 5.64 Å.

Results and discussion

Our investigations show that ΔE_{tr}^0 may be smaller than zero or bigger than zero. The middle values of the ΔE_{tr}^0 in the region of the temperatures (15–50) °C are 1.3 kJ mol^{-1}

(cation I), $+4.9$ kJ mol^{-1} (cation III), $+6.1$ kJ mol^{-1} (cation IV).

In case of the cation II we observed a change of the sign of ΔE_{tr}^0 with temperature: the negative hydration at (15–30) °C became the positive hydration at temperatures higher than 35 °C. The change of the sign of hydration in case of the nonorganic ion K$^+$ was observed in [3] at 45 °C. Perhaps it can be explained by changing of the water structure with temperature.

Positive hydration became stronger in cases of lengthening of the carbon chain and the increase of the ion radius.

Comparison of values of the hydration parameters at 25 °C for nonorganic and surfactant ions are presented in Fig. 1. In the region of the positive hydration the nonorganic ions Li$^+$, Na$^+$ and surfactant ions III, IV are situated. In the region of the negative hydration there are the nonorganic ions: Rb$^+$, Cs$^+$, K$^+$ and surfactant ions: I, II. Point 4′ corresponds to the ion II in the solution of the dezilpyridinium chloride and at point 4 – in the solution of the dezilpyridinium bromide.

From Fig. 1 we can obtain the value of ΔS_{II} which is difficult to estimate from experiment.

The type of the hydration must influence the micellization and CMC because these phenomena are determined by the water–surfactant interaction. In paper [6] there were data about the CMC of the dezilpyridinium chloride. These values were minimum near the (35–40) °C. In our experiments near this temperature the type of the hydration changes. Furthermore, we can say that the CMC decreases with temperature at the negative hydration and increases at the positive hydration. Increase of the temperature stimulates the micellization at the negative hydration.

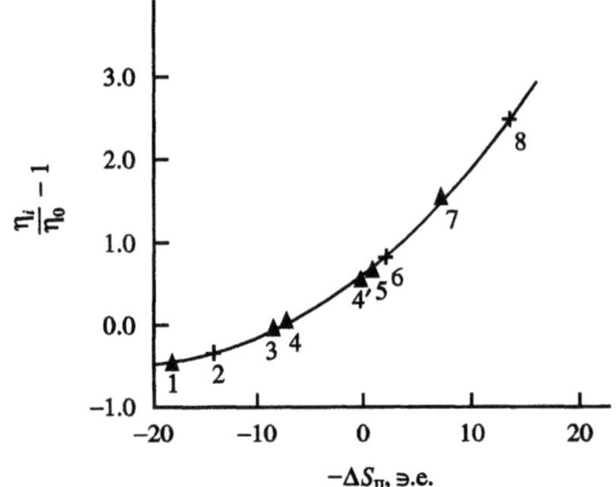

Fig. 1 The relative local viscosity vs the changing of the water entropy at 25 °C with cations: 1 – $C_{16}H_{33}N^+(CH_3)_3$, 2 – Cs$^+$, 3 – K$^+$, 4 – $C_{10}H_{21}N^+C_5H_5$, 5 – $C_{12}H_{25}N^+C_5H_5$, 6 – Na$^+$, 7 – $C_{16}H_{33}N^+C_5H_5$, 8 – Li$^+$

Progr Colloid Polym Sci (1998) 110:1–3
© Steinkopff Verlag 1998

Besides Fig. 1 we can estimate the ΔS_{II} at 25 °C. For the solution of the dezilpyridinium chloride this value is about 0.083 kJ mol^{-1} K^{-1} and it is close to the entropy of micellization 0.078 kJ mol^{-1} K^{-1} at 25 °C from [6]. The positive value of ΔS_{II} corresponds to the negative hydration which the parameter ΔE_{tr}^{0} shows at this temperature.

References

1. Samoilov O Ya (1957) Structure of the electrolyte aqueous solutions and hydration of ions. Moscow Izd Akad Nauk SSSR
2. Krestov GA (1973) Thermodynamics of the ion processes in solutions. Leningrad, Khimia
3. Gurikov Yu V (1992) J Phys Chem 66: 1257–1262 (in Russian)
4. Kochurova NN, Surkov KN, Rusanov AI (1995) Obschaya Khimia 65:1276–1278 (in Russian)
5. Rid R, Prausniz J, Shervud T (1982) Properties of Gases and Liquids. Leningrad, Khimia
6. Mehrian T, Keizer A, Korteweg AJ, Lyklema J (1993) Colloid Surf A: Physicochem Eng Aspects 71:255–267

Progr Colloid Polym Sci (1998) 110:4–7
© Steinkopff Verlag 1998

A.O. Ivanov
S.V. Bulytcheva

Evolution of colloidal fractal aggregates: diffusion-limited mathematical model

Prof. A.O. Ivanov (✉) · S.V. Bulytcheva
Department of Mathematical Physics
Urals State University
Lenin Av. 51
620083 Ekaterinburg
Russia

Abstract A mathematical model of the growth and the evolution of internal structure of a separate aggregate in a diluted colloid has been developed. The model includes the aggregation kinetic equation and a mass transfer equation, describing the diffusional transport of a colloidal particles to the aggregate. The attachment of single particles to the aggregate skeleton has been described on the basis of coexisting into penetrating media conception and with the help of the non-linear mass exchange terms in the diffusion equation. The self-similar solution of the model and the aggregate growth rate have been obtained under the condition of a slow aggregate growth. The aggregate structure resembles a fractal cluster and the aggregate growth is described by a classical kinetic-limited growth rate.

Key words Aggregation – kinetics – fractal cluster – colloids

A particle coagulation process in colloidal dispersions represents a reason for the formation and subsequent growth of the colloidal aggregates, the internal structure of which is not spatially homogeneous. The general observations concerning the chemical and physical peculiarities of the interparticle interaction obviously lead to a conception of the colloidal aggregates as the natural examples of the so-called "fractal clusters" [1, 2]. Alongside with the computer modelling [1–4], of great interest are the analytical models, describing the evolution of the internal aggregate structure and the aggregate growth rate. In present paper we should focus our attention on both the problem of developing of such a model and the self-similar solutions and principles of evolution of the large aggregate.

Mathematical model

A mathematical model of the growth of a separate spatially inhomogeneous colloidal aggregate may be formulated on the basis of the conception of coexisting into-penetrating media under the following assumptions:

– All colloidal particles are divided in the two main classes: aggregated particles inside the aggregate volume and "free" particles inside and outside the aggregate. Due to the interparticle bonds, the hydrodynamic mobility of the aggregated particles is neglected as compared with the diffusional motion of the "free" particles. The aggregate boundary motion $\Sigma(t)$ is limited by the rate of diffusional transport of "free" non-aggregated particles from the bulk of a colloid to the aggregate surface.

– A diluted colloidal system is considered. The evolution of a spontaneously formed aggregate is dependent on the kinetics of attachment of the single particles to the aggregate skeleton. The probability of such attachment is proportional to the aggregated particle concentration and is sufficiently large in comparison with the probability of the single particle–particle coagulation.

– The aggregate structure is described by the volume concentration $\varphi(t, \mathbf{r})$ of the aggregated particles, coexisting with the "free" particles of the concentration $\rho(t, \mathbf{r})$ inside and $\sigma(t, \mathbf{r})$ outside the aggregate. The last quantities are defined over the volume which is not occupied by the aggregated particles.

– Inside the aggregate the φ- and ρ-particles are considered as coexisting into-penetrating media. The intertransformation of these media is convenient to describe with the help of non-linear mass transfer terms under the condition when both the processes of particle attachment to the aggregate and breaking from the aggregate are taken into account.

According to these assumptions the mass transfer equation, describing the diffusional transport of the "free" ρ-particles and its transformation in φ-particles, may be developed from the mass balance law:

$$\frac{\mathrm{d}}{\mathrm{d}t}\int_V\left(1-\frac{\varphi}{\varphi_\mathrm{m}}\right)\rho\,\mathrm{d}v = -\int_S\mathbf{j}\cdot\mathrm{d}s - \int_V \text{mass exchange}\,\mathrm{d}v\,,\tag{1}$$

where φ_m is the maximal aggregated particle density and $\mathbf{j}$ represents the diffusional flux density of the "free" particles over the surface which is not occupied by the φ-particles. Using the definition for $\mathbf{j}$ and the Gauss theorem, we obtain

$$\int_S\mathbf{j}\cdot\mathrm{d}s = -D\int_S\nabla\rho\cdot\mathrm{d}s = -D\int_V\nabla\left[\left(1-\frac{\varphi}{\varphi_\mathrm{m}}\right)\nabla\rho\right]\mathrm{d}v\,,\tag{2}$$

where D is the single-particle diffusion coefficient. For the aggregated particles the mass balance law has the simple form

$$\frac{\mathrm{d}}{\mathrm{d}t}\int_V\varphi\,\mathrm{d}v = \int_V \text{mass exchange}\,\mathrm{d}v\,.\tag{3}$$

Going to the differentials, we get the diffusional equation, describing the concentrational profiles of the ρ-particles inside the aggregate volume:

$$\frac{\partial}{\partial t}\left[\left(1-\frac{\varphi}{\varphi_\mathrm{m}}\right)\rho\right] = D\nabla\left[\left(1-\frac{\varphi}{\varphi_\mathrm{m}}\right)\nabla\rho\right] - \frac{\partial\varphi}{\partial t}\,.\tag{4}$$

The last term describes the evolution of the aggregate structure. Let us consider the three types of the aggregation kinetics. The A type kinetics takes into account the possibilities of the attachment of "free" particles to the aggregate skeleton and the breaking of particles from the aggregate in a classical way when these possibilities are considered to be proportional to the local values of the corresponding φ- and ρ-particles concentrations:

$$\frac{\partial\varphi}{\partial t} = a\rho\varphi - b\varphi\varphi_\mathrm{m}\,,\quad a, b \approx \text{const}\,.\tag{5}$$

Here the effective coefficients a and b evidently depend on the chemical and physical properties of the colloidal system and the solvent.

The B type kinetics takes into account that the breaking is complicated in the presence of the interparticle bonds between the aggregated particles:

$$\frac{\partial\varphi}{\partial t} = a\rho\varphi - b\varphi(\varphi_\mathrm{m} - \varphi)\,.\tag{6}$$

In the last C-type kinetics we assume that the probabilities are proportional to the local numbers of the aggregated and "free" particles:

$$\frac{\partial\varphi}{\partial t} = a\rho\varphi\left(1-\frac{\varphi}{\varphi_\mathrm{m}}\right) - b\varphi(\varphi_\mathrm{m} - \varphi)\,.\tag{7}$$

Outside the aggregate only the "free" particles are considered:

$$\partial\sigma/\partial t = D\,\Delta\sigma\,,\quad \sigma(\infty) = \sigma_\infty\,.\tag{8}$$

The following conditions would be fulfilled at the aggregate boundary $\Sigma(t)$:

$$D\frac{\partial\sigma(t,\Sigma)}{\partial\mathbf{n}} - D\left[1-\frac{\varphi(t,\Sigma)}{\varphi_\mathrm{m}}\right]\frac{\partial\rho(t,\Sigma)}{\partial\mathbf{n}}$$
$$= v\left[a\sigma(t,\Sigma)\,\varphi(t,\Sigma) - b\varphi(t,\Sigma)\,\varphi_\mathrm{m}\right]\,,\tag{9}$$

where v is the length dimension coefficient which may be considered approximately to be equal to the particle radius. The flux consistency condition (9) must be supplemented with the boundary motion condition:

$$\varphi(t,\Sigma)\frac{\mathrm{d}\Sigma}{\mathrm{d}t} = v\left[a\sigma(t,\Sigma)\,\varphi(t,\Sigma) - b\varphi(t,\Sigma)\,\varphi_\mathrm{m}\right]\,.\tag{10}$$

Hence, the "free" particle concentration at the aggregate boundary is determined by the third-order boundary condition taking into account the balance between the diffusional transport of the particles and the kinetics of its aggregation.

Self-similar problem

Let us investigate the spherically symmetric solutions of the problem for large values of an aggregate. Introducing new self-similar spatial variable $s = r/\Sigma(t)$ and assuming that this spatial scaling will result in self-similar quasi-stationary concentration profiles, we get the problem for determining the functions $y(s) = \varphi(s)/\varphi_\mathrm{m}$ and $\rho(s)$:

$$\frac{\mathrm{d}}{\mathrm{d}s}\left[s^2(1-y)\frac{\mathrm{d}\rho}{\mathrm{d}s}\right] + fs^3\frac{\mathrm{d}}{\mathrm{d}s}[\varphi_\mathrm{m}y + \rho(1-y)] = 0\,,$$
$$0 < s < 1\,,\tag{11}$$

$$-Pfs\frac{\mathrm{d}y}{\mathrm{d}s} = y(\rho - K)\,,\quad 0 < s < 1\,,\quad \text{A kinetics}\,,\tag{12}$$

$$-Pfs\frac{dy}{ds} = y\rho - Ky(1-y), \quad 0 < s < 1, \quad \text{B kinetics}, \tag{13}$$

$$-Pfs\frac{dy}{ds} = y(1-y)(\rho - K), \quad 0 < s < 1, \quad \text{C kinetics}, \tag{14}$$

$$\frac{d}{ds}\left(s^2\frac{d\sigma}{ds}\right) + fs^3\frac{ds}{ds} = 0, \quad 1 < s < \infty, \tag{15}$$

$$f = \frac{\Sigma}{D}\frac{d\Sigma}{dt} \approx \text{const}, \quad P = \frac{D}{a\Sigma^2} \approx \text{const}, \quad K = \frac{b}{a}\varphi_{\mathrm{m}} \ll 1. \tag{16}$$

The defined parameters f and P may be approximately considered as constant in the case of a large aggregate and a slow growth. These parameters have the meanings of the relations between the character time periods: $\tau_D = \Sigma^2/D$ – diffusion relaxation time period, $\tau_A = \Sigma/(d\Sigma/dt)$ – relative aggregate growth time period and $\tau_a = 1/a$ – effective aggregation time period.

The problem (11)–(16) is supplemented by the corresponding number of the boundary conditions:

$$\frac{d\sigma(1)}{ds} - [1 - y(1)]\frac{d\rho(1)}{ds} = Sy(1)[\sigma(1) - K],$$

$$f = S[\sigma(1) - K], \tag{17}$$

$$y(0) = 1, \quad \frac{dy(0)}{dy} = 0, \quad \sigma(\infty) = \sigma_\infty, \quad S = \frac{v\varphi_{\mathrm{m}}a\Sigma}{D}.$$

Droplet-like solutions

In the case of very slow aggregation kinetics $fP = \tau_a/\tau_A \ll 1$ from the set of Eqs. (11)–(17) it follows the approximate solution $y(s) \approx 1$, $\varphi(s) \approx \varphi_{\mathrm{m}}$. It means that all the particles inside the aggregate are bonded (aggregated). In this case the internal structure of an aggregate is spatially homogeneous and the subsequent evolution is determined only by the attachment of single particles to the aggregate surface. The aggregates of this type occur under the phase separation of colloidal systems [5, 6] and are characterized by a surface tension, remaining the liquid droplet placed in a vapour. The growth rate of such an aggregate follows from the boundary condition (17)

$$\frac{d\Sigma}{dt} = \frac{va(\sigma_\infty - K)}{1 + (v\,a\varphi_{\mathrm{m}}/D)\Sigma} \tag{18}$$

and takes the classical diffusion-limited form for large aggregate $\Sigma \gg D/va\varphi_{\mathrm{m}}$.

Fractal-like solution

For the case of a spatially inhomogeneous aggregate we obtain the following approximate solutions:
Kinetics of the A type

$$\varphi(s) = \varphi_{\mathrm{m}}\exp(-\varphi_{\mathrm{m}}s^2/10P), \tag{19}$$

$$\frac{d\Sigma}{dt} = \frac{va(\sigma_\infty - K)}{1 + (v\,a\varphi_{\mathrm{m}}/D)\Sigma\exp(-a\varphi_{\mathrm{m}}\Sigma^2/10D)}.$$

Kinetics of the B type

$$\varphi(s) = \varphi_{\mathrm{m}}\exp\left(-\frac{\varphi_{\mathrm{m}}fs^2}{10Pf + 5K}\right), \tag{20}$$

$$\frac{d\Sigma}{dt} = \frac{va(\sigma_\infty - K)}{1 + (v\,a\varphi_{\mathrm{m}}/D)\Sigma\exp(-a\varphi_{\mathrm{m}}\Sigma^2/10D)}.$$

Kinetics of the C type

$$\varphi(s) = \varphi_{\mathrm{m}}/[1 + (\varphi_{\mathrm{m}}/10P)s^2\exp(\varphi_{\mathrm{m}}s^2/10P)]^{-1}, \tag{21}$$

$$\frac{d\Sigma}{dt} = \frac{va(\sigma_\infty - K)[1 + (\varphi_{\mathrm{m}}a\Sigma^2/10D)\exp(\varphi_{\mathrm{m}}a\Sigma^2/10D)]}{1 + (va\varphi_{\mathrm{m}}/D)\Sigma + (\varphi_{\mathrm{m}}a\Sigma^2/10D)\exp(a\varphi_{\mathrm{m}}\Sigma^2/10D)}.$$

This kind of the solution remains a fractal cluster and contains a central dense "core" and a surrounding zone of the decreasing φ-particle concentration, in which the aggregate becomes more and more friable. The aggregated particle concentration profiles for all kinetics (19)–(21) are presented in Fig. 1 in comparison with the numerical solutions of the problem (11)–(17). As may be easily seen from expressions (19) and (20), the difference between solutions for the A and B kinetics is negligibly small. For the C type kinetics the concentration of the aggregated

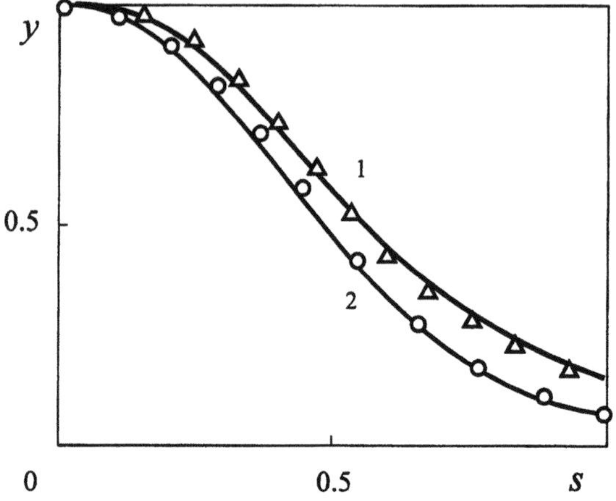

Fig. 1 The aggregated particle concentration profiles inside the aggregate for A and B kinetics (curve 1) and C kinetics (curve 2), dots – numerical solutions of the problem (11)–(17): $\varphi_{\mathrm{m}} = 0.5$, $\sigma_\infty = 0.05$, $K = 0.0001$

Progr Colloid Polym Sci (1998) 110:4–7
© Steinkopff Verlag 1998

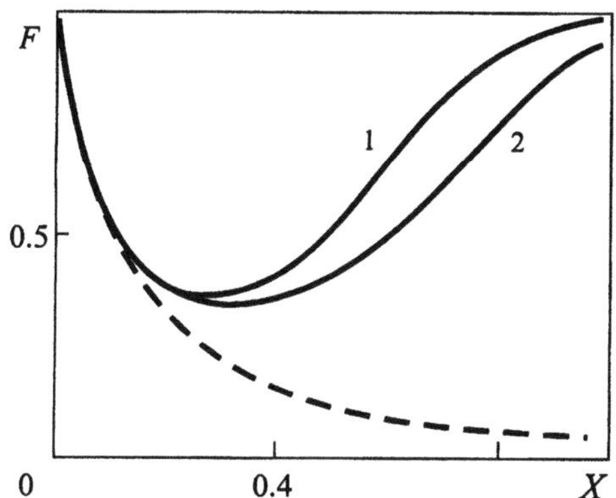

Fig. 2 Dimensionless aggregate growth rate $F = [va(\sigma_\infty - K)]^{-1} \times (\mathrm{d}\Sigma/\mathrm{d}t)$ vs. dimensionless aggregate radius $X = \varphi_\mathrm{m}va\Sigma/D$: curve 1 – kinetics A and B, curve 2 – kinetics C, dashed line – diffusion-limited growth rate (18)

particles decreases more rapidly when going out of the aggregate "core".

The aggregate growth rate for all kinetics as functions of the dimensionless aggregate size are demonstrated in Fig. 2 in comparison with the diffusion-limited growth rate (18). The distinctive feature of the fractal-like aggregate is connected with the transformation of the aggregate growth rate from the classical diffusion-limited law to the kinetic-limited one:

$$\mathrm{d}\Sigma/\mathrm{d}t = va(\sigma_\infty - K) \tag{22}$$

when the aggregate size exceeds the value $\Sigma > D/\varphi_\mathrm{m}av$. For the A and B aggregation kinetics, this transformation begins for smaller aggregates than for C kinetics. The monotonous decreasing of the growth rate for droplet-like aggregate is caused with the limitation of a process by a diffusional transport of the "free" particles to the aggregate surface. In the case of inhomogeneous aggregate the boundary motion is connected with the kinetics of the particle attachment to the aggregate boundary. Therefore, the aggregate growth results in the boundary motion to the region of the larger concentration of the "free" particles and has to be accompanied with the increasing of the growth rate.

Conclusion

We have discussed the partial differential model describing the growth and the evolution of internal structure of a separate colloidal aggregate for three types of aggregation kinetics. In the case of slow aggregate growth, the two kinds of self-similar solutions have been found:

– The first one is characterized by a constancy of the aggregated particles inside an aggregate and by a classical diffusion-limited growth rate. Such aggregates arise during a phase separation process in colloidal systems and are similar to the liquid drops in a supersaturated gas;

– The second kind of solution reminds a fractal cluster and contains a central dense "core" and a surrounding zone, in which the aggregate becomes more and more friable. In this region the concentration of the aggregated particles decreases similarly to the fractal clusters. The model predicts the exponential behaviour instead of the power dependence. The growth rate is characterized by a transformation from a diffusion-limited growth rate to a kinetic-limited one during the aggregate growth.

By using the results of the model on the aggregate internal structure and growth rate, the kinetics of a colloid aggregation may be investigated under the condition when both the aggregate distribution over size and the particle balance law in the colloid will be taken into account.

Acknowledgements The present research was carried out within the financial support of the Russian Basic Research Foundation (Grant No. 95-03-32038).

References

1. Witten TA, Sander LM (1981) Phys Rev Lett 47:1400
2. Vicsek T (1989) Fractal Growth Phenomena. World Scientific, Singapore
3. Pastor-Satorras R, Rubi JM (1995) Phys Rev E 52:5602–5609
4. Meakin P (1992) Physica A 187:1–17
5. Buyevich Yu A, Ivanov AO (1992) Physica A 190:276–294
6. Bushmanova SV, Ivanov AO, Buyevich YuA (1994) Physica A 202:175–195

Progr Colloid Polym Sci (1998) 110:8–11
© Steinkopff Verlag 1998

Dynamic light scattering by preserved skimmed cow milk: A comparison of two-colour and three-dimensional cross-correlation experiments

C. Sinn
R. Niehüser
E. Overbeck
T. Palberg

Dr. C. Sinn (✉) · R. Niehüser
E. Overbeck · T. Palberg
Johannes-Gutenberg-Universität
Institut für Physik
Staudingerweg 7
D-55099 Mainz
Germany

Abstract We investigate the dynamic light scattering of preserved skimmed cow milk and compare the performance of a standard two-colour (2C) cross-correlation setup with the recently developed three-dimensional (3D) cross-correlation setup. Undiluted milk could only be investigated using the 3D setup because of too low signal-to-noise ratio of the 2C setup. A tenfold diluted milk, however, could be investigated by both setups, and comparable results for the dynamics of the sample were obtained thus demonstrating the equivalence of the decorrelation schemes. On the other hand, we found that the dynamics of the milk is strongly altered upon dilution. This finding elucidates the benefits of the new 3D setup for the investigation of highly turbid samples.

Key words Dynamic light scattering – multiple scattering – cross-correlation – cow milk

Introduction

Currently, there is a growing interest in dense suspensions of colloidal particles where different direct interactions combine with hydrodynamic interactions to result in relevant influences on structure and dynamics. Related topics range from crystallization kinetics to the microscopic understanding of rheological behaviour. However, in dense samples, light scattering experiments are complicated due to multiple scattering events, which may even dominate the scattered intensity. Only in some rare cases can the turbidity of the samples be reduced by a certain match of the refractive index of the particles and the suspension medium. Dynamic light scattering (DLS) measurements have been performed by decorrelation of multiple scattering applying setups with a different degree of complexity [1–4]. Here we compare the performance of two state-of-the-art setups, both realized in our laboratory, i.e. the two-colour (2C) and the three-dimensional (3D) setup. As an illustrative example we investigate the dynamics of cow milk, because of its high reproducibility and its close relation to possible industrial applications of cross-correlation methods.

Experimental details

The experimental setup of the 2C experiment follows closely the description given in the original publication [2] and a recent performance report [5], respectively, while the setup of the 3D experiment has been already described elsewhere [6]. Consequently, we give here only a very short summary of the working principles. Both experiments measure the sample dynamics by two independent light scattering experiments, though applying the same scattering vector $q_1 = q_2$. While in the 2C setup the two experiments are distinguished by two different colours, the 3D setup switches to the third dimension thereby separating the light paths. These basic principles may be understood from Fig. 1, which shows a wavevector scheme for the respective DLS arrangements. Decorrelation of multiple scattering is performed by cross-correlating the signal fluctuations of both experiments [7]. The main

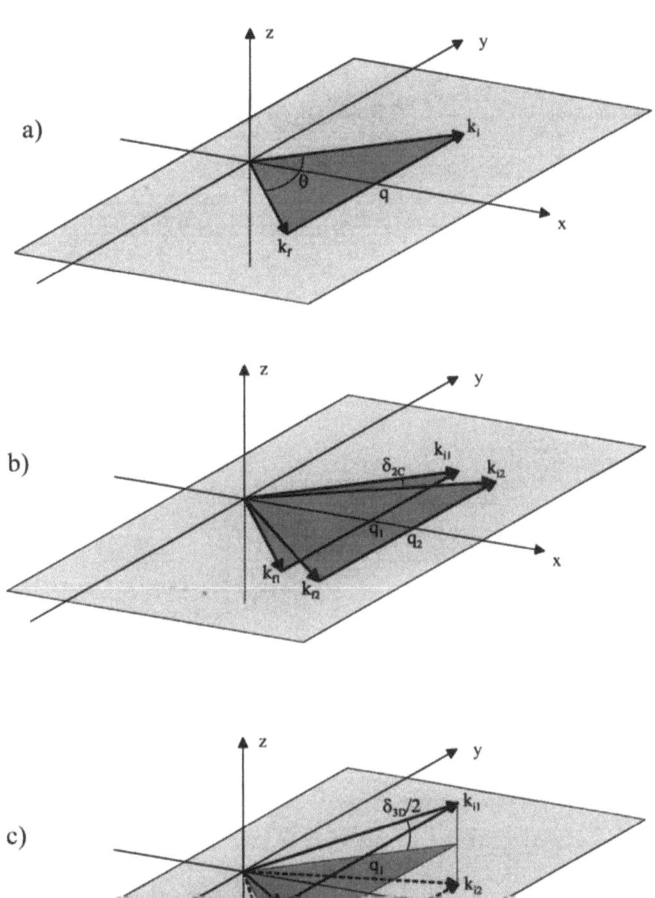

Fig. 1 "Evolution" of dynamic light scattering: (a) Conventional setup, with incoming and scattered wavevectors $\mathbf{k}_i$ and $\mathbf{k}_f$, respectively. (b) 2C setup; the incoming and scattered wavevectors both include a difference angle δ_{2C}; $\lambda_1 = 514.5$ nm, $\lambda_2 = 488.0$ nm. (c) 3D setup; the wavevectors include a difference angle δ_{3D} and share the common wavelength $\lambda = 790.0$ nm

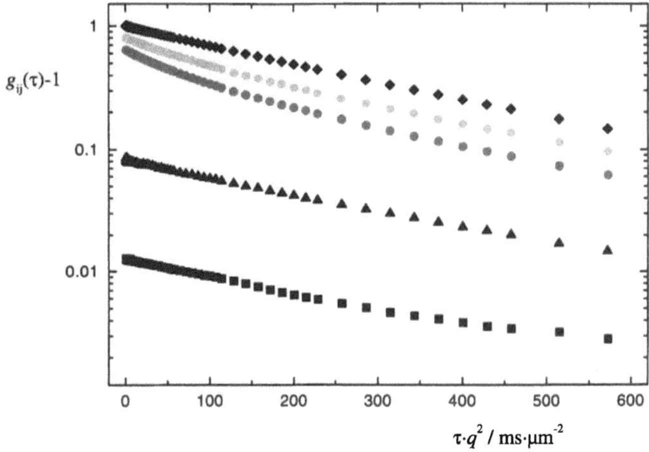

Fig. 2 Comparison of 3D and 2C DLS correlation functions applying a sample with $c/c_0 = 0.1$. An identical value of the scattering vector, $q = 16.7\ \mu\text{m}^{-1}$, is obtained by choosing $\theta_{2C} = 60°$ and $\theta_{3D} = 104°$. Auto-correlation: 3D (diamonds, $\lambda = 790.0$ nm); 2C (circles; $\lambda = 514.5$ nm: light grey, $\lambda = 488.0$ nm: dark grey). Cross-correlation: 3D (triangles); 2C (squares)

performance difference of the 2C and the 3D setup stems from the different wavelengths used. The mean wavelength of the 2C experiment is $\lambda = 501.3$ nm, whereas the 3D setup works at the longer wavelength of $\lambda = 790.0$ nm. As the scattering cross section in the Rayleigh–Debye–Gans limit varies as $1/\lambda^4$, the turbidity of the samples differs by a factor of six approximately, meaning a ratio of transmission coefficients $I_{2C}/I_{3D} \approx 2 \times 10^{-3}$. Utilising the 3D experiment one is therefore able to investigate samples of higher turbidity compared to the 2C experiment.

The milk under study was given by happy Hessian cows and, following a certain treatment, supplied by Schwälbchen Molkerei Jakob Berz AG, Bad Schwalbach, Germany, in tetragonal packages. We used the skimmed supply with a fat content of 0.3%; the original milk (c_0)

was diluted by demineralised water in order to obtain samples with different concentration c and turbidity. No care was taken to prevent air contact. The milk has been stored in a refrigerator for approx. 24 h after dilution before starting the measurements, which were performed at room temperature.

Results and discussion

Let us first discuss differences and similarities of both setups for one chosen sample. Figure 2 shows auto-correlation functions $g_{11}(\tau)$ and cross-correlation functions $g_{12}(\tau)$ obtained from a tenfold diluted milk sample ($c/c_0 = 0.1$), measured with both setups under a scattering vector of $q = 16.7\ \mu\text{m}^{-1}$. Multiple scattering contributions to the auto-correlation function increase with decreasing wavelength, as clearly seen in the wavelength sequence of the $g_{11}(\tau)$ data in Fig. 2. The higher signal-to-noise ratio (i.e. the intercept β of $g_{ij}(\tau)$) of the 3D experiment in auto-correlation as well as in cross-correlation is both due to small multiple scattering contributions and an excellent signal-to-noise ratio obtained by using single-mode fibres [8].

Figure 3 shows only the cross-correlation results for both methods, now for four different sample concentrations as indicated. An increase in concentration is accompanied by a decrease in intercept, as expected. This behaviour is more pronounced for the 2C setup, still again due to the wavelength dependence, leading even to the "passing" of 3D by 2C. It becomes obvious that using the

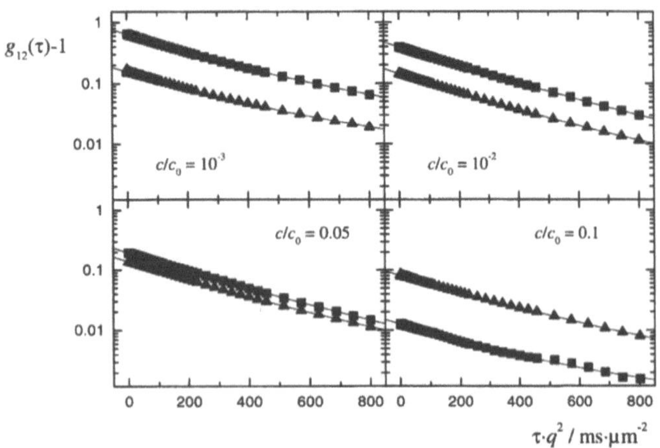

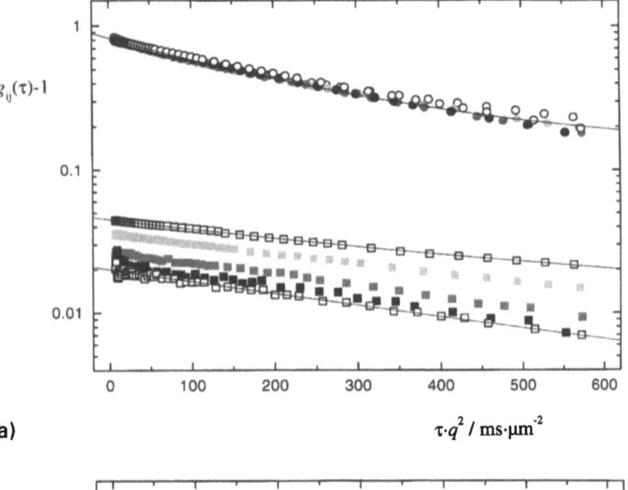

(a)

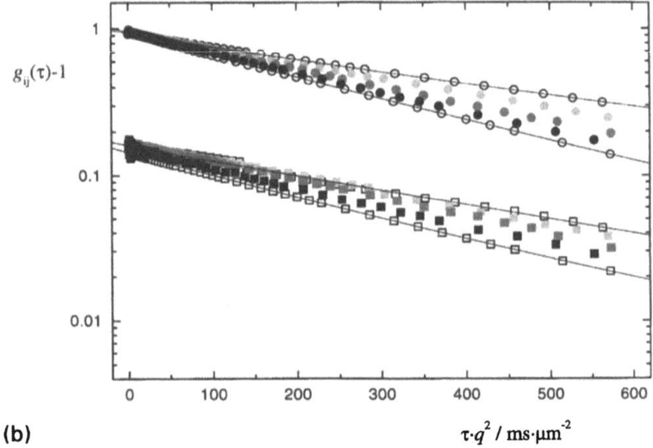

(b)

Fig. 3 Cross-correlation functions obtained from both, the 3D setup (triangles) and the 2C setup (squares) applying four samples with their concentration indicated in the plots. The drawn lines are fits to a second order cumulant expansion of the cross-correlation function. The droplet radii obtained from the first cumulant as determined by the 3D experiment are (in order of increasing concentration): $R = \{126, 116, 116, 127\}$ nm. The droplet radii determined by the 2C experiment are (in the same order): $R = \{119, 119, 118, 128\}$ nm

3D setup we are able to investigate by far more concentrated samples than by using the 2C setup, thus confirming the expectation in the experimental section. The drawn lines are fits of a second order cumulant expansion to the data according to $\ln[g_{12}(\tau) - 1] = \ln\beta - K_1\tau + \frac{1}{2}K_2\tau^2$. We checked that a higher order cumulant fit does not change the results significantly. From the first cumulant, the droplet radii can be determined by using $R = k_B T q^2 / 3\pi\eta K_1$. The results are given in the legend of Fig. 3 and demonstrate very clearly the equivalence of both decorrelation methods, as the radii agree within the error of 5%.

Because of the signal-to-noise limitation mentioned we were not able to investigate undiluted cow milk with the 2C setup. However, with the 3D setup we employ rectangular cuvettes, the thickness of which we can make very small (approx. 1 mm). Multiple scattering contributions are reduced concomitantly as long as multiple scattering events take place outside the scattering volume, and the measurements can be extended to still more turbid samples. This method is by far more complicated if cylindrical cuvettes are employed, as is the in the two-colour scheme, because optical aberrations have to be considered in detail. This additional feature of the 3D setup enabled us to investigate even the dynamics of undiluted milk.

In Fig. 4 we present the data of angular dependent measurements obtained with the 3D setup for undiluted ($c = c_0$; Fig. 4a) and diluted ($c/c_0 = 10^{-2}$; Fig. 4b) milk. The undiluted milk shows a clear non-exponential behaviour in $g_{11}(\tau)$, which is almost absent in $g_{12}(\tau)$, indicating

Fig. 4 (a) Correlation functions of undiluted milk ($c/c_0 = 1$) obtained with the 3D setup. Auto-correlation functions (circles) and cross-correlation functions (squares) are shown. Parameter is the scattering angle (36° and 104° (open symbols), 54°, 72° and 90° (increasing grey scale)). The drawn lines are fits to a second order cumulant expansion of the correlation functions; only the extremal values are shown. Droplet radii as obtained from the first cumulant of the cross-correlation function are (in order of increasing angle): $R = \{266, 252, 252, 266, 217\}$ nm; the experimental error is approximately 5%. (b) Correlation functions of diluted milk ($c/c_0 = 10^{-2}$) obtained with the 3D setup. Parameter is again the scattering angle, symbols are the same as in Fig. 4a. Droplet radii as obtained from the first cumulant of the cross-correlation function are (in order of increasing angle): $R = \{174, 171, 149, 124, 116\}$ nm; the experimental error is approximately 5%

that the non-exponential behaviour in auto-correlation is due to multiple scattering. The decreasing intercept with increasing angle in cross-correlation is due to a methodical artefact. Note, that the droplet radii (approx. $R = 250$ nm) of the cross-correlation measurements are angular independent within the experimental errors.

In clear contrast, the diluted milk shows a strong angular dependence of $g_{11}(\tau)$ as well as $g_{12}(\tau)$. We interpret this finding as an increase in polydispersity of cow milk upon dilution. In addition, the droplet radii are found significantly smaller (approx. $R = 150$ nm) than for undiluted milk. We conclude that cow milk cannot be diluted without significantly affecting its composition as measured by its droplet dynamics.

Conclusion

We reported measurements of the dynamics of preserved skimmed cow milk with two different methods to decorrelate multiple scattering. While we could demonstrate that both methods measure the dynamics without distortions due to multiple scattering, only by applying the recently introduced technique of the 3D setup using semiconductor lasers we were able to investigate undiluted milk. However, the dynamics of cow milk is drastically altered upon dilution. This finding might illustrate the importance of our development for the application of dynamic light scattering techniques to highly turbid samples.

References

1. Phillies GDJ (1981) Phys Rev 24: 1939–1963
2. Drewel M, Ahrens J, Podschus U (1989) J Opt Soc Amer A 7:206–210
3. Overbeck E, Sinn C, Palberg T (1997) Progr Colloid Polym Sci 104:117–120
4. Aberle LB, Wiegand S, Schröer W, Staude W (1997) Progr Colloid Polym Sci 104:121–125
5. Segrè PN, van Megen W, Pusey PN, Schätzel K, Peters W (1995) J Mod Opt 42:1929–1952
6. Overbeck E, Sinn C, Palberg T, Schätzel K (1997) Coll Surf A 122:83–87
7. Schätzel K (1991) J Mod Opt 38: 1845–1865
8. Ricka J (1993) Appl Opt 32:2860–2875

Progr Colloid Polym Sci (1998) 110:12–15
© Steinkopff Verlag 1998

Attraction between likely charged colloidal macroions?

H. Löwen

H. Löwen*
Institut für Theoretische Physik II
Heinrich-Heine-Universität Düsseldorf
Universitätsstraße 1
D-40225 Düsseldorf
Germany

*Institut für Festkörperforschung
Forschungszentrum Jülich
D-52425 Jülich
Germany

Abstract The distance-resolved effective force between two spherical highly charged colloidal macroions is calculated within the primitive model of strongly asymmetric electrolytes using computer simulations. For parameters corresponding to typical experimental samples, a repulsive force is obtained. Possibilities for an effective attraction induced by very high coupling and small distances between the macroionic surfaces. We also discuss briefly the influence of glass plates on the effective interaction.

strong coupling between the macroions or by a geometric confinement of the macroions between glass plates are briefly discussed.

Key words Charged colloids – effective interaction – screening – computer simulation

Introduction

The total effective force between two highly charged colloidal particles ("macroions") in solution with their microscopic counterions is expected intuitively to be repulsive since the screening by the thermal counterions is imperfect, in general. Nevertheless, there is a long debate in the literature about a possible attraction between macroions (or – equivalently – an "overscreening" by the counterions), for recent references and reviews see e.g. [1–3]. Recent measurements [4–7] reveal that the presence of a charged planar glass plate may influence the nature of the effective interactions between macroions drastically: Apparently it is repulsive in the bulk and gets attractive near the glass plates.

The aim of the present paper is to understand and to calculate within a simple microscopic model (the so-called primitive model of strongly asymmetric electrolytes) the nature of these forces theoretically. Using extensive Monte-Carlo simulations, we obtain the distance-resolved forces between the macroions. For parameters typically encountered in experiments, there is no indication for an attractive force. This may change, however, for extremely

Theory of the effective force

We consider N_m macroions with bare charge Ze and diameter σ confined in a volume V corresponding to a finite number density $\rho_m = N_m/V$ at a temperature T. The macroion density can conveniently be expressed in terms of their volume fraction $\phi = \pi\rho_m\sigma^3/6$. The counterions carry an opposite charge $-qe$ and possess a number density $\rho_c + N_c/V$ which is determined by global charge neutrality to be

$$\rho_c = Z\rho_m/q \,. \tag{1}$$

Since the system is invariant with respect to charge inversion $Z \to -Z$ and $q \to -q$, it is sufficient to consider positive Z and q. Within the "primitive model" one assumes the following pair interaction potentials $V_{mm}(r)$, $V_{mc}(r)$, $V_{cc}(r)$ between macroions and counterions, r denoting the corresponding interparticle distance and ε the dielectric constant of the solvent:

$$V_{\mathrm{mm}}(r) = \begin{cases} \infty & \text{for } r \le \sigma \,, \\ \frac{Z^2 e^2}{\varepsilon r} & \text{for } r > \sigma \,, \end{cases} \tag{2}$$

$$V_{\mathrm{mc}}(r) = \begin{cases} \infty & \text{for } r \le \sigma/2 \,, \\ \frac{Zqe^2}{\varepsilon r} & \text{for } r > \sigma/2 \,, \end{cases} \tag{3}$$

$$V_{\mathrm{cc}}(r) = \frac{q^2 e^2}{\varepsilon r} \,. \tag{4}$$

The effective macroion forces contain two parts, one stemming from the direct Coulomb repulsion between the macroions (2) and the other from the macroion– counterion interaction (3). This implies that the force $\mathbf{F}_i$ acting on the ith macroion located at $\mathbf{R}_i$ can be written as

$$\mathbf{F}_i = \mathbf{F}_i^{(\mathrm{m})} + \mathbf{F}_i^{(\mathrm{c})} \,, \tag{5}$$

where the direct macroionic part $\mathbf{F}_i^{(\mathrm{m})})$ is pairwise

$$\mathbf{F}_i^{(\mathrm{m})} = - \nabla_{\mathbf{R}_i} \sum_{j=1;\, j \ne i}^{N_{\mathrm{m}}} V_{\mathrm{mm}}(|\mathbf{R}_i - \mathbf{R}_j|) \,, \tag{6}$$

$\{\mathbf{R}_j;\, j = 1, \ldots, N_{\mathrm{m}}\}$ being the given macroion positions. The second contribution to the total effective force, $\mathbf{F}_i^{(\mathrm{c})}$, is the canonically counterion-averaged force from the macroion–counterion-interaction (3):

$$\mathbf{F}_i^{(\mathrm{c})} = - \left\langle \sum_{j=1}^{N_{\mathrm{c}}} \nabla_{\mathbf{R}_i} V_{\mathrm{mc}}(|\mathbf{R}_i - \mathbf{r}_j|) \right\rangle_{\mathrm{c}} \,. \tag{7}$$

Here, $\{\mathbf{r}_j;\, j = 1, \ldots, N_{\mathrm{c}}\}$ are the counterion positions and the canonical average $\langle \cdots \rangle_{\mathrm{c}}$ over an $\{\mathbf{r}_j\}$-dependent quantity $\mathscr{A}$ is defined via the classical trace

$$\langle \mathscr{A}(\{\mathbf{r}_k\}) \rangle_{\mathrm{c}} = \frac{1}{Z} \frac{1}{N_{\mathrm{c}}!} \frac{1}{\Lambda^{3N_{\mathrm{c}}}} \int_V \mathrm{d}^3 r_1 \cdots \int_V \mathrm{d}^3 r_{N_{\mathrm{c}}}$$
$$\times \mathscr{A}(\{\mathbf{r}_k\}) \exp\left[-\frac{V_{\mathrm{c}}}{k_{\mathrm{B}} T} \right] \,, \tag{8}$$

where k_{B} is the Boltzmann constant, Λ the de Broglie thermal wavelength of the counterions and

$$V_{\mathrm{c}} = \sum_{n=1}^{N_{\mathrm{m}}} \sum_{j=1}^{N_{\mathrm{c}}} V_{\mathrm{mc}}(|\mathbf{R}_n - \mathbf{r}_j|) + \frac{1}{2} \sum_{i,j=1;\, i \ne j}^{N_{\mathrm{c}}} V_{\mathrm{cc}}(|\mathbf{r}_i - \mathbf{r}_j|) \tag{9}$$

is the total counterionic part of the potential energy. Furthermore, the classical partition function

$$Z = \frac{1}{N_{\mathrm{c}}!} \frac{1}{\Lambda^{3N_{\mathrm{c}}}} \int_V \mathrm{d}^3 r_1 \cdots \int_V \mathrm{d}^3 r_{N_{\mathrm{c}}} \exp\left[-\frac{V_{\mathrm{c}}}{k_{\mathrm{B}} T} \right] \tag{10}$$

guarantees the correct normalization $\langle 1 \rangle_{\mathrm{c}} = 1$.

At this stage let us emphasize a couple of points.

(i) Since the counterions are averaged out, the effective forces acting on the macroions do not depend on the counterion positions $\{\mathbf{r}_j\}$. They do, however, depend parametrically on the positions of the macroions $\{\mathbf{R}_j\}$. While the repulsive $\mathbf{F}_i^{(\mathrm{m})}$ is explicitly given and clearly a pairwise interaction, the nontrivial part $\mathbf{F}_i^{(\mathrm{c})}$ contains in general *many-body forces* of arbitary order (triplet, quadruplet, etc.) due to nonlinear counterion screening [8]. It is only in linearized theories (like Debye–Hückel approaches) that these forces are pairwise, too.

(ii) As it can be checked immediately, the effective force $\mathbf{F}_i$ can be derived as a gradient of an effective *potential energy*, i.e. $\mathbf{F}_i = - \nabla_{\mathbf{R}_i} V_{\mathrm{eff}}(\{\mathbf{R}_j\})$ with

$$V_{\mathrm{eff}}(\{\mathbf{R}_j\}) = \sum_{i,j;\, i<j}^{N_{\mathrm{m}}} V_{\mathrm{mm}}(\mathbf{R}_i - \mathbf{R}_j) - k_{\mathrm{B}} T Z \,. \tag{11}$$

This quantity clearly differs from the counterion-averaged total potential energy

$$U(\{\mathbf{R}_j\}) := \sum_{i,j;\, i<j}^{N_{\mathrm{m}}} V_{\mathrm{mm}}(|\mathbf{R}_i - \mathbf{R}_j|) + \langle V_{\mathrm{c}} \rangle_{\mathrm{c}} \,. \tag{12}$$

(iii) An alternative expression for the counterion-induced forces $\mathbf{F}_i^{(\mathrm{c})}$ can be obtained by using the inhomogeneous equilibrium counterion density profile in the field of fixed macroions which is defined via

$$\rho^{(0)}(\mathbf{r}; \{\mathbf{R}_j\}) = \left\langle \sum_{n=1}^{N_{\mathrm{c}}} \delta(\mathbf{r} - \mathbf{r}_n) \right\rangle_{\mathrm{c}} \,. \tag{13}$$

Then $\mathbf{F}_i^{(\mathrm{c})}$ can be rewritten as

$$\mathbf{F}_i^{(\mathrm{c})}(\{\mathbf{R}_j\}) = - \int_V \mathrm{d}^3 r \, \rho^{(0)}(\mathbf{r}; \{\mathbf{R}_j\}) \, \nabla_{\mathbf{R}_i} V_{\mathrm{mc}}(|\mathbf{r} - \mathbf{R}_i|) \,. \tag{14}$$

(iv) The definition of the effective forces differs from those derived from the potential of mean force. While the effective force represent the full bare interaction, the latter incorporates also correlations between the macroions, see e.g. [9].

Equations (5)–(6) are the exact expressions for the effective forces in the framework of classical statistical mechanics. The counterion average, however, can in general not performed analytically. In principle, there are two different routes to follow: First, one can use (approximative) density functional theory to obtain $\rho^{(0)}(\mathbf{r}; \{\mathbf{R}_j\})$ and then use (14) to get $\mathbf{F}_i^{(\mathrm{c})}(\{\mathbf{R}_j\})$. This strategy was employed in Ref. [8]. However, since the exact free energy functional is only known approximatively, one gets an uncontrolled error at least if the counterion density profile is strongly inhomogeneous near the macroionic surfaces.

The second route which we shall follow in this paper is to calculate the counterion average (6) by "exact" computer simulation. The drawback here is that – due to the enormous number of counterions – one can only treat relatively small system sizes. In the following we consider only *two* macroions ($N_{\mathrm{m}} \equiv 2$). The two macroions are placed symmetrically along the room diagonal of a cubic box such that the centre of the cube coincides with the center-of-mass of the two particles. The box length L is determined by the macroion density $L = (N_{\mathrm{m}}/\rho_{\mathrm{m}})^{1/3}$. The center-of-mass distance between the two macroionic spheres is denoted by $r \equiv |\mathbf{R}_2 - \mathbf{R}_1|$. Periodic boundary conditions resulting in periodically repeated image charges were taken. Hence our macroion configuration corresponds to a body-centered cubic (BCC) crystal distorted in $(1\,1\,1)$ direction. Due to symmetry, $\mathbf{F}_1 = -\mathbf{F}_2$ and the total effective force depends only on the macroion distance r. From symmetry consideration, the direction of $\mathbf{F}_1$ is along the room diagonal, i.e. parallel to $\mathbf{R}_1 - \mathbf{R}_2$. We therefore consider only the magnitude of the force projecting it onto the room diagonal

$$F(r) = \mathbf{F}_1(r) \cdot (\mathbf{R}_1 - \mathbf{R}_2)/r \qquad (15)$$

which is the key quantity of the paper. If $F(r)$ is positive, then the effective force is repulsive for the given separation r while it is attractive if $F(r)$ becomes negative.

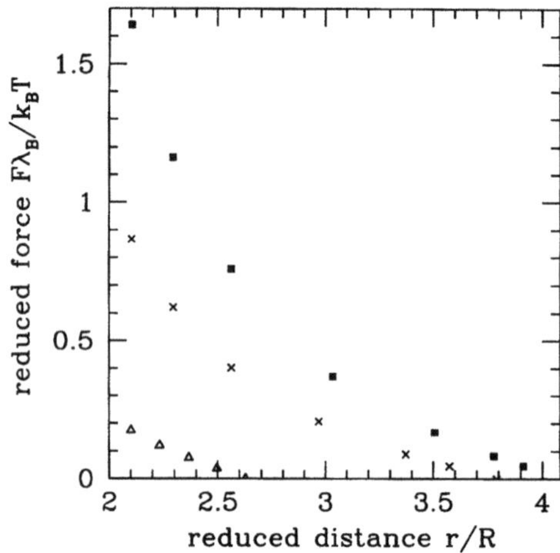

Fig. 1 Effective macroion force $F(r)$ in units of $k_{\mathrm{B}}T/\lambda_{\mathrm{B}}$ versus reduced distance r/R. $\lambda_{\mathrm{B}} = e^2/\varepsilon k_{\mathrm{B}}T$ is the Bjerrum length and $R \equiv \sigma/2$ is the macroion radius. The statistical error of the Monte-Carlo data is smaller than the size of the symbols. The parameters are as follows: (a) crosses: $Z = 200$, $\phi = 0.1$. (b) open triangles: $Z = 100$, $\phi = 0.3$. (c) full squares: $Z = 300$, $\phi = 0.08$

Results from computer simulation

Let us now summarize and discuss computer simulation results, some of them are published in Refs. [10–12].

(i) Monte Carlo results for $F(r)$ are shown in Fig. 1. The parameters are typical for experimental samples and chosen as in Ref. [8]: $q = 1$, $T = 300\,K$, $\varepsilon = 78$, $\sigma = 106$ nm. As is clear from Fig. 1, the forces are always *repulsive* (i.e. $F(r) > 0$) over the broad range of distances explored. Of course, in an undistorted configuration the forces are exactly zero due to symmetry of the macroionic configuration.

(ii) For extremely high couplings, the computer simulation data exhibit large fluctuations in the force and the corresponding statistical error is huge [13]. Nevertheless, the force becomes much less repulsive if the distance between the macroions is of the order of few Bjerrum lengths. It is still unclear whether the net force gets attractive but in principle large fluctuations in the counterionic Stern-layer around the macroionic surfaces can give rise to an attraction [14]. At this stage let us discuss and compare the effective forces in a different geometric set-up, namely that between rigid rod-like polyelectrolytes and that between charged planar plates. In the first case, a counterionic attraction was found recently [15, 16] if the rod-radius is microscopic. Although this is realistic for DNA-strands, it does not seem to be important for charged macroionic spheres since their radius is much larger. In the second case (charged plates), an effective attraction was found for

large couplings [17–19]. One may argue that the case of spheres and planes is simply related performing a Derjaguin-like approximation for nearly touching spheres. This is not that simple, however, since the interaction is still long-ranged and the topology of two spheres is different from that of two planes.

(iii) For suspensions with added salt and asymmetric macroions (i.e. macroions of different radii and different charges) there is again repulsion for moderate couplings [12]. However, in the presence of a small neutral cage, the interaction between a particle and a neutral plane can become attractive. This is expected due to counterion exclusion by the neutral cage [12].

Experiments and discussion

As regards recent experiments discussed already in the Introduction, one carefully has to distiguish whether the experimental samples considered were in the bulk or confined between plates. It now seems to be established that there is no direct hint towards a macroionic attraction in bulk systems which is in agreement with our theoretical findings. In a confined system, however, Kepler and Fraden [4] observed a peak in the pair correlation function for high dilution revealing an attraction. Grier and coworkers [5–7, 20] systematically investigated the effect of confinement on the interaction. While the interaction appears to be repulsive in a multilayer system far away from the plates, it becomes attractive for two macroions close to the plates. This becomes apparent by the observation of metastable crystallites near the charged surfaces but also from dynamical measurements of the diffusion of two nearby particles pinned by tweezers.

Theoretically, it is still unclear whether there is any mechanism leading to attraction near the plates. A possible explanation is the presence of additional counterions stemming from the highly charged plates which give rise to strong fluctuations and correlations absent in the bulk system. More theoretical work is needed to explore this interesting question further.

References

1. Arora AK, Tata BVR (eds) (1996) Ordering and Phase Transitions in Charged Colloids. VCH Publishers Inc., New York
2. Löwen H (1997) Physica A 235:129
3. Stevens MJ, Falk ML, Robbins MO (1996) J Chem Phys 104:5209
4. Kepler GM, Fraden S (1994) Phys Rev Lett 73:356
5. Crocker JC, Grier DG (1996) Phys Rev Lett 73:352
6. Crocker JC, Grier DG (1996) Phys Rev Lett 77:1897
7. Larsen AE, Grier DG (1996) Phys Rev Lett 76:3862
8. Löwen H, Hansen JP, Madden PA (1993) J Chem Phys 98:3275
9. Chu X, Wasan DT (1996) J Colloid Interface Sci 184:268
10. D'Amico I, Löwen H (1997) Physica A 237:25
11. Löwen H, D'Amico I (1997) J Phys Condens Matter 9:8879
12. Allahyarov EA, Löwen H, Trigger SA (1998) Phys Rev E 57
13. Allahyarov E, D'Amico I, Löwen H, to be published
14. Rouzina I, Bloomfield VA (1996) J Phys Chem 100:9977
15. Groubech-Jensen N, Maschl RJ, Bruinsma RF, Gelbart WM (1997) Phys Rev Lett 78:2477
16. Ha BY, Liu AJ (1997) Phys Rev Lett 79:1289
17. Guldbrand L, Jonsson B, Wennerstrom H, Linse P (1984) J Chem Phys 80:2221
18. Stevens M, Robbins MO (1990) Europhys Lett 12:81
19. Valleau JP, Ivkov R, Torrie GM (1991) J Chem Phys 95:520
20. Larsen AE, Grier DG (1997) Nature 385:230

Progr Colloid Polym Sci (1998) 110:16–20
© Steinkopff Verlag 1998

E. Lécolier
A. Mourchid
P. Levitz

Rheological properties of anisotropic charged colloidal suspensions

E. Lécolier (✉) · A. Mourchid
P. Levitz
Centre de Recherche sur la
Matière Divisée (CRMD)
Centre National de la Recherche
Scientifique (CNRS)
Université d' Orléans
1B, rue de la férollerie
45071 Orléans
France

Abstract In this paper, we have investigated the viscoelastic properties of aqueous clay suspensions. The complex modulus has been measured on a very extended scale ranging from 10^{-5} to 10^2 rad s^{-1}. For this purpose, we used two sets of experiments: a controlled shear rate and a controlled stress rheometer. The data show that the relaxation mechanisms are very slow, and the noticeable variation of the viscoelastic moduli is located in the very narrow range of our frequency scale. We have pointed out that the elastic modulus evolves monotically with the particle concentration and follows the power law: $G' = A(C - C_0)^\alpha$. Our results show that only the threshold C_0 varies with the ionic strength I. This power law holds in the concentration regime where the suspensions are either optically isotropic or birefringent viscoelastic solids. The observation of flow birefringence of isotropic gels leads to conclude that the Laponite suspensions are composed of micro-sized nematic clusters. This conclusion is in good agreement with the large relaxation times measured and with the hierarchical structure of these suspensions.

Key words Laponite – rheology – clay suspension – complex fluid – viscoelastic properties

Introduction

The rheological properties of clay suspensions have been intensively studied over the last decades. Clay suspensions present a wide variety of rheological behavior: Newtonian fluid, flocculated dispersion or thixotropic gel [1–3]. These properties are of prime importance for both industrial use and fundamental research. Although clay suspensions are known to present interesting rheological behaviors, it is still a matter of research.

In our laboratory, we have studied Laponite suspensions, a synthetic clay manufactured by Laporte Industry. This synthetic hectorite-type clay has a high chemical purity and a small anisotropy of shape compared to other natural clay materials. Moreover, suspensions of Laponite have emerged as one of the best material for fundamental studies of colloidal clay dispersions [4–12]. Relations between thermodynamics, structural and mechanical properties of Laponite dispersions were recently discussed in the literature. The occurrence of a mechanical sol–gel transition and an isotropic–nematic transition constitutes one of the most important results [5, 9]. This behavior raises several questions concerning a possible micro phase separation, the temporal evolution and the different geometric scales involved in the structure of the gel suspensions.

This paper deals with rheological properties of Laponite suspensions in the linear regime in order to clarify some of these questions.

Experimental section

The Laponite suspensions were prepared by dispersing the clay powder in aqueous solutions. The pH of the aqueous solutions was fixed at 10 with NaOH to avoid congruent dissolution of Laponite [13]. The ionic strength was fixed by adding NaCl to the water at pH = 10. Laponite is dispersed in water as individual platelike sheets having a thickness of 10 Å and a diameter ranging between 250 and 300 Å. The Laponite particles bear a structural negative charge. The charge is balanced by Na^+ counterions located all around the sheets in the so-called ionic double layer. Positive and negative charges can appear on the edge due to dissociation of amphoteric surface acids. But with a pH fixed at 10, the dissociation of -MgOH and -LiOH are limited and then we can consider the particle as negatively charged. The samples were stirred with an homogenizer at high speed (5000–7000 rpm) for 10 min, sealed and stored for one week under nitrogen at room temperature prior to any measurements. The samples were not studied after one month from the date of their preparation to avoid chemical problems due to ageing process of the suspensions. This protocol allowed to obtain homogeneous and transparent dispersions with clay concentrations, C, up to 3% (w/w), and ionic strength, I, ranging from 10^{-4} to 10^{-2} M. Above $I = 2 \times 10^{-2}$ M, the dispersions flocculate. In order to prepare suspensions with higher clay concentrations, we used osmotic stress method. This method allows to prepare suspensions with solid concentration up to 12% (w/w), and to define correctly the ionic strength and then the screening Debye length which plays a major role in colloidal systems.

The rheological measurements were divided into two sets:

(1) *Controlled shear rate rheometry*: the measurements were made using Haake RV20 rheometer at room temperature and controlled humidity level. A vane was immersed in the dispersion and rotated at constant low speed. Then, the stress arising in the materials was recorded as a function of time. The measurements were performed during several hours. The geometry of the vane consisted of six blades located at equal angles around a cylindrical shaft. The dimensions of the blades were 22 mm diameter and 16 mm height. This geometry prevented the delicate problem of wall slip. As the material was disturbed by the immersion of the vane, the gel was allowed to relax during about 3 h before starting up the experiment.

(2) *Controlled stress rheometry*: a Carri-med CSL 50 was used to obtain the frequency variation of viscoelastic modulus of the suspensions in the range of 10^{-2}–10^2 rad s^{-1}. The sample was placed between a cone and a plate which were sealed within a thermostated housing with constant humidity at 20 °C. Two types of experiments were performed. In the first one, we were following the restructuration of the gel which was disturbed during the setting up of the sample between the cone and the plate. A weak sinusoidal stress was applied to the sample and the complex viscoelastic modulus, G^*, was recorded as a function of time. Once G', the real part of the complex modulus, and G'', the imaginary part, have reached a plateau (typically after 3 h), we considered that the suspension has recovered its equilibrium gel state. In the second experiment, the evolution of G' and G'' as a function of the frequency was measured in the linear elastic regime of the sample.

Measurements of the stress relaxation function were achieved either for fixed C with I variable, or for C variable with I constant. The purpose was to apply a finite deformation and to measure the resulting stress during the period of application of this deformation and after ceasing it. The first part of the curve is an almost linear increase of the stress with time. The slope of the curve for $t = 0$ leads to the determination of the elastic modulus at high frequency, $G'_{\omega \to \infty}$, according to the following equation [14]:

$$G' = \frac{R_C^2 - R^2}{2\Omega R_C^2} \frac{d\tau}{dt},\qquad(1)$$

where R and R_C are the inner and outer radii, respectively, of the geometry used, and Ω is the angular velocity of the vane. The second part of the stress–time curve, obtained after the cessation of the sample shearing, represents the relaxation function of the gel.

Results

Figure 1 shows the resulting shear stress for a clay dispersion of 3% (w/w) concentration and $I = 2 \times 10^{-3}$ M. The relaxation function decreases very slowly and shows no evidence of complete relaxation even after several hours. This is indicative of a viscoelastic solid behavior. Similar comportment was observed for different clay concentrations and ionic strengths in the viscoelastic phase.

The relaxation part of the curves exhibits a non-exponential variation. However, it was possible to model the curves for all the suspensions studied with a stretch exponential decay with a relaxation time, t_0, of about 10 000 seconds as indicated by the expression:

$$\exp[-(t/t_0)^a],\qquad(2)$$

where the value of the exponent, a, is nearly 0.45. This observation implies that the relaxation time, t_0, and the exponent, a, do not change strongly with C and I unlike

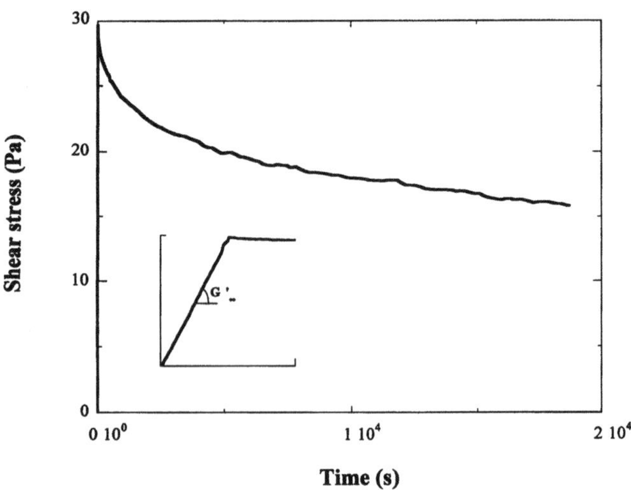

Fig. 1 Transient evolution of the shear stress for a viscoelastic suspension of Laponite of $C = 3\%$ w/w and $I = 2 \times 10^{-3}$ M

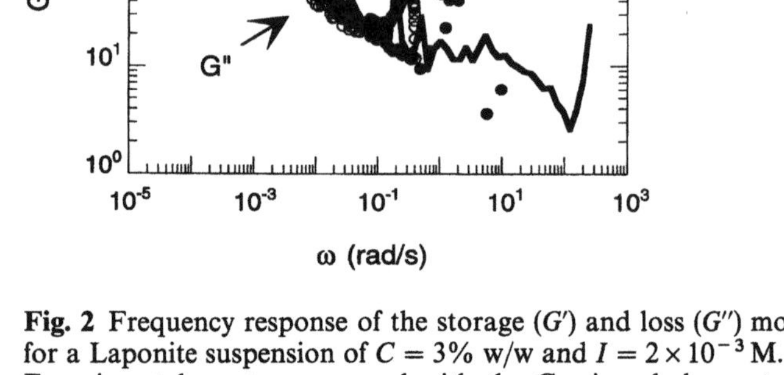

Fig. 2 Frequency response of the storage (G') and loss (G'') moduli for a Laponite suspension of $C = 3\%$ w/w and $I = 2 \times 10^{-3}$ M. (—) Experimental spectra measured with the Carri-med rheometer. (● and ○) The computed spectra of G' and G'' obtained by using the Fourier transform of the transient shear stress

the amplitude of the measured stress which increases with these parameters.

It is interesting to check the validity of the rheological data by probing the analogy between the results of the two rheometers. Then we used the Fourier transform analysis of the relaxation function. In the linear regime, this method permits to check if there is any classical problem with the rheology experiments like slipping at the interface sample/geometry in the case of cone/plate. Another interesting point of this method is that we can extend the domain frequency of the viscoelastic moduli spectra by using a Fourier transform of the transient stress recorded during a long time.

The method applied is numerical calculation of Fourier transform of the shear stress and shear rate. The viscoelastic modulus: $G^* = G' + iG''$ is equal to the ratio of $\hat{\tau}$ on $\hat{\gamma}$, where $\hat{\tau}$ is the Fourier transform function of the derivative of the shear stress with respect to time, and, $\hat{\gamma}$ is the Fourier transform of the shear rate.

A typical result of this method is presented in Fig. 2. The transformation shown was done on the relaxation function results obtained with the same sample as in Fig. 1. The viscoelastic modulus determined by the Fourier transform analysis is compared to the viscoelastic modulus spectra measured with the Carri-med in an oscillatory frequency sweep. Comparison between the two data is very good as the G' and G'' spectra obtained from the Fourier transform method recover the direct experimental data over three orders of magnitude of the frequency scale: from 10^{-2} to 10 rad s^{-1}. The interval of the frequency was lowered for the calculated G' and G'' to few 10^{-5} rad s^{-1}. We can see Fig. 2 that G' varies slowly with the frequency.

The G'' shows an increase as the frequency is decreased but no maximum seems to be reached.

The method confirms our previous results [4] concerning the weak evolution of G' and G'' with the frequency indicating that the samples are viscoelastic solids.

From the transient stress measured we can extract modulus of the suspensions as mentioned above. The experiments was done on a large number of suspensions with different clay concentrations and at fixed ionic strength (10^{-2} M). The Carri-med was also used to determine the elastic modulus G' for different particle concentrations and different ionic strengths. We shown before that with the aid of the rheology a transition from Newtonian liquid to viscoelastic solid was observed. This transition is induced by increasing the clay concentration or the ionic strength, and the transition point C_0 is given by

$$G' = A(C - C_0)^{\alpha} . \tag{3}$$

We realized that the elastic modulus is only sensitive to this transition point C_0. All the G' data collected on the Laponite suspensions, by using the Haake and the Carri-med rheometers, over several years, for different clay concentrations and ionic strengths can be plotted versus $(C - C_0)$. We can see on Fig. 3 that all the experimental points are setting on a master curve, representing the power law of Eq. (3). It results that the amplitude A and the exponent $\alpha(= 2.35)$ are independent of the clay concentration and of the ionic strength. This result suggests that the mechanical state of the suspensions of Laponite can be fixed by adapting the distance to the transition point C_0. This parameter is only dependent on the ionic strength.

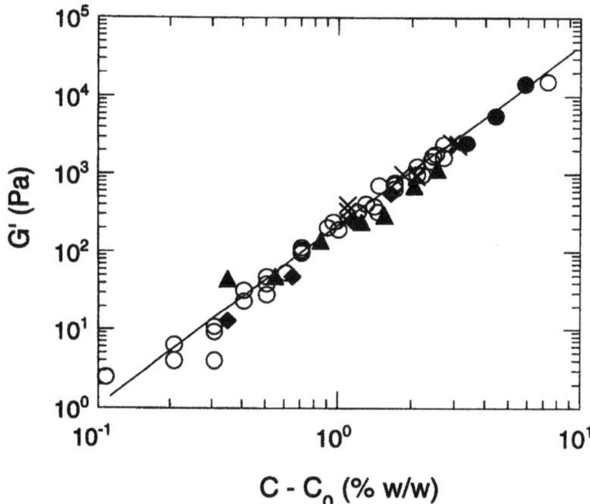

Fig. 3 Variation of the elastic modulus (G') as a function of $(C - C_0)$. (o) $I = 10^{-2}$ M and $C_0 = 0.3\%$ w/w. (▲) $I = 6 \times 10^{-3}$ M and $C_0 = 0.45\%$ w/w. (●) $I = 5 \times 10^{-3}$ M and $C_0 = 0.6\%$ w/w. (♦) $I = 2 \times 10^{-3}$ M and $C_0 = 1.35\%$ w/w. (×) $I = 10^{-4}$ M and $C_0 = 1.85\%$ w/w. The continuous curve follows Eq. (3)

Discussion and conclusion

Gelation of clay colloid suspensions of Laponite and its related structure was the subject of many publications. Some models were proposed in order to explain the phenomenon of Laponite gelation. The classical mechanism proposed by van Olphen [1] is a micro-flocculation promoted by attraction between the oppositely charged edges and faces of the clay plate-like particles and reinforced by van der Waals bonds, yielding to the "house of cards" structure. However, it is now admitted that this mechanism of gelation is favored only at high ionic strengths when the electrostatic repulsions are weak [2, 15]. The flocculation of particles is avoided at low ionic strength. This was evidenced by small angle X-ray scattering measurements, TEM observations of cryofractured gels and osmotic pressure determination.

Recently, it was pointed out, by studying the textures of the dispersions that the gel state in the phase diagram of Laponite consists of a non-birefringent gel at low clay concentration and a birefringent one at high solid concentration [9]. Another study of the static light scattering (SLS) at very small angles and the rheology measurements, showed that the gel phase of Laponite splits into two different gel states made of micro-sized aggregates [10]. The first state, obtained at low clay concentration, has a fractal dimension of 1 and quadratic variation of the yield stress with the solid concentration. At higher clay concentration, the fractal dimension of the gel is 1.8 and the yield stress is a cubic function of the solid concentration.

The rheology data reported in this paper do not show a difference in the evolution of the elastic modulus with the particles concentration in the gel state instead of the two distinct regimes, occurring by increasing the clay concentration, reported previously [9, 10]. However, there are some similarities with the results presented here. The length scale deduced from SLS data of about $1 = \mu$m, in ref. [10], is compatible with the relaxation time obtained from the relaxation function of the shear stress shown in Fig. 1. This observation is supported by the form of the spectra of G' and G'' which exhibit an evolution more pronounced at very low frequencies (10^{-5}–10^{-4} rad s^{-1}, see Fig. 2). This is a signature of the existence of a mechanism of relaxation with a long time value and compatible with the presence of micro-sized objects inside the suspensions [12, 16]. At this stage of discussion, the nature of these micro-sized objects should be examined. For this, a gel prepared with particle concentration near the transition point was allowed to flow in a pipette placed between crossed polarizers. At rest the suspension did not show any permanent birefringence, and a laser beam was not transmitted. When the suspension was starting to flow with a low average shear rate (~ 10 s^{-1}) two side intense bands, separated by a dark central band, were transmitted. When the flow was stopped, it took few seconds for the side bands to disappear. This experiment was similar to the one used by Langmuir [17] for the study of the flow birefringence in some suspensions of colloidal particles like bentonite clay. Two arguments suggest that the anisotropic objects are not constituted of single particles. (i) The ratio of the shear rate on the coefficient of diffusion of a single particle leads to a very low Peclet number, $\sim 10^{-4}$, and seems to indicate that individual particles of Laponite cannot become oriented in these conditions. (ii) The time it takes the side bands to disappear when the flow stops is very long compared to the rotational time of a single particles of Laponite ($\sim 10^{-4}$ s^{-1}). These observations, leading to conclude that the Laponite suspensions are composed of micro-sized nematic clusters, are compatible with our previous results obtained from the analysis of the SAXS data and from numerical calculations [4].

The origin of the ordering of particles was discussed in detail in a previous publication. However there is still an indetermination on the mechanism of the finite size of this ordering which prevent the obtaining of a nematic liquid crystal. The rheological study exposed here permits to locate the point of the transition from the Newtonian liquid to the viscoelastic solid state by means of Eq. (3). The meaning of this equation and the values of its parameters (the amplitude and the exponent) are still obscure. Their understanding supposes the knowledge of the structure of the suspensions and of the interparticle potential of interaction over a long range as the length scale involved in the rheology is very large. The relaxation behavior observed by the study of the transient decay of the shear stress shows long relaxation times, compatible with the existence of micro-sized nematic clusters. The analysis of the decay function was well fitted with a stretch exponential variation and suggests that the system is highly polydisperse and that some hierarchy is controlling the phenomenon of the relaxation of the shear stress [18]. As it was pointed out above, the Fourier transform formalism used in this study extended the frequency domain of the rheological moduli towards the very low value. This frequency scale seems to be more appropriate for the mechanical study of the viscoelastic clay suspensions. Interrelation and/or interconnection of the microdomains should be responsible of the appearance of the rigidity of the suspensions at this very large length scale. It is worth noticing that the elastic and the viscous moduli evolve as $G' = G'_0 + \Delta G(\omega)$, and $G'' = \Delta G(\omega)$, respectively, in the frequency range of 10^{-5}–10^2 rad s^{-1}. This behavior indicates that a relaxation mechanism occurs with very long relaxation time.

References

1. van Olphen H (1977) An introduction to Clay Colloid Chemistry. Wiley, New York
2. Rand B, Pekenc E, Goodwin JW, Smith RW (1980) J Chem Soc Faraday Trans 76:225
3. Sohm R, Tadros ThF (1989) J Colloid Interface Sci 132:62
4. Mourchid A, Delville A, Lambard J, Lécolier E, Levitz P (1995) Langmuir 11:1942
5. Mourchid A, Lécolier E, Van Damme H, Levitz P (1998) Langmuir, to be published
6. Mourchid A, Delville A, Levitz P (1995) J Chem Soc Faraday Discuss 101:275
7. Dijkstra M, Hansen JP, Madden PA (1995) Phys Rev Lett 75:2236
8. Ramsay JDF (1986) J Colloid Interface Sci 109:441
9. Gabriel J-CP, Sanchez C, Davidson P (1996) J Phys Chem 100:11 139
10. Pignon F, Piau J-M, Magnin A (1996) Phys Rev Lett 76:4857
11. Willenbacher N (1996) J Colloid Interface Sci 182:501
12. Kroon M, Wegdam GH, Sprik R (1996) Phys Rev E 54:6541
13. Thompson DW, Butterworth JT (1992) J Colloid Interface Sci 151:236
14. Alderman NJ, Meeten GH, Sherwood JD (1991) Non-Newtonian Fluid Mech 39:291
15. Norrish K (1954) Discuss Faraday Soc 18:120
16. Tassin J-F, Robert G, Benyahia L (1996) Les Cah Rhéo 15:238
17. Langmuir I (1938) J Chem Phys 6:873
18. Bellini T, Mangtegazza F, Piazza R, Degiorgo V (1989) Europhys Lett 10(5):499

Progr Colloid Polym Sci (1998) 110:21–24
© Steinkopff Verlag 1998

A. Fernández-Nieves
C. Richter
F.J. de las Nieves

Point of zero charge estimation for a TiO$_2$/water interface

A. Fernández-Nieves
F.J. de las Nieves (✉)
Universidad de Almería
Departamento de Física Aplicada
La Cañada de San Urbano s/n
04120 Almería
Spain

C. Richter
DLR-Plataforma Solar de Almería
Department of Solar Chemistry
04080 Almería
Spain

Abstract The point of zero charge of a certain commercial titanium dioxide has been determined. This is crucial in determining the colloidal stability that our TiO$_2$/water dispersion exhibits, which greatly influences the efficiency of the detoxification process where this interface is employed. The method used is based on the semiconductor properties of titanium dioxide and on the slow OH$^-$ consumption that has been observed in this type of interfaces. The electrophoretic mobility of the TiO$_2$ dispersion was measured at different pHs, establishing for the point of zero charge a value of 7.5.

Key words Titanium dioxide – point of zero charge – isoelectric point

Introduction

In the last few years, solar water detoxification has received a lot of attention. The removal of persistent non biodegradable organic chemicals from the water is a very important and pressing ecological problem. The methods used presently have moved from processes involving phase transfer, such as active carbon adsorption, to processes that achieve the destruction of the contaminants [1]. Solar photocatalytic detoxification has proven to be a very efficient technique in this fields. Titanium dioxide has been the most employed catalyst for this kind of study [1–4].

Two experimental settings have been proposed for this process, which basically differ in the way the TiO$_2$ is employed: as a flat layer over which the waste water passes, or as a colloidal suspension. As established in [2, 3], the second facility gives the best results.

It is obvious that for a given quantity of TiO$_2$, the area the colloidal suspension offers, plays a very important role in the detoxification process, since the reactions that originate the destruction of the toxic organic compounds take place on or near the catalyst surface. The necessity of a rapid removal of the titanium dioxide from the water, once it has been purified, is another important variable that may be enhanced if we are able to destabilize our colloidal suspension, increasing then its sedimentation rate. Finally, the possibility of reusing the TiO$_2$ also makes attractive the knowledge of the conditions under which our colloidal suspension is stable or not.

All these facts point to the necessity of characterizing, in a colloidal sense, the dispersion formed by the TiO$_2$ particles immersed in water. Variables, such as the point of zero charge (PZC) and the isoelectric points (IEP) in the presence of different substances, are very important. They will enable us to gain information that will do the detoxification process more efficiently. This is the main objective of this work: the estimation of the PZC of a colloidal dispersion formed by TiO$_2$ particles in water.

Preliminary aspects

For oxide colloidal systems, the H$^+$ and OH$^-$ are considered to be the potential determining ions. The PZC is defined as the solution pH value for which the free surface charge of the colloidal particles is zero.

Our idea for estimating the PZC has its origin in the slow OH$^-$ consumption found by several authors. Normally, a distinction is made between a fast consumption of potential determining ions and a forward slow wastage of those ions. Janssen and Stein [4] point out that for their TiO$_2$ particles, which are very similar to the ones we are using, this consumption is due to the suspension contamination with CO$_2$, in accordance with their infrared spectroscopy results. Trimbos and Stein [5] have studied this process with the ZnO/water interface, neglecting the presence of CO$_2$; but their experiments are carried out in a nitrogen atmosphere. They found that after a primary consumption of H$^+$ and OH$^-$, a slow consumption of the first or second ion occurred, depending on whether the pH was greater or lower than 8.9, respectively. They proposed that the cause of one or other process is the presence of free holes or electrons close to the surface hydroxyl groups, ZnOH. These authors found a square root dependence between the consumed concentration of ions and the time. Explicitly;

$$[\text{OH}^-] - [\text{H}^+] = A\sqrt{t} \tag{1}$$

with A, a pH-dependent constant that takes values within the interval 0 and $2 \times 10^{-8}\text{M}/\sqrt{\text{min}}$ and which is positive or negative depending on whether the consumption is of H$^+$ and OH$^-$, respectively.

Our colloidal particles are semiconductors, in spite of what, the latter described effect might take place. Additionally, a consumption due to the presence of CO$_2$ could also happen if its presence is not clearly avoided.

These two events led us to believe that our dispersion might experience a decrease in solution pH due to a slow consumption of OH$^-$, despite the starting value being basic. If measurements of electrophoretic mobility are done while the pH is dropping, as will be in our case, we might observe a change in its sign, that will coincide with a change in the particles electrokinetic sign charge. The pH for which this might occur, will be the IEP of our suspension under the presence of the base's cation employed in raising the pH to its starting value. Now, if this starting pH is not too high, specific adsorption of this ion may be neglected, which makes possible the equality of the zero mobility pH and the PZC of our colloidal particles.

This procedure would then be valid, provided the PZC of our dispersion is in between the starting pH, which cannot be high, and the final pH that may be due to either of the two effects mentioned. The advantage is that it is a very easy way to estimate the PZC of colloidal suspensions that are sensitive to the described possible causes.

Experimental

Materials

Degussa titanium dioxide P25 obtained by the aerosol technique was employed in this work, as received. XRD was used to establish that the anatase : rutile percentages in our sample are between 70 : 30 and 80 : 20. BET surface area was found to be around 55 m^2/g. TEM measurements showed irregularly shaped particles and high polydispersity, in accordance with PCS.

The NaOH employed was of analytical grade. The water was ultrapure with a conductivity lower than 1 μS cm^{-1} (ATAPA SA, Spain)

Electrophoretic mobility measurements

These measurements were carried out with a Zetamaster-S (Malvern Instruments). Colloidal dispersions were prepared in a concentration of 0.3 g/l of titanium dioxide, sonicating for 10 min and letting sedimentation for 30 min. The supernatant was taken and used for the experimental measurements. Each mobility value is the average of ten data, changing the suspension sample twice, using as experimental error the standard deviation of these measurements.

Results and discussion

Figure 1 shows the encountered spontaneous pH variation of our TiO$_2$ suspension. The starting pH is 9, which is low enough as to neglect the possible specific adsorption of Na$^+$ ions, finding a stationary value of 6.5, around 12 h from the start of the experiment. Notice that this drop was found after the rapid potential determining ions consumption, which Janssen and Stein found during the first 15 min from the preparation of the suspension. In Fig. 2, the measurements of electrophoretic mobility during the drop of the solution pH are presented, versus the pH. A zero mobility is observed at around 7.5, which as explained before, may be considered as a good estimation of our colloidal particles PZC.

The cause of this pH drop seems to be as proposed by Trimbos and Stein: the presence of free holes in the vicinity of surface hydroxyl groups. The reason for thinking so, may be understood by observing Fig. 3. In it we are representing the H$^+$ concentration versus time together with a square root fitting of the experimental points. This fit corresponds to the following equation:

$$[H^+] = B\sqrt{t} \tag{2}$$

with B $= 0.58 \times 10^{-8}$ M/$\sqrt{\text{min}}$.

Progr Colloid Polym Sci (1998) 110:21–24
© Steinkopff Verlag 1998

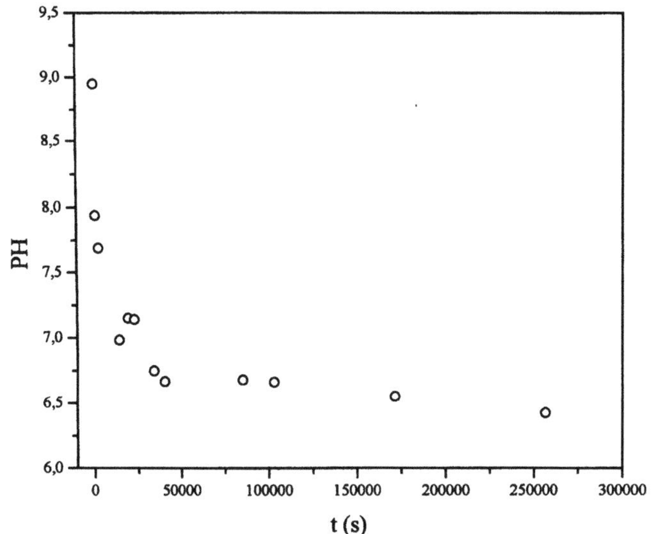

Fig. 1 pH spontaneous variation versus time for the TiO_2 suspension

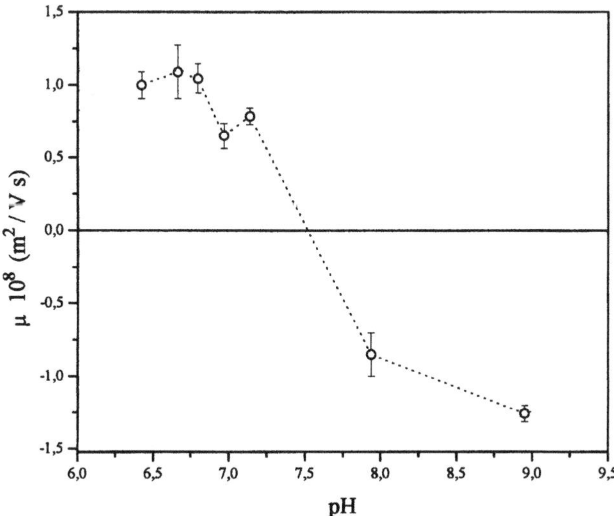

Fig. 2 TiO_2 particles' electrophoretic mobility versus the solution pH

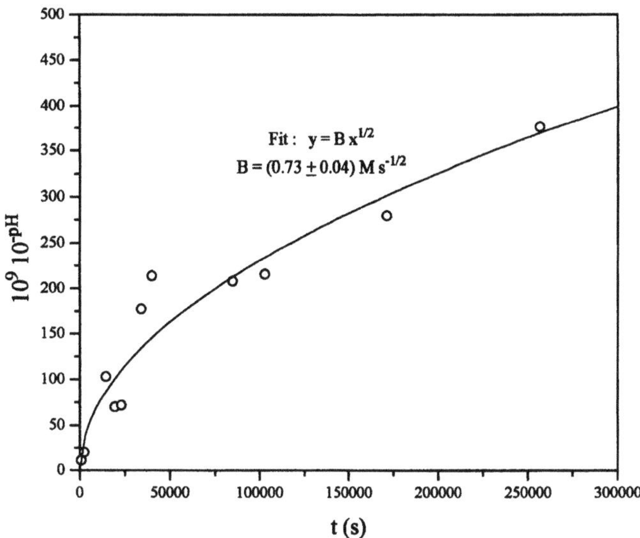

Fig. 3 H^+ concentration against time. Also a fit of the points to a square root function is shown

In viewing our fit, it seems that the obtained results are in good agreement with those of Trimbos and Stein. Anyway, we cannot assure that the CO_2 contribution to the slow OH^- consumption is not present.

The mechanism that might govern the observed behaviour is what we introduce in the following paragraphs.

Probably the process that is taking place is similar to what occurs in a PN union [6], with what is concerned with the flux of electrons from the electrolyte (from the OH^-) to the semiconductor. The driving force for this to occur is the greater Fermi energy level of the solution electrolyte, compared to that of the semiconductor, i.e., the difference in the solution and semiconductor electrochemical potential of the electrons. The TiO_2, even though is a type n semiconductor, acts as a type p in this process, since it is receiving electrons from outside.

Another important fact is that, as in the semiconductors in a PN union, our TiO_2 forms an inside space charge region. In this zone an electric field is generated, which causes the potential energy of the electrons to change. As a consequence, the energy bands are bent (see Fig. 4a).

The electron flux from the solution to the semiconductor, causes the pH to decrease, reducing thus the particles surface charge. Electrons are then accommodated during the process into the semiconductor, while some of them are leaving its surface. Obviously, since the pH is dropping, the electron flux from the solution electrolyte is greater than the electron flux leaving the TiO_2 surface. While this is taking place, the spatial charge region inside the semiconductor is reducing its charge (see Fig. 4b).

When the PZC is reached, the surface free charge is zero and the energy bands flat. The Fermi energy levels must still not be equal, since equilibrium has not yet been established (Fig. 4c).

Below pH 7.5, the particles surface charge as well as the inside charge zone of TiO_2 change in sign. As a consequence the energy bands are bent downwards, since the electron potential energy is reduced (Fig. 4d).

At a sufficiently large time, both Fermi energy levels come to a similar value, causing the electron flux to stop. This occurred around 12 h from the start of the experiment, at a pH around 6.5.

Fig. 4 Schematic representation of the proposed mechanism for the observed slow OH$^-$ consumption in our TiO$_2$/water interface. E_F: Fermi energy level; E_c: Energy associated to the floor of the semiconductor conduction band; E_v: Energy associated to the top of the semiconductor valence band

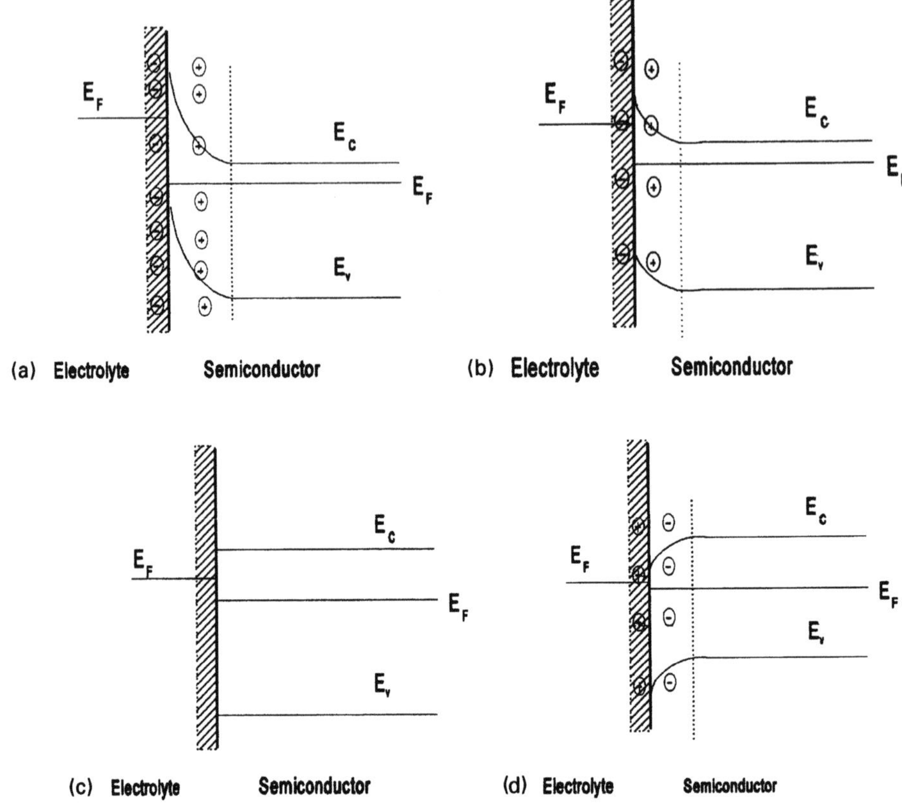

Finally, the proposed mechanism justifies the observed pH drop, an object of future work for its validation with a quantitative support. In ref. [7] the electrolyte/semiconductor interface is considered to have a behaviour which is in agreement with the one proposed here. We remark again that the effect of the CO$_2$ is probably superimposed to the one described, which the authors consider as the dominant mechanism for the observed OH$^-$ slow consumption.

The procedure followed for the determination of the PZC, has been useful for our semiconducting particles of titanium dioxide and may also be a good method for other semiconductor suspensions.

Acknowledgments One of us, A. Fernández-Nieves, wishes to express his gratitude to the Plataforma Solar of Almería (PSA) for the grant which enabled him to initiate this work. The financial support provided by the Commisión Interministerial de Ciencia y Tecnología under proyect MAT96-1035-C03-03 is also gratefully acknowledged.

References

1. Minero C, Pelizetti E, Malato S, Blanco J (1996) Solar Energy 5
2. Malato S, Blanco J, Richter C, Vincent M, (1996) Solar Energy 5:56
3. Blanco J, Malato S, Bahnemann D, Bockelman D, Weichgrebe D, Goslich R, Carmona F, Martínez F (1994) 7th Int Symp on Solar Thermal Concentrating Technologies, Moscow, Russia
4. Janssen MJG, Stein HN (1986) J Colloid Interface Sci 2:109
5. Trimbos HFA, Stein HN (1980) J Colloid Interface Sci 2:77
6. Shalimova KV (1975) Physics of Semiconductors. Mir, Moscow
7. Grätzel M (1989) Colloidal Semiconductors. In Serpone N, Pelizzetti E (Eds) Photocatalysis: Fundamentals and Applications. Wiley Interscience, Canada

Progr Colloid Polym Sci (1998) 110:25–28
© Steinkopff Verlag 1998

J. Stellbrink
B. Abbas
J. Allgaier
M. Monkenbusch
D. Richter
C.N. Likos
H. Löwen
M. Watzlawek

Structure and dynamics of star polymers

Dr. J. Stellbrink (✉) · B. Abbas · J. Allgaier
M. Monkenbusch · D. Richter
IFF-Neutronenstreuung II
Forschungszentrum Jülich
D-52425 Jülich
Germany

C.N. Likos · H. Löwen*
IFF-Theorie II
Forschungszentrum Jülich
D-52425 Jülich
Germany

M. Watzlawek
*Institut für Theoretische Physik II
Heinrich-Heine-Universität Düsseldorf
Universitätsstraße 1
D-40225 Düsseldorf
Germany

Abstract The structural and dynamical properties of an 18-arm poly(isoprene) star polymer ($M_n^{arm} = 7360\ g\,mol^{-1}$) are investigated by experiment and theory. Light scattering techniques reveal the existence of static and dynamic long range correlations ($\xi \approx 100$ nm) above the overlap concentration c^*, which are an indication for a structural glass transition at higher concentrations. Combining statistical-mechanical theories and small angle neutron scattering (SANS), it is shown that the effective pair potential between star polymers is exponentially decaying for large distances and crosses over, at a density-dependent corona diameter, to an ultra-soft logarithmic repulsion for small distances. Finally, neutron spin echo spectroscopy (NSE) is used to investigate the influence of interstar entanglements on the internal dynamics of the star polymer. Hence, star polymers can actually be viewed as hybrids between polymer-like entities and colloidal particles establishing an important link between these different domains of physics.

Key words Star polymer – colloids – neutron scattering – light scattering – liquid state theory – Monte Carlo simulations

Introduction

Regular star polymers consist of a well-defined number f of flexible polymer chains tethered to a central microscopic core. By enhancing this functionality (or arm number) f which governs the interpenetrability of two stars, one can continuously switch from unbranched polymer chains ($f = 1, 2$) to a colloidal sphere ($f \gg 1$). Hence, star polymers can actually be viewed as hybrids between polymer-like entities and colloidal particles establishing an important link between these different domains of physics. Moreover, star polymer solutions reveal quite a number of novel structural and dynamical properties which occur neither in single-chain polymers nor in suspensions of colloidal spheres, for recent reviews see [1, 2].

Here we present a comprehensive study of the structural and dynamical properties of star polymer solutions, in particular in the concentration range above the overlap concentration $c^* = (3/(4\pi R_g^3))x(M_w/N_A)$. Using neutron (SANS, NSE) and light scattering techniques (SLS, PCS), combined with viscometry, it is possible to investigate a spatial and temporal range spanning from microscopic to macroscopic length and time scales. The experimental work is compared to new theoretical approaches using a recently proposed pair potential [3] as a description of the particle interaction. In the theoretical investigations, we use standard liquid-state theory techniques to obtain accurate data for the structure and thermodynamics of the fluid phase and density-functional theory to investigate the phase diagram of the system as a function of concentration and arm number.

Experimental

The partially labeled 18-arm polyisoprene (PI) star was prepared by anionic polymerization following an established procedure [4, 5]. The synthesis of the arms started with deuterated PI and secondary butyl lithium as initiator and proceeded with protonated PI. The still living polymer chains were coupled to the linking agent octadeca-chlorosilan. The result of the synthesis is a labeled 18-arm star with a near-monodisperse molecular weight distribution, which has a protonated core and a deuterated shell. Light scattering experiments were performed on a standard set-up (ALV-125 compact goniometer, Ar^+-Laser $\lambda = 514.5$ nm, ALV5000E correlator). The SANS and NSE experiments were performed at the FRJ-2 reactor of the Forschungszentrum Jülich. Using fully deuterated methylcyclohexane the solvent and the deuterated shell scatter neutrons in the same way and in the experiments only the protonated core is visible (For the light scattering experiments protonated methylcyclohexane was used). Samples covering polymer volume fractions ϕ in the range $5 \times 10^{-4} \leq \phi \leq 0.33$ were investigated.

Results and discussion

In a recent dynamic light scattering study [6] of PI star solutions ($f = 18$) we find two dynamical regimes: (i) Polymer like relaxation dynamics at short times with a collective response following the renormalisation group theory [7, 8] independent of the polymer architecture. (ii) At long times above c^* a very well separated second slow dynamic regime reveals itself in star polymer solutions, see Fig. 1. This slow diffusive process is connected to low angle excess scattering and indicates long range, slow density fluctuations, with $\xi_{slow} \gg R_g$, reminiscent of features of the glass process [9]. Comparing the reduced diffusion coefficient D_0/D_{slow} and the reduced macroscopic viscosity η/η_0 ($D_0 =$ diffusion coefficient at infinite dilution, η_0 solvent viscosity), it is found that both show the same scaling law above c^*, $\sim c^{2.1}$, but the ratio of the diffusion coefficients is two orders of magnitude larger than the ratio of the viscosities. These experimental results resemble properties that have been recently reported for a colloidal PMMA system [10]. Those authors interpreted the slow process as a structural relaxation. This was mainly concluded from the fact that long time diffusion and viscosity show exactly the same ϕ-dependence, only when the diffusion coefficient is measured at q_m, the peak position of the static structure factor. For the star polymer solution $q_m \approx 0.05 \text{ Å}^{-1}$ is much larger than the q values covered in our experiment. Thus, the difference observed here cannot be explained in

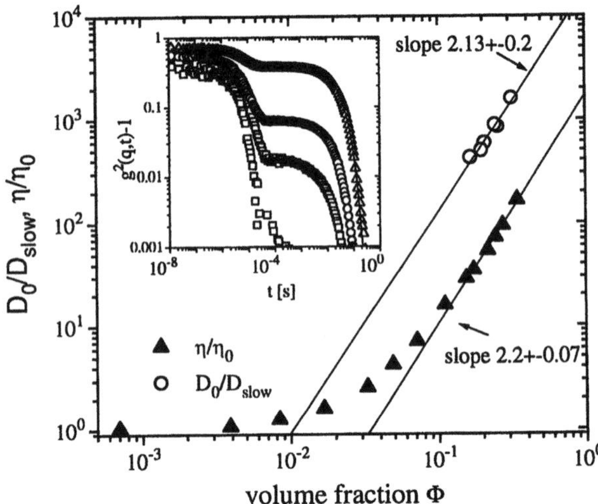

Fig. 1 The reduced diffusion coefficient D_0/D_{slow} is plotted against polymer volume fraction ϕ together with the reduced macroscopic viscosity η/η_0 ($D_0 =$ diffusion coefficient at infinite dilution, η_0 solvent viscosity). The inset shows the intensity autocorrelation functions for the star polymer at $c \geq c^*$ together with that of a linear polymer

this way but must relate in a different way to the colloidal character of the star solution.

For describing the SANS data we make use of a pair potential [11] $V(r)$ for star polymers which reads as follows [3]:

$$V(r)k_BT = \begin{cases} (5/18)f^{3/2}[-\ln(r/\sigma) + (1 + \sqrt{f}/2)^{-1}] \ (r \leq \sigma); \\ (5/18)f^{3/2}(1 + \sqrt{f}/2)^{-1}(\sigma/r) \\ \quad \times \exp[-\sqrt{f}(r - \sigma)/2\sigma] \ (r > \sigma). \end{cases}$$

$$(1)$$

The potential $V(r)$ is an interpolation between a Yukawa-form, suitable for $r > \sigma$, and a logarithmic behavior, appropriate for $r < \sigma$, shifted by a constant which is chosen in such a way that the potential is smooth at $r = \sigma$. The logarithmic form of the interaction sets in when two stars are separated by such a distance that the outermost blobs of the stars in the Daoud–Cotton model [1, 12] fully overlap, i.e., $\sigma/2$ is the distance from the center of the star to the center of the outermost blob. By geometry, the latter has a radius $R_b = \sigma/\sqrt{f}$. We now take the pair potential given by Eq. (1) and apply the Rogers–Young (RY) closure [13] and associated Monte Carlo simulations to obtain information about the pair structure of the liquid, in particular the center-to-center structure factor $S(q)$ of the stars.

In attempting to fit the experimental data for the total scattering intensity $I(q)$ with the theoretical predictions based on an analytic pair potential, we must take into consideration the fact that the star size itself has a

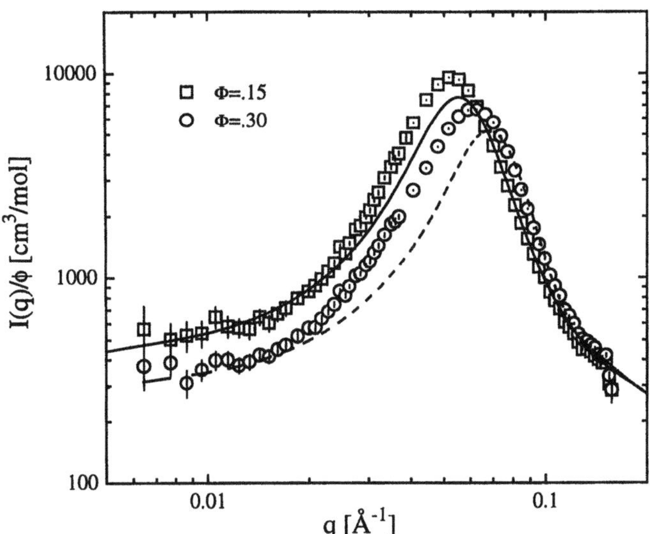

Fig. 2 Experimental (points) vs. theoretical results (lines) for the total scattering intensity $I(q)/\phi$ of 18-arm stars at various volume fractions ϕ. From top to bottom $\phi = 15$ and 30%

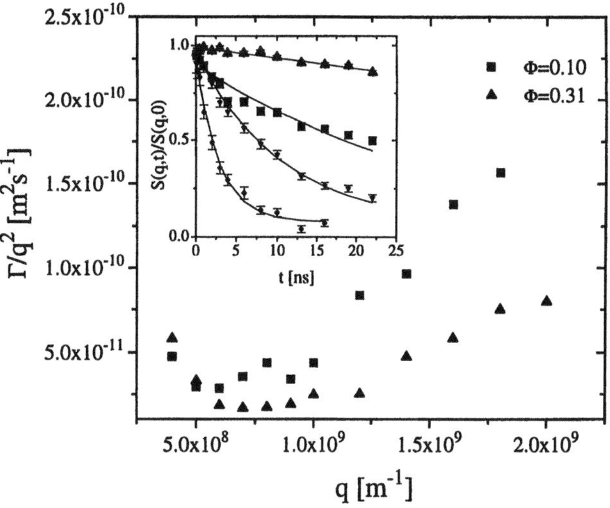

Fig. 3 The reduced relaxation frequencies Γ/q^2 show a minimum when plotted against scattering vector q. The insert show $S(q, t)$ obtained at a polymer volume fraction $\phi = 0.31$, from top to bottom: $q = 0.06, 0.12, 0.16, 0.2 \, \text{Å}^{-1}$

dependence on the concentration ϕ, as was earlier found experimentally [14, 15]. Accordingly, the length σ must also have a ϕ-dependence, $\sigma = \sigma(\phi)$. We fixed this dependence as follows: for the lowest concentration considered, $\phi = 2\%$ we obtain σ by optimizing the agreement between the theoretical prediction and the experimental results. For the potential given by Eq. (1), we obtain in this way $\sigma(\phi = 2\%) = 96 \, \text{Å}$. At the same concentration, the experimentally measured radius of gyration is $R_G(\phi = 2\%) = 76.1 \, \text{Å}$ [14, 15]. This fixes once and for all the ratio $\sigma/R_G \equiv \tau = 1.26$. For all other concentrations ϕ we set $\sigma(\phi) = \tau R_G(\phi)$, where $R_G(\phi)$ is read off from the experimental results [14, 15]. In this way we can say that our fit contains *no adjustable parameters* since σ does not vary arbitrarily with ϕ, but rather in a way dictated by the measured values of the size of the star. Moreover, the value $\sigma = 96 \, \text{Å}$ yields a theoretical prediction $70.63 \, \text{Å}$ for the star radius which is within two error bars from the experimental result. In Fig. 2 we show representative results for $\phi = 15$, and 30%. It can be seen that the fit is quite satisfactory for the whole range of concentrations. In particular, the compressibility of the solution, being proportional to $I(q \to 0)$ is given correctly for all concentrations, as well as the general shape and wavenumber q_{max} at which the scattering intensity displays a maximum. The height of the peak is underestimated by the theory and the agreement worsens somewhat as the concentration grows. However, at high values of ϕ the decoupling between form- and structure factors implied in writing down $I(q) = V_W P(q) \, S(q)$ becomes questionable and this is a possible source of discrepancies between theory and

experiment. We emphasize that our logarithmic-Yukawa potential is the first that gives semi-quantitative agreement between theory and experiment for such a wide range of concentrations. Earlier attempts to fit the experimental results with a hard sphere – Yukawa interaction, for example, failed at and beyond the overlap concentration ϕ^* [16]. Indeed, the existence of a "soft core", such as the logarithmic term in our potential is crucial at high concentrations where the stars start interpenetrating.

Due to the applied matching conditions neutron spin echo experiments (NSE) observe the collective motion of the star centers if we measure close to the structure factor peak (for our star polymer this occurs around $q_m \approx 0.05 \, \text{Å}^{-1}$ in the concentration regime of interest, see previous section). With increasing scattering vector more and more contributions from internal segmental motions inside the core become visible. The reduced relaxation frequencies Γ/q^2, where Γ is obtained from the initial slope of $S(q, t)/S(q, 0)$, show a minimum at the structure factor peak, see Fig. 3. This minimum is a characteristic feature of star polymer solutions. In dilute solution it arises due to the special molecular architecture of the star polymers [17]. Above c^* intermolecular interactions come into play and the observed phenomenon resembles the *deGennes narrowing* known from simple liquids. The depth of the minimum increases with increasing concentration. Moreover, in the high q-range a strong concentration effect is unexpectedly found, which might result from entanglements in the star shell affecting the internal motions of the core. For illuminating this hypothesis we have investigated the dynamics of a smaller star with size equal to the

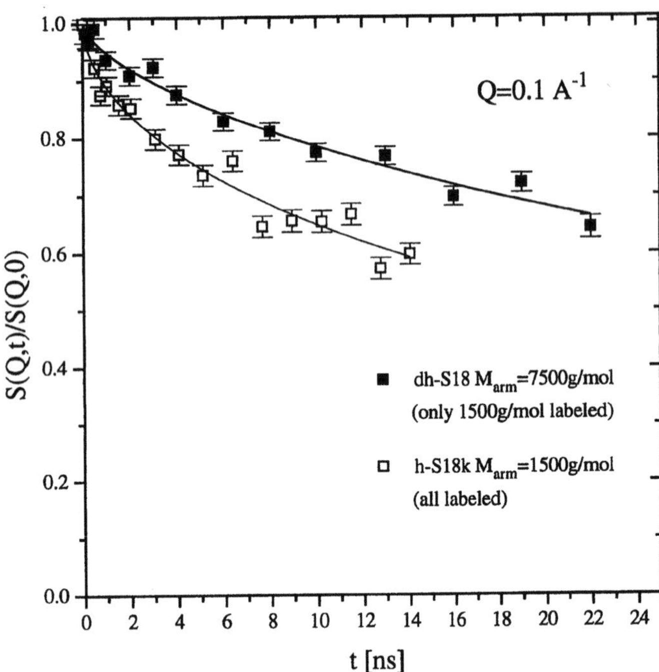

Fig. 4 NSE spectra for two star polymers with M_w^{arm} well below (open symbols) and above (filled symbols) the entanglement molecular weight of polyisoprene $M_e \approx 5000$ g/mol; solid line: fit to a stretched exponential, see text

protonated core of the former one. Both stars have been dissolved separately in a matrix of a fully deuterated star ($M_n^{arm} = 7360$ g mol^{-1}, $\phi_{matrix} = 0.26$). Preliminary results from these experiments are shown in Fig. 4. As can be clearly seen, entanglements slow down the dynamics as expected. Both spectra have been fitted, after correction for

resolution and background, to a stretched exponential decay (Kohlrausch–Williams–Watts function); $S(Q, t)/S(Q, 0) = A \exp(-(\Gamma t)^\beta)$. Whereas the characteristic decay times $\tau = 1/\Gamma$ of both spectra are well separated, the stretching exponents are very similar, 0.60 ± 0.03 for h-S18k and 0.67 ± 0.04 for dh-S18. Thus, the lineshape seems to be not affected by the presence of entanglements.

Conclusions

Structure and dynamics of star polymer solutions above the overlap concentration c^* reveal features of both (i) solutions of linear homopolymers and (ii) colloidal systems. The polymer aspect concerns the collective short time dynamics following the renormalisation group description, which is shown to be valid independent of the polymer architecture, and the influence of entanglements on the internal segmental motions. In addition long range density fluctuations give rise to an *ultra slow* diffusive mode observed by PCS indicating a nearby concentration-driven glass transition. Combining statistical-mechanical theories and SANS it is shown that the effective pair potential between star polymers is exponentially decaying for large distances and crosses over, at a density-dependent corona diameter, to an ultra-soft logarithmic repulsion for small distances. The latter two results stand for the colloidal aspects and underline the importance of star polymers as a link between polymer and colloidal properties. Star polymers can be considered as a new class of colloids, which we would like to call *ultra soft* colloid.

References

1. Grest GS et al (1996) Adv Chem Phys XCIV:67
2. Gast AP (1996) Langmuir 12 4060
3. Likos CN et al (1998) Phys Rev Lett 80:4450
4. Allgaier J, Young RN, Efstratiadis V, Hadjichristidis N (1996) Macromolecules 29:1794
5. Hadjichristidis N, Fetters LJ (1980) Macromolecules 13:191
6. Stellbrink J, Allgaier J, Richter D (1997) Phys Rev E 56:3772
7. Shiwa Y (1987) Phys Rev Lett 58:2102
8. B. Nystroem, Roots J (1990) J Polym Sci Part C: Polym Lett 28:101
9. Fischer EW (1993) Physica A 201:183
10. Segre PN, Meeker SP, Pusey PN, Poon WCK (1995) Phys Rev Lett 75:958
11. Triplet forces start to become relevant only if three spheres of diameter σ exhibit a triple overlap within their coronae, corresponding to densities $\rho > 2\rho^*$
12. Daoud M, Cotton JP (1982) J Physique (France) 43:531
13. Rogers FA, Young DA (1984) Phys Rev A 30:999
14. Willner L et al (1984) Macromolecules 27:3821
15. Jucknischke O (1995) Doctoral Thesis, Westfälische Wilhelms-Universität Münster
16. Abbas B (1996) Diploma Thesis, Westfälische Wilhelms-Universität Münster
17. Richter D et al (1990) Macromolecules 23:1845
18. Stellbrink J, IFF-Neutronenstreuung II, Forschungszentrum Jülich, D-52425 Jülich, Germany

Progr Colloid Polym Sci (1998) 110:29–33
© Steinkopff Verlag 1998

R. Pastor-Satorras
J.M. Rubí

Fractal properties of cluster of colloidal magnetic particles

R. Pastor-Satorras
Department of Earth, Atmospheric
and Planetary Sciences
Massachusetts Institute of Technology
Cambridge, Massachusetts 02139
USA

Prof. Dr. J.M. Rubí (✉)
Department de Física Fonamental
Facultat de Física
Universitat de Barcelona
Av Diagonal 647
E-08028 Barcelona
Spain

Abstract We have studied the
properties of clusters of colloidal
magnetic particles generated from
a 2D aggregation model with dipolar
interparticle interactions. Particles
diffuse off-lattice, experiencing dipolar 1 with the already attached
particles until either they stick to the
cluster or wander far away and are
removed. Our results are interpreted
in terms of a fractal dimension that is
a monotonically decreasing function
of the temperature, varying between
a definite value close to 1 at $T = 0$,
and the limit $T \to \infty$, corresponding to
free diffusion-limited aggregation. By
analyzing orientational correlation
functions, an ordered state is found at
low temperatures; this state is
destroyed by the fractal disorder
generated at high T. Our study could
be relevant in understanding
aggregation of dipolar colloids and
phase transitions in Langmuir
monolayers.

Key words Magnetic particles –
ferrofluids – cluster aggregation –
dipolar interactions

Introduction

Fractal growth phenomena [1] have been a particularly
active field of physics in the last decade, due to the potential applications to many disciplines, in particular to the
physics of colloids. In the context of computer models, the
most noteworthy are the diffusion-limited aggregation
(DLA) model [2] and the cluster–cluster aggregation
model [3], which indeed can describe the fractal structure
of colloidal aggregates [4].

Most often the above-mentioned models assume very
short-range interparticle interactions; usually a hard-core
potential plus an infinite well on the surface of the particles. They are therefore appropriate to describe aggregation with interactions strongly decaying with distance.
However, they fail to represent aggregation in the presence
of long-range forces. Some examples of such kind of processes are the aggregation of particles subject to dipolar
forces. Actual experiments have been conducted in ferro-

fluids and the so-called magnetic holes [5]; part of
their interest lines in the extreme simplicity of the experimental setup, which can be easily performed with magnet-
ized microspheres [6]. On the other hand, there is a
considerable interest in the study of phase transitions
in Langmuir monolayers [7]. Far from equilibrium,
the condensed phase grows at the expense of the liquid
phase, forming clusters of different shapes. The phospholipids constituting the Langmuir monolayer experience repulsive dipolar interactions, which must play a major role in determining the morphology of the condensed
aggregates.

Some authors have extended the classic models in
order to take interparticle interactions into account. The
modifications proposed so far are either deterministic
[8, 9] or random [10–13], using Monte Carlo or Langevin
dynamics methods. However, because the implementation
of long-range interactions is extremely time-consuming,
only modest-sized aggregates can be grown (up to 128
particles in [9, 14, 15]). Therefore, the results do not allow

categorical conclusions to be made about the fractal properties of the dipolar clusters.

Our purpose in this paper is to extend the particle–cluster aggregation model in order to include fully anisotropic attractive dipolar interparticle interactions. Our algorithm is considerably fast, allowing us to generate clusters up to 10 000 particles at zero temperature, in a reasonably short time. In the next section we describe the technical details of the algorithm. Then we discuss the evolution as a function of temperature of the fractal dimension and the order induced on the orientation of the dipoles by their reciprocal interactions. Our conclusions are discussed finally.

Cluster formation algorithm

We consider the two-dimensional aggregation process of magnetic particles of diameter d and magnetic moment $\boldsymbol{\mu} = \mu\mathbf{u}$, with μ being the magnetic moment strength and $\mathbf{u}$ a unit vector oriented along its direction. The dipolar energy between two particles i and j, located at the positions $\mathbf{r}_i$ and $\mathbf{r}_j$, respectively, is $\varepsilon_{ij} = \mu^2 E_{ij}$, E_{ij} being the dimensionless dipolar energy

$$E_{ij} = \{\mathbf{u}_i \cdot \mathbf{u}_j - 3(\mathbf{u}_i \cdot \mathbf{r}_{ij})(\mathbf{u}_j \mathbf{r}_{ij})/r_{ij}^2\}/r_{ij}^3 , \tag{1}$$

and $\mathbf{r}_{ij} = \mathbf{r}_i - \mathbf{r}_j$.

The simulation starts with a seed particle located at the origin of coordinates, bearing a randomly oriented tridimensional vector $\mathbf{u}_1$, parallel to the plane of growth. The following particles are released from a random position on a circle of radius R_{in} centered on the seed. The particles have assigned a vector $\mathbf{u}_i$ oriented at random. Each particle undergoes a random walk until it either contacts the cluster or moves away from the origin a distance greater than R_{out}. In this case, the particle is removed and a new one is released from the circle surrounding the seed. We have used the values $R_{\mathrm{in}} = 2R_{\mathrm{max}} - 5$ and $R_{\mathrm{out}} = 2R_{\mathrm{max}}$, where R_{max} is the maximum radius of the cluster. The random walk experienced by the incoming particles is affected by the interactions exerted by the particles already attached to the cluster. We have taken this fact into account by using a Metropolis algorithm inspired by Refs. [14, 15]. Suppose that the cluster is composed by $n - 1$ particles, placed at the points $\mathbf{r}_i, i = 1, \ldots, n - 1$. At some time t the incoming particle occupies the position $\mathbf{r}_n$ and has an interacting energy $E = \sum_{i=1}^{n-1} E_{ni}$. At time $t + 1$ we compute a new position $\mathbf{r}_n$; the particle arrives there by means of jump of length d in a direction chosen at random. The movement to $\mathbf{r}_n$ is performed rigidly, without changing $\mathbf{u}_n$. The energy experienced in the new position is E^*, and the total change in the energy due to the movement is $\Delta E = E^* - E$. If $\Delta E < 0$,

then the movement is accepted; if $\Delta E > 0$, we compute the quantity $p = \exp(-\Delta E/T_r)$, where

$$T_r = \frac{d^3 k_B T}{\mu^2} . \tag{2}$$

(T_r is a reduced temperature, related to the intensity of the interaction and the actual temperature T.) In this latter case, the movement is accepted with probability p. After every accepted movement, the moment of the random walker is oriented along the direction of the total field on its position. This fact indeed assumes that the relaxation time for the orientation of particles is very short in comparison with the movement of the center of mass. The particle sticks to the cluster when it overlaps one or more particles already incorporated. After sticking, the newly attached particle undergoes one last relaxation.

Fractal properties of the clusters

The purpose of this section is to analyze the fractal properties of the clusters generated using the prescription outlined above. In Fig. 1 we have represented typical clusters

Fig. 1 Typical 2D dipolar clusters of 1000 particles, generated for several values of T_r. (a) $T_r = 0$, $D = 1.20 \pm 0.02$. (b) $T_r = 10^{-3}$, $D = 1.35 \pm 0.04$. (c) $T_r = 10$, $D = 1.74 \pm 0.02$. (c) Pure DLA cluster, grown in the limit $T_r \to \infty$, $D = 1.71 \pm 0.01$

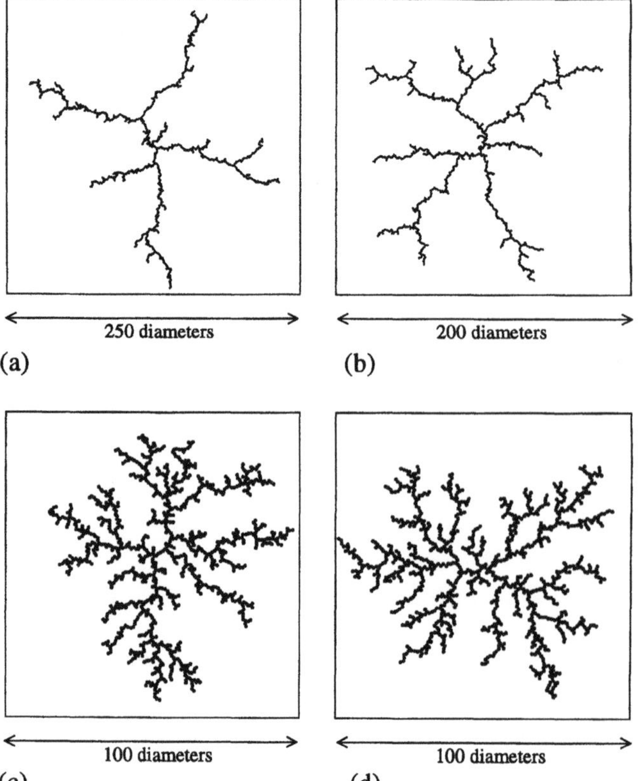

of 1000 particles, grown at four different values of T_r. The effects of temperature can be seen by comparing with a pure non-interacting DLA cluster, as shown in Fig. 1d. At low T_r, Fig. 1a, the dipolar clusters have a lesser branched and more open structure, that is to say, they have a smaller ratio of bifurcation. Even though the clusters may seem to be anisotropic, they still possess spherical symmetry: When collapsing an ensemble of clusters at the same T_r, we recover a perfectly symmetric structure. When increasing the temperature, the branching of the clusters increases correspondingly; for $T_r = 10$ (the largest value simulated), Fig. 1c, the clusters are completely indistinguishable from true DLA.

In order to quantify this temperature dependence, we have computed the fractal dimension D of the clusters, determined from a log–log plot of the radius of gyration as a function of the number of particles N [1],

$$R_g(N) = \left(\frac{1}{N} \sum_{i=1}^{N} (\mathbf{r}_i - \mathbf{r}_{\text{c.m.}})^2 \right)^{1/2} \sim N^{1/D} . \tag{3}$$

The algorithm was checked by computing the dimension of an ensemble of 100 clusters of 1000 particles, generated in the limit $T_r \to \infty$ (pure DLA). The value computed was $D = 1.71 \pm 0.01$; we recover, within the error bars, the well-known result $D = 1.715 \pm 0.004$ [16].

Figure 2 shows a plot of D as a function of T_r for the whole range of values analyzed. It seems to exhibit a smooth increasing behavior when increasing T_r, the values of D ranging between the limits corresponding to $T_r \simeq 0$ (aggregation with dipolar forces of infinite strength or zero temperature) with $D = 1.13 \pm 0.01$ and $T_r = \infty$ (aggregation with no interactions or infinite temperature) with $D = 1.71 \pm 0.01$. We can compare our results with the colloidal aggregation experiment described in [15],

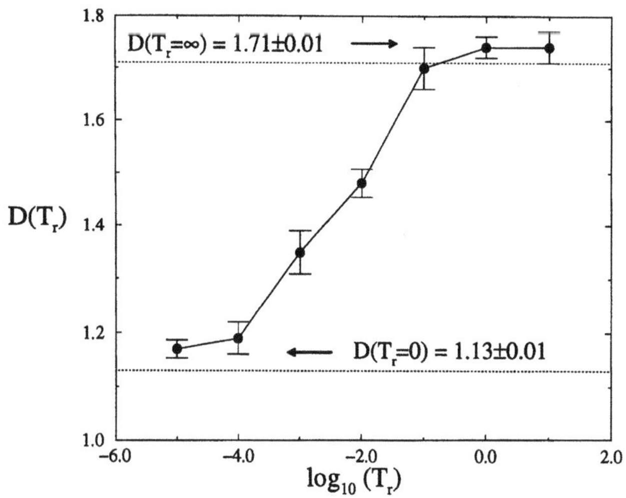

Fig. 2 Fractal dimension D as a function of T_r for dipolar clusters

where magnetic particles were employed for which $T_r^{-1} \simeq 1360$ at room temperature. In our simulations, the fractal dimension obtained for clusters grown at $T_r \simeq 10^{-3}$ is $D = 1.35 \pm 0.04$, a value clearly different from the one computed at $T_r = 0$.

The shape of Fig. 2 can be explained as follows. Given the expression (1) for the dipolar energy, it is energetically favorable for an incoming particle to stick at the tip of a branch, contributing thus to its growth, rather than sticking on any side and splitting it. At low temperatures, therefore, the most probable scenario is that of a cluster with a very small branching ratio and a low fractal dimension. This in consistent with the fact that, at $T_r = 0$, the fractal dimension seems to decrease when increasing N; we cannot even reject the possibility of a dimension equal to 1 in the limit $N \to \infty$. On the other hand, at large temperature the growth of a given branch and its split are equally likely events. The efffect of dipolar interactions is overcome by the thermal disorder and we recover the original DLA model with no interactions. At intermediate values of T_r, there would be a competition between growth and splitting, ruled by the thermal disorder as well as the dipolar interactions. We should then expect a continuous change in the geometry of the cluster and, therefore, a smooth dependence of D on the temperature.

The fractal dimension we have considered so far is a macroscopic property of dipolar aggregates. However, their composing particles also bear a rigid magnetic dipole, which confers a microscopic structure to the clusters. In order to obtain information about this structure, we have analyzed the relative orientation between pairs of dipoles. To this end, we define the function $g_T(\theta)$ as the probability density that the relative angle formed by the directions of a pair of dipoles randomly chosen from a cluster at temperature T_r is included in the interval $[\theta, \theta + d\theta]$; θ is defined to be normalized to the interval $[0, \pi]$. In practice, if $n(T_r, \theta)$ is the number of pairs with a relative angle between θ and $\theta + \Delta\theta$, for a fixed angular increment $\Delta\theta$, then we have

$$g_T(\theta) = \frac{2}{N(N-1)} \frac{1}{\Delta\theta} n(T_r, \theta) , \tag{4}$$

N being the number of particles in the cluster. g_T is a measure of the order of dipoles on the cluster. In a completely disordered distribution, all relative orientations are equally probable and, therefore, we have $g(\theta) = 1/\pi$. On the other hand, in a distribution in which all the dipoles point in the same direction (for example, in the presence of a strong magnetic field) every pair forms a relative angle of zero radians, and thus $g(\theta) = \delta(\theta)$.

Figure 3 depicts $g_T(\theta)$ computed from several ensembles of clusters at different values of T_r, between 0 and 10.

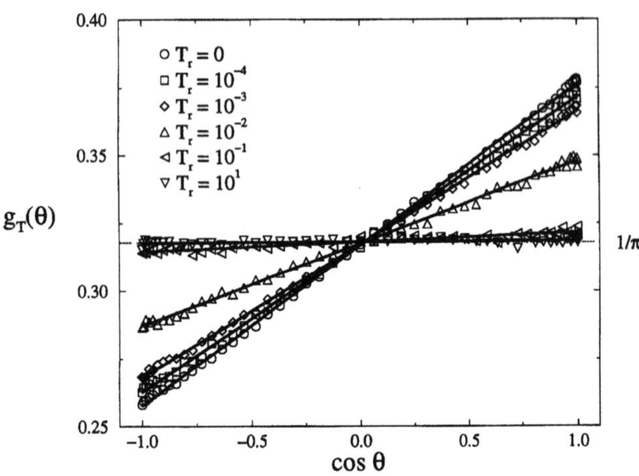

Fig. 3 Orientational correlation $g_T(\theta)$ as a function of $\cos\theta$, computed for dipolar clusters at different values of T_r. At high temperatures the slope is almost zero. Solid lines are least-squares fittings

Numerically, we observe that $g_T(\theta)$ fits extremely well to the function

$$g_T(\theta) = a + b(T_r)\cos\theta .\tag{5}$$

The normalization condition of g_T implies obviously $a = 1/\pi$, as observed. On the other hand, $b(T_r)$ seems to be a decreasing function of temperature, between the limits $b(0) = 0.0594 \pm 0.0005$ and $b(\infty)\sim 0$. The slope $b(T_r)$ can be seen as a measure of the degree of internal order in the orientation of the dipoles in the cluster. In the low-temperature limit (in the absence of any thermal disorder) the orientation of the dipoles is strongly correlated, and this fact is reflected in a nonzero slope b. As temperature raises $g_T(\theta)$ decreases continuously. In the high-temperature limit dipoles show a large disorder, imposed by the intrinsic fractal geometry, and the function g_T is almost flat (all relative orientations are equally probable). Once again, the apparently continuous variation of g_T hints towards a smooth dependence of the geometry of the clusters on the temperature.

Conclusions

We have investigated the effects of dipolar interactions in particle–cluster aggregation in two dimensions. The relevant parameter in the model is the dimensionless temperature T_r defined in Eq. (2), which relates the real temperature T and the strength of the magnetic interactions μ. At low temperatures we observe clusters with a less branched and more open structure than free DLA, and correspondingly, a fractal dimension close to 1. This is an effect of the dipolar interaction on the local growth-site probability distribution [1]. The growth of a given branch is a more likely event than its split, resulting in an enhancement of the screening of the inner regions. At high temperatures the fractal dimension raises its value until it reaches the limits of free DLA. The plot of D as a function of T_r, Fig. 2, seems to show a smooth behaviour, resulting from the competition between dipolar attractive forces and thermal disorder. The internal structure of the clusters, probed through the function g_T, seems also to show a smooth transition between an ordered state at low temperature, with long-range correlations between dipoles, and a disordered state at high temperature, in which all relative orientations are equally probable. This loss of order is explained as the effect of the geometrical disorder induced by the fractal character of the clusters. Our results could be relevant to understand the processes of cluster aggregation in dipolar colloids. Even though our results are relative to attractive dipolar interactions, they could also be applied to study cluster growth in Langmuir monolayers; the similarity with experimental clusters has already been pointed out in Ref. [13], in the particular case isotropic dipolar interactions (potential decaying with distance as r^{-3}).

As a final remark, we would point out that, even though all our results are consistent with a geometry continuously varying with the temperature, they are also compatible with the different scenario, in which there is sharp crossover between the low-temperature dimension $D_0 \simeq 1.13$ and the high-temperature dimension $D_\infty \simeq 1.71$. From Fig. 2, the crossover temperature could be estimated to be $T_c \cong 10^{-3.5} \sim 3 \times 10^{-4}$. The presence of the sharp fall in D would then be smoothed out by finite-size effects, unavoidable for the cluster sizes we are considering. Further work should be done in order to elucidate this possibility, especially by simulating aggregates larger than those actually available with our present computer resources.

Acknowledgements RPS benefited from a scholarship grant from the Ministerio de Educación y Cultura (Spain). JMR acknowledges financial support by CICyT (Spain), Grant no. PB92-0895.

References

1. Vicsek T (1992) Fractal Growth Phenomena, 2nd ed. World Scientific, Singapore

2. Witten TA, Sander LM (1981) Phys Rev Lett 47:1400

3. Meakin P (1983) Phys Rev Lett 51:1119; Kolb M, Botet R, Jullien R (1983) Phys Rev Lett 51:1123

4. Weitz DA, Oliveria M (1984) Phys Rev Lett 52:1433
5. Skjeltorp AT (1983) Phys Rev Lett 51:2306
6. Helgesen G, Skjeltorp AT (1991) J Appl Phys 69:8277
7. Möhwald II (1993) Rep Prog Phys 56:653
8. Block A, von Bloh W, Schellnhuber HJ (1991) J Phys A 24:L1037
9. Eriksson AB, Jonson M (1989) Phys Rev B 40:884
10. Ansell GC, Dickinson E (1985) Chem Phys Lett 122:594
11. Meakin P, Muthukumar M (1989) J Chem Phys 91:3212; Meakin P (1990) J Colloid Int Sci 134:235
12. Pastor-Satorras R, Rubí JM (1995) Phys Rev E 51:5994
13. Indivery G, Levi AC, Gliozzi A, Scalas E, Möhwald H (1996) Thin Solid Films 284–285:106
14. Mors PM, Botet R, Jullien R (1987) J Phys A 20:L975
15. Helgcscn G, Skjeltorp AT, Mors PM, Botet R, Jullien R (1988) Phys Rev Lett 61:1736
16. Tolman S, Meakin P (1989) Phys Rev A 40:428

Progr Colloid Polym Sci (1998) 110:34–36
© Steinkopff Verlag 1998

H. Gau
W. Mönch
S. Herminghaus

Coalescence dynamics of ordered breath figures

H. Gau · W. Mönch · S. Herminghaus (✉)
Max-Planck-Institut für Kolloid- und
Grenzflächenforschung
Rudower Chaussee 5
D-12489 Berlin-Adlershof, Germany

Abstract Breath figures are the characteristic droplet patterns which appear when one breathes onto a cold surface. Since the droplets are formed by nucleation, these patterns are usually disordered. In the present study we examine, for the first time, the growth of ordered breath figures which are generated by providing a substrate with a structured wettability. We are able to introduce disorder in a controlled manner and study its influence on the coalescence dynamics. The experimental results are compared with theoretical predictions based on geometrical considerations at which significant discrepancies were found. We conclude that the coalescence behavior of ordered breath figures is very sensitive to the microscopic dynamics taking place at the aqueous interfaces between the droplets.

Key words Breath figures – dewetting – coalescence

When a liquid condenses onto a cold surface, characteristic droplet patterns are generated which are called breath figures. Since deposition proceeds via nucleation processes, these patterns are intrinsically disordered, and previous studies have thus been restricted exclusively to disordered structures [1, 2]. However, since droplet arrays with a certain degree of order may occur, e.g., as the late stage of spinodal dewetting [3], the investigation of coalescence and ripening of ordered droplet patterns is desirable. In the present work, we present a study of coalescence dynamics of ordered breath figures. With the help of a special preparation procedure, we are able to examine the time-dependent evolution of hexagonal, ordered monodisperse droplet arrays during further adsorption of liquid. We are able to vary the degree of order and study its influence upon the coalescence dynamics.

In highly ordered hexagonal pattern, geometrical arguments show immediately that the dominating coalescence event should be the coalescence of 4 droplets at a time (so-called 4-fold coalescence) as opposed to the 2-fold coalescence usually observed in disordered structures. Upon introducing disorder, also 3-fold coalescences will occur before the droplets are large enough to geometrically enforce 4-fold coalescences. The experimental results are compared with theoretical predictions based on geometrical considerations which rest on two simple assumptions made invariably in the theory of breath figures: (a) the center of mass is conserved during a coalescence event, and (b) geometrical overlap between two droplets is necessary and sufficient for coalescence to occur.

The ordered patterns were prepared as follows. We evaporated a few Å of a water soluble salt through an appropriate mask onto a cured polydimethylsiloxane (PDMS) surface. When the substrate is cooled below the dew point by means of a Peltier element, water adsorbs on the hygroscopic salt patches only. Hence, a droplet pattern is formed reflecting dimensions and symmetry of the mask used. The salt, in our case $CaCl_2$, dissolves easily in water. Therefore, the droplets come to sit on a uniform and smooth surface (roughness <1 nm, as measured with

AFM). The time dependent evolution of the droplet pattern during further condensation was observed by light microscopy and video recording. The degree of order of the array can be varied by complete evaporation (heating of the surface) and subsequent readsorption of water. Due to contact line pinning, the salt clusters which are left upon evaporation do not form a perfectly ordered lattice. Therefore, readsorption of water leads to a droplet array with a certain disorder, which can be characterized quantitatively by calculation of the pair correlation function $P(x)$ (Fig. 1b).

Further adsorption of water leads to coalescences of droplets, which finally lead to a complete loss of order in the late condensation stage. The time resolution was sufficient to distinguish all coalescence events from each other.

In order to characterize the impact of disorder on the coalescence dynamics, we plotted the number of 4-fold (N_4) versus 3-fold (N_3) coalescence events per unit area for the different droplet arrays in Fig. 2a. Displayed are three orbits, each of which represents a single experimental run and starts at the origin of the N3/N4 plane at $t = 0$. For a highly ordered lattice, mostly 4-fold coalescences occur as seen in curve 1. Only in the very beginning of the evolution, some 3-fold coalescence events occur due to isolated defects of the lattice. Upon increasing disorder, 3-fold coalescences become more and more important (curves 2 and 3), due to the fact that the droplets coalesce earlier than for highly ordered patterns. The slopes of the curves give the ratio of the numbers of 4-fold and 3-fold coalescences.

Fig. 1 (a) Highly ordered array of droplets the periodicity is 10 μm. (b) Pair correlation function $P(x)$ of 2 different arrrays with different degree of disorder. The dotted curve represents $P(x)$ of the highly ordered array in Fig. 1a

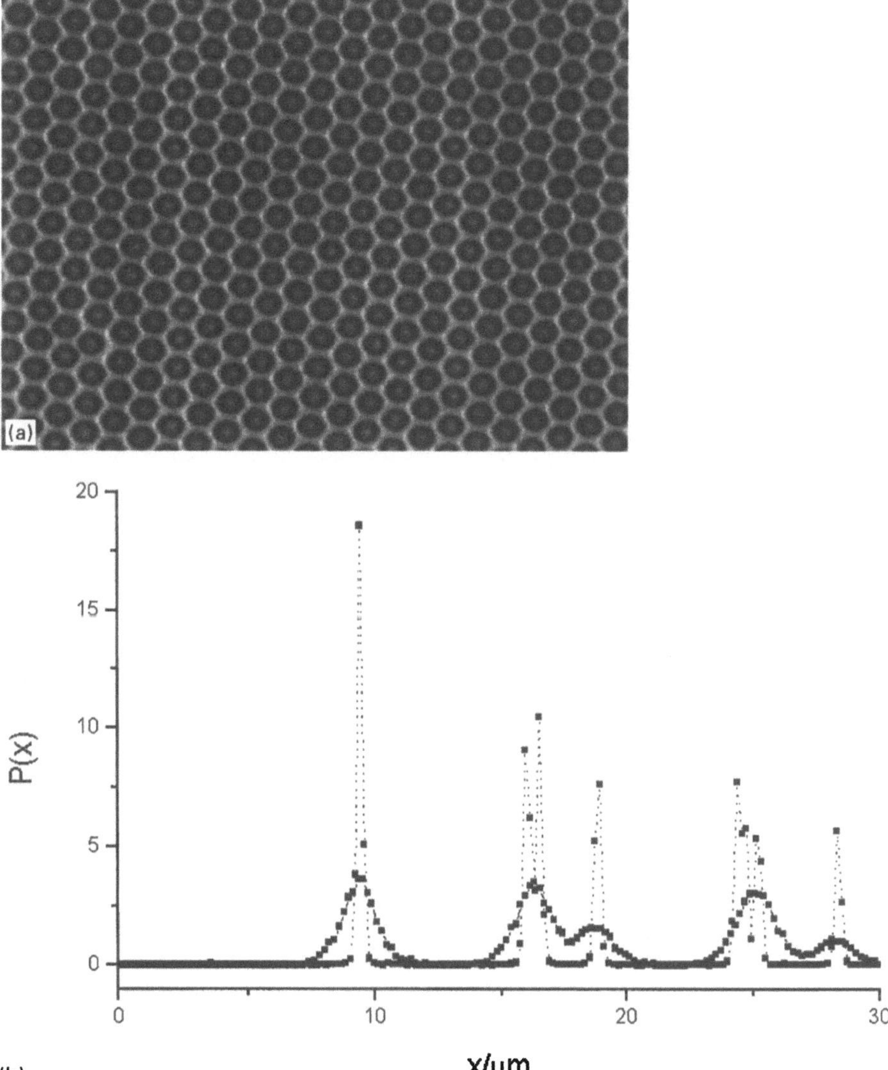

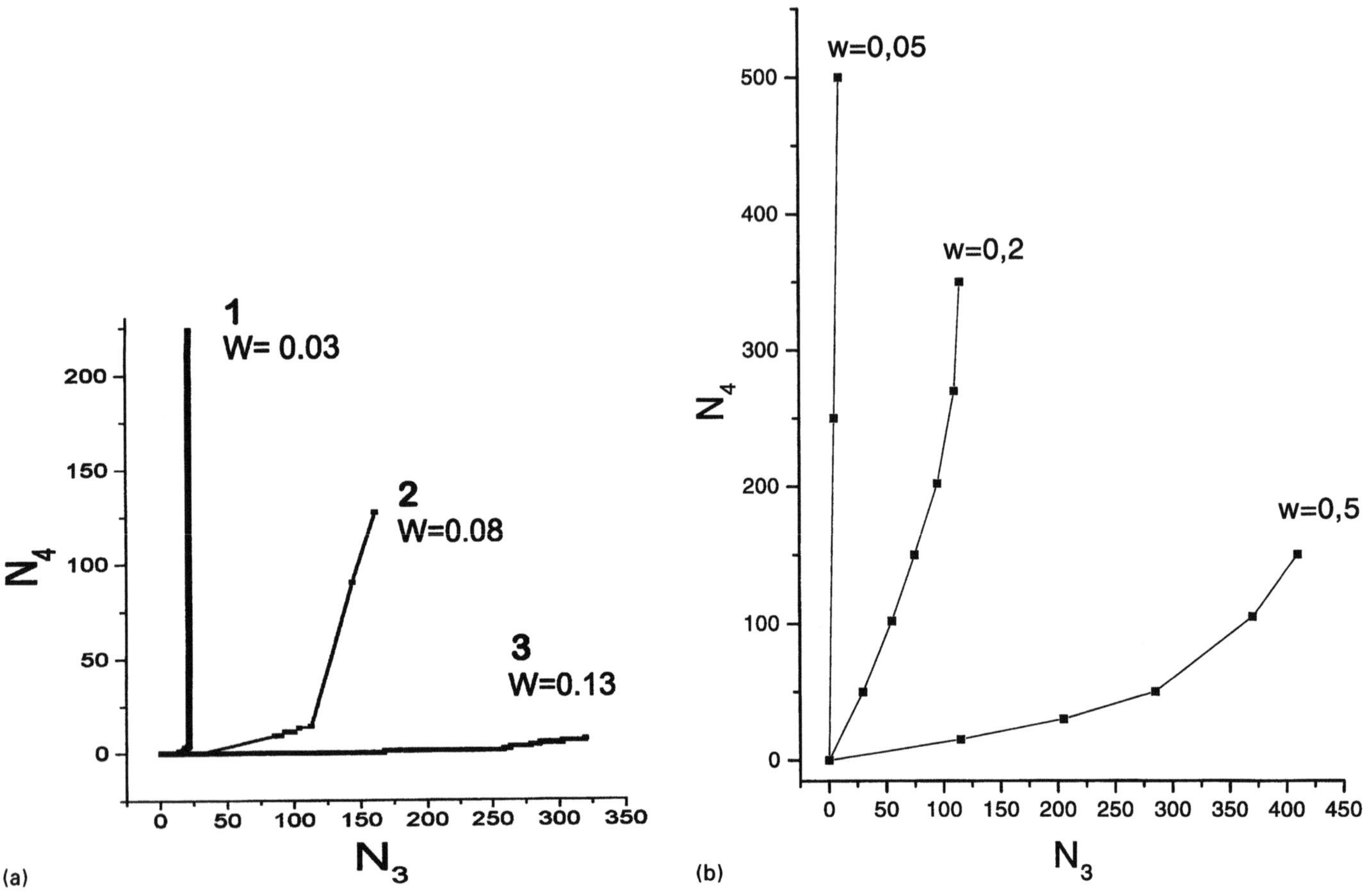

Fig. 2 (a) $4(N_4)$- and $3(N_3)$-fold coalescences of three arrays with different disorder. w denotes the width of $P(x)$ of the nearest neighbor for each lattice. (b) Calculated curves of a geometrical based theory. In contrast to Fig. 2a, 3-fold coalescences occur at a higher degree of disorder

It is instructive to compare the data with theoretical predictions, based on the assumptions that (a) the center of mass is conserved during coalescence and (b) a geometrical overlap of two droplets is necessary and sufficient for coalescence to occur. The results of the calculation are shown in Fig. 2b for three different degrees of disorder. Obviously, the suppression of 4-fold coalescence (in favour of 3-fold coalescence) by the disorder is substantially stronger than predicted by the theory. This may be understood qualitatively by observing that in order for a coalescence to occur, the layer system water/air/water which is present where two droplets are about to touch each other, must undergo a transition to forming a liquid bridge. This requires a certain amount of energy, in particular when polar surfaces are involved, and may thus require time, just as dewetting does, to which it is formally equivalent. The energy barrier may even be enhanced by the presence of amphiphilic impurities which may have segregated to the surface, for what reason we are about to repeat the experiments under conditions of specially purified atmosphere.

So far, we can conclude that the coalescence behaviour of ordered breath figures is not only qualitatively different from what is observed with usual (disordered) ones, but is also very sensitive to the microscopic dynamics taking place at the aqueous interfaces between the droplets.

References

1. Beysens D, Knobler CM (1986) Phys Rev Lett 57:1433

2. Briscoe BJ, Galvin KP (1991) Phys Rev A 43:1906

3. Bischof J, Scherer D, Herminghaus S, Leiderer P (1996) Phys Rev Lett 77:1536

Progr Colloid Polym Sci (1998) 110:37–40
© Steinkopff Verlag 1998

The interaction between colloidal particles and differently treated glass surfaces studied with evanescent wave scattering

D. Rudhardt
C. Bechinger
P. Leiderer

D. Rudhardt (✉) · C. Bechinger
Lehrstuhl Prof. Dr. P. Leiderer
Fakultät für Physik
University of Konstanz
Postfach 5560 M676
D-78434 Konstanz
Germany
E-mail: daniel.rudhardt@uni-konstanz.de

Abstract Using total internal reflection microscopy (TIRM) the interaction between a charged 3 μm polystyrene sphere which acts as a probe particle and substrates with different surface charges are investigated. As substrates we used fused silica surfaces and glass plates which were coated with a monolayer of hexamethyldisilazane (HMDS). The first is known to provide highly polar surfaces, whereas the substrate coated with HMDS is unpolar. Our results demonstrate that the curves for the obtained potentials strongly vary in both cases which makes the technique suited to distinguish between surfaces of different polarity. The obtained potentials are compared with theoretical predictions.

Key words Evanescent waves – TIRM – electrical double layer – particle-wall-interactions

Introduction

When light is subjected to total internal reflection at the interface of a medium with lower refractive index an evanescent wave is produced which penetrates the medium with the lower refractive index. The intensity of the evanescent wave decays exponentially with increasing distance perpendicular to the surface on a length scale which is typically in the order of the wavelength of the incident light. For this reason scattering experiments with evanescent waves are suitable to investigate systems close to a surface [1–3]. When an object which can scatter light, e.g. a colloidal sphere is brought into the region of the evanescent field, the sphere can couple to the photons of the incident beam and frustrated total reflection will occur. Since the scattering intensity of the colloid is proportional to that of the evanescent wave, it decreases also exponentially with the separation distance z between the colloidal sphere and the surface [4, 5] and allows to calculate z. If the system is in equilibrium the particle-wall-interaction potential can be calculated from the probability distribution of z and the use of the Boltzmann distribution. This technique which is called Total Internal Reflection Microscopy (TIRM) was first used by Prieve et al. [6]. Since then it is established as a powerful method to measure the interaction between single colloidal spheres and surfaces [7–9].

In this work we present data which show the influence of differently charged walls on the interaction with a charged colloidal sphere.

Experimental

Here we want to describe our TIRM-setup only in brief, for details we refer to the literature [6]. The cell used in our experiments consists of two fused silica glass plates (Suprasil, Steeg & Reuter) separated by a spacer ($d = 1$ mm). In some experiments the silica plates which are known to be very polar due to OH terminating bonds were covered with a monolayer of hexamethyldisilazane (HMDS). Silylation [10] was achieved by exposing the glass plates to the vapor phase of HMDS for several hours. After this process the surface energy of the glass surface is lowered, i.e. it becomes unpolar. Therefore the contact angle of water on the HMDS treated glass plate is about 65° in contrast to a value of only 5° before the silylation.

After assembling the cell, a BK7 prism is matched to the lower glass plate. Then the cell was connected to a closed circuit which contained the colloidal suspension. We used electrically stabilized polystyrene microspheres with a diameter of 3 μm which are suspended in water (IDC spheres). To guarantee that there is only a single particle in the field of view during our experiments, only highly diluted suspensions with a particle density less than 0.5 mm^{-3} were used. The suspension was then pumped through this circuit which also contained a vessel of ion exchanger and an electrical conductivity probe to control the ionic strength in the suspension. This method [11] allowed us to perform measurements at controlled ionic strengths. The lowest ion concentration which is stable during measuring time was $c_{NaCl} = 1$ μmol/l.

Red light of a HeNe laser ($\lambda = 632.8$ nm, 10 mW) is coupled through this prism into the cell and is totally reflected at the glass plate/water interface. The penetration depth of the evanescent wave can be varied by changing the angle of incidence of the laser beam at the glass/water interface. In the experiments described here, the penetration depth was kept constant at about 2.1 μm, being small compared to the separation distance of the glass plates.

The scattered intensity of the colloidal spheres illuminated by the evanescent field was collected by a microscope objective (Plan L50x, Leitz) and measured with a photomultiplier tube (Hamamatsu). To obtain sufficient data and to minimize statistical errors during each measurement the scattering intensity of a particle is sampled over 1200 s at a sampling rate of 50 Hz. In order to avoid systematic errors, several experiments under the same conditions (salt concentration, angle of incidence and temperature) but on different particles were recorded and averaged. The data points were then stored on a PC. From the raw data the particle-wall-interaction potential as a function of the particle distance was calculated. For details regarding the TIRM method and data evaluation we refer to the literature [5–7, 12].

Results and discussion

The inset of Fig. 1 shows a typical particle-wall-potential curve as a function of z which was obtained with uncoated silica plates. Towards larger distances one can see that the potential increases linearly. In this region the dominant force on the particle is gravity and the slope of the potential is given by

$$\frac{G}{kT} = \frac{(\rho_P - \rho_W)Vg}{kT}, \tag{1}$$

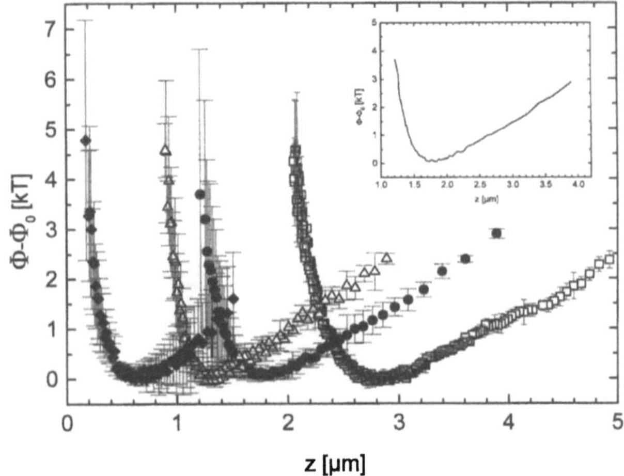

Fig. 1 Potentials for a charged particle above a fused silica surface measured at different salt concentrations $c_{NaCl} = 3.5$ μM (open squares), 9.2 μM (solid circles), 12.9 μM (open triangles), and 111 μM (solid diamonds). The inset shows a typical potential curve obtained with TIRM

where G is the weight of the particle, k the Boltzman constant, T the temperature of the suspension, ρ_P and ρ_W the density of the particle and water, g the gravity constant and V the volume of the sphere. By inserting the density of polystyrene and water and the temperature of the system in Eq. (1) we obtain a particle diameter which is about the value given by the manufacturer.

When the particle comes closer to the surface the potential strongly increases due to repulsion effects. This is due to the electrostatic interaction between the negatively charged polystyrene sphere and the fused silica surface which is also negatively charged when it is in contact with water due to the dissociation of polar surface groups [13]. The repulsive part of the potential can be fitted to an exponential curve. The obtained fit curves and parameters are discussed below.

To vary the interaction between the particle and the wall, we performed measurements like above at different salt concentrations (Fig. 1). With increasing salt concentration the surface charge of both the sphere and the wall are more and more screened which then should lead to a decreased electrostatic repulsion. This should reflect itself in a systematic change in the potential curves.

In our experiments we varied the salt concentration c_{NaCl} by more than a factor of 30 between 3.5 and 111 μM, the results are plotted in Fig. 1. With increasing salt concentration the minima of the potential curves are shifted closer to the wall because the electrostatic repulsion decreases and the particle is pressed closer to the wall by gravity. One can also see that with increasing salt concentration the increase towards smaller particle-wall distances gets stronger, indicating that larger c_{NaCl} makes the potential more hard-core like. This is in agreement with the behavior of a screened electrostatic interaction between the wall and the particle.

Accordingly, one would also expect the potential curves to be influenced by the surface charge of the glass plate used in our experiments. Therefore we repeated the experiments above with HMDS coated silica plates which are expected to posses a rather uncharged surface. To compare the results with those of the uncoated silica surfaces, we plotted in Fig. 2 two of the resulting potential curves of HMDS coated (solid circles) and two of the non coated silica (open circles) surfaces at the same salt concentrations, i.e. $c_{\mathrm{NaCl}} = 111\ \mu\mathrm{M}$ and $9.4\ \mu\mathrm{M}$, respectively. On the first glance, the curves for the unpolar surfaces are very similar to the results for the silica substrates. Again we observe the linear increase towards larger distances and a strong increase towards smaller z. As expected, we find the slope of the gravitational contribution to be identical in both cases. However, the minima of the potentials of the HMDS-covered surface are at smaller z values compared to the uncoated silica surfaces. This can be seen more clearly in Fig. 3 where the minima positions of HMDS-covered and uncovered silica surfaces for different salt concentrations are plotted. For all measurements we found the minima position of the HMDS covered surfaces to be smaller than that of the non-coated silica surfaces. As above, the explanation of this shift can be easily explained when taking into account that the electrostatic repulsion between the particle and the glass substrate is lowered when the glass is made unpolar by the HMDS coating. A more detailed analysis of Fig. 2 shows also that at small distances the increase in the repulsive part of the potential is steeper in the case of the covered surface which shows again that the repulsive interaction in this case is more hard-core like.

The results show that the interaction potentials are dominated by gravity and electrostatic repulsion. In the following we want to investigate these interactions more quantitatively. For particle-wall distances several times larger than the Debye length κ^{-1} the double-layer potential between a sphere and a flat wall is given by an exponential according to the Derjaguin approximation [14]. This yields to a total potential of the form

$$\Phi(z) = A\,\mathrm{e}^{-\kappa z} + G z\,, \qquad (2)$$

where A is a constant and G the weight of the particle in water. With z_0 as the position of the potential minimum this can be translated [6, 7] into

$$\frac{\Phi(z) - \Phi(z_0)}{kT} = \frac{G}{\kappa kT}\,\mathrm{e}^{-\kappa(z - z_0)} + \frac{G}{kT}(z - z_0) - \frac{G}{\kappa kT}\,. \qquad (3)$$

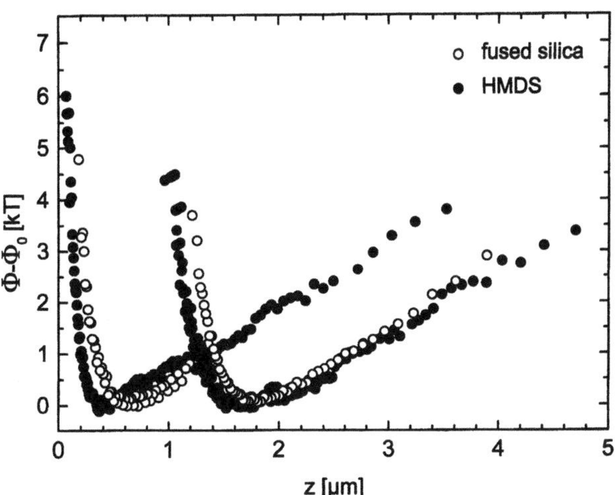

Fig. 2 Comparison of the particle-wall-potentials obtained for HMDS covered (solid circles) and bare fused silica surfaces (open circles). The potentials for the two systems are measured at the same salt concentration $c_{\mathrm{NaCl}} = 9.4\ \mu\mathrm{M}$ and $c_{\mathrm{NaCl}} = 111\ \mu\mathrm{M}$, respectively

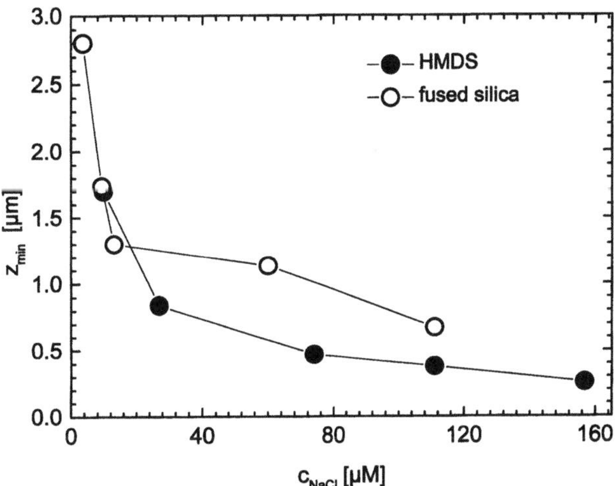

Fig. 3 Positions of the potential minima for the fused silica surface (open circles) and the HMDS covered surface (solid circles) as a function of the salt concentration

Indeed this potential form has been experimentally observed in some cases [6, 7].

As one example a potential curve taken from Fig. 1 (open squares) together with a fit according to Eq. (3) (solid line) is shown (Fig. 4). κ was chosen to be the only fit parameter. One obtains the screening length to be $\kappa^{-1} = 0.26\ \mu\mathrm{m}$. This deviates by 40% from the Debye length calculated from the measured electrical conductivity of the suspension ($\kappa^{-1} = 0.16\ \mu\mathrm{m}$). This deviation can not be explained by experimental errors which are also shown in Fig. 4.

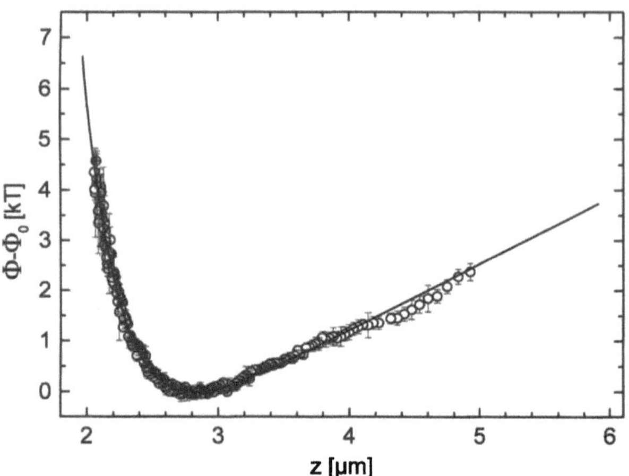

Fig. 4 The symbols correspond to the potential curve for the suspension with $c_{\text{NaCl}} = 3.5\ \mu M$ from Fig. 1 and the solid line is the fit according to Eq. (3). When κ is used as fit parameter good agreement with experiment is found, however, the value obtained for the screening length $\kappa^{-1} = 0.26\ \mu m$ is 40% higher than the expected value from an independent electrical conductivity measurement

The finding that the increase of the repulsive part of the potential is smaller than expected is also found by other authors [12]. The comparison of experimental data and theoretical predictions for other curves leads to deviations discussed above. In particular, we do not find the deviation in κ to depend on the salt concentration, i.e. the mean distance between the particle and the wall. Therefore, even if the attractive van der Waals interaction with its strong distance dependence is additionally considered it cannot explain this deviation. We also do not believe that hydrodynamic interactions are responsible for the deviations in κ because they only affect the dynamics of the particle but not the equilibrium potential. Since these deviations are not yet understood, investigations of other surfaces with different properties are needed.

Conclusions

The interaction potential between a $3\ \mu m$ negatively charged polystyrene sphere suspended in water and a charged fused silica surface is measured with evanescent wave scattering as well as the interaction with a unpolar HMDS covered fused silica surface. It was shown that TIRM is a suitable method to resolve the differences in the potentials of these two systems. By making the surface unpolar the mean separation distance between the sphere and the glass surface is decreased compared to the system with a polar fused silica surface and the increase of the repulsive part gets steeper by reducing the surface charge. The potential curves are compared with theoretical prediction and show a good agreement. Deviations between the obtained and the predicted screening lengths found also by other authors are not yet understood.

Acknowledgments We acknowledge helpful discussions with R. Klein. This work is supported by the Deutsche Forschungsgemeinschaft (SFB 513).

References

1. Lan KH, Ostrowsky N, Sornette D (1986) Phys Rev Lett 57:17
2. Polverari M, v.d. Ven TGM (1995) J Colloid Interface Sci 173:343
3. Walz JY, Suresh L (1996) J Chem Phys 103:10714
4. Chew H, Wang DS, Kerker M (1979) Appl Opt 18:2679
5. Prieve DC, Walz JY (1993) Appl Opt 32:1629
6. Prieve DC, Luo F, Lanni F (1987) Faraday Discuss Chem Soc 83:297
7. Prieve DC, Bike SG, Frej NA (1990) Faraday Discuss Chem Soc 90:209
8. Frej NA, Prieve DC (1993) J Chem Phys 98:7552
9. Flicker SG, Bike SG (1993) Langmuir 9:257
10. Cooper BE (1978) Chem Ind 21:794
11. Palberg T, Härtl W, Wittig U, Versmold H, Würth M, Simnacher E (1992) J Phys Chem 96:8180
12. Liebert RB, Prieve DC (1995) Biophys J 69:66
13. Hunter RJ (1988) The Zeta Potential in Colloid Science, ch. 7. Academic Press, London
14. Verwey EJ, Overbeck JT (1948) The Theory of Stability of Lyophobic Colloids, Elsevier, Amsterdam

Progr Colloid Polym Sci (1998) 110:41–45
© Steinkopff Verlag 1998

R. Bubeck
S. Neser
C. Bechinger
P. Leiderer

Structure and dynamics of two-dimensional colloidal crystals in confined geometry

R. Bubeck · S. Neser
Dr. C. Bechinger (✉) · P. Leiderer
Fakultät für Physik
University of Konstanz
D-78434 Konstanz
Germany

Abstract The properties of two-dimensional (2D) colloidal crystals have been widely investigated during the last 20 years, but only little attention has been paid to effects of finite system size. We use video-microscopy to study small 2D crystals consisting of approximately 10 to 100 particles confined in hexagonal, circular and square regions. The observed crystal structures are strongly dominated by the shape of those regions and anisotropic diffusion coefficients, indicating a reduced mobility perpendicular to the walls.

Key words Two-dimensional crystallization – colloidal crystals – confinement – restricted geometry – finite-size effects

The fascinating melting and freezing properties of two dimensional (2D) systems motivated a lot of scientific work during the past years [1–3]. In 1973 Kosterlitz and Thouless proposed a two step melting scenario where melting is mediated by the dissociation of paired lattice faults [1]. In the following years this theory was refined by Halperin, Nelson and Young [2]. In contrast to the 3D case where both translational and bond orientational order are lost at the melting temperature T_m, the KTHNY theory predicts a new intermediate hexatic phase where translational order is lost but, some bond orientational order is still present. Experimental data are available for many different 2D systems, like electrons on helium, noble gases adsorbed on graphite and colloidal spheres confined between two walls or at an interface [3]. The great advantage of colloidal systems are their convenient time (milliseconds) and length scales (microns) which allow the observation of single particle trajectories and the variety of available interaction potentials.

In this paper we present experimental data on the structure and dynamics of laterally confined 2D colloidal systems consisting of large superparamagnetic spheres lying on a smooth polymer substrate. A similar system has been already used by Zahn et al. [4] for the investigation of melting in large 2D systems. The lateral confinement is obtained by patterning the surface. An external magnetic field **B** perpendicular to the 2D plane is used to generate a magnetic moment **M** within the particles leading to a repulsive pair potential of the form $V \propto M^2/r^3$, where r is the distance between two particles. In this system the plasma parameter Γ, defined as ratio between the magnetic energy U and $k_B T$, depends on both the external magnetic field B and the temperature T. For reasons of convenience, in this system usually temperature is kept constant and phase transitions are observed as a function of the magnetic field B. It is noticeable, that the well known interaction allowed a comparison of the experiment and theory without any free parameter! [4].

The sample cell (Fig. 1a) is formed by two circular fused silica plates which are fixed at a distance of 1 mm by an O-ring. To reduce the sticking of the particles, a smooth 3–4 μm thick film of poly(methyl-methacrylate) (PMMA) was spin coated on top of the bottom silica plate. To realize the lateral confinement we then applied thin structured copper foils onto the PMMA substrate. These foils which are commercially available in a variety of sizes and geometries as grids for Transmission Electron Microscopy (TEM), consist of a large number (20–100) of identical

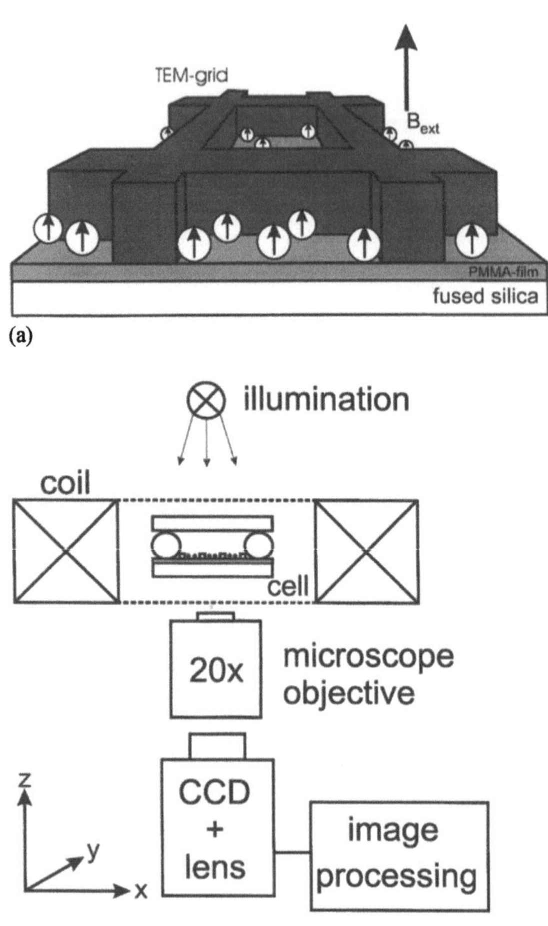

Fig. 1 Experimental setup. (a) Magnification of the patterned silica plate at the bottom of the cell: paramagnetic colloidal particles with induced magnetic moments are lying on a PMMA film. A TEM grid serves as lateral confinement for the particles. (b) Inverted microscope consisting of NPLAN 20×0.4 objective, CCD-camera with $f = 135$ mm lens and Köhlerian illumination unit. A coil is used to generate a magnetic field inside the cell

holes with very narrow mechanical tolerances separated by small bridges. The large number of identical structures greatly improves statistics. To fix a grid we heated it up to a temperature of $150\,^\circ$C and pressed it onto the PMMA-film. By this process the grid is then glued onto the PMMA film which is only slightly distorted by this procedure. Consequently, the walls of the compartments are formed by the bridges of the grid, whereas the bottom of the confined areas consists of PMMA.

After insertion of the colloidal suspension (DynaBeads 4.5 μm, Fa. Dynal, Lot-No. A44610, $\Phi_0 \approx 1 \times 10^{-4}$) which was additionally stabilized with 0.2 wt% Triton X-100, gravity causes the particles to sediment onto the bottom plate. Particles which sediment onto the bridges perform a Brownian random walk and therefore are also trapped in the holes within a few minutes. This results in the forma-

tion of a 2D colloidal system whose particle density can be tuned via the initial volume fraction Φ_0. Since the displacement of a particle a tenth of its diameter against gravity would increase its potential energy by about $18\,k_{\rm B}T$ the assumption for the system to be 2D is well fulfilled.

Our experimental setup is shown in Fig. 1b. The cell is placed in the center axis of a copper coil which can generate magnetic fields of up to 12 mT. The particles were imaged with an inverted microscope which consists of a Leica NPLAN 20×0.4 objective, a CCD camera equipped with a $f = 135$ mm lens and a Köhlerian illumination unit. This setup provides a homogeneously illuminated field of view with 550×450 microns in size. Data are taken with an image processing system connected to a computer which allows to calculate particle trajectories $\mathbf{r}(t) = (x(t), y(t))$ with a spatial resolution of 0.5 μm and a temporal resolution of 20 ms. In the absence of an external magnetic field the particles behave like hard spheres. To characterize the dynamical behaviour of the system we consider the mean square displacement

$$\Delta r^2(\tau) = \langle (\mathbf{r}_i(t + \tau) - \mathbf{r}_i(t))^2 \rangle_{t,\,i} \tag{1}$$

where τ denotes the considered time interval and $\mathbf{r}_i(t)$ is the position vector of the ith particle at the time t, which – in the absence of external forces – shows a linear increase in time and can be described by the usual 2D diffusion law

$$\Delta r^2(\tau) = 4D\tau , \tag{2}$$

where D denotes the diffusion constant. If external lateral forces are present nonlinear terms may appear: e.g. when the 2D plane is not exactly horizontal, gravity may superimpose a term $\propto \tau^2$ in $\Delta r^2(\tau)$ (corresponding to a constant drift velocity). To avoid this, before each measurement the apparatus is calibrated until the square component of a fit to the measured $\Delta r^2(\tau)$ is negligible within the resolution of the apparatus.

For a system where no boundaries are present, i.e. no TEM grid is glued onto the PMMA film, we find a diffusion constant $D = 0.035(1)\ \mu\text{m}^{-2}\,\text{s}^{-1}$ parallel to the surface corresponding to about 36% of the free diffusion constant D_0 as calculated with the Stokes–Einstein equation (see Fig. 2). This result is consistent with the picture drawn by Faucheux et al. [5] who predicted and measured a reduction of the free diffusion up to a third of D_0 in the case of colloidal particles moving on a flat surface.

When applying a magnetic field the behaviour of the mean square displacement changes significantly. In Fig. 2 we plotted the mean square displacements as a function of time for different magnetic fields. For large values of $\tau\ \Delta r^2(\tau)$ still behaves linearly with a slope which decreases with increasing magnetic field strength. This dependence of $D_{\rm long}^{\rm self}$ on B is also shown in the inset of Fig. 2, where one can see that the decrease is strongest for small fields while

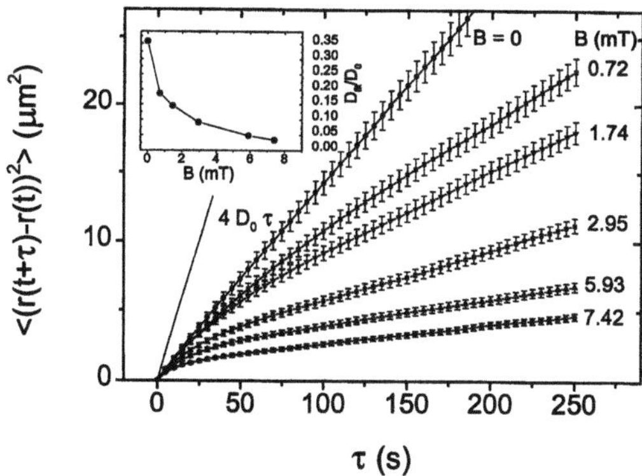

Fig. 2 Mean square displacements of ≈ 600 particles which are part of a infinite system: Increasing the magnetic field reduces the long-time self-diffusion and increases the localization of the particles. In the inset the long time self diffusion coefficient $D_{\text{long}}^{\text{self}}$ is plotted over the applied magnetic field B

$\Delta r^2(\tau)$ approaches a constant for large values of B. For small τ's there is a cross-over regime where the slope of $\Delta r^2(\tau)$ approximately increases to the value found when no magnetic field was present. This experimental observation is consistent with the well known crossover from short-time to long time self diffusion: On short time scales a particle can diffuse freely ($D = D_0^{\text{self}}$) in the cage formed by its nearest neighbours. With increasing time $\Delta r^2(\tau)$ cannot increase linearly anymore because the particle starts to sample the energy well of the surrounding spheres and a deformation of the cage would be neccessary for a further movement in the same direction ($D = D_{\text{long}}^{\text{self}}$). With increasing field strength the interaction becomes stronger and thus $D_{\text{long}}^{\text{self}}$ decreases until finally – in the solid phase – it becomes extremely unlikely that a particle leaves its cage. Therefore the mean square displacement becomes a constant and $D_{\text{long}}^{\text{self}}$ approaches zero.[1] A qualitatively similar behaviour has been found by Zahn et al. [4] investigating the same colloidal particles at the water air interface within a small water droplet. After this brief discussion of the behaviour of the extended system, the effects of finite system sizes will be considered: In contrast to infinite 2D systems where the hexagonal lattice is the energetically preferred structure, in finite systems the equilibrium structure for a given number of particles strongly depends on the size and geometry of the confinement. To get these equilibrium structures we slowly increased the external magnetic field up to values of 12 mT with a rate of approximately 1 mT per hour. This small rate is needed because of the slow dynamics for our large spheres, i.e. the

diffusion constant without applied magnetic field of $D = 0.035\ \mu\text{m/s}$ corresponds to a time of 150 s for a particle to diffuse a distance of its own diameter. This fact leads to long structural relaxation times of several hours. In the following we show the crystal structures in the case of high magnetic fields for different confinement geometries.

Hexagonal symmetry

An example for hexagonal confinement is shown in Fig. 3a. The orientation of the whole lattice is aligned to that of the confinement region and one clearly observes the hexagonal symmetry. Consequently, the particles are arranged in hexagonal "shells" around the center particle. For infinite 2D systems the hexagonal lattice is distinguished as the equilibrium structure of the crystalline phase [6]. Therefore, one would expect that confining particles to a hexagonal region should lead to the formation of a triangular lattice. However, this only holds for some *magic* numbers where the colloidal crystal and the confinement region are commensurate. One can easily calculate that if S denotes the number of hexagonal shells, a crystal of $1 + 3S(S + 1)$ particles fits exactly into a hexagonal shape, otherwise distortions will occur [7]. In the example shown in Fig. 3a, there are 4 shells with a total number of 59 particles. However, since a perfect crystal requires 61 instead of 59 particles, slight distortions in the fourth shell are found.

Square symmetry

In this case, due to the competition of the intrinsic hexagonal symmetry and the extrinsic square symmetry the dependence of structure formation on the particle number N is even more pronounced. This can be demonstrated easily when comparing two systems of consecutive particle numbers (Fig. 4a and b). In Fig. 4a is considered a system of 16 particles in a square of 90 μm sidelength. The particles are found to arrange themselves in a 4×4 square lattice, i.e. the structure is only determined by the square shaped confinement. This, however, is only observed when the number of particles N (16 in this case) is commensurate to a square symmetry. For the case of 17 particles a couple of different structures are found. In Fig. 4b the additional particle is simply added at the bottom edge, while in Fig. 4c it is inserted at an interstitial site, hence considerably

---

[1] The mean square displacement in a two dimensional crystal diverges logarithmically with system size. However, for the sizes of systems considered in this work this divergence can be neglected.

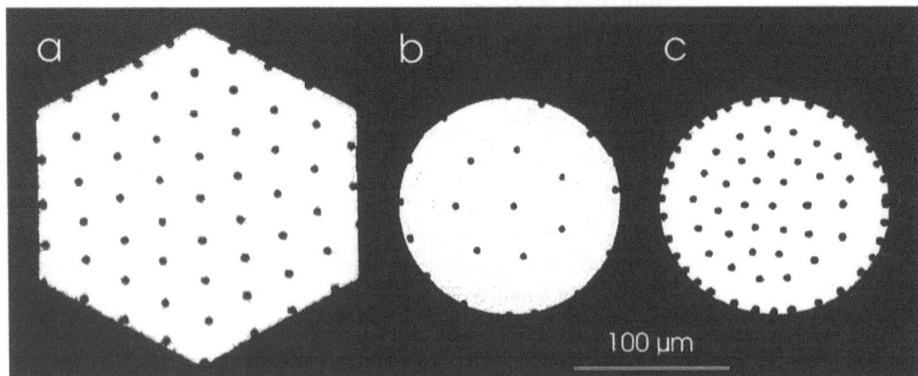

Fig. 3 Experimentally observed ground states for different geometries. (a) hexagonal ordering in a hexagonal confinement region. (b) Formation of one inner shell in a circular system of 21 particles. (c) Formation of three inner shells in a system of 71 particles in a circular confinement. On the right side, three particles chains which are marked with arrows locally distort the structure. These particle chains are formed by two or more coagulated particles and are alligned parallel to the direction of B when the field is turned on. Because a chain has a larger magnetic moment M than a single particle it slightly distorts the lattice

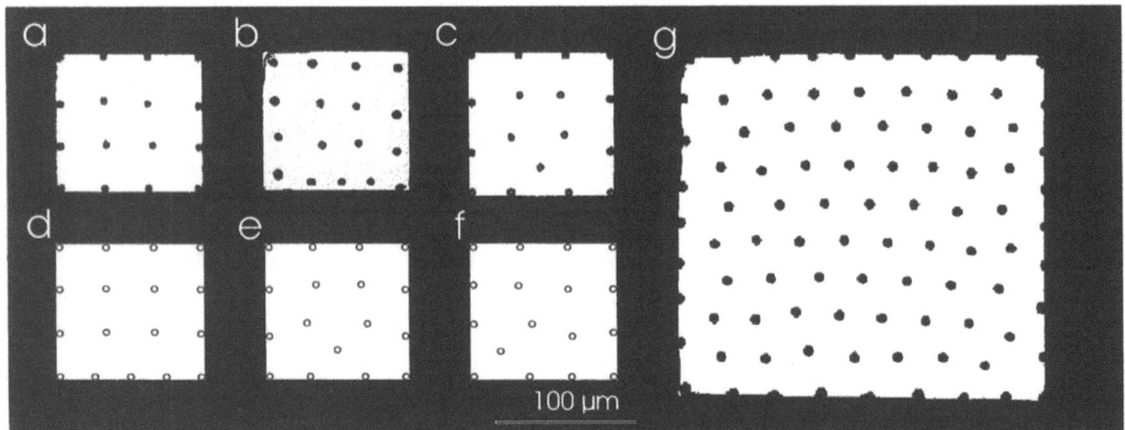

Fig. 4 Experimental results for a square confinement of 90 μm sidelength: (a) 16 particles arrange in a square 4×4 lattice. In (b) and (c) different arrangements for a system with $17 = 16 + 1$ particles are shown: (b) the additional particle moved to the bottom edge of the boundary, whereas in (c) a distorted square lattice of 16 particles with an occupation of a subsquare is formed. The system (b) is slightly distorted due to imperfections of the PMMA film at the edges of the boundary. (d)–(g) Results for 17 particles achieved by simulated annealing

distorting its environment. In order to find out whether this states are degenerated or not we calculated their corresponding energies and additionally performed a simulated annealing computer calculation to derive the ground states. The results are shown in Figs. 4d–f. While Fig. 4d and e correspond to the experimental results of Fig. 4b and c, we found no experimental match for f. The energies of Fig. 4d–f are given as 1216, 1223 and 1222 in arbitrary units, respectively. So Fig. 4d is the ground state and Fig. 4e and f are metastable states with an energy being about 0.5% larger than that of the ground state. This small difference might explain that we experimentally observe more than one structure (Fig. 4b and c).

For larger particle numbers the influence of the boundary condition gets weaker. In Fig. 4g we have plotted a picture of a square containing 98 particles. Here the system does not adapt a square lattice but has hexagonal structure with one lattice line aligned to the boundary. It is noticeable that for this system size the hexagonal lattice is superior even with the wall-induced distortions visible on the left and right boundary of Fig. 4g.

Circular symmetry

As a last example we show systems with circular confinement regions, where, unlike as in the systems with hexagonal or square boundaries, the rotational symmetry of the confinement region is larger than the intrinsic symmetry of the lattice. Consequently, in this case the

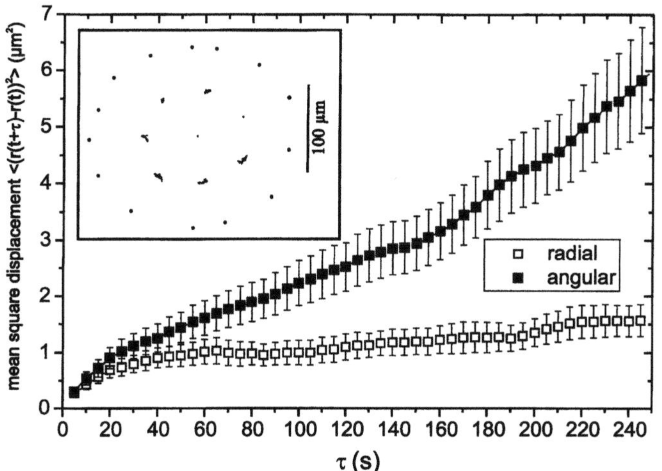

Fig. 5 Radial and angular components of the mean square displacements calculated by trajectories of 22 followed over 100 pictures in an interval of 5 s. The trajectories of the particles are shown in the inset. During the measurement two of the inner particles stuck to the substrate. This two particles and those which are in contact with the boundary (●) have not been included into the calculation

problems of commensurability are much smaller. Fig. 3b and c show two systems with circular confinement and 21 and 71 particles, respectively. As can be seen the particles arrange in circular shells centered within the confinement region. This is consistent with computer simulations for 2D systems with both dipole–dipole [8] and Coulomb interactions [9] where the systems consist of shells which are filled consecutively with increasing particle number N, in analogy to a classical periodic system of elements.

Finally, after the discussion of static properties we want to present some first results about dynamics: In the inset of Fig. 5 we plotted the center-of-mass positions of the particles taken from 100 different video frames taken in an interval of 5 s. From these trajectories we derived the angular and radial components of the mean square displacement and plotted the results in Fig. 5. It can be seen that the confinement results in a strongly anisotropic diffusion indicating that the mobility of the particles on the shells is by a factor of two larger than in the radial direction.

In conclusion, we performed experiments on the structure of 2D colloidal crystals in the presence of additional lateral confinement. It has been shown that the boundary conditions may completely dominate the structure and dynamics of small crystals.

The rather good agreement of our experimental results with computer simulations shows the reliability of the used superparamagnetic colloidal system as a model system for investigations of finite size effects in 2D. An interesting topic for future reasearch is the freezing behaviour of small confined 2D crystals because in this case the sequence in which the system aquires bond orientational and translational order might be reversed.

We gratefully acknowledge financial support by the Deutsche Forschungsgemeinschaft (SFB 513) and valuable discussions with K. Zahn and G. Maret.

References

1. Kosterlitz JM, Thouless DJ (1973) J Phys C 8:1181
2. Nelson DR, Halperin BI (1979) Phys Rev B 19:2457
3. Strandburg KJ (1988) Rev Mod Phys 60:161
4. Zahn K (1997) Phys Rev Lett 79:175
5. Faucheux LP, Libchaber AJ (1994) Phys Rev E 49:5158
6. Bonsall L, Maradudin AA (1977) Phys Rev B 15:1959
7. Lubachevsky BD, Graham RL (1997) Discrete Comput Geom 18:179
8. Cebers A, Zhuk V (1987) Fizikas un tehnisko zinātņu sērija 1:112
9. Bedanow VM, Peeters FM (1994) Phys Rev B 49:2667

Progr Colloid Polym Sci (1998) 110:46–49
© Steinkopff Verlag 1998

Qi-huo Wei
C. Bechinger
D. Rudhardt
P. Leiderer

Structure of two-dimensional colloidal systems under the influence of an external modulated light field

Qi-huo Wei · Dr. C. Bechinger (✉)
D. Rudhardt · P. Leiderer
Fakultät für Physik
Lehrstuhl Prof. Dr. P. Leiderer
University of Konstanz
D-78434 Konstanz
Germany
E-mail: Clemens.Bechinger@uni-konstanz.de

Abstract The presence of a modulated laser field can induce crystallization of a colloidal liquid where the particles interact via screened Coulomb repulsion. This phenomenon is called laser-induced freezing (LIF). In this paper, we present experimental results on LIF which were performed under controlled particle interaction potentials. This was achieved by defined ion concentration conditions during our experiments. We observed distorted and almost perfect hexagonal structures as well as a modulated liquid as a function of the periodicity of the modulated laser field.

Key words Laser-induced freezing and melting – colloidal dispersion

Introduction

Since the pioneering work by Ashkin and co-workers on optical forces acting on small dielectric particles [1], there has been an enormous interest in the field of particle manipulation with light fields. One example are optical tweezers which allow to trap and manipulate single or several particles with one or more intense laser beams. Light forces can also be used to measure interaction forces between colloidal particles and motor molecules [2–5], to probe the elasticity of single polymer like DNA [6], or to investigate properties of membranes [7, 8].

It has been also demonstrated, that by creating an extended light intensity pattern, e.g. by interfering two or more laser beams, the structure of many colloidal particles can be manipulated [9, 10]. For colloids with effectively hard sphere interaction, one can organize the particles to form any structure, even a two-dimensional fivefold symmetrical and three-dimensional ones, according to the structure of intensity antinodes where the particles are trapped [9, 10]. For strongly interacting charge-stabilized colloidal particles where the interparticle interaction is a screened Coulomb potential, it has been shown experimentally by Chowdhury et al. [9] that a two-dimensional colloidal liquids starts to crystallize when exposed to a periodic light pattern created by two interfering laser beams. When the wave vector of the modulation potential is chosen to coincide the location of the first peak of the structure factor of colloidal liquids (or the periodicity $d = \sqrt{3}a/2$, with the mean interparticle separation), a dominantly hexagonal order is observed. This effect is called laser-induced freezing (LIF) [9, 11].

Later, density functional theory and Monte-Carlo simulations confirmed the existence of LIF and also predicted that this freezing transition changes from a first order to second order one via a tricritical point. Furthermore, it is expected that a colloidal crystal can re-melt (LIM) when the external field exceeds some critical value [12, 13]. This, however, has not been proven experimentally, yet.

LIF is anyway a result of many-body effects, although one can understand it in the following way, that the external potential induces the alignment of the particles along rows, whereas the interparticle-screened Coulomb repulsion leads to an equal distribution of particles within a single row (see Fig. 2c) and to the registration of particles in neighboring rows. As shown by density functional theory [12], LIF is, in fact, the excitation of the density modulation modes in colloidal liquids with one specific external modulation potential. In this paper we perform experiments to study the structure when the periodicity (or the wave vector) of the external modulation potential deviates from the above value ($\sqrt{3}a/2$). To control the ion concetration we exployed a continuous deionization technique [14] which allows us to adjust different salt concentrations. In the following we present the results on LIF and the induced structures.

Experimental

The sample cell is composed of two microscopic cover glasses whose spacing can be adjusted from several mm to about 20 μm. After assembling the cell, it was connected to a closed circuit which contained the colloidal suspension. We used charge-stabilized surfactant-free polystyrene sulfate particles from IDC with a diameter of 3 μm. The particle concentration was about 1.5×10^7/ml, but due to sedimentation the actual particle concentration in the cell is assumed to be somewhat higher. The suspension was then pumped through this circuit which also contained a vessel of ion exchanger and an electrical conductivity probe to control the ionic strength in the suspension. This method allowed us to perform measurements at different ionic strengths [14].

Figure 1 shows schematically the setup used in our experiments. The beam of an argon ion laser (TM$_{00}$ mode, $\lambda = 514$ nm, $I_{max} = 2.6$ W) is split into two parallel beams of equal intensity by means of two beam splitters (BS1, BS2) and two mirrors (M1, M2). The distances s of the parallel beams can be adjusted by the position of the mirror M2 which is mounted on a motor controlled translation stage. After passing the lens L, the two beams are overlapped inside the sample cell where they produce interference fringes. The spacing of the interference fringes d is controlled by beam spacing s through

$$d = \lambda/2 \sin(\theta/2) = f\lambda/s \,, \tag{1}$$

where θ is the angle between the laser beams and f the focus length of the lens L. The sample cell with the colloidal suspension is illuminated with white light (not shown in Fig. 1) and imaged with a microscope objective (magnification 40) on a CCD camera. In order to prevent the camera to be damaged by the intense laser light, the transmitted and scattered laser light is blocked by a filter. The obtained data were recorded on tapes through a video system which was connected to a computer for further analysis.

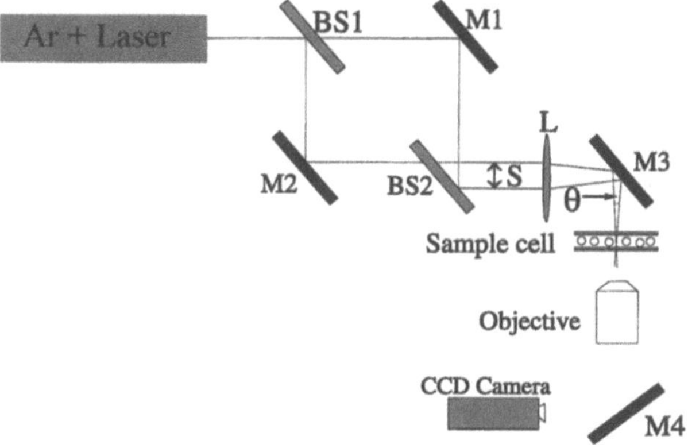

Fig. 1 The optical setup used during the experiments. M1, M2, M3, M4 are mirrors, BS1, BS2 beam splitters, L is a lens. The position of mirror M2 which can be changed by a motor-controlled translation stage determines the fringe spacing d

Results

When the laser is switched on and the colloidal suspension is subjected to an intererence pattern, the radiation pressure causes the particles to be pushed towards the bottom glass plate and a two-dimensional system is produced. Since glass surfaces are known to be negatively charged when immersed in water, the particles are prevented from sticking to the glass surface by electrostatic repulsion [15]. Due to the difference in the refraction indices of PS ($n_p = 1.59$) and water ($n_w = 1.33$) the particles are drawn into the intensity maximum of the interference grid which can be considered as an external periodic potential. The form of this potential $V(x)$ can be written as [11]

$$V(x) = V_0 \cos(2\pi x/d)\,, \tag{2}$$

where $V_0 = [3n_w Pr^3(n^2 - 1)/c\sigma_0^2(n^2 + 2)][j_1(\pi r/d)/2\pi r]$, with P being the laser power, c the light velocity in vacuum, $n = n_p/n_w$, j_1 the first-order spherical Bessel function, r the particle diameter, and σ_0 the waist radius of the laser beam in the sample. Due to the Gaussian shape of the interfering laser beams, V_0 has also an Gaussian envelope. To minimize this effect [9] which would complicate the analysis we expanded the interference region to an area of about 300 μm in diameter.

Figure 2a shows a typical configuration of particles in the sampe cell when no interference pattern is present. The particles are arranged as expected for a colloidal liquid and interact only via a screened Coulomb potential. The corresponding Fourier transformation which is plotted in Fig. 2b consists of two ring-shaped areas, the larger one being due to an illustration artefact, whereas the smaller (and darker) one confirms – due to the absence on any

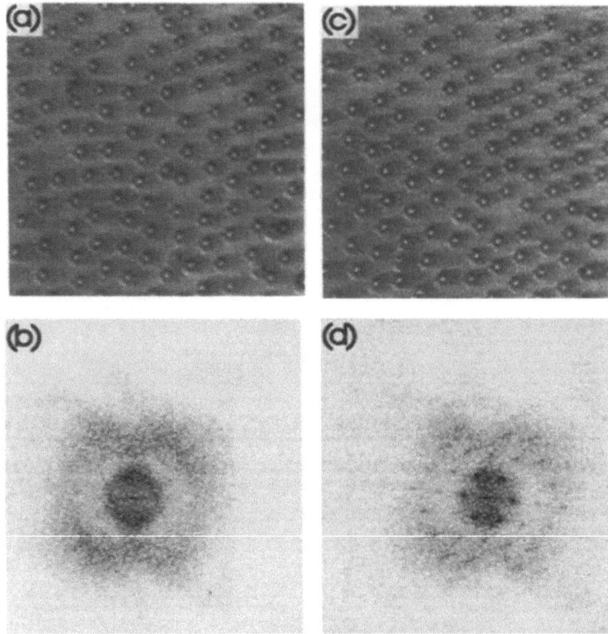

Fig. 2 Microscopic pictures and corresponding Fourier transformation of colloidal structure (a), (b) in absence of laser field, and (c), (d) when exposed to an interference pattern (laser intensity 200 mW). The direction of the interfering fringes is vertical

distinct features – the liquid structure of Fig. 2a. When the laser is turned on and the interference pattern (the fringes are aligned vertically) interacts with the particles, the structure changes and starts to crystallize. This can be seen in Fig. 2c, where the laser light intensity is 200 mW, corresponding to a potential depth of 1.9 $k_B T$, and the fringe spacing $d = \sqrt{3}a/2$, with $a = 10\ \mu m$ being the average particle distance determined from Fig. 2a. Under these conditions the interference fringes are commensurate with a hexagonal lattice which can be also seen in the Fourier transformation in Fig. 2c.

However, when the modulation periodicity $d \neq \sqrt{3}a/2$, deviation from a hexagonal symmetry are expected. Figure 3 shows several structures formed under different fringe spacing conditions, the laser light intensity was kept constant at a value of 200 mW as above. The ion concentration, i.e. the ionic conductivity during the experiments was kept contants at a value of 0.5 $\mu S/cm$. When d is increased (from the left to the right) the corresponding Fourier transformations clearly indicate that a change of the induced structure from a crystalline (Fig. 3a and b) into a liquid-like structure (Fig. 3g and h) occurs. With a_1 being the mean distance of particles along a row (parallel to the interference fringes) we can define the parameter $k = d/a_1$. As mentioned above, the close packed hexagonal lattice corresponds to $k = \sqrt{3}/2 = 0.866$. When k is smaller than that value, as being the case in Fig. 3a and b, where k was chosen to be 0.55, the particles in adjacent rows are so close that a crystal with almost quadratic symmetry (which can be also considered as a hexagonal lattice distorted in the vertical direction) is observed. In fact, for $k = 0.5$ we

Fig. 3 Microscopic pictures and corresponding Fourier transformation for different fringe spacing. The ratio of the fringe spacing to the mean particle separation a_1 is 0.55 in (a) and (b), 0.91 in (c) and (d), 1.0 in (e) and (f), and 1.2 in (g) and (h). The laser intensity is 200 mw, the direction of the fringes is vertical

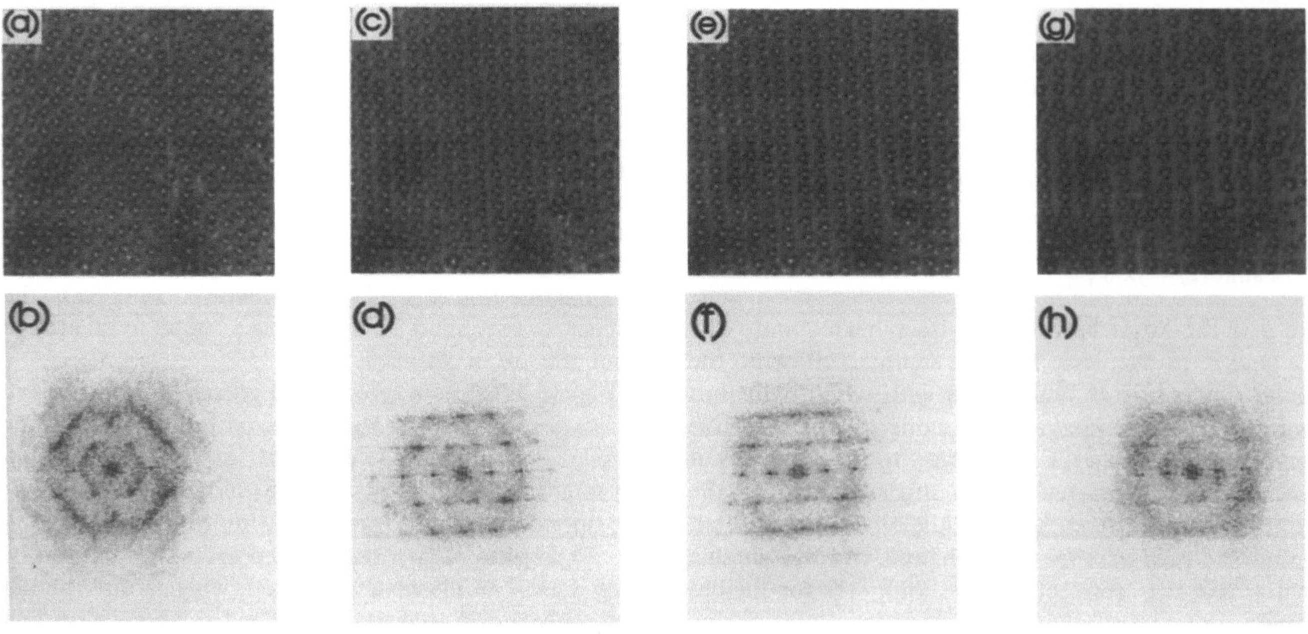

Progr Colloid Polym Sci (1998) 110:46–49
© Steinkopff Verlag 1998

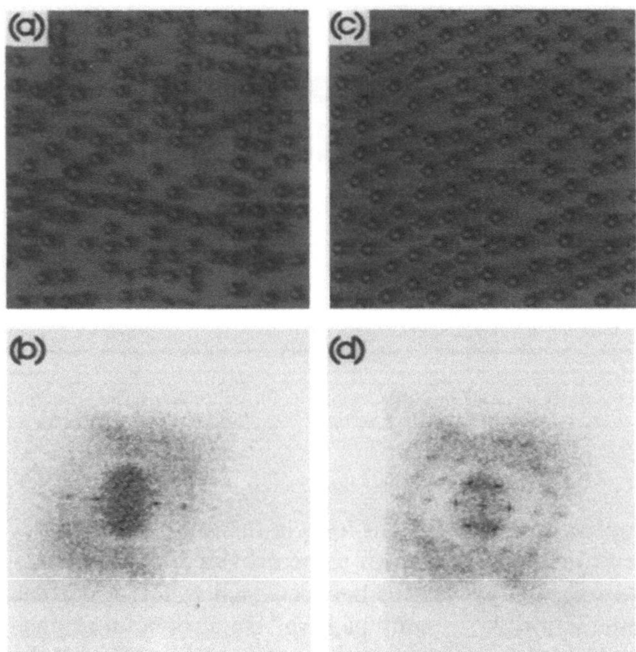

Fig. 4 Microscopic pictures and corresponding Fourier transformation of colloids for two different ion concentrations. The conductivity of the colloidal suspension is 2.6 and 0.5 μS/cm respectively, in (a) and (b) and (c) and (d). The laser intensity is 200 mW in both cases

observed a structure with exact quadratic symmetry. The structure in Fig. 3c and f with $k = 0.91$, being very close to 0.866, is nearly a hexagonal lattice. With k increased further to 1.0, we obtain the structure shown in Fig. 3e and g which is still a hexagonal lattice but now distorted in the horizontal direction. However, as can be seen from the spots in the corresponding Fourier transformation (Fig. 3f), registration between neighboring rows (i.e. repulsive interaction between particles of adjacent rows) still occurs. Figure 3g and h finally, show the structure for $k = 1.2$. The spots in Fig. 3h only correspond to a particle density modulation along vertical direction, but the registration between neighboring rows is lost because their interaction is smaller than the thermal energy. The obtained structure is a modulated liquid.

Finally, we want to demonstrate the effect on the light-induced structures when changing the ionic concentration in the system. We found that for low particle density or high ion concentration where the system is far from the freezing condition, the colloidal liquids never freeze to a crystalline phase, even at very high light intensities. This can be seen from Fig. 4, where the particle concentrations are nearly the same (about 9% higher in Fig. 4a), but the ion concentration is changed. The measured conductivity is 2.6 and 0.5 μS/cm, respectively, for Fig. 4a and c. In Fig. 4a we only observe the alignment of particles along the interference fringes, however, no order within rows and no registration between them are found. This can be also seen in the corresponding Fourier transformation in Fig. 4b, which is characteristic of a modulated liquid. In contrast to this, after the ionic concentration was decreased, the colloidal suspension is crystallized under the influence of the same periodic light potential (Fig. 4c and d). This is in agreement to theoretical calculations [13].

In summary, we have studied the phase transitions of colloids under the influence of a periodic light field. We observed the light-induced freezing transition of the system when the fringe spacing and the particle concentration is chosen properly. Additionally, we also observed strong deviations of the induced crystal structure from a perfect hexagonal symmetry when the fringe spacing is veried. Finally, we demonstrated the influence of the salt concentration on LIF.

Acknowledgments Financial support by the Deutsche Forschungsgemeischaft, Sonderforschunggsbereich 513 is gratefully acknowledged. One of the authors (Q.H.W.) would like to acknowledge the research fellowship support from the Alexander von Humboldt Foundation.

References

1. Ashkin A (1970) Phys Rev Lett 24:156; Ashkin A, Dziedzic JM (1975) Science 187:1073
2. Kepler GM, Fraden S (1994) Phys Rev Lett 73:356
3. Crocker JC, Grier DG (1996) Phys Rev Lett 73:352
4. Ashkin A, Schütze K, Dziedzic JM, Euteneuer U, Schliwa M (1990) Nature 348:346
5. Block SM, Goldstein LSB, Schnapp BJ (1990) Nature 348:348
6. Austin RH, Brody JP, Cox EC, Duke T, Volkmuth W (1997) Phys Today 33 and reference therein
7. Bariv R, Moses E (1994) Phys Rev Lett 73:1392
8. For a review, see Annu Rev Biophys Biomol Struct (1994) 23:247
9. Chowdhury A, Ackerson BJ, Clark NA (1985) Phys Rev Lett 60:833; Ackerson BJ, Chowdhury AH (1987) Faraday Discuss Chem Soc 83:309
10. Burns MM, Fournier J-M, Golovchenko JA (1990) Science 249:749
11. Loudiyi K, Ackerson BJ (1992) Physica 184A:1; 26
12. Chakrabati J, Krishnamurthy HR, Sood AK (1994) Phys Rev Lett 73:2923
13. Chakrabati J, Krishnamurthy HR, Sood AK, Sengupta S (1995) Phys Rev Lett 75:2233
14. Palberg T, Härtl W, Wittig U, Versmold H, Würth M (1992) J Phys Chem 96:8081
15. Prieve DC, Luo F, Lanni F (1987) Faraday Discus Chem Soc 83:297

Progr Colloid Polym Sci (1998) 110:50–53
© Steinkopff Verlag 1998

R. van Roij
J.-P. Hansen

Phase separation in suspensions of repelling charged colloids

R. van Roij (✉)
H.H. Wills Physics Laboratory
Royal Fort, Tyndall Avenue
Bristol BS8 1TL
UK

J.-P. Hansen
Department of Chemistry
University of Cambridge
Lensfield Road
Cambridge CB2 1EW
UK

Abstract We use the simplest possible density functional theory to show that the purely repulsive screened-Coulomb (or DLVO) interaction between charged colloidal particles is compatible with gas–liquid, gas–solid and solid–solid coexistence in colloidal suspensions of low ionic strength of about 10^{-6} mol l^{-1}. The cohesive energy of the condensed phase, which produces a Van der Waals loop in the total free energy, is shown to be provided by the attractions between each colloidal particle and "its own" cloud of counterions. This finding may (partially) resolve the ongoing debate on attractions between like-charged particles.

Key words Van der Waals theory – like-charge attraction – DLVO-theory

Introduction

Attractive interactions between particles are well-known to be essential for the occurrence of gas–liquid coexistence in the phase diagram of simple fluids. This relation between micoscopic interactions and macroscopic phase behaviour was first demonstrated in 1873 by Van der Waals [1]. His theory is easily explained in terms of his expression for the Helmholtz free energy, denoted F_{vdW} here, as a function of the density ρ and temperature T. Van der Waals assumed it to be given by

$$\frac{F_{vdW}}{V} = k_B T \rho \left(\log\left(\frac{\rho b}{1 - \rho b}\right) - 1 \right) - a\rho^2 . \tag{1}$$

with V the total volume and k_B the Boltzmann constant. In Eq. (1) the parameter b is a typical molecular volume (and thus describes short-ranged repulsions), while a is a measure for the strength of the long-ranged attractions. In Fig. 1 we plot F_{vdW} as a function of ρ for a temperature above and below the critical temperature $k_B T_c = 8a/27b$. Below T_c a common tangent construction reveals that coexistence of a low-density phase (a gas) and a high-density phase (a liquid) is more favorable than a single homogeneous fluid phase; above T_c no such phase separation occurs. As $T_c = 0$ if $a = 0$, Van der Waals' theory implies that *there is no gas–liquid coexistence in the phase diagram of purely repulsive particles*. Although this conceptually important statement has been refined and quantified considerably in the course of the 20th century, it is still accepted to be essentially correct.

In fact, the current belief in this statement is so strong that recent experimental observations of gas–liquid [2, 3] and gas–solid [4] coexistence in low-salt suspensions of charged colloidal particles have been interpreted as proof of the existence of long-ranged *attractions* between colloids of the *same* charge-sign! This nature of the interaction between colloidal particles is not only counterintuitive, but also in contradiction with the standard and well-established DLVO theory (which predicts purely repulsive screened-Coulomb interactions [5]). Attempts to formulate alternatives for the DLVO-theory, as for instance by Sogami and Ise [6], have been criticized by others [7, 8], resulting in a sometimes furious debate in the literature.

In this contribution we attempt to settle the controversy by showing that the observed coexistence of a dilute

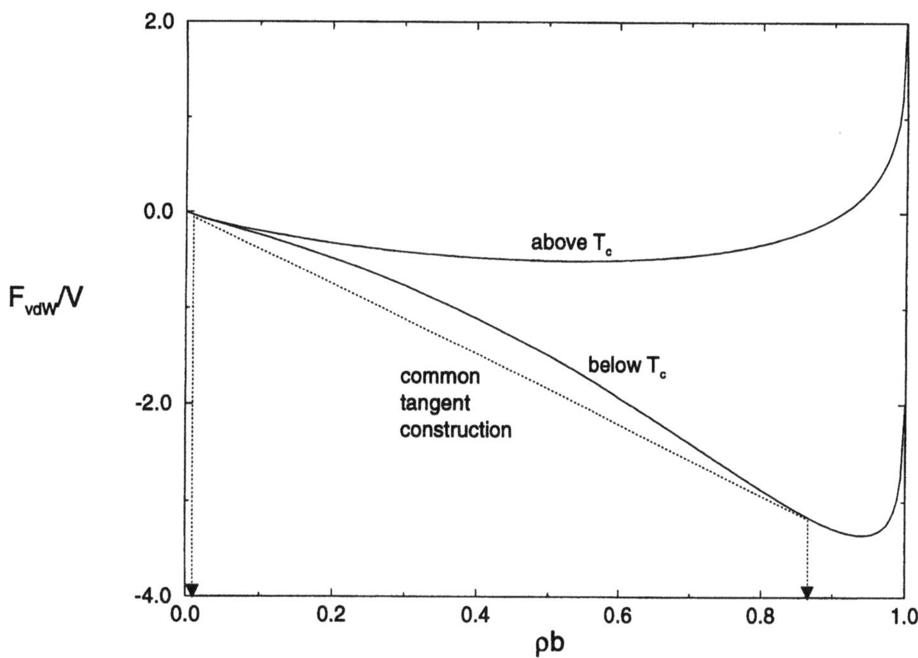

Fig. 1 Van der Waals' Helmholtz free energy, given in Eq. (1), as a function of density above and below the critical temperature T_c. Below T_c, a common tangent construction determines the coexisting gas and liquid densities

and a dense phase in colloidal suspensions is yet compatible with purely repulsive DLVO-interactions between the colloids. The catch is that the internal (free) energy of a single dressed colloidal particle (i.e. a colloidal particle with its surrounding cloud of co- and counterions) *cannot* be considered as an irrelevant constant, since it depends nontrivially on the colloidal density. This is qualitatively different from the analogous free energy contribution in simple fluids, as internal molecular states are not (or hardly) affected by the density of the fluid. Using a simple density functional theory, which reproduces the DLVO-potential between two dressed colloidal particles, we show that the density dependence of the internal free energy of a dressed colloidal particle is such that it causes a "Van der Waals loop" at low salt concentrations, and hence coexistence between a dilute and a dense phase. As this loop is driven by the Coulombic *attractions* between the colloidal particle and the surrounding counterions, the condensation phenomenon is yet caused by attractions, although not by attractions between two dressed colloidal particles. Part of this work has been published recently [9], to which we refer for more details.

Effective Hamiltonian of colloidal suspension

As a simple model for a colloidal suspension we consider a thermodynamic system of volume V and dielectric constant ε, suspending (i) N_m identical macroionic hard spheres of radius R carrying a charge $-Ze$ (with $-e$ the

electron charge), (ii) $N_c = ZN_m$ point-like counterions of charge $+e$, and (iii) N_s pairs of point-like added salt ions of charge $\pm e$. The total Hamiltonian $H = H_m + H_\mu + H_{m\mu}$ of this system consists of the bare Hamiltonians H_m and H_μ of the macroions (m) and microions (μ), respectively, and their interaction energy $H_{m\mu}$. The resulting Helmholtz free energy F at fixed inverse temperature $\beta = 1/k_B T$ is formally given by [9–12]

$$\exp(-\beta F) = \text{Tr}_m \text{Tr}_\mu \exp(-\beta H) \equiv \text{Tr}_m \exp(-\beta H_m^{\text{eff}}),$$

(2)

where the trace $\text{Tr}_{m(\mu)}$ is short for the phase space integral over all macro(micro)ion degrees of freedom. The effective Hamiltonian is written as $H_m^{\text{eff}} = H_m + F'$, with

$$\exp(-\beta F') = \text{Tr}_\mu \exp(-\beta(H_\mu + H_{m\mu})).$$ (3)

Clearly, F' is the free energy of an inhomogeneous fluid of microions in the external field of fixed macroions. It is well-known that F' and the density profiles of the microions can be determined from density functional theory [13]. Using the simplest possible free energy functional – which leads to the linearized Poisson–Boltzmann equation for the density profiles – we showed in Ref. [9] that the charge density outside a colloidal particle decays with distance r from its center as $\exp(-\kappa r)/r$, where the screening parameter κ is given by

$$\kappa^2 = \frac{4\pi e^2(Zn_m + 2n_s)}{\varepsilon k_B T},$$ (4)

with the number densities $n_m = N_m/V$ and $n_s = N_s/V$. The resulting effective Hamiltonian, which depends on the colloid's center-to-center separations r_{ij}, was shown to be

$$H_m^{\text{eff}} = K_m + \frac{1}{2} \sum_{\substack{i \neq j = 1}}^{N_m}$$

$$\left[v_{\text{hs}}(r_{ij}) + \frac{Z^2 e^2}{\varepsilon} \left(\frac{\exp(\kappa R)}{1 + \kappa R} \right)^2 \frac{\exp(-\kappa r_{ij})}{r_{ij}} \right] + F_0 , \qquad (5)$$

with the colloidal kinetic energy K_m, the hard-sphere potential v_{hs}, the Yukawa-like DLVO pair-interaction, and the coordinate-independent term

$$F_0 = F_{\text{id}}(N_s, V, T) + F_{\text{id}}(ZN_m + N_s, V, T) - \frac{Z^2 e^2}{2\varepsilon} \frac{N_m \kappa}{1 + \kappa R} . \qquad (6)$$

Here F_{id} is the ideal gas contribution to the free energy of the microions, where the released counterions and the positive added salt ions are assumed to be identical species. The final term of Eq. (6) represents the internal free energy of the dressed colloids, as discussed in the introduction. It is of the order of $-Z^2 e^2 \kappa/\varepsilon$ per macroion, which is the Coulombic energy of charges $-Ze$ and $+Ze$ separated by the range $1/\kappa$ of the ionic charged cloud. Naively, one could argue that F_0 is an irrelevant additive constant to the effective Hamiltonian, which can be neglected. The key point of our study is, however, that F_0 depends non-trivially on the density n_m, and can thus be a relevant contribution to the total Helmholtz free energy F. It is easily checked that the dependence of F_0 on n_m is negligible if $2n_s \gg Zn_m$, i.e. at high salt concentrations or at low Z. Under such conditions F_0 is indeed an irrelevant constant. In contrast, this dependence is very strong if $2n_s \ll Zn_m$. In the latter case $\kappa \propto n_m^{1/2}$ and hence the final term of F_0/V behaves as $-n_m^{3/2}$. This term has a similar effect as the $-a\rho^2$ term of F_{vdW}/V in Eq. (1): it drives a "van der Waals loop" to which a common tangent construction can be applied, yielding phase coexistence. The power 3/2 is typical for theories of long-ranged Coulombic interactions [15], while the power 2 is typical for shorter-ranged potentials (with a finite second virial coefficient). The crossover from the high to the low salt regime takes place at salt concentrations of the order of micromoles/liter, which corresponds precisely to the experimental conditions of two-phase observations [2–4].

Helmholtz free energy and phase diagram

So far we have discussed the effective Hamiltonian (5) and its coordinate-independent term F_0, but the actual phase diagrams can only be inferred from the total free energy F.

Obviously, it follows from Eqs. (2) and (5) that $F = F_0 + F_1$, where F_1 is the free energy of a classical system of N_m structureless particles interacting with the hard-sphere plus screened-Coulomb pair-potential. Here we approximate F_1 variationally using the Gibbs–Bogolyubov inequality, where the reference fluid and solid phase are, respectively, the hard-sphere fluid and the Einstein crystal [14]. Once F is known, phase coexistences are determined by imposing equal osmotic pressure and chemical potential (both for the colloids and the added salt) in the two phases.

We present a typical phase diagram of an aqueous colloidal suspension at room-temperature, and thus set $e^2/\varepsilon k_B T = 7.2$ Å. We also take into acount the dissociation of water at pH = 7, as the finite concentration of H^+ and OH^- ions provides additional screening. We take the colloidal parameters that were recently used in experiments by Larsen and Grier [16], and set the diameter $D = 2R = 652$ nm and the charge number $Z = 7300$. The phase diagram, shown in Fig. 2, is represented in the η–n_s plane, with the colloidal packing fraction $\eta = (4\pi/3) R^3 n_m$. At $n_s > 3 \times 10^{-6}$ mol l^{-1}, the only coexistence is that of the fluid (F) and the face-centered-cubic solid (FCC). This coexistence, which is well-known to occur in purely repulsive systems, is called "narrow" because the coexisting densities are very similar. In contrast, the coexistence at $n_s < 3 \times 10^{-6}$ mol l^{-1} are "broad", and are due to the van der Waals loop in F_0. The critical point, indicated by $\times$, occurs in the FCC-solid phase, so that there is a (small) regime of isostructural FCC-FCC coexistence, which is

Fig. 2 Room temperature phase diagram of aqueous monodisperse suspensions of colloidal particles of diameter $D = 652$ nm and chargenumber $Z = 7300$, as a function of the colloidal packing fraction η and the salt concentration n_s. The two possible phases are fluid (F) and face-centered-cubic solid (FCC). The critical point is indicated by $\times$, the triple points by $\triangle$

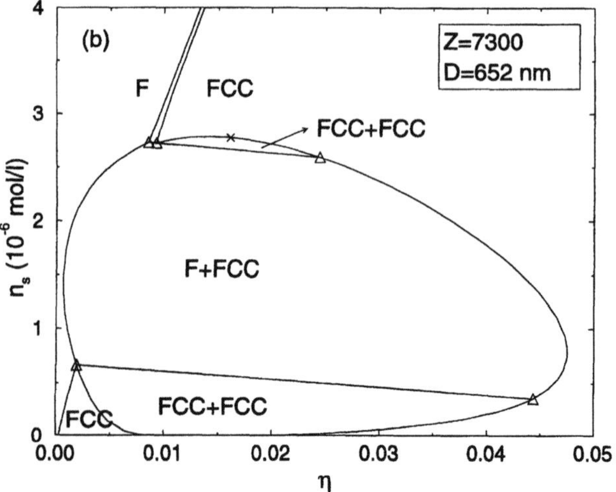

Progr Colloid Polym Sci (1998) 110:50–53
© Steinkopff Verlag 1998

intersected by the narrow F-FCC transition to yield a broad F-FCC coexistence. The triple points are indicated by Δ. At extremely low salt concentrations below $n_s \approx 6 \times 10^{-7} \, \mathrm{mol\,l^{-1}}$ a reentrant phenomenon occurs, yielding another F-FCC-FCC triple-point below which the narrow F-FCC coexistence preempts a broad FCC-FCC coexistence.

We note that Larsen and Grier observed unexplained long-lived metastable crystals at $\eta \approx 0.02$ and $n_s \leq 2.5 \times 10^{-6} \, \mathrm{mol\,l^{-1}}$, which they interpreted as evidence of colloidal attraction [16]. Our results suggest, however, that one should expect a broad F-FCC coexistence at such thermodynamic state-point, so that metastable FCC-crystals do not melt as rapidly as those in narrow coexistence with a fluid. We stress, once more, that the broad coexistence is *not* due to colloid–colloid attraction, but rather to colloid–counterion attraction.

In Ref. [9] we also showed that the critical point can be shifted into the fluid phase, yielding gas–liquid coexistence, by increasing the total charge Z. The critical point disappears completely for sufficiently small values of Z, leaving the narrow fluid–solid coexistence the only one in the phase diagram.

Conclusion

We have presented a theoretical explanation for the experimental observation of coexistence of dilute and dense phases in colloidal suspensions at low salt concentrations. The theory is based on the simplest possible density functional, which not only reproduces the purely repulsive DLVO-potential but also gives the internal free energy of a colloidal particle with its cloud of microions. The density dependence of this latter term at low salt concentrations drives a Van der Waals loop, which leads to phase coexistence without the presence of attractions between the colloidal particles. This implies that one should not interpret the experimental observation of gas–liquid or gas–solid coexistence in colloidal suspensions as a proof of the existence of colloid–colloid attractions, as has often been done [2–4, 6, 16, 17].

Acknowledgements It is a pleasure to thank Thierry Biben, Emmanuel Trizac, Raul Leote de Carvallo, and Marjolein Dijkstra for numerous discussions. This work was made possible by financial support from the "Région Rhône-Alpes".

References

1. Van der Waals JD (1873) PhD Thesis, Hoogeschool te Leiden, The Netherlands
2. Ito K, Hiroshi H, Ise N (1994) Science 263:66; Tata BVR, Yamahara E, Rajamani PV, Ise N (1997) Phys Rev Lett 78:2660
3. Tata BVR, Rajalakshmi M, Arora AK (1992) Phys Rev Lett 69:3778; Palberg T, Würth M (1994) Phys Rev Lett 72:786
4. Ise N, Smalley MV (1994) Phys Rev B 50:16722
5. Verwey EJW, Overbeek JThG (1948) Theory of the Stability of Lyotropic Colloids. Elsevier, Amsterdam
6. Sogami I, Ise N (1984) J Chem Phys 81:6329
7. Overbeek JThG (1987) J Chem Phys 87:4406
8. Woodward CE (1988) J Chem Phys 89:5140
9. van Roij R, Hansen JP (1997) Phys Rev Lett 79:3082
10. Evans R, Hasegawa M (1981) J Phys C 14:5225
11. Grimson MJ, Silbert M (1991) Mol Phys 74.397
12. Löwen H, Hansen JP, Madden PA (1993) J Chem Phys 98:3275
13. Evans R (1979) Adv Phys 28:143
14. Shih WH, Stroud D (1983) J Chem Phys 79:6254; Shih WY, Aksay IA, Kikuchi R (1987) J Chem Phys 86:5127
15. Fisher ME, Levin Y (1993) Phys Rev Lett 71:3826 and references therein
16. Larsen AE, Grier DG (1997) Nature 385:230
17. Ise N, Yoshida H (1996) Acc Chem Res 29:3

Progr Colloid Polym Sci (1998) 110:54–60
© Steinkopff Verlag 1998

Contrast variation and optical polydispersity in colloidal suspensions

A.J. Banchio
G. Nägele
A. Ferrante
R. Klein

A.J. Banchio · G. Nägele (✉)
A. Ferrante · R. Klein
Fakultät für Physik
Universität Konstanz
P.O. Box 5560
D-78457 Konstanz
Germany
E-mail: gerhard.naegele@uni-konstanz.de

Abstract We present theoretical results on the effect of refractive index variations on static and dynamic light scattering in strongly correlated charge-stabilized suspensions which are optically and size polydisperse. The equilibrium microstructure and the short-time dynamics of these systems are calculated using a HNC integral equation scheme, and a core-shell model of optical polydispersity is discussed. Our calculations show that, close to an index matching point, the scattered intensity $I(k)$, the measurable structure factor $S_M(k)$, and the measurable hydrodynamic function $H_M(k)$ become very sensitive to the refractive index contrast with respect to the solvent. Various definitions of index matching points are considered. The incoherent part of the scattered intensity is particularly strong close to these points. The anomalous behaviour of $I(k)$ and $H_M(k)$ in the corresponding range of the solvent refractive index can be semi-quantitatively described by a simple approximative scheme. Consequences for scattering experiments aimed to determine size distributions or structural properties near index matching are discussed.

Key words Optical polydispersity – contrast variation – index matching – light scattering

Introduction

There has been considerable interest in recent years in studying the structural and dynamic properties of model systems of charge-stabilized spherical colloidal particles [1, 2]. Major tools for investigating the microstructure and diffusion of such systems are static and dynamic light scattering (SLS and DLS), respectively. Many colloidal suspensions which have been investigated are polydisperse with respect to the particle sizes and to the pair interaction. As a consequence, various theoretical investigations have been carried out to assert the influence of size- and interaction polydispersity on the static and dynamic scattering functions probed in light scattering [3–9], and good agreement between theory and experiment has been achieved for static and short-time dynamical properties

[3–6, 8, 10]. Besides being polydisperse in size and interaction, colloidal particles often exhibit an internal optical structure reflecting the form of particle synthetization. In other words, the particle refractive index $v(\mathbf{r})$ depends on the position $\mathbf{r}$ inside the particle. Even for optically homogeneous particles, the refractive index can depend on the diameter σ, i.e. $v = v(\sigma)$. We will refer to this additional type of polydispersity arising from a non-constant $v(\mathbf{r}, \sigma)$ as optical polydispersity. Very frequently, the various types of polydispersity are correlated with each other. Well-known examples of optically polydisperse colloids are TPM-coated silica spheres, which show a radially symmetric core-shell structure $v(r)$ of two different dielectric materials [10, 11], and partially crystalline fluorinated polymer particles [12]. The fluorinated particles are optically anisotropic and can therefore be used to measure rotational diffusion using depolarized DLS [13, 14].

In the interpretation of light scattering experiments on colloidal suspensions, it is often assumed that all particles have the same refractive index v and hence show no internal optical structure. For this idealized situation, quantities like the measurable static structure factor $S_M(k)$ and the measurable hydrodynamic function $H_M(k)$ are independent of the refractive index contrast $v - v_s$ between particles and suspending solvent. For optically polydisperse particles, this simplifying assumption is justified only when the refractive index distribution inside a particle is significantly smaller than the average contrast with the solvent. Strong optical contrast, however, gives rise to unwarranted multiple scattering of light, which invalidates the assumption of dominant single scattering underlying the conventional interpretation of scattering data. To avoid multiple scattering in more concentrated systems, the optical contrast is varied, e.g. by changing the composition and/or temperature of the solvent so that according to some criterion (cf. below), an optimal index matching between particles and solvent is achieved. Close to such an index matching point, the optical structure of the particles becomes important and, as we will show, it has to be accounted for in a correct interpretation of scattering data.

The aim of this work is to provide a well-founded theoretical analysis of contrast variation on static and short-time dynamic light scattering experiments in strongly correlated suspensions of particles, which are optically, size and interaction polydisperse. For this purpose, we will calculate and discuss the scattered intensity $I(k)$, the measurable static structure factor $S_M(k)$, and the hydrodynamic function $H_M(k)$. We show the sensitivity of these quantities to the refractive index contrast close to an index matching point. In this context, various definitions of optimal index matching are discussed, and a simple decoupling approximation is proposed which gives a semi-quantitative description of the anomalous behaviour of $S_M(k)$ and $H_M(k)$ close to index matching. In this article, we restrict ourselves to the core-shell model of optical polydispersity, and to charge-stabilized suspensions. A full account of our investigations, which encompasses further models of optical polydispersity, and hard sphere suspensions, will be given in a forthcoming article [15].

Light scattering in optically polydisperse suspensions

A polydisperse system of spherical colloidal particles can be regarded as an m-component mixture. The particles forming a component α, with $\alpha = 1, \dots, m$, are assumed to be identical, whereas particles belonging to different components can differ in terms of diameter σ_α, interaction and scattering properties. The average scattered intensity $I(k)$, measured in SLS experiments, is given, up to an irrelevant multiplicative constant, by [1, 2]

$$I(k) = N\overline{f^2}(k)\, S_M(k)\,, \tag{1}$$

where it is assumed that single scattering events prevail. Here, $N = \sum_{\alpha=1}^{m} N_\alpha$ is the total number of spheres in the volume V, and

$$S_M(k) = \frac{1}{\overline{f^2}(k)} \sum_{\alpha,\beta=1}^{m} (x_\alpha x_\beta)^{1/2}\, f_\alpha(k) f_\beta(k)\, S_{\alpha\beta}(k) \tag{2}$$

is the measurable static structure factor at wave number k. The function $S_M(k)$, normalized so that $S_M(k \to \infty) = 1$, is a linear superposition of partial static structure factors $S_{\alpha\beta}(k)$, weighted by the partial molar fractions x_α, and by the scattering amplitudes $f_\alpha(k)$ of α-type particles. The partial static structure factors $S_{\alpha\beta}(k)$ are related to the radial distribution function $g_{\alpha\beta}(r)$ of an $\alpha - \beta$ pair of particles as [2]

$$S_{\alpha\beta}(k) = \delta_{\alpha\beta} + n(x_\alpha x_\beta)^{1/2} \int \mathrm{d}r\, e^{ik \cdot r}\, [g_{\alpha\beta}(r) - 1]\,, \tag{3}$$

with $n = N/V$. Moreover,

$$\overline{f^n}(k) = \sum_{\alpha=1}^{m} x_\alpha f^n(k) \tag{4}$$

is the nth moment of the distribution of scattering amplitudes.

The key quantity determined in dynamic light scattering (DLS) is the electric field autocorrelation function $I(k, t)$. For polarised single scattering, this quantity reads, up to a multiplicative factor [2],

$$I(k, t) = N\overline{f^2}(k)\, S_M(k, t)\,, \tag{5}$$

and the measurable dynamic structure factor $S_M(k, t)$ is given by the right hand side of Eq. (2) with $S_{\alpha\beta}(k)$ replaced by its dynamic generalization, i.e. by the dynamic partial structure factors $S_{\alpha\beta}(k, t)$. In a short-time DLS experiment, one determines the initially exponential decay of $I(k, t)$, which is described for a polydisperse suspension by

$$I(k, t) = I(k)\, e^{-k^2 D_M(k)t}\,. \tag{6}$$

From the initial decay rate, the k-dependent measurable diffusion coefficient $D_M(k)$ is obtained. This coefficient can be expressed as [2, 10]

$$D_M(k) = \frac{H_M(k)}{S_M(k)}\,, \tag{7}$$

i.e. in terms of the static quantity $S_M(k)$, and of the so-called measurable hydrodynamic function $H_M(k)$, with [4]

$$H_M(k) = \frac{1}{\overline{f^2}(k)} \sum_{\alpha,\beta=1}^{m} (x_\alpha x_\beta)^{1/2}\, f_\alpha(k) f_\beta(k)\, H_{\alpha\beta}(k)\,. \tag{8}$$

The partial hydrodynamic functions $H_{\alpha\beta}(k)$ contain the configuration-averaged effect of the solvent-mediated hydrodynamic interactions on the collective short-time diffusion of α- and β-particles. For the statistical mechanical definition of $H_{\alpha\beta}(k)$ in terms of an equilibrium average involving hydrodynamic mobility tensors, consider Refs. [2, 4]. Without hydrodynamic interactions, $H_{\alpha\beta}(k) = D^0(\sigma_\alpha)\delta_{\alpha\beta}$, where $D^0(\sigma) = k_\mathrm{B}T/(3\pi\eta\sigma)$ denotes the Stokesean diffusion coefficient of an isolated sphere in a fluid with shear viscosity η and temperature T.

Contrary to $S_{\alpha\beta}(k)$ and $H_{\alpha\beta}(k)$, which are purely statistical mechanical quantities, it is apparent from Eqs. (1), (2) and (8) that $I(k)$, $S_\mathrm{M}(k)$ and $H_\mathrm{M}(k)$ depend also on the scattering properties through the dependence of $f_\alpha(k)$ on the size, and on the internal optical structure of the particles. Within the Rayleigh–Gans–Debye regime, which we assume to apply, the scattering amplitude of an α-sphere with radially symmetric refractive index $v_\alpha(r)$ is given by [11]

$$f_\alpha(k) = \frac{1}{v_\mathrm{s}} \int_{V_\alpha} \mathrm{d}\mathbf{r}\, e^{i\mathbf{k}\cdot\mathbf{r}} \left[v_\alpha(r) - v_\mathrm{s} \right] = f_\alpha^\mathrm{h}(k) + f_\alpha^\mathrm{inh}(k) , \tag{9}$$

where

$$f_\alpha^\mathrm{h}(k) = \frac{\overline{v_\alpha} - v_\mathrm{s}}{v_\mathrm{s}} B(k, \sigma_\alpha) , \tag{10}$$

$$f_\alpha^\mathrm{inh}(k) = \frac{1}{v_\mathrm{s}} \int_{V_\alpha} \mathrm{d}^3 r\, e^{i\mathbf{k}\cdot\mathbf{r}} \left[v_\alpha(r) - \overline{v_\alpha} \right] . \tag{11}$$

Here, r is the distance from the particle center, V_α is the particle volume, $\overline{v_\alpha} = (\int_{V_\alpha}\mathrm{d}^3 r\, v_\alpha(r))/V_\alpha$ is the volume averaged refractive index of an α-type particle, and $B(k, \sigma) = (\pi\sigma^3/2)\, j_1(k\sigma/2)/(k\sigma/2)$, where j_1 is the first order spherical Bessel function.

In Eq. (9), $f_\alpha(k)$ has been split into a homogeneous part, $f_\alpha^\mathrm{h}(k)$, representing scattering from an optically homogeneous particle of refractive index $\overline{v_\alpha}$, and into an inhomogeneous part, $f_\alpha^\mathrm{inh}(k)$, arising from the internal optical structure of an α-particle. The latter part is, up to the trivial prefactor v_s^{-1} in Eq. (11), independent of v_s, i.e., it does otherwise not depend on the optical contrast with the solvent. Notice that $f_\alpha^\mathrm{inh}(k \to 0) = 0$, expressing the fact that the internal structure is not resolved in the long wavelength limit. When the internal refractive index distribution $v(r)$ is quite narrow as compared to the average difference (contrast) to v_s, then $f_\alpha(k) \approx f_\alpha^\mathrm{h}(k)$, and the particles act like homogeneous spheres with a size-dependent refractive index $\overline{v_\alpha}$. In the opposite regime of nearly index matched particles, where $|v_\alpha - v_\mathrm{s}|/v_\mathrm{s}$ is small, the internal optical structure of the particles strongly influences the shape of $I(k)$, $S_\mathrm{M}(k)$ and of $H_\mathrm{M}(k)$. We will discuss this

interesting point in the last section, which contains our numerical results.

A particularly simple approximation for $S_\mathrm{M}(k)$ is obtained by disregarding the size- and interaction polydispersity, and by accounting for optical polydispersity only. In this so-called decoupling approximation, $S_\mathrm{M}(k)$ is expressed as [1, 2, 5, 6]

$$S_\mathrm{M}(k) \approx S_\mathrm{D}(k) \equiv X(k) + [1 - X(k)]S^\mathrm{I}(k; \bar{\sigma}) , \tag{12}$$

where

$$X(k) = 1 - \frac{\bar{f}(k)^2}{\overline{f^2}(k)} \tag{13}$$

is referred to as the incoherent scattering contribution, and $S^\mathrm{I}(k, \sigma)$ is the structure factor of the reference system of N ideally monodisperse particles of size $\bar{\sigma} = \sum_{\alpha=1}^m x_\alpha\sigma_\alpha$. In general, $S_\mathrm{M}(k)$ agrees well with its decoupling approximation $S_\mathrm{D}(k)$ for hard spheres and charge-stabilized particles only for small size polydispersities $s \leq 0.05$. Here, s is the relative standard deviation of the particle size distribution. For nearly index matched particles, however, the decoupling approximation is applicable within good accuracy for significantly larger size-polydispersities, as our results will demonstrate. This arises from the fact that close to index matching, optical polydispersity becomes most important.

We can introduce a simple decoupling approximation for the partial hydrodynamic functions $H_{\alpha\beta}(k)$, by accounting for size polydispersity only in the free-particle Stokesean diffusion coefficients $D^0(\sigma_\alpha)$. With regard to potential and hydrodynamic interactions, the spheres are treated as ideally monodisperse of common diameter $\bar{\sigma}$. This approximations amounts to replace $H_{\alpha\beta}(k)$ by

$$H_{\alpha\beta}(k) \approx \delta_{\alpha\beta}D^0(\sigma_\alpha) + \delta_{\alpha\beta}[D_\mathrm{s}^\mathrm{I} - D^0(\bar{\sigma})] + (x_\alpha x_\beta)^{1/2}H_\mathrm{d}^\mathrm{I}(k) \tag{14}$$

and hence $H_\mathrm{M}(k)$ by

$$H_\mathrm{M}(k) \approx H_\mathrm{D}(k) = \frac{1}{\overline{f^2}(k)} \sum_{\alpha=1}^m x_\alpha f_\alpha^2(k)\, D^0(\sigma_\alpha) + [D_\mathrm{s}^\mathrm{I} - D^0(\bar{\sigma})]$$

$$+ [1 - X(k)]\, H_\mathrm{d}^\mathrm{I}(k) . \tag{15}$$

Here $H^\mathrm{I}(k) = D_\mathrm{s}^\mathrm{I} + H_\mathrm{d}^\mathrm{I}(k)$ denotes the hydrodynamic function of the ideally monodisperse reference system, and it is split into a k-independent self-part D_s^I, and into a distinct part $H_\mathrm{d}^\mathrm{I}(k)$. For vanishing hydrodynamic interactions, the decoupling approximation of $H_\mathrm{M}(k)$ reduces to the exact result $H_\mathrm{M}^0(k) = \sum_\alpha x_\alpha f_\alpha^2(k)\, D^0(\sigma_\alpha)/\overline{f^2}(k)$. The decoupling approximation is justified for systems with weak hydrodynamic interactions dominated by the Stokesean friction, where the particles are nearly identical with respect to potential and hydrodynamic interactions.

In the remainder of this section, for polydisperse systems we will discuss various possibilities of defining the point of maximal partial index matching with respect to the suspending solvent. Quite often, the index matching point is defined by a value $v_s = v_s^{(1)}$ of the solvent refractive index, which is equal to the volume- and component-averaged particle refractive index $\bar{v}$. Explicitly,

$$v_s^{(1)} = \bar{v} \equiv \sum_{\alpha=1}^{m} x_\alpha \frac{1}{V_\alpha} \int_{V_\alpha} \mathrm{d}^3 r \, v_\alpha(r) . \tag{16}$$

We refer to this definition of index matching as IMP1. For particles which are optically polydisperse but otherwise identical, the forward scattering is minimized when $v_s = v_s^{(1)}$. The intensity is then given by $I(k=0) = N\overline{f^2}(0) \, S_\mathrm{D}(0)$, with $S_\mathrm{D}(0) = 1$ and $X(0) = 1$.

Alternatively, the index matching point can be defined as that value $v_s = v_s^{(2)}$, where (IMP2)

$$X(k=0; v_s^{(2)}) = 1 \tag{17}$$

or, equivalently, where $\bar{f}(k=0; v_s^{(2)}) = 0$. Clearly, $v_s^{(2)}$ can be different from $v_s^{(1)}$ for particles which are different not only in their optical properties. The definitions IMP1 and IMP2 have the advantage that no structural information is needed for calculating $v_s^{(1)}$ and $v_s^{(2)}$.

As a third possibility, we can define the index matching point by a value $v_s = v_s^{(3)}$, where the solid-angle integrated total intensity,

$$I_\mathrm{tot} = \int_{4\pi} \mathrm{d}\Omega_k \, I(k; v_s^{(3)}) , \tag{18}$$

becomes minimal (IMP3). Experimentally, $v_s^{(3)}$ can be determined from turbidimetric measurements. For a theoretical computation of $v_s^{(3)}$, information on the pair correlation functions is needed. All three definitions (IMP1-3) of index matching agree with each other only in the exceptional case of homogeneous colloidal spheres with narrow refractive index distribution, and which are identical otherwise.

If one is interested only in an estimate of the range of values v_s, where the optical contrast and therefore multiple scattering is small, any of the three definitions is equally good. For the example of a core–shell model, we will exhibit that $S_\mathrm{M}(k)$ and $H_\mathrm{M}(k)$ show large shape variations when v_s is in a critical range of values located around the $v_s^{(i)}$.

Model system and theoretical treatment

As an example of a colloidal system with related size, interaction and optical polydispersity, we consider a core–shell model characterized by a radially symmetric refractive index dependence of the form [11]

$$v_\alpha(r) = \begin{cases} v_1, & 0 \le r \le \sigma_\alpha/2 - d , \\ v_2, & \sigma_\alpha/2 - d < r \le \sigma_\alpha/2 . \end{cases} \tag{19}$$

According to Eq. (11), it follows for the inhomogeneous part that

$$f_\alpha^{\mathrm{inh}}(k) = \frac{v_2 - \overline{v_\alpha}}{v_s} B(k, \sigma_\alpha) + \frac{v_1 - v_2}{v_s} B(k, \sigma_\alpha - 2d) . \tag{20}$$

In order to mimic TPM-coated charged silica particles suspended in DMF [10], in our calculations we use $v_1 = 1.451$, $v_2 = 1.458$, and $d = 6$ nm for the thickness of the TPM coating. We describe the distribution of particle diameters by an optimized m-component histogrammatic representation of an unimodal Schulz distribution $p_s(\sigma; \bar{\sigma}, s)$ characterized by the mean diameter $\bar{\sigma} = 124$ nm, and by the relative standard deviation $s = 0.15$ (for details cf. Refs. [4, 10]).

The effective pair potential $u_{\alpha\beta}(r)$ between two particles of components α and β is given by their hard-core repulsion, plus a screened Coulomb potential of the DLVO form [2]

$$\frac{u_{\alpha\beta}(r)}{k_\mathrm{B}T} = L_\mathrm{B} Z_\mathrm{eff}^2 \left(\frac{e^{\kappa\sigma_\alpha/2}}{1 + \kappa\sigma_\alpha/2} \right) \left(\frac{e^{\kappa\sigma_\beta/2}}{1 + \kappa\sigma_\beta/2} \right) \frac{e^{-\kappa r}}{r} . \tag{21}$$

Here, $L_\mathrm{B} = 1.93$ nm is the Bjerrum length, and $Z_\mathrm{eff} = 220$ is the effective particle valency which we choose to be constant and to be the same for all particles, in accordance with a recent study where experimental and theoretical work have been combined [10]. We only consider salt-free suspensions of strongly correlated particles with univalent counter-ions. Then, the screening parameter κ is determined through $\kappa^2 = 4\pi L_\mathrm{B} n Z_\mathrm{eff}$. According to Eq. (21), size and interaction polydispersity are related, since $u_{\alpha\alpha}(r) < u_{\alpha\beta}(r) < u_{\beta\beta}(r)$ for $\sigma_\alpha < \sigma_\beta$. The partial static correlation functions $S_{\alpha\beta}(k)$ and $g_{\alpha\beta}(r)$ needed to determine $S_\mathrm{M}(k)$ and $H_\mathrm{M}(k)$, respectively, are calculated using an m-component HNC integral equation scheme [2, 10]. For $s = 0.15$, it is sufficient to account for $m = 5$ components. With regard to our calculations of $H_{\alpha\beta}(k)$ and $H^1(k)$, we employ the assumption of pairwise additive hydrodynamic interactions and a far-field expansion of the hydrodynamic two-body mobility tensors as described and justified by Nägele et al. [4]. In decoupling approximation, the static correlation functions $S^1(k; \bar{\sigma})$ and $g^1(r; \bar{\sigma})$ of the reference system are calculated in HNC using Eq. (21) with $\sigma_\alpha = \bar{\sigma}$ for $\alpha = 1, \ldots, m$. The volume fraction $\phi = (\pi/6) n \sum_{\alpha=1}^{m} x_\alpha \sigma_\alpha^3$ is kept fixed in our calculations at a value $\phi = 0.063$.

Results and discussion

For the core–shell model, we obtain the following values of the index matched v_s according to the index matching point criteria IPM1-3: $v_s^{(1)} = 1.4529$, $v_s^{(2)} = 1.4528$ and $v_s^{(3)} = 1.4523$. The differences in the $v_s^{(i)}$ appear in the fourth decimal point, since v_1 differs from v_2 only at the third decimal point.

Figure 1 shows the second moment, $\overline{f^2}(k)$, the measurable static structure factor, $S_M(k)$, and the average intensity, $I(k) = N\overline{f^2}(k)\,S_M(k)$, for $v_s = 1.4500$ and for larger values of v_s up to the value $v_s = v_s^{(2)}$ characterizing the index matching point IMP2. At $v_s = 1.4500$, the particles–solvent contrast is so pronounced that $f_\alpha(k) \approx f_\alpha^h(k)$, i.e., the α-type particles can be considered as homogeneous spheres of refractive index $\overline{v_\alpha}$. For this situation, $\overline{f^2}(k)$ is monotonously decaying in the k-interval displayed in the figure.

With v_s approaching $v_s^{(2)}$, the incoherent part $f_\alpha^{inh}(k)$ increasingly contributes to $f_\alpha(k)$. From Eq. (20), together with $v_2 > \overline{v_\alpha} > v_1$, it can be deduced that, contrary to $f_\alpha^h(k)$, $f_\alpha^{inh}(k)$ is negative in the k-interval depicted in Fig. 1. Due to partial cancellation of the two contributions to $f_\alpha(k)$, the second moment of $\overline{f^2}(k)$ and thus $S_M(k)$ and $I(k)$ become

Fig. 1 Second moment $\overline{f^2}(k)$ of the scattering amplitude distribution (in arbitrary units), measurable static structure factor $S_M(k)$, and average scattered intensity $I(k)$ (in arbitrary units) versus wavenumber k for various solvent refractive indices v_s as indicated in the figure. The dashed curves correspond to the index matching point IMP2, where $X(k=0) = 1$ and $S_D(k=0) = 1$

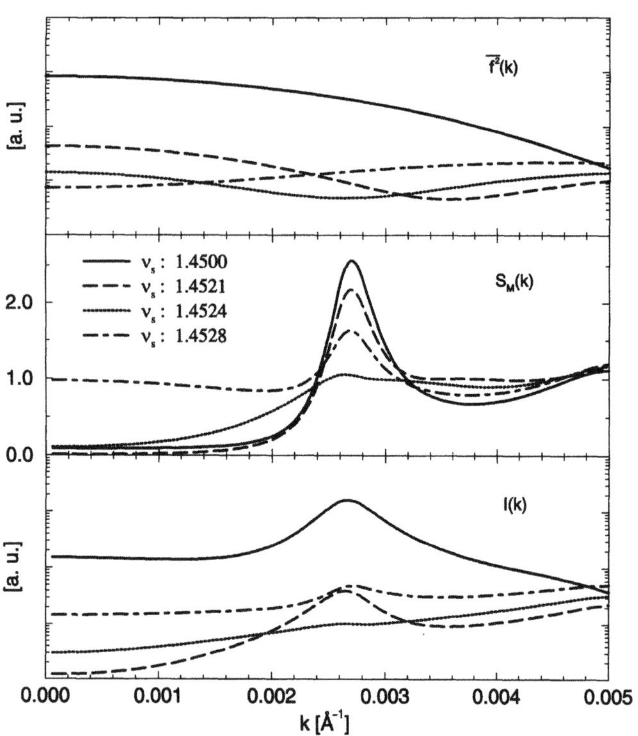

very sensitive to the values of v_s close to the matching point values $v_s^{(i)}$, with $i = 1, 2, 3$. The strong dependence on v_s is particularly pronounced at $k = 0$ and around the principal peak of $S_M(k)$.

Notice the overall decrease in the scattering intensity $I(k)$ as compared to the case $v_s = 1.4500$. Our corresponding findings for the normalized measurable hydrodynamic function $H_M(k)$ are displayed in Fig. 2. Apparent is the similarity in the shapes of $H_M(k)/H_M^0(k)$ and of $S_M(k)$ in dependence on k and v_s. In particular, analogous to $S_M(k)$ also $H_M(k)/H_M^0(k)$ behaves strongly anomalous in the index matching regime of v_s-values.

At the matching points IMP1-3, the comparison between $S_M(k)$ and its decoupling approximation $S_D(k)$ is depicted in Fig. 3. The decoupling approximation reproduces quite well the shape of $S_M(k)$ for a polydisperse system with $s = 0.15$ in the index matching regime, in particular at small k. The position k_m of the principal peak of $S_M(k)$ is somewhat overestimated in decoupling approximation, and there are also slight differences in the peak heights of $S_D(k)$ and $S_M(k)$.

Analogous to what has been found for the measurable structure factor, there is also good overall agreement between $H_M(k)$ and $H_D(k)$ in the index matching regime. This fact is illustrated in Fig. 4, where the case $v_s = v_s^{(2)}$ with $X(k = 0, v_s^{(2)}) = 1$ has been selected. Figure 4 further includes the hydrodynamic function of the ideally monodisperse system, and the incoherent scattering function $X(k)$.

In Figs. 5a and b, we plot $S_M(k)$ and $S_D(k)$, both at $k = 0$ and at the peak position k_m of $S_M(k)$, as functions of the solvent refractive index v_s. Figure 5a nicely demonstrates

Fig. 2 Normalized measurable hydrodynamic function $H_M(k)$ corresponding to Fig. 1. Here, $H_M^0(k)$ is the hydrodynamic function in the absence of hydrodynamic interactions

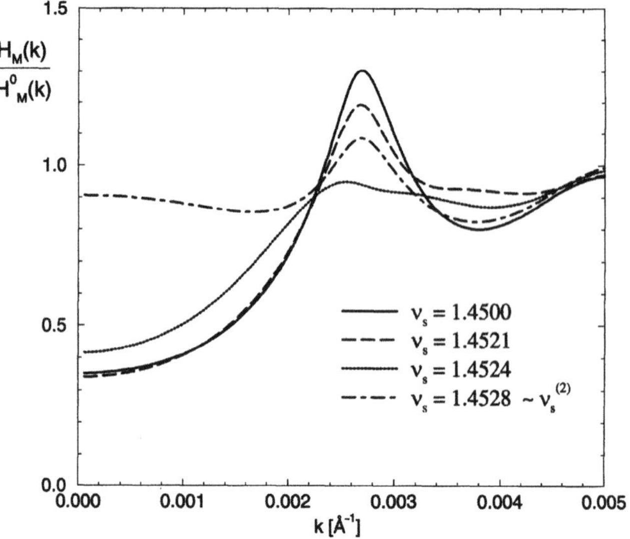

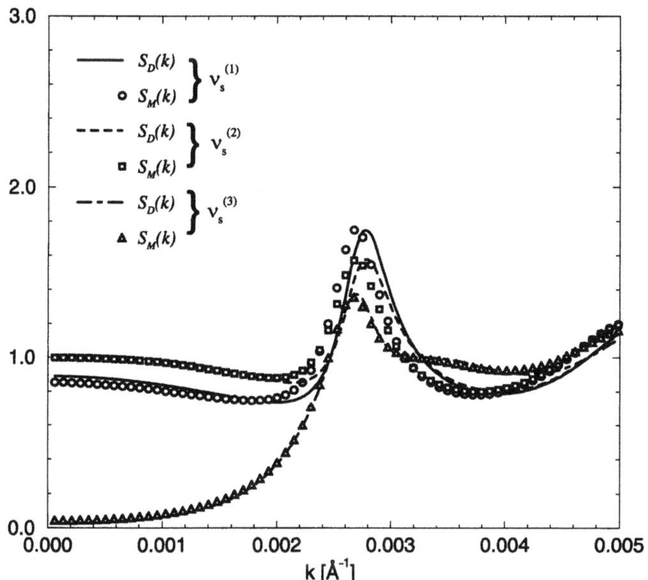

Fig. 3 Measurable static structure factor $S_M(k)$ at the index matching points IMP1-3. For comparison, the corresponding results of $S_D(k)$ in decoupling approximation are included in the figure

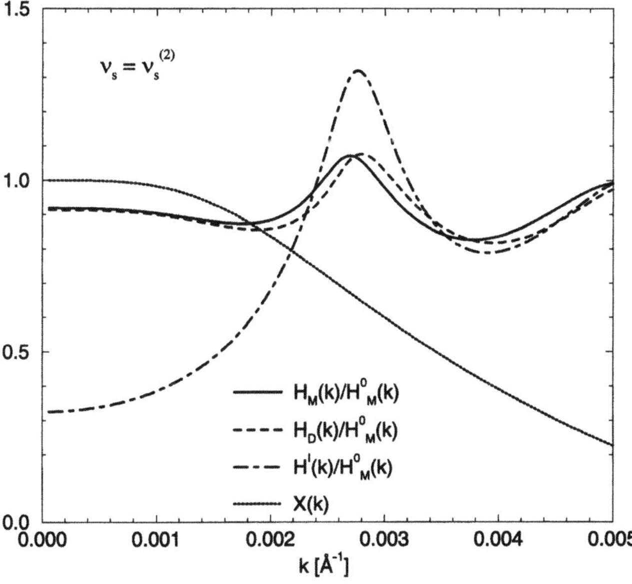

Fig. 4 Normalized hydrodynamic function at the index matching point IMP2. Comparison between decoupling approximation results $H_D(k)$ and $H_M(k)$, the latter calculated using a m-component histogrammatic representation of the Schulz size distribution. Further included in the figure is the hydrodynamic function $H^I(k)$ of the monodisperse reference system employed in the decoupling approximation, and the incoherent scattering function $X(k)$

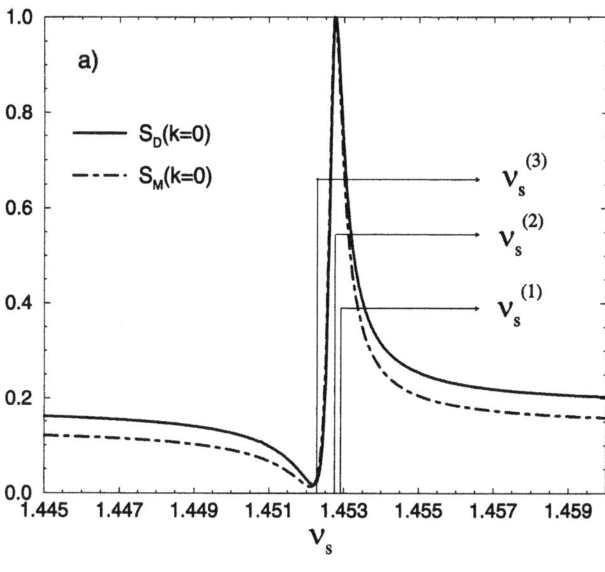

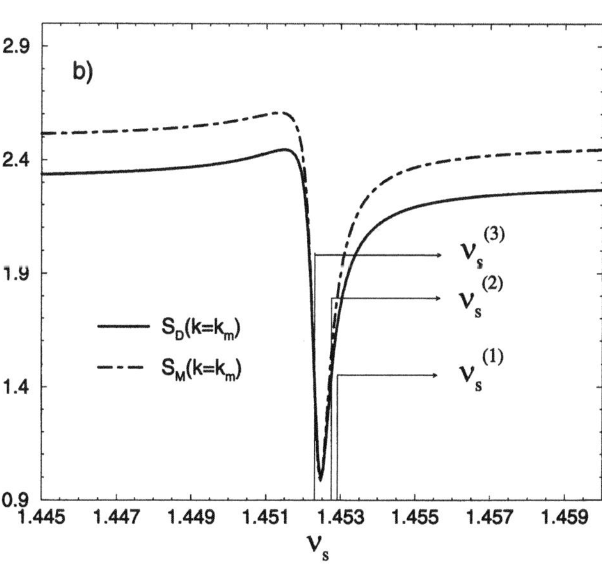

Fig. 5 Dependence of $S_M(k)$ and $S_D(k)$ on v_s for (a) $k = 0$ and (b) $k = k_m$. Here, k_m denotes the wavenumber where $S_M(k)$ attains its maximum. The double arrow marks the range of critical index matching. Notice the strong variations of the measurable static structure factor in the critical v_s-range

the enormous variation of $S_M(k = 0)$, and the nearly perfect agreement with the decoupling approximation in the regime of critical index matching located by the $v_s^{(i)}$. Figure 5b shows the corresponding finding at $k = k_m$. Suffi-

ciently outside the matching regime, indicated by the grey double arrow, $S_M(k)$ is only weakly dependent on v_s, and it deviates here more strongly from $S_D(k)$. Since $0 \leq X(k) \leq 1$ and, usually, $S^I(k = 0) \leq 1$, it follows that $S_D(k = 0) \leq 1$. The maximal value of one for $S_D(k = 0)$ and hence for $S_M(0) \approx S_D(0)$ is thus reached at $v_s \approx v_s^{(2)}$ where $X(0) = 1$ (cf. Fig. 5). The minimum of $S_M(0)$ and of $S_M(k_m)$ occurs around $v_s \approx v_s^{(3)}$, and this is, according to Eq. (18), consistent with a minimized total scattering intensity I_{tot}.

Summarizing, we have theoretically analyzed the influence of contrast variation on the statics and on the measurable hydrodynamic function of polydisperse suspensions of charged core–shell particles. Since the definition of an optical index matching point is not unique in the presence of correlated size and optical polydispersity, three distinct definitions (IMP1-3) have been examined. We have shown that the simple decoupling approximation provides a semi-quantitatively good description of $S_M(k)$ and $H_M(k)$ in the critical regime of v_s values. This regime of optimized index matching can be conveniently localized using, e.g., the simple criterion IMP2 where only the knowledge of $\overline{f^2}(k = 0)$ is needed. In the critical v_s-regime, one can not determine the width of the size-polydispersity, mean hydrodynamic radius and the effective charge of optically inhomogeneous particles on the symplifying assumption of a homogeneous sphere model. This assumption would lead to wrong results. In order to avoid experimentally the critical v_s-regime with its strong sensitivity on optical polydispersity, calculations of $S_D(k)$ and $X(k)$ based on the decoupling approximation can prove to be very helpful. However, if one is particularly interested in the critical regime, then optical polydispersity has to be accounted for in full detail.

Acknowledgments A.J.B. acknowledges financial support by the DAAD. We further acknowledge useful discussions with Prof. A. Vrij and Dr. B. Weyerich, and financial support by a BMBF-DECHEMA project on concentrated colloidal dispersions.

References

1. Pusey PN (1991) In: Hansen J-P, Levesque D, Zinn-Justin J (eds) Liquids Freezing and Glass Transition: II, Les Houches Sessions 1989. North-Holland, Amsterdam, pp 763–942
2. Nägele G (1996) Phys Rep 272:215
3. Krause R, D'Aguanno B, Mendez-Alcaraz JM, Nägele G, Klein R, Weber R (1991) J Phys C 3:4459
4. Nägele G, Kellerbauer O, Krause R, Klein R (1993) Phys Rev E 47:2562
5. Nägele G, Zwick T, Krause R, Klein R (1993) J Colloid Interface Sci 161:347
6. van Veluwen A, Lekkerkerker HNW, de Kruif CG, Vrij A (1988) J Chem Phys 89:2810
7. D'Aguanno B, Klein R (1991) J Chem Soc Faraday Trans 87:379
8. Pusey PN, Fijnaut HM, Vrij A (1982) J Chem Phys 77:4270
9. van Beurten P, Vrij A (1981) J Chem Phys 74:2744
10. Phalakornkul JK, Gast AP, Pecora R, Nägele G, Ferrante A, Mandl-Steininger B, Klein R (1996) Phys Rev E 54:661
11. Philipse AP, Smits C, Vrij A (1989) J Colloid Interface Sci 129:335
12. Degiorgio V, Piazza R, Bellini T (1994) Adv Colloid Interface Sci 48:61
13. Degiorgio V, Piazza R, Jones RB (1995) Phys Rev E 52:2707
14. Watzlawek M, Nägele G (1997) Physica A 235:56
15. Banchio AJ, Nägele G, Ferrante A, submitted

Progr Colloid Polym Sci (1998) 110:61–65
© Steinkopff Verlag 1998

Dynamic light scattering in turbid suspensions: An application of different cross-correlation experiments

C. Urban
P. Schurtenberger

C. Urban · P. Schurtenberger (✉)
Institut für Polymere
ETH Zürich
CH-8092 Zürich
Switzerland
E-mail: pschurt@ifp.mat.ethz.ch

Abstract The ability to characterize colloidal suspensions by means of dynamic light scattering is limited to systems with negligible contributions from multiple scattering. For larger particle sizes with high scattering contrast this immediately limits the technique to very low concentrations. A very interesting solution of this problem is to suppress multiple scattering in dynamic light scattering experiments using various cross-correlation schemes. Based on these considerations we have constructed a so-called 3D cross-correlation experiment with which we are able to characterize extremely turbid suspensions. We have tested the feasibility of these experiments with well defined model systems such as suspensions of monodisperse and bimodal latex particles with relatively high volume fractions. The results are very promising and demonstrate unambiguously that such systems can be quantitatively characterized by means of dynamic light scattering methods without having to resort to high dilution.

Key words Dynamic light scattering – multiple scattering – cross – correlation – turbid suspensions – concentrated colloid suspensions

Introduction

Dynamic light scattering (DLS) is one of the most popular experimental techniques in the study of colloidal suspensions. It provides a measure of the time scale for fluctuations in the index of refraction of a complex fluid. These are typically caused by density fluctuations, which in the case of colloidal particles are caused by the diffusive motion of the particles. DLS probes these fluctuations on the length scale of the inverse of the magnitude of the scattering vector, q^{-1}. However, its application to many systems of industrial relevance has often been considered to be too complicated due to the very strong multiple scattering in undiluted solutions. One quite recent approach to this problem, which has already been implemented in commercial products, works in the limit of very strong multiple scattering, where a diffusion model can be used in order to describe the propagation of the light across the sample [1]. Unfortunately, this so-called "diffusing wave spectroscopy" does not generally allow to determine important features such as polydispersity of the particles due to the inherent averaging over a large range of spatial length scales. A very interesting opposite approach aims at sufficiently suppressing contributions from multiple scattering from the measured photon correlation data [2].

The general idea is to isolate singly scattered light and suppress undesired contributions from multiple scattering in a dynamic light scattering experiment (for details see [2]). This can be achieved by performing two scattering experiments simultaneously on the same scattering volume (with two mutually incoherent laser beams, initial wave vectors $\mathbf{k}_{i1}$ and $\mathbf{k}_{i2}$, and two detectors positioned at final wave vectors $\mathbf{k}_{f1}$ and $\mathbf{k}_{f2}$) and cross-correlating the signals seen by the two detectors. If both experiments then have the same scattering vector $\mathbf{q}$, but use different scattering geometries, and in the absence of multiple scattering, the corresponding relations between the dynamic structure factor $S(\mathbf{q}, \tau)$ and the measured auto-correlation ($G_{11}^{(1)}(\tau)$) and cross-correlation ($G_{12}^{(1)}(\tau)$) function, where $^{(1)}$ indicates singly scattered photons, can then be written as

$$G_{11}^{(1)}(\tau) = I_1^{(1)^2}\left(1 + \beta|S(\mathbf{q}, \tau)|^2\right),\qquad(1a)$$

$$G_{12}^{(1)}(\tau) = I_1^{(1)}I_2^{(1)}\left(1 + \beta_{12}|S(\mathbf{q}, \tau)|^2\right),\qquad(1b)$$

where I_1 and I_2 are the average scattered intensities seen by detectors 1 and 2, respectively, and the intercept of the cross-correlation function $G_{12}^{(1)}(\tau)$ is given by

$$\beta_{12} = \beta\exp\left(-\delta q^2 R^2/4\right)\exp\left(-\delta x^2/R^2\right).\qquad(2)$$

The intercept β detected in the auto-correlation experiment primarily depends on the detection optics, and the additional terms describe the reduction of the intercept due to phase mismatch ($\delta q = |\delta\mathbf{q}|$ describes the match of the scattering vectors $\mathbf{q}_1$ and $\mathbf{q}_2$) and misalignment ($\delta x = |\delta\mathbf{x}|$ is the spatial match of the scattering volumes). For single scattering, both auto- and cross-correlation therefore yield the same information. However, in the case of multiple scattering, the situation changes. In the auto-correlation experiment the multiply scattered photons contribute to $G_{11}(\tau)$ as well and make an interpretation of the data and a deduction of $S(\mathbf{q}, \tau)$ very difficult if not impossible. However, in the cross-correlation experiment outlined above only singly scattered light will produce correlated intensity fluctuations on both detectors. In contrast, multiply scattered light will result in uncorrelated fluctuations that contribute to the background only due to the fact that it has been scattered in a succession of different $\mathbf{q}$ vectors, and the contributions from multiple scattering to the signal are suppressed by a factor of order $(R\delta k_j)^{-1}$ [2]. This leads to the following expression for $G_{12}(\tau)$

$$G_{12}(\tau) \approx I_1 I_2 + \beta_{12} I_1^{(1)} I_2^{(1)} |S(\mathbf{q}, \tau)|^2,\qquad(3)$$

where I_j is the average intensity summed over all contributions (singly and multiply scattered photons) measured at detector j, and $I_j^{(1)}$ contains the singly scattered light only. Such a cross-correlation experiment should thus provide us with $S(\mathbf{q}, \tau)$ even for turbid suspensions, and multiple scattering will be visible in the decreasing intercept only.

A particularly interesting scheme is the so-called 3D cross-correlation experiment, in which the two incident and the two detected light paths are placed at an angle $\delta/2$ above and below the plane of symmetry of the scattering experiment (see Fig. 1A). The initial ($\mathbf{k}_{i1}, \mathbf{k}_{i2}$) and final ($\mathbf{k}_{f1}, \mathbf{k}_{f2}$) wave-vector pairs are rotated by some angle about the

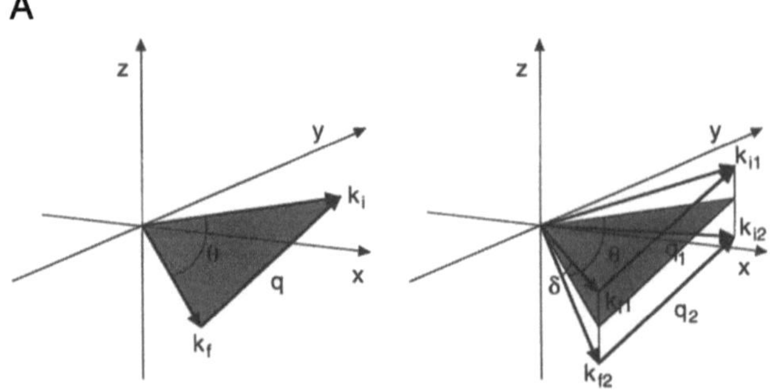

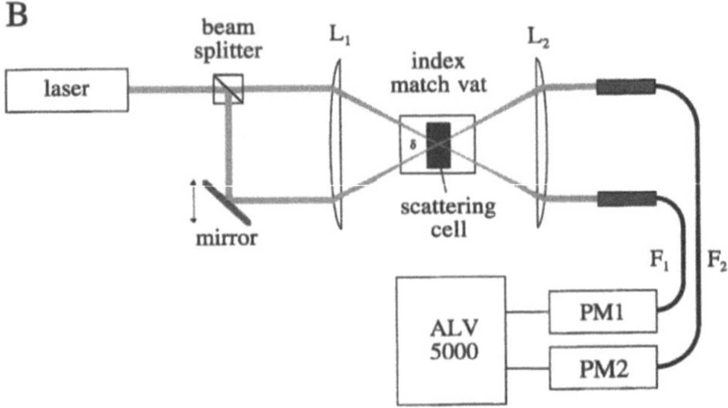

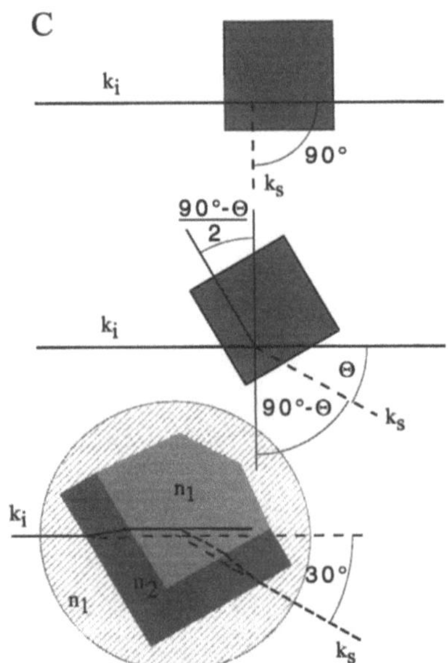

Fig. 1 Schematic description of the wave vector arrangement in auto-correlation and 3D cross-correlation experiments (A), an outline of the experimental set-up for 3D cross-correlation (B), and a description of the θ–2θ geometry used in order to work with rectangular scattering cells and minimized optical path length (C)

common scattering vector $\mathbf{q}_1 = \mathbf{q}_2$ for the two scattering processes 1-1 and 2-2, whereas the two other scattering processes 1-2 and 2-1 detected in this experiment have different scattering vectors. These additional scattering processes will therefore contribute to the background only, which means that the maximum intercept for the 3D experiment is only one quarter of the value obtained for auto-correlation or other cross-correlation schemes such as the two-colour method, i.e. $\beta_{12,\text{ideal}} = 0.25$ [2]. Several research groups [3–5] have very recently demonstrated the feasibility of such an experiment and clearly shown that dynamic light scattering need not be restricted to dilute suspensions of small particles, but that it can be used to successfully characterize extremely turbid suspensions. This opens up a completely new field of suspension characterization using dynamic light scattering experiments combined with novel cross-correlation schemes in numerous areas of soft condensed matter research and technology. Here we describe the construction of such a spectrometer and demonstrate it's capability to efficiently suppress contributions from multiple scattering.

Experimental set-up

A schematic layout of the experimental set-up is shown in Fig. 1B. We use either an argon-ion laser (Coherent Innova 300-8) operating at a wavelength of $\lambda_0 = 488$ nm, or a HeNe laser (Uniphase 1145P) at $\lambda_0 = 633$ nm as the light source. The laser beam is split into two parallel beams, which are then focused onto the scattering cell by the lens L_1 (planar convex, focal length 90 mm, diameter 63 mm, Spindler and Hoyer). This lens has been chosen in order to allow for a sufficiently large δ required for an efficient suppression of contributions from multiple scattering. In our current set-up we have chosen values of $R \approx 50$ μm for the beam waist at the scattering volume and $\delta = 18°$, which results in a suppression of double scattering by approximately a factor of 6×10^{-3} (the efficiency for suppressing higher order contributions is much higher, for details see [2]). The scattering cell is immersed in an index match vat (ALV 88/ALV, without flat entrance and exit windows, outer diameter 85 mm). Water is used as the index matching fluid. The vat temperature is controlled via a heat exchanger coil which is connected to a thermostat. An identical lens (L_2) is used for the detection side, and the scattered light with wave vectors $\mathbf{k}_{f1}$ and $\mathbf{k}_{f2}$ is collected and guided by a combination of a GRIN lens and a single mode fiber (F_1, F_2, OZ Optics) [6, 7]. The end of each fiber is connected via standard SMA fiber connectors to a home-built housing for a photomultiplier tube (PM; H5783P-01, Hamamatsu), and the PM signal is processed by two amplifier/discriminators (C3866 photon counting unit, Hamamatsu) and fed into a digital autocorrelator (ALV 5000, ALV).

For turbid colloidal suspensions, due to the strong attenuation of the singly scattered intensity which essen-

tially decays like $I^{(1)} \sim \exp(-t/L)$, where t is the optical path across the cell for singly scattered light and L is a characteristic length scale (photon transport mean free path length), which may easily be less than 1 mm for dense colloidal suspensions, the size of the scattering cell becomes important. In order to avoid a strong reduction of the singly scattered light and the subsequent reduction in signal (see Eq. (3)), the light path should be of the order of L. In principle this could be achieved best by using flat cells. However, since these cells will then be used in general with beams at non-normal incidence angles, this will lead to considerable beam deflection and displacement. We therefore implemented a different approach outlined in Fig. 1C. We use square cells with 10 mm path length that are of superior optical quality for experiments at low scattering angles, and position them such that the scattering volume is located in a corner of the cell to have short optical path lengths. For experiments at scattering angles different from 90° we turn the cell in a so-called θ–2θ geometry, where the sample is rotated by half the rotation angle $(90° - \theta)$ in order to recover a symmetrical situation in which the displacement and deflection of the incident and scattered beam almost cancel.

Results and discussion

We have performed measurements with a bimodal dispersion of polystyrene spheres in order to demonstrate the capability of the technique to suppress contributions from multiple scattering in turbid colloidal suspensions. We have prepared two suspensions of highly monodisperse latex particles with radii of $R_1 = 23$ nm and $R_2 = 274$ nm, respectively, which we both carefully and individually characterized by static and dynamic light scattering experiments. The corresponding q-dependence of the scattering intensity obtained with these dispersions is shown in Fig. 2. The two suspensions were then mixed and the relative concentrations were adjusted such (weight ratio $c_1/c_2 = 8.77$) that the two particle components equally contribute to the total scattering intensity $I(q)$ of the mixture at a scattering angle of 48°. Due to the very pronounced particle form factor of the larger spheres in the accessible q-range, their relative contribution to $I(q)$ varies considerably with scattering angle, which in principle should be measurable in a classical particle sizing experiment with dynamic light scattering. At $\theta = 61°$, the large spheres contribute approx. 65% and the small spheres approx. 35% to the overall intensity. The corresponding two distinct populations are clearly visible in a dynamic light scattering experiment when performing auto-correlation measurements on dilute suspensions (transmission $T = 0.95$) and using Contin [8, 9] in order to perform the

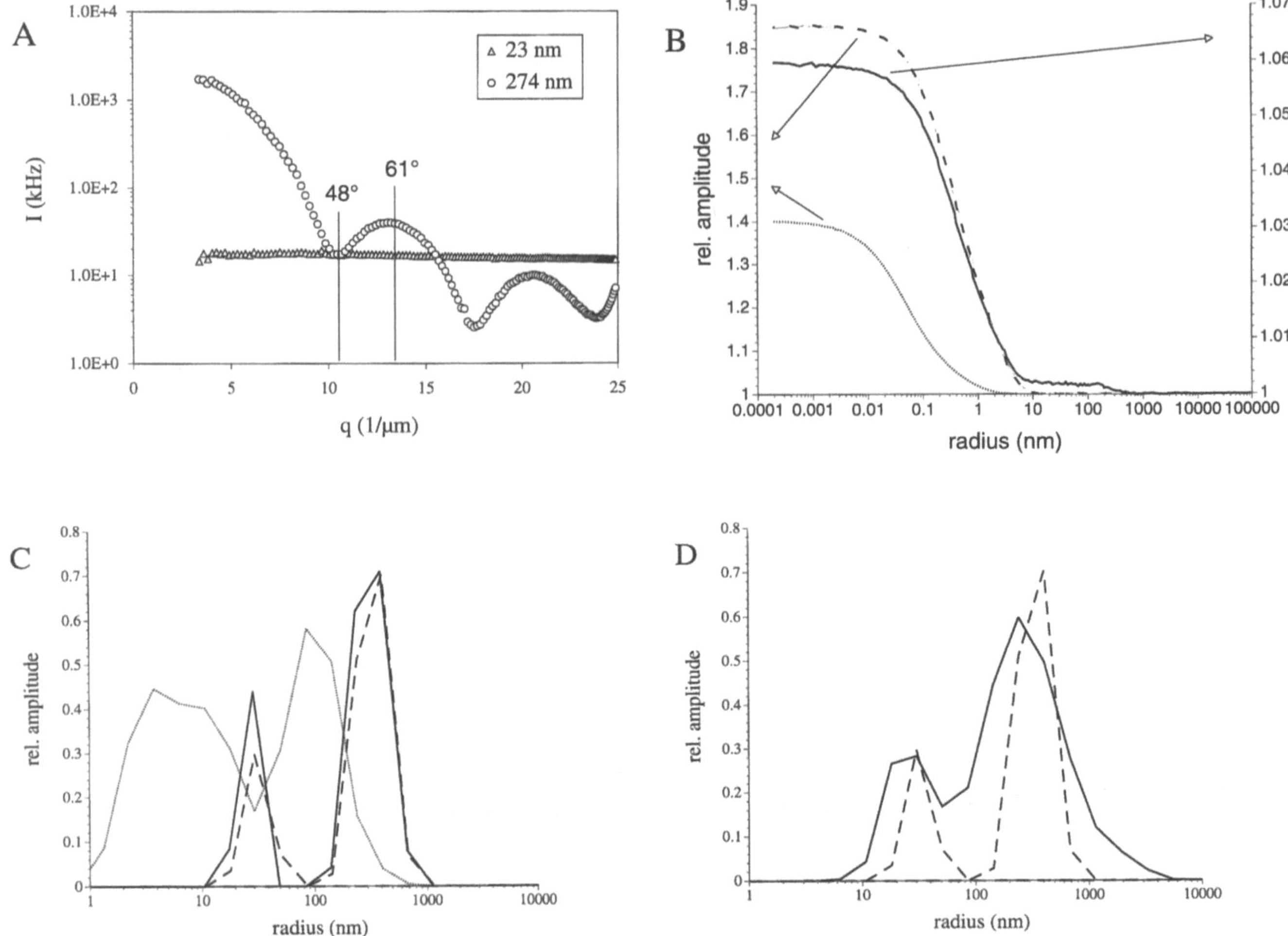

Fig. 2 Results from sample characterization and auto- and cross-correlation experiments with dilute and concentrated mixtures of two different polystyrene spheres with radii of $R_1 = 23$ nm and $R_2 = 274$ nm. (A) Results from static light scattering experiments with the individual monodisperse particle suspensions ($\circ$: $R = 23$ nm; $\triangle$: $R = 274$ nm). The stock solutions were diluted and the relative concentrations were adjusted to the same weight ratio as used in the experiments with mixtures. (B) A comparison of the measured auto-correlation function with a dilute (dashed line) and concentrated (dotted line) mixture and the cross-correlation function obtained with the 3D instrument from a concentrated mixture (solid line). (C) A comparison of the calculated intensity-weighted distribution of hydrodynamic radii (dashed line, based on the measurement of the individual dilute suspensions) with the results obtained from an application of Contin to the auto-correlation functions measured with dilute (solid line) and concentrated (dotted line) mixtures. (D) A comparison of the calculated intensity-weighted distribution of hydrodynamic radii (dashed line) with the results obtained from an application of Contin to the cross-correlation measured with concentrated mixtures (solid line) on the 3D set-up

inverse Laplace transformation required to obtain the size distribution (see Fig. 2B and C). We are in fact able to quantitatively recover the appropriate relative amplitudes in the intensity weighted size distribution from Contin as shown in Fig. 2C.

However, the situation completely changes for higher particle concentrations. We have also made auto-correlation measurements for an undiluted mixture at a total latex concentration of 0.21% by weight. Although the colloid concentration is still relatively low and the collective diffusion coefficient measured by dynamic light scat-

tering is only slightly influenced by contributions from interparticle interactions, the mixture is now completely turbid with a transmission of $T = 0.05$. This has a very pronounced effect on the results obtained from auto-correlation measurements. As shown in Fig. 2B, the intercept of the auto-correlation function has decreased considerably, we observe strong distortions at short lag times τ, and the characteristic decay times are much smaller now. While the decrease in intercept is mainly caused by the fact that the coherence length of the laser is shorter than the photon transport mean free path length in the sample, the shift of the characteristic correlation times measured in an auto-correlation experiment is a genuine effect of multiple scattering [3, 10] and for example severely limits the ability to recover the correct distribution of diffusion coefficients or particle hydrodynamic radii in particle sizing experiments with large particles such as emulsion droplets to low concentrations only. This is also evident from Fig. 2C, which shows the result from Contin when applied to the auto-correlation function measured with the turbid suspension.

In contrast to the results obtained from auto-correlation experiments, the use of a 3D cross-correlation set-up allows us to very efficiently suppress any effects from multiple scattering. This is shown in Fig. 2B and D, where the cross-correlation function and the corresponding distribution of hydrodynamic radii from Contin have been plotted. The intercept is much lower due to the combined effect of the intrinsically lower value for the ideally aligned instrument $\beta_{12,\text{ideal}} = 0.25$ and the actual phase mismatch and misalignment, which results in an intercept of $\beta_{12} \approx 0.14$ for our instrument, and the effect of multiple scattering (see Eq. (3)). However, the actual distribution of decay times remains unchanged as is already clearly visible from a comparison of the auto-correlation function from the dilute sample and the cross-correlation function obtained from the turbid sample (see Fig. 2B). Figure 2D demonstrates that the use of a 3D cross-correlation scheme allows us to determine the correct distribution of decay times (or hydrodynamic radii) even for highly turbid suspensions. The current value of $\beta_{12} \approx 0.14$ is high enough that we obtain correlation functions of sufficient accuracy in order to successfully apply inverse Laplace transformation programs such as Contin after measurements of typically 60–300 s. This opens up a wide field of new applications ranging from particle sizing in industrially relevant systems such as emulsions to the determination of the dynamics of concentrated model suspensions that can not be optically matched.

Acknowledgement We gratefully acknowledge financial support from KTI (Kommission für Technologie und Innovation) and Nestec SA.

References

1. Weitz DA, Zhu JX, Durian DJ, Pine DJ (1992) In: Chen S-H, Huang JS, Tartaglia P (ed.) Structure and Dynamics of Strongly Interacting Colloids and Supramolecular Aggregates in Solution. Kluwer Academic Publishers, Dordrecht, pp 731–748
2. Schätzel K (1991) J Mod Opt 38:1849–1865
3. Overbeck E, Sinn C, Palberg T (1997) Progr Colloid Polym Sci 104:117–120
4. Aberle LB, Wiegand S, Schröer W (1997) Progr Colloid Polym Sci 104:121–125
5. Schurtenberger P (1997) In: Windhab E-J, Wolf B (eds) Proc 1st Int Symp on Food Rheology and Structure. Vincentz Verlag, Hannover, pp 84–90
6. Ricka J (1993) Appl Opt 32:2860–2875
7. Gisler T, Rüger H, Egelhaaf SU, Tschumi J, Schurtenberger P, Ricka J (1995) Appl Opt 34:3546–3553
8. Provencher SW (1979) Makromol Chem 180:201–209
9. Provencher SW (1982) Comput Phys Commun 27:213–227
10. Stieber F, Richtering W (1995) Langmuir 11:4724–4727

Progr Colloid Polym Sci (1998) 110:66–69
© Steinkopff Verlag 1998

S.H. Behrens
M. Semmler
M. Borkovec

Aggregation in sulfate latex suspensions: The role of charge for stability predictions

S.H. Behrens (✉) · M. Borkovec
Department of Chemistry
Clarkson University
Potsdam, NY 13699-5814
USA

M. Semmler
Swiss Federal Institute of Technology
ETHZ-ITO
Grabenstrasse 3
CH-8952 Schlieren
Switzerland

Abstract It is commonly reported that the classical aggregation theory by Derjaguin, Landau, Verwey and Overbeek (DLVO) fails to predict quantitatively the dependence of colloid stability on the electrolyte concentration of the solution and on the particle size. A connection of such problems to the typical surface charge density of model particles was indicated by a recent study of pH-dependent aggregation. The present investigation compares the mobility and aggregation behavior of differently charged sulfate latex particles, measured by electrophoresis and dynamic light scattering. We find a much closer agreement of observed and predicted stability for particles of very low surface charge. A new picture of the limitations of DLVO-theory follows from the analysis of theoretical energy profiles when combined with results from direct force measurements, and is strongly supported by our present findings.

Key words Aggregation – DLVO-theory – light scattering – electrophoresis – sulfate latex

Colloidal aggregation has been a subject of increasingly elaborate scientific studies for a long time. In spite of the tremendous impact such studies have already had on applied sciences like materials engineering, environmental science or biochemistry, their potential of promoting novel applications is still far from being exhausted. On the other hand, with the availability of uniform, well characterized particles, it has become possible to elucidate the exact mechanisms of aggregation by comparing stability measurements to quantitative theoretical predictions. As it turns out, our current understanding of the particle interactions inducing or prohibiting colloidal aggregation is often unsatisfactory.

A prime example is the relatively simple case of charged particles dispersed in an aqueous solution of monovalent salt. This situation is commonly addressed in terms of the theory presented 50 years ago by Derjaguin, Landau, Verwey and Overbeek [1]. This so-called DLVO-theory describes interactions of charged surfaces across a liquid as an interplay of van der Waals forces and electrostatic double-layer forces, the latter being calculated on the basis of the Poisson–Boltzmann equation [2]. Its greatest success in the field of colloid stability was to explain the existence of the experimentally observed regimes of slow and fast aggregation and to deduce the empirically established Schulze–Hardy rule. However, when used for quantitative predictions of stability in the slow aggregation regime, the theory badly fails, claiming a much stronger dependence of aggregation (and deposition) rates on electrolyte concentration than was ever observed experimentally [2–4]. Moreover, stability is predicted to depend sensitively on the primary particle size, a trend which could not be confirmed by measurements either [5].

The reasons for these well-known discrepancies between theory and experiment have been the issue of a long and controversial scientific debate [6]. Dynamic processes other than pure diffusion have been suggested to be a major source of deviations [7], but additional non-DLVO forces, too, have been considered [8]. Some progress has been reported in incorporating into the existing theory the effects of physical roughness and charge heterogeneities on the particle surfaces [4, 9]. In any case, the DLVO framework would not be expected to give an adequate picture whenever ion correlations become important (especially in polyvalent salt solution), since electrostatic forces are treated on a mean field level only [10]. While the search for solutions to the deficiencies of DLVO-theory is going on, surprising news has lately come from direct force measurements using the surface force apparatus or the atomic force microscope: the forces probed between smooth, charged surfaces agree quantitatively with DLVO predictions down to surface separations of a few nanometers [11].

The paradox of seeing the theory confirmed by directly measured interparticle forces, but not by the observed resulting dynamics, was addressed in a very recent study on pH-dependent aggregation of different model colloids [12]. It was suggested that the traditionally reported deficiencies in predicting slow aggregation rates were related to the large surface charges of the particles typically used in aggregation experiments. Inducing aggregation in such systems requires fairly high electrolyte concentration – the critical coagulation concentration (ccc), crossover between slow and fast aggregation regime, usually exceeds 0.1 mol/l for monovalent salt. Then, the barrier in the DLVO-potential that two aggregating particles would have to overcome in the slow regime, attains its maximum at surface separations of less than one nanometer. This is problematic, because stability predictions strongly depend on the energy profile around this maximum position [13], while in view of the results from direct probing techniques, no realistic description can be expected from DLVO-theory for such minute distances. If, on the other hand, the surface charges (and the corresponding coagulant concentrations) are sufficiently low ($\sim$0.5–2 mC/m^2), then the energy barrier lies at a few nanometers. Here the DLVO picture is more reliable, and quantitative stability predictions should be correct. For particles with pH-dependent charge, the aggregation theory was seen to become accurate, indeed, when charge densities were reduced [12]. Yet, for constant charge systems, no such predictive success has ever been reported to our knowledge.

An example of such systems shall be considered in the present study. We have investigated two samples of monodisperse sulfate latex with different surface charge density (Table 1). Prior to experiments, they were dialyzed extensively against deionized water (nanopure quality). Figure 1 shows the electrophoretic mobility of these particles as a function of electrolyte concentration measured on a laser Doppler velocimetry setup. A difference in charge is clearly reflected by the difference in mobility. The experimental mobility data were translated into zeta potentials

Table 1 Characterization of the samples. The primary particle size is given in terms of the number averaged diameter and its coefficient of variation (CV) obtained from Transmission Electron Microscopy by the manufacturer. The surface charge density was inferred from electrophoretic mobility measurements. In the case of the 215 nm particles, the given value agrees with the one obtained by the manufacturer in conductometric titration. The separation of the DLVO-energy maximum from the surface is given for the electrolyte concentrations corresponding to a stability ratio of $W \approx 10$)

Diameter [nm]	CV [%]	Charge density [mC/m^2]	CCC [mol/L]	Barrier position [nm]
215	3.9	20	0.4	0.74
180	3.2	7	0.04	1.3

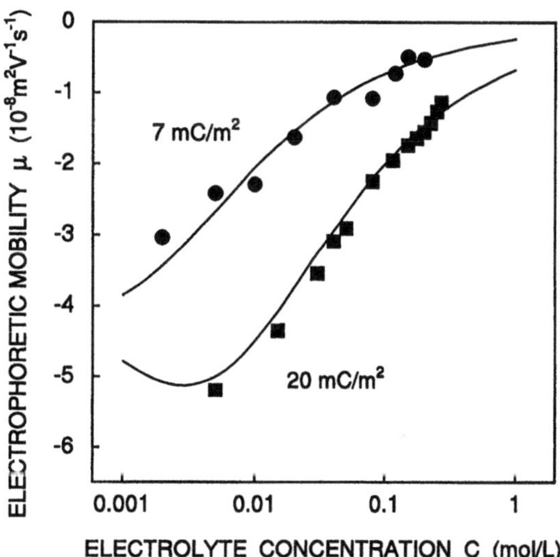

Fig. 1 Electrophoretic mobility of differently charged sulfate latex particles in potassium chloride solution. The lines represent theoretical calculations and were generated by transforming zeta potentials (identified to a first approximation with the diffuse layer potential) into mobilities using the O'Brien and White procedure. (The electrolyte concentration C is related to the number concentration n entering Eq. (1) via the Avogadro constant.)

through the method proposed by O'Brien and White [14]. The result is shown in Fig. 2 (points) together with a theoretical calculation (lines). For the calculated curves, we used the Gouy–Chapman relation [2]

$$\sigma = \sqrt{8\varepsilon\varepsilon_0 nk_B T} \cdot \sinh\left(\frac{e\zeta}{2k_B T}\right) \tag{1}$$

to relate the charge density σ (values indicated in the figures) to the surface potential, identified in a first approximation with the zeta potential ζ. In Eq. (1) $\varepsilon\varepsilon_0$ denotes the total electric permittivity, n is the number concentration of salt in the bulk, e the protonic charge and $k_B T$ the thermal energy. Corrections for surface curvature were seen to be

S.H. Behrens et al.
Aggregation of sulfate latex

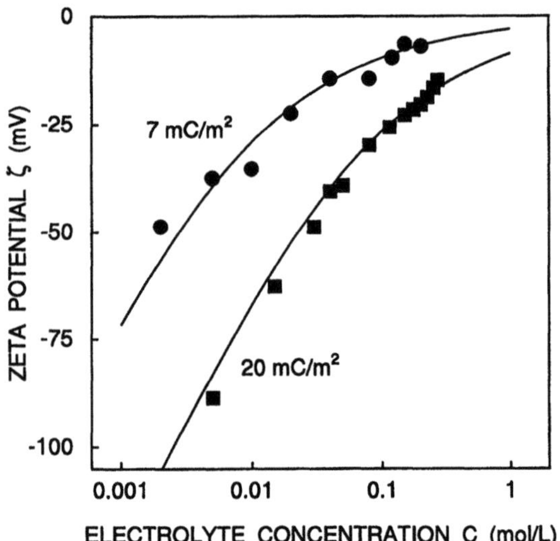

Fig. 2 The zeta potentials corresponding to Fig. 1. Here, the experimental mobility points were calculated back into zeta potentials according to O'Brien and White [14]

of minor importance for the cases considered. On account of the strong acidity of the sulfate groups, we assume the charge density to be constant. Transformation of the theoretical zeta potentials for a given charge into mobilities via the O'Brien and White procedure results in the theoretical curves of Fig. 1. The minimum in the prediction for the strongly charged particles is related to the extremum in the mobility dependence on zeta potential [14].

Dynamic light scattering was used to measure the colloid stability of our samples.

The results will be discussed in terms of the ratio

$$W = k_{\text{fast}}/k \tag{2}$$

of the dimer formation rate in the fast regime and the rate in the actual conditions under consideration. Theoretical aggregation rate constants were calculated from the modified Fuchs expression [2]

$$k = 4\pi \left\{ \int_{2a}^{\infty} \frac{\exp\left[V(r)/k_{\text{B}}T\right]}{r^2 D(r)} \, dr \right\}^{-1}, \tag{3}$$

$D(r)$ being the relative diffusion coefficient for two spheres of radius a at center-to-center-distance r. The effect of hydrodynamic interaction has been included in $D(r)$ according to Honig [15]. $V(r)$ stands for the net interaction energy of a pair of colloid particles

$$V(r) = V_{\text{vdW}}(r) + V_{\text{dl}}(r), \tag{4}$$

composed of a (non-retarded) van der Waals attraction and a repulsive double-layer term. The Hamaker constant entering the van der Waals potential was obtained from Lifshitz theory and has a value of $1.3 \times 10^{-20} J \ (\simeq 3.2 \ k_{\text{B}}T$

at 25 °C) for polystyrene across water [16]. The electrostatic part has been calculated from the non-linear Poisson–Boltzmann equation and can be expressed in terms of Jacobi elliptic functions [17]. All the contributions were taken in the Derjaguin approximation relating the interaction energy for two spheres to the corresponding energy density of two half-spaces. This approximation should be rather accurate for all the situations considered, since the Debye length is always much smaller than the radius of our particles [18]. We would like to stress that our theoretical predictions of the stability ratios involve no adjustable parameters.

A comparison of these predictions with the experimental stability data is shown in Fig. 3; the position of the

Fig. 3 Stability data from dynamic light scattering versus DLVO prediction for sulfate latex. Experimental stability ratios W were obtained from the measured initial increase in hydrodynamic radius r_h as $W = [(dr_{\text{h}}/dt)_{\text{fast}}/(dr_{\text{h}}/dt)]_{t\to 0}$; the theoretical values were calculated according to Eqs. (2)–(4) using the charge densities that fit to the electrophoresis data of Fig. 1. (a) Measured points for a typical (highly charged) latex in sodium perchlorate solution were taken from [19]. The striking failure of DLVO-theory in predicting the proper slope has been reported many times before. (b) The same plot for particles of lower charge (measured in potassium chloride solution). Note the closer agreement of theory and experiment

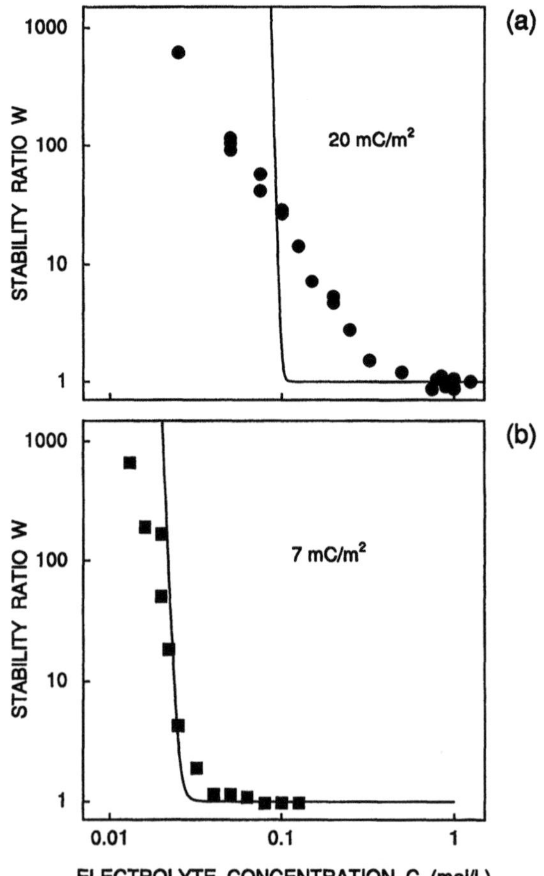

Progr Colloid Polym Sci (1998) 110:66–69
© Steinkopff Verlag 1998

DLVO potential barrier corresponding to theoretical stabilities of $W \approx 10$ is given in Table 1. The situation of the more strongly charged latex, depicted in Fig. 3a, is very typical: for a common model colloid, the stabilization at low electrolyte concentrations takes place much more gradually than predicted by DLVO-theory, the observed slope of the stability curve in the slow aggregation regime comes nowhere near to the expected one. The discrepancy with respect to the ccc is usually not seen this way, because Hamaker constants are commonly adjusted to yield the experimental ccc-value; variation of the Hamaker constant essentially produces a parallel shift of the theoretical curve, but the problem of the disagreeing slopes always remains [4]. As argued before, we cannot actually expect the theory to work well for these particles, because of the small separation of the DLVO energy barrier from the surface associated with the high charge (Table 1). Let us now consider the more weakly charged latex in Fig. 3b. Here the agreement between theory and experiment is clearly much better. In the sense of our argumentation, even this sample would be too strongly charged still; it does not meet the criterion (barrier maximum at separations of several nanometers) for colloids suitable to confirm DLVO aggregation theory, either. If our reasoning is correct, we may however conjecture a smooth transition from the charge regime where the theory fails into the regime where it is bound to work, and expect the quality of the prediction to be not quite as bad for our weakly charged particles as for the common ones. This is exactly the message of Fig. 3b.

In conclusion, the present study corroborates the idea, that DLVO-theory, proven to describe accurately the forces between charged surfaces at separations of some nanometers or more, also works for colloid aggregation, provided that the same range of distances encompasses all the length scales involved (Debye length, energy barrier position). This condition is only met for very low charge densities (of at most ~ 2 mC/m^2; the typical charge densities in constant charge systems such as sulfate latex are higher by at least one order of magnitude). Synthesis and handling of such weakly charged systems is necessarily difficult, because it only takes very little partial decharging – by traces of adsorbing, charged impurities say – to initiate aggregation. We suspect that this is the reason why no aggregation studies on colloids of this kind have come to our knowledge before.

We are indebted to R. Emmerzael, H. Holthoff, P. Schurtenberger and H. Sticher for various kinds of support and thank IDC, Portland for preparing the weakly charged sample. This work was funded by the Swiss National Science Foundation.

References

1. Derjaguin BV, Landau L (1941) Acta Physicochim USSR 14:633–662; Verwey EJW, Overbeek JThG (1948) Theory of the Stability of Lyophobic Colloids. Elsevier, New York
2. Russel WB, Saville DA, Schowalter WR (1989) Colloidal Dispersions. Cambridge Univ Press, New York
3. Lips A, Willis E (1973) Trans Farad Soc 69:1226–1236; Zeichner GR, Schowalter WR (1979) J Colloid Interface Sci 71:237–253; Elimelech M, Gregory J, Jia X, Williams R (1995) Particle Deposition & Aggregation. Butterworth-Heinemann, Oxford
4. Kihira H, Ryde N, Matijevic E (1992) Chem Soc Faraday Trans 88:2379–2386
5. Reerink H, Overbeek JThG (1954) Discuss Faraday Soc 18:74–84; Overbeek JThG (1982) Adv Colloid Interface Sci 16:17–30; Tsuruta LR, Lessa MM, Carmona-Ribeiro AM (1995) J Colloid Interface Sci 175:470–475
6. Swanton SW (1995) Adv Colloid Interface Sci 54:129–208
7. Shulepov SYu, Dukhin SS, Lyklema J (1995) J Colloid Interface Sci 171:340–350; Shulepov SYu (1997) J Colloid Interface Sci 189:199–207
8. Einarson MB, Berg JC (1992) J Colloid Interface Sci 155:165–172; Israelachvili J, Wennerström H (1996) Nature 379:219–225
9. Shulepov SYu, Frens G (1996) J Colloid Interface Sci 182:388–394
10. Kjellander R, Akesson T, Jönsson B, Marčelja S (1992) J Chem Phys 97:1424–1431
11. Israelachvili JN, Pashley RM (1983) Nature 306:249–250; Horn RG, Clarke DR, Clarkson MT (1988) J Mater Res 3:413–416
12. Behrens SH, Borkovec M, Schurtenberger P (1998) Langmuir 14:1951–1954
13. Gisler T, Borkovec M (1993) Langmuir 9:2247–2249
14. O'Brien RW, White LR (1978) J Chem Soc Faraday Trans II 77:1607–1626
15. Honig EP, Roebersen GJ, Wiersema PH (1971) J Colloid Interface Sci 36:97–109
16. Bowen WR, Jenner F (1995) Adv Colloid Interface Sci 56:201–243
17. Ninham BW, Parseggian VA (1971) J Theor Biol 31:405–428
18. Carnie SL, Chan DYC, Stankovich J (1994) J Colloid Interface Sci 165:116–128
19. Holthoff H, Egelhaaf SU, Borkovec M, Schurtenberger P, Sticher H (1996) Langmuir 12:5541–5549

Progr Colloid Polym Sci (1998) 110:70–75
© Steinkopff Verlag 1998

J.L. Burns
Y.D. Yan
G.J. Jameson
S. Biggs

A comparison of the fractal properties of salt-aggregated and polymer-flocculated colloidal particles

J.L. Burns* · Dr. S. Biggs (✉)
Department of Chemistry
The University of Newcastle
University Drive
Callaghan NSW 2308
Australia

Y.D. Yan · G.J. Jameson
* Department of Chemical Engineering
The University of Newcastle
University Drive
Callaghan NSW 2308
Australia

Abstract The aggregation of colloidal polystyrene latex particles has been investigated using small-angle static light scattering. Aggregate structures were found to be mass fractal in nature for both the salt-aggregated and depletion-flocculated systems. The measured fractal dimensions were found to be highly dependent on the salt concentration, the concentration of non-adsorbing polymer and the initial particle concentration in solution. The fractal dimensions observed for the depletion-flocculated aggregates, particularly at low polymer concentrations, were significantly larger than those observed for the salt-induced aggregates. Interestingly, for both aggregation systems, an increase in the initial particle concentration resulted in lower fractal dimensions.

Key words Fractal dimension – salt aggregation – depletion flocculation – colloids – light scattering

Introduction

Mandelbrot [1] was the first to use the term *fractal* to express that many common structures could be characterized by a non-integer (fractal) dimensionality. Fractal objects are self-similar objects, meaning that the structure of the object is invariant to a change of scale. Colloidal aggregates form an important class of fractal objects with the fractal structure and scaling properties of particle aggregates receiving considerable attention in recent years. It is possible to assign a mass fractal dimension, d_F, to an aggregate to describe its space-filling capacity or density of its structure. The mass density, $\rho(R)$, of a fractal aggregate of radius R, can be related to its mass fractal dimension by the following equation [2]:

$$\rho(R) \propto R^{d_F - 3},\tag{1}$$

where $1 < d_F < 3$ for a three-dimensional object.

Static light scattering provides a useful technique to measure the fractal dimension of aggregates in an aggregating colloidal system. In a static light scattering experiment, a beam of light is directed onto the aggregating system and the scattered intensity, $I(Q)$, is measured as a function of the magnitude of the scattering vector, Q, with

$$Q = \frac{4\pi n_0}{\lambda_0} \sin\left(\frac{\theta}{2}\right),\tag{2}$$

where n_0 is the refractive index of the dispersion medium, λ_0 is the wavelength of the incident light *in vacuo* and θ is the scattering angle. For fractal aggregates made up of monodisperse primary particles which satisfy the Rayleigh–Gans–Debye criteria, the mass fractal dimension can be determined using the power-law scattering equation [2]

$$I(Q) \propto Q^{-d_F},\tag{3}$$

provided that $1/R \ll Q \ll 1/r_0$, where r_0 is the primary particle radius.

The structure of aggregates formed from colloidal particles is highly dependent on the conditions used in the aggregating system. The aggregation of colloidal particles by the addition of inert electrolytes has been extensively studied [3–8]. Two limiting regimes of aggregation have been established to characterize these salt-induced aggregations, namely, diffusion-limited cluster–cluster aggregation (DLCA) and reaction-limited cluster–cluster aggregation (RLCA). For charge-stabilized particles, the height of the repulsive energy barrier between two approaching particles determines the aggregation regime and hence the aggregate structure. In the absence of any potential energy barrier between the colliding particles, loose, open structures are formed rapidly giving rise to the DLCA regime. Typically, these cluster–cluster aggregates have fractal dimensions in the range 1.75–1.80. In contrast, the existence of significant electric double-layer repulsion between the approaching particles will mean that the particles and clusters will have to collide many times before becoming permanently attached. The result is an increase in the fractal dimension over the diffusion-limited case with $d_F = 2.1$–2.2 and this is known as the RLCA regime.

The addition of non-adsorbing polymer to a colloidal dispersion can also promote phase separation and this is commonly referred to as depletion flocculation. The depletion attraction is the result of the exclusion of the polymer molecules from between the particles upon close particle approach, which is due to configurational entropy restrictions. This creates a pure solvent reservoir between the particles and an osmotic pressure gradient, which causes an attractive interaction potential. Many studies have been presented on depletion-flocculated systems [9–15], with a range of theories proposed to explain the mechanism by which the particles flocculate in the presence of non-adsorbing polymer [16–20]. However, there are only very limited studies reported on the structural properties of aggregates formed by a depletion–flocculation process, which is in direct contrast to the extensively studied salt-induced aggregations.

This paper aims to provide a comparison of the aggregate structures formed from a monodisperse colloidal dispersion aggregated in the respective presence of a simple electrolyte and a non-adsorbing polymer. Monodisperse polystyrene latex particles were aggregated with KNO_3 and non-adsorbing poly(acrylic acid), and the aggregations were followed using small-angle static light scattering. The light scattering curves confirmed the mass fractal nature of the aggregates and hence mass fractal dimensions could be determined for a wide range of aggregating conditions. To support the findings of the fractal data, sedimentation measurements were also performed.

Experimental

Latex particles

Polystyrene latices were synthesized by the surfactant-free emulsion polymerization method of Goodwin et al. [21]. The polystyrene latex particles were prepared at a temperature of 80 °C and at a stirrer speed of 200 rpm. Styrene monomer (100 g, Aldrich) and ammonium persulfate initiator (0.565 g, Aldrich) were reacted in 910 cm^3 of water for 24 h. After this, the latex was dialyzed against Millipore water. The size distribution of the latex particles was determined using transmission electron microscopy and the sample was found to have an average particle diameter of 330 ± 10 nm.

Fractal dimension determination

Small-angle static light scattering measurements, used to determine the fractal dimensions of the latex particle aggregates, were performed using a Malvern Mastersizer S, fitted with a 633 nm He–Ne laser. The salt aggregations were investigated over the concentration range of 0.4–1.5 M KNO_3 at pH 6, to allow both the RLCA and DLCA regimes to be observed. Depletion flocculation was studied at poly(acrylic acid) $(M_w = 2.5 \times 10^5$, Aldrich) levels of 1 to 20 g/l, without the addition of salt. These solutions were maintained at pH 10 to ensure that the polyacid did not adsorb onto the surface of the latex particles. Samples were gravity-fed into the light scattering cell and the scattered intensity was measured simultaneously at a range of scattering angles. The mass fractal dimension of the latex particle aggregates was then obtained by taking the absolute value of the slope of the plot of $\log I(Q)$ vs $\log Q$, when the aggregate size had grown sufficiently large.

Sedimentation studies

The measurement of sedimentation volume provides a useful means for comparing aggregate densities produced under different aggregating conditions. For the salt-aggregated systems, samples were prepared at concentrations ranging from 0.1 to 1.5 M KNO_3. For the depletion–flocculated systems, the range of concentrations of the non-adsorbing polyacid was 1–20 g/l. Samples were prepared by mixing a solution of the flocculant of a given concentration with a known volume of the latex dispersion and then shaking for 5–10 min. Each sample was then allowed to aggregate at room temperature before photographs of representative samples were taken.

Results and discussion

The aggregation of the colloidal latex particles was investigated over a wide range of electrolyte concentrations to allow both regimes of aggregation (RLCA and DLCA) to be observed. This also allowed observation of the fractal behavior of the aggregates at intermediate salt concentrations. The experiments were conducted at three different particle concentrations for each salt concentration examined. The mass fractal dimensions were found to range from 1.78 to 2.20, which correspond well to known literature values of 1.8 and 2.1 for diffusion-limited cluster–cluster aggregation and reaction-limited cluster–cluster aggregation, respectively. The observed trends of the fractal dimensions with ionic strength and particle concentration are plotted in Fig. 1.

Clearly, the structure of aggregates formed from the latex particles differs according to the prevailing conditions within the aggregating system. At any given particle concentration, more open structures (i.e. lower fractal dimensions) were formed at higher salt concentrations, whilst more tightly packed compact structures (i.e. higher fractal dimensions) were formed at low salt concentrations. At intermediate salt concentrations, the "effective" fractal dimensions were found to be intermediate to the theoretical values predicted by the DLCA and RLCA models. Furthermore, the aggregate structures were found to become more compact as the latex particle concentration was decreased, that is, the fractal dimension was observed to increase with a decrease in the number of initial particles in the aggregating system.

According to the DLVO theory of colloid stability, the aggregation behavior of aqueous colloids is determined by the potential energy barrier to coagulation between two colloidal particles, which in turn is controlled by the salt concentration in solution. The primary effect of salt on a charge-stabilized aqueous dispersion is to suppress the extent of the electric double layer. This leads to a reduction in the repulsive potential energy and hence the system becomes more unstable. At low salt levels, a potential energy barrier is present resulting in a reduced particle sticking efficiency. Only if the colliding particles or aggregates have sufficient thermal energy can they surmount the energy barrier, and hence a number of collisions may be required before entering the deep primary minimum. At high salt levels, there is no energy barrier between the particles and hence all collisions result in irreversible aggregation. Therefore, the absence of any energy barrier to coagulation indicates the DLCA regime (low d_F), whilst the presence of a significant barrier leads to the RLCA regime (high d_F).

In our study, three particle concentrations were examined, with each being investigated over a wide range of salt concentrations. Our work reveals that there is a noticeably larger variation in the fractal dimension with particle concentration at lower salt concentrations, whereas at high salt levels the fractal dimension does not vary significantly with changes in the particle concentration. The latter can be attributed to the fact that all particle collisions would result in permanent contact, hence reconfiguring of the aggregates was effectively prevented. In contrast, under slow growth conditions (i.e. low salt levels) particles could undergo a number of collisions before they became permanently locked into position. In this case, a lower particle concentration and hence a slower aggregation rate will result in an increase in the time for reconfiguration of the partially formed aggregates to occur. As a result, denser structures are to be expected at lower particle concentrations.

The structure of aggregates formed from the depletion flocculation of the polystyrene latex particles in the presence of the non-adsorbing poly(acrylic acid) (PAA) was also found to be mass fractal. Hence, the fractal dimension of the aggregates was measured over a wide range of PAA concentrations (1–20 g/l) and at the same three particle concentrations investigated above. The resulting fractal dimensions were found to range from 2.02 to 2.99 and the results are shown in Fig. 2. Clearly, the values at low PAA concentrations are significantly larger than those of 1.78–2.20, which were found for the salt-induced aggregates.

As can be seen from Fig. 2, the fractal dimension of the latex particle aggregates was largely influenced by the amount of non-adsorbing PAA in solution. Specifically, the fractal dimension increased as the concentration of the free bulk polymer was decreased. Interestingly, the fractal dimension approached a value of almost 3 when the polyacid concentration was just sufficient to induce

Fig. 1 The dependence of the mass fractal dimension of aggregates of polystyrene latex particles with changes in the electrolyte concentration measured at the following particle concentrations: (▲) 0.0020% w/w; (◇) 0.0035% w/w and (■) 0.0070% w/w

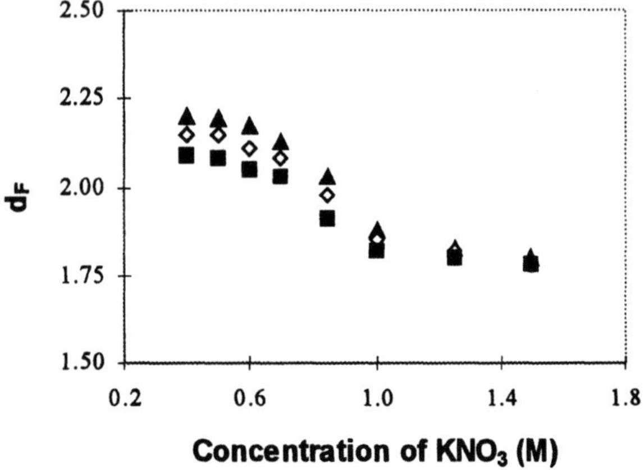

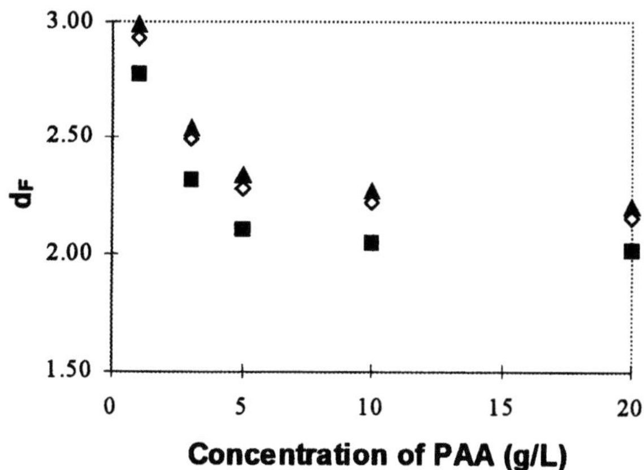

Fig. 2 The dependence of the mass fractal dimension of aggregates of polystyrene latex particles with changes in the non-adsorbing poly (acrylic acid) concentration measured at the following particle concentrations: (▲) 0.0020% w/w; (◇) 0.0035%w/w and (■) 0.0070% w/w

Fig. 3 Photographs of the sediment volumes of polystyrene latex particles aggregated in the presence of (a) non-adsorbing PAA and (b) KNO_3. The respective concentrations of PAA and KNO_3 increase from left to right

flocculation of the latex particles. The observed increase in the fractal dimension as the concentration of PAA was decreased may be explained by a reduction in the particle sticking efficiency, this being the result of weaker depletion forces between the latex particles at lower polymer concentrations. This increased probability of particle escape will allow for the aggregate to reconfigure to achieve a higher density of particle packing. At high polymer levels the fractal dimension of the flocculated particles exhibited relatively small variations with changes in the PAA concentration. This leveling off of the fractal dimension curves at high PAA concentrations may be the result of only small increases in the depletion energy of interaction as the polymer concentration becomes relatively high.

The fractal dimension of the depletion-induced flocs was found to be also largely dependent on the initial number of particles in the solution. At a given level of

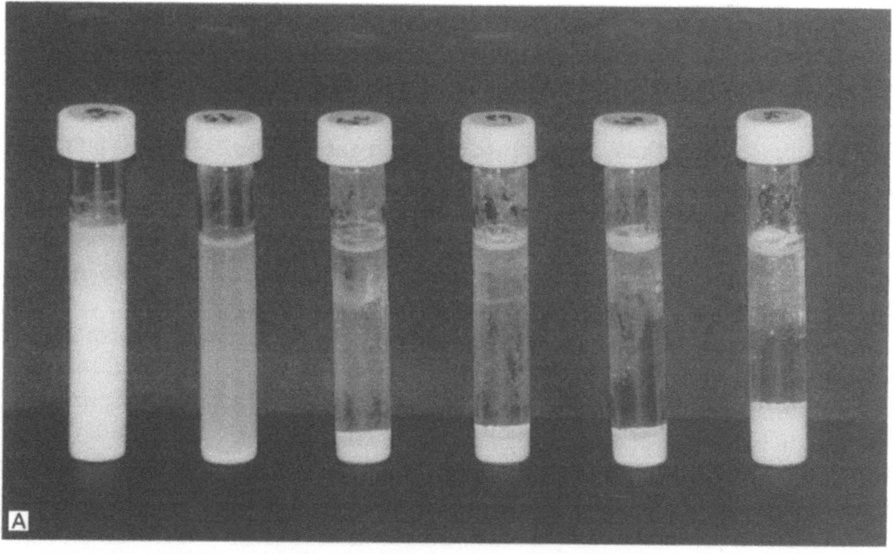

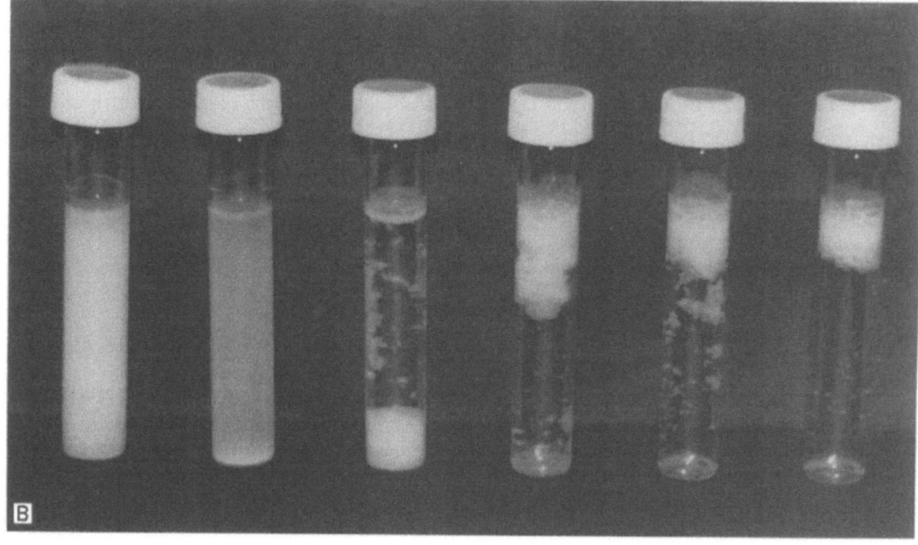

PAA, denser aggregate structures (i.e. higher fractal dimensions) were observed at lower initial particle concentrations. Similar to the salt aggregation case, lower latex particle concentrations will lead to a reduction in the particle collision frequency and as a result there will be more time for particles attached to the edge of the aggregate structure to "roll around" as the structure develops. This will necessarily lead to denser aggregate structures at a given free polymer concentration. Unlike the salt aggregations, the particle concentration effect was significant at all flocculant levels. This is most likely due to depletion flocculation being a secondary energy minimum effect and hence there are only relatively weak interparticle attractions within the aggregates.

The aggregate structural information obtained from the fractal studies was compared with the observed aggregate volumes. For the depletion-flocculated system (Fig. 3a) the volume occupied by the flocculated material was found to increase with an increase in the bulk polymer concentration. As expected from earlier discussions, when the level of polymer is increased, the fractal dimension and hence the compactness of the aggregate structures will decrease and thus a more open aggregate phase is to be expected. Overall, the aggregate volumes were observed to be extremely compact, particularly at low PAA concentrations, in line with the very high fractal dimensions obtained from the light scattering measurements. Therefore, the observed sediment volumes of the aggregates do show good correlation with the fractal dimension data.

Of particular interest were the aggregate volumes obtained for the salt-induced aggregations (Fig. 3b). At high salt levels, where the aggregate fractal dimensions were observed to fall below 2, the aggregates were seen to float rather than settle in the sample vials. This is indicative of very open aggregate structures with extremely low mass densities as suggested by the measured fractal dimensions. As the concentration of salt was lowered, the aggregates began to settle in the sample vials, but nevertheless, the aggregates still had a very open appearance. This also

implies a relatively low aggregate density, which again supports the fractal dimension data.

Conclusions

Small-angle static light scattering was used to investigate the aggregate structures formed for salt-coagulated and depletion-flocculated systems. For the salt aggregations, the measured mass fractal dimensions ranged from 1.78 to 2.20, which are in good agreement with the predicted values of the diffusion-limited cluster-cluster aggregation and reaction-limited cluster-cluster aggregation regimes, respectively. In general, it was found that increasing the salt concentration had the effect of reducing the fractal dimension, indicating the formation of more open aggregate structures. Fractal dimensions measured on the depletion-flocculated systems were found to be highly dependent on the concentration of the non-adsorbing polymer, resulting in values from 2.02 to 2.99. An increase in the polyacid concentration resulted in a reduction in the fractal dimension, which is indicative of an increase in the depletion energy of the system. Clearly, the fractal dimensions observed for the depletion-flocculated aggregates at low polymer concentrations are significantly higher than those observed for the salt-induced aggregates. In general, for both aggregation systems an increase in the initial particle concentration resulted in more open aggregate structures (i.e. lower fractal dimensions). This was attributed to a reduction in the time available for reconfiguring of particles in the aggregates due to an increased particle collision frequency. The aggregate structural information inferred from the light scattering experiments for both the salt and depletion aggregations was in good correlation with the sedimentation results.

Acknowledgements J.B. wishes to thank the Australian Research Council for the Australian Postgraduate Award (Industry) and Jetflote Pty. Ltd. for the financial support of this work. The authors also acknowledge the Centre for Multiphase Processes, a Special Research Centre of the Australian Research Council.

References

1. Mandelbrot BB (1982) The Fractal Geometry of Nature. Freeman, San Francisco
2. Teixeira J (1988) J Appl Crystallogr 21:781–785
3. Cannell DS, Aubert C (1986) Phys Rev Lett 56:738–741
4. Wilcoxon JP, Martin JE, Schaefer DW (1989) Phys Rev A 39:2675–2688
5. Lin MY, Lindsay HM, Weitz DA, Ball RC, Klein R, Meakin P (1989) Nature 339:360–362
6. Carpineti M, Ferri F, Giglio M, Paganini E, Perini U (1990) Phys Rev A 42:7347–7354
7. Zhou Z, Chu B (1991) J Colloid Interface Sci 143:356–365
8. Amal R, Raper JA, Waite TD (1992) J Colloid Interface Sci 151:244–257
9. Cowell C, Li-in-on R, Vincent B (1978) J Chem Soc, Faraday Trans 1 74:337–347
10. Clarke J, Vincent B (1981) J Colloid Interface Sci 82:208–216
11. Cowell C, Vincent B (1982) J Colloid Interface Sci 87:518–526
12. Vincent B, Clarke J, Barnett KG (1986) Colloids Surf A 17:51–65
13. Vincent B, Edwards J, Emmett S, Jones A (1986) Colloids Surf A 18:261–281
14. Jones A, Vincent B (1989) Colloids Surf A 42:113–138

15. Milling A, Vincent B, Emmett S, Jones A (1991) Colloid Surf A 57:185–195
16. Asakura S, Oosawa F (1954) J Chem Phys 22:1255–1256
17. Feigin RI, Napper DH (1980) J Colloid Interface Sci 75:525–541
18. de Gennes PG (1982) Macromolecules 15:492–500
19. Scheutjens JMHM, Fleer GJ (1982) Adv Colloid Interface Sci 16:341–359
20. Vincent B (1990) Colloids Surf A 50:241–249
21. Goodwin JW, Hearn J, Ho CC, Ottewill RH (1974) Colloid Polym Sci 252:464–471

Progr Colloid Polym Sci (1998) 110:76–79
© Steinkopff Verlag 1998

M. Ballauff

Analysis of latex particles by small-angle X-ray scattering: The isoscattering point revisited

M. Ballauff
Polymer-Institut
Universität Karlsruhe
Kaiserstrasse 12
D-76128 Karlsruhe
Germany

Abstract A discussion of the analysis of composite latex particles by small-angle X-ray scattering (SAXS) including contrast variation is given. For monodisperse core-shell particles marked isoscattering points will become visible in which all scattering curves measured as function of contrast cross each other. The relation of this isoscattering point to fine details of the radial structure and to polydispersity is discussed. It will become apparent that the isoscattering point may be used for a detailed analysis of the internal structure of latex particles.

Key words Small-angle X-ray scattering – contrast variation – isoscattering point – latex

Introduction

Small-angle X-ray scattering is a powerful tool for the analysis of the structure of latex particles [1–4]. This is due to the fact that practically all polymers used in emulsion polymerization have sufficient contrast to the dispersion medium water which leads to sufficient SAXS-intensities. Often the contrast between the polymers used for synthesizing composite particles is quite high and SAXS can therefore be used to analyze the structure of the particles. On the other hand, the overall contrast can still be matched by adding sucrose or glycerol to the latex thus allowing to measure the SAXS-intensities at different contrast [2–4]. In this communication we wish to discuss the analysis by contrast variation in more detail using model calculations. Particular attention will be paid to the problem of non-uniformity. The results will be compared to a recent study of poly(styrene) particles covered by a non-ionic surfactant [5].

Theory

We consider a system of N monodisperse spherical particles dispersed in the scattering volume. The scattering intensity $I(q)$ as function of q, the magnitude of the scattering vector is given by [6]

$$I(q) = NI_0(q)S(q) , \qquad (1)$$

where $I_0(q)$ is the scattering intensity of a single particle at infinite dilution and $S(q)$ denotes the structure factor. The alterations effected by $S(q)$ can be removed by proper extrapolation to vanishing concentration. In addition to this, for nearly isometric objects its influence is restricted to small scattering angles. At higher scattering angles it can be neglected (cf. the discussion of this point in Ref. [1]).

The local scattering length density $\rho(\mathbf{r})$ can be rendered as the product of a shape function $T(\mathbf{r})$ and the local scattering length density $\rho_p(\mathbf{r})$ inside the particle [7, 8, 2]:

$$\rho(\mathbf{r}) = T(\mathbf{r})\rho_p(\mathbf{r}) + \rho_m(1 - T(\mathbf{r})) . \qquad (2)$$

Here the shape function $T(\mathbf{r})$ is 1 if the point $\mathbf{r}$ is inside of the latex particle and 0 otherwise. The quantity ρ_m denotes the electron density of the dispersion medium. With this definition, the average electron density $\bar{\rho}$ follows as

$$\bar{\rho} = \frac{1}{V_p} \int T(\mathbf{r})\rho_p(\mathbf{r})\,d\mathbf{r} . \qquad (3)$$

The difference $\bar{\rho} - \rho_m$ is the contrast and the match point is defined by $\bar{\rho} = \rho_m$. The local excess electron density may be split into a part depending the contrast $\bar{\rho} - \rho_m$ and into a function $\Delta\rho(\mathbf{r})$ independent of contrast [7, 2]

$$\rho(\mathbf{r}) - \rho_m = T(\mathbf{r})[\bar{\rho} - \rho_m] + T(\mathbf{r})\Delta\rho(\mathbf{r}) . \qquad (4)$$

From this definition of the function $\Delta\rho(\mathbf{r})$ it is evident that

$$\int T(\mathbf{r})\Delta\rho(\mathbf{r})\,\mathrm{d}\mathbf{r} = 0 . \qquad (5)$$

For particles with spherical symmetry as e.g. core-shell latex particles the scattering intensity may be calculated from [7, 2]

$$I_0(q) = B^2(q) , \qquad (6)$$

with

$$B(q) = B_0(q) + \varepsilon(q) , \qquad (7)$$

where

$$B_0(q) = (\bar{\rho} - \rho_m)\,4\pi \int_0^R T(r)\,\frac{\sin(qr)}{qr}\,r^2\,\mathrm{d}r \qquad (8)$$

and

$$\varepsilon(q) = 4\pi \int_0^R T(r)\Delta\rho(r)\,\frac{\sin(qr)}{qr}\,r^2\,\mathrm{d}r \qquad (9)$$

The quantity $B_0(q)$ presents the scattering amplitude of a homogeneous sphere whereas $\varepsilon(q)$ solely refers to the variation of the electron density inside the sphere. $B_0(q)$ will vanish for $\tan(q^*R) = q^*R$ and $I_0(q^*) = \varepsilon^2(q^*)$. Hence, in case of well-defined particles with spherical symmetry the isoscattering points present a prominent feature of the scattering curves as function of contrast.

Until recently [1, 2, 11, 12], this isoscattering point first discussed by Kawaguchi [9, 10] and by Vrij and coworkers [7] has not attracted much attention. The main reason for this is the fact that this feature is often smeared out by polydispersity of size or contrast (see below). Therefore, it is interesting to elaborate under which condition isoscattering points may become visible in SAXS experiments and how to use this feature for structural analysis.

In the following we shall discuss the isoscattering point, a ensemble of monodisperse and polydisperse particles without interactions, i.e., $S(q) = 1$. Since latexes exhibit always a finite breadth of the size distribution, the effect of polydispersity will be considered in detail [13].

Model calculation and comparison with experimental data

The isoscattering point is discussed first for a monodisperse core-shell sphere the radial electron density of which

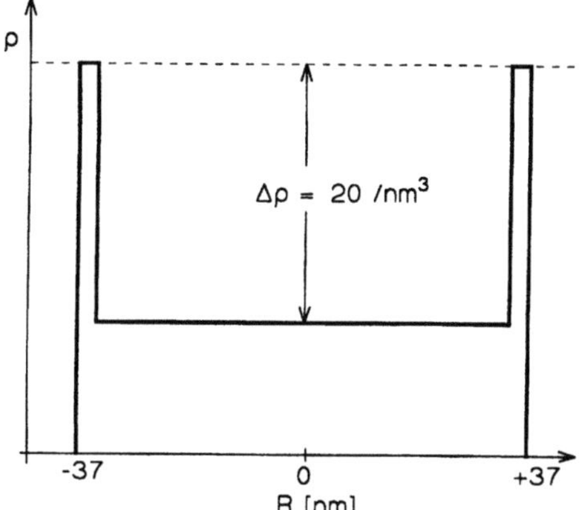

Fig. 1 Radial electron density used in the model calculation

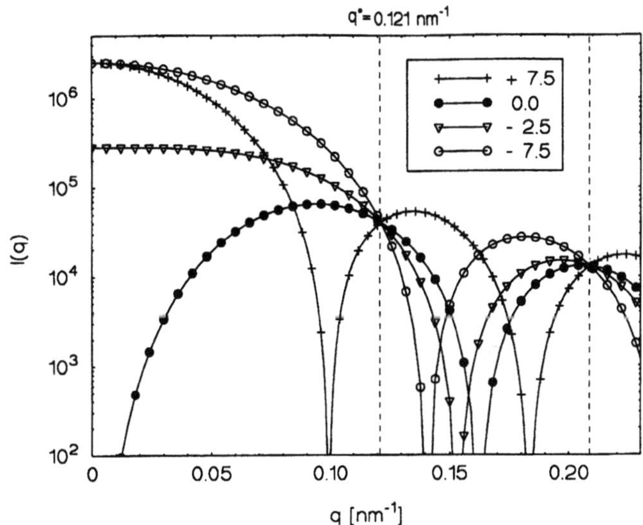

Fig. 2 Isoscattering point for a monodisperse core-shell sphere. The electron density used for this model calculation is displayed in Fig. 1. The inset gives the contrast $\bar{\rho} - \rho_m$ (electrons/nm³)

is displayed in Fig. 1. There is a shell of three nanometers thickness in which the electron density is higher by 20 electrons/nm³.

Figure 2 displays the SAXS-intensities $I_0(q)$ calculated for the radial electron density shown in Fig. 1 [13]. The isoscattering points are clearly visible. Furthermore, the calculation show that forward scattering for a monodisperse particle will vanish at zero contrast in accord with the above deductions. As a consequence of this, the radius of gyration will increase or decrease rapidly as function of contrast in the vicinity of the match point. It is thus evident that this quantity is difficult to measure by SAXS for latex systems in the vicinity of the match point.

Figure 2 also shows that the isoscattering points are features of $I_0(q)$ which are located in a q-range easily accessible by conventional SAXS-equipment ($q > 0.08$ nm). A careful study of composite particles by contrast variation may therefore serve to elucidate their internal structure without the necessity to explore the region of smallest q-values.

Effect of size polydispersity

For vanishing concentration, Eq. (1) can be generalized to polydisperse systems by adding up the contributions $I_{0,i}(q)$ of all species i weighted by their respective number density N_i:

$$I_0(q) = \sum_i N_i I_{0,i}(q) \,. \tag{10}$$

First of all, polydispersity has a profound influence on the form amplitude $B_0(q)$ and will therefore smear out the deep minima or zeros of the scattering curves. As an example, the upper curve in Fig. 3 gives $B_0^2(q)$ of a polydisperse homogeneous spheres with a radius of 37 nm and a polydispersity of 9%. In case of inhomogeneous particles polydispersity will obscure the isoscattering points as well since these features are located directly at the q^*-values of minima of the form part $B_0^2(q)$. Therefore, the crossing

points will become apparent only in the vicinity of the match point where the influence of $B_0^2(q)$ is small.

Another problem is given by polydispersity of internal structure leading to an ensemble of particles with different average contrast. In this case forward scattering does not vanish as is immediately obvious from Eq. (8). This can be demonstrated when considering the same system as discussed in Fig. 1 but now with a size distribution which was assumed to be Gaussian with a standard deviation of 9%. The thickness of the shell having a higher electron density (see Fig. 1) has been kept constant. Therefore, the particles differ not only with regard to size but also with regard to $\bar{\rho}$. The result of this calculation is displayed in Fig. 3. It refers directly to the scattering intensities of latex particles exhibiting a well-defined core-shell structure [1, 2, 4].

Figure 3 points to the importance of contrast variation: It shows that only the first isoscattering point is clearly visible; at higher q this feature is no longer discernible except for low contrast. If contrast is high, the variation of contrast leads only to shift of the scattering curves parallel to the ordinate. Measurements at sufficiently low contrast will, however, lead to scattering curves which differ markedly from the results and clearly reveal the existence of an isoscattering point.

This result is immediately evident when considering that low contrast masks the contribution $B_0^2(q)$ and the contribution $\varepsilon(q)$ governs the measured intensity. Figure 3 also points to the obvious experimental difficulties caused by the much lower scattering intensities around the match point. Hence, it may be advantageous to perform

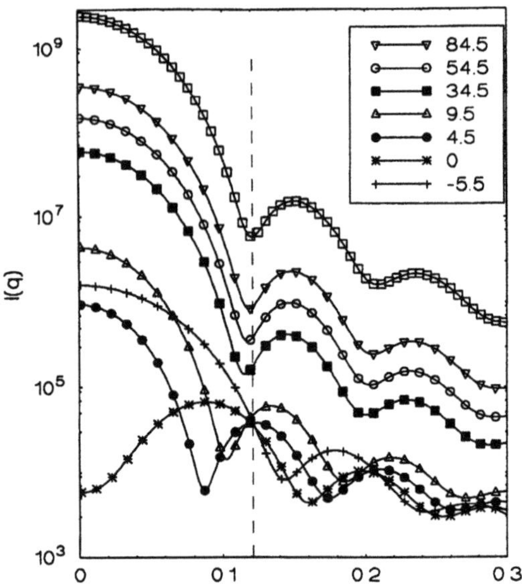

Fig. 3 Isoscattering point for a system of polydisperse spheres having a Gaussian size distribution with a standard deviation of 9%. The electron density shown in Fig. 1 was assumed to scale linearly with the size of the spheres. The inset gives the contrast $\bar{\rho} - \bar{\rho}_m$ (electrons/nm^3). The uppermost curve refers to the scattering intensity of a homogeneous spheres with diameter 37 nm. The dashed line marks the isoscattering point which coincides with the minimum of the scattering intensity $B_0^2(q)$ (cf. Eq. (8)) of the homogeneous sphere

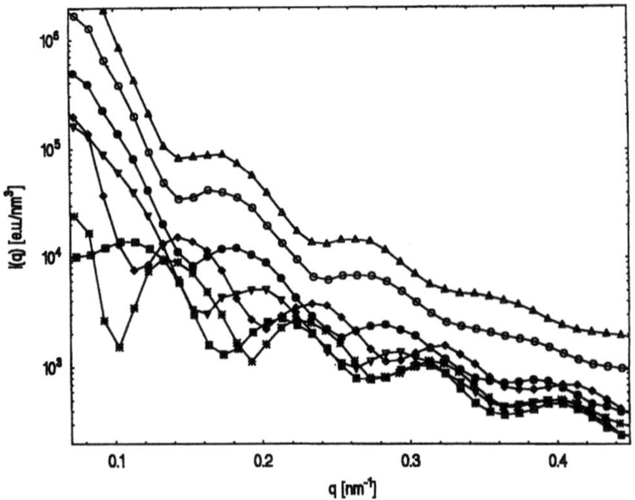

Fig. 4 Scattering curves of the covered PS-latex (103 mg Triton per g PS) measured at different contrast by addition of glycerol. For clarity the points were linked by lines. The curves refer to the following concentrations by weight of glycerol: (◇) 0%; (∗) 6.4%; (■) 13.4%; (▽) 20.4%; (●) 27.3%; (○) 41.3%; (△) 55.3%. The data have been taken from Ref. [5]

Progr Colloid Polym Sci (1998) 110:76–79
© Steinkopff Verlag 1998

SAXS-measurements at low but finite contrast; a comparison of $I_0(q)$ obtained at high contrast is sufficient for the analysis of the internal structure.

An experimental example of contrast variation is shown in Fig. 4 [5]. Here poly(styrene) latex particles covered by the non-ionic surfactant Triton X-405 have been analyzed by SAXS. The contrast had been adjusted by adding different amounts of glycerol to the latex. A critical examination showed that measurements using sucrose for raising ρ_m leads essentially to the same result [5]. The intensities shown in Fig. 4 demonstrate indeed that $I_0(q)$ measured at high contrast (uppermost curves in Fig. 4) are only shifted parallel to the ordinate as one would expect for homogeneous particles. An analysis solely based on these intensities would certainly fail to detect the inhomogeneity of the radial electron density. At low contrast, however, the side maxima are shifted in a characteristic manner which allows to determine the core-shell structure due to the adsorption of the surfactant [5].

Acknowledgment Financial support by the Deutsche Forschungsgemeinschaft is gratefully acknowledged.

References

1. Ballauff M, Bolze J, Dingenouts N, Hickl P, Pötschke D (1996) Macromol Chem 197:3043
2. Dingenouts N, Ballauff M (1993) Acta Polymer 44:178
3. Dingenouts N, Kim YS, Ballauff M (1994) Makromol Chem Rapid Commun 15:613
4. Dingenouts N, Kim YS, Ballauff M (1994) Colloid Polym Sci 272:1380
5. Bolze J, Hörner KD, Ballauff M (1996) Langmuir 12:2906
6. Glatter O, Kratky O (eds) (1982) Small Angle X-Ray Scattering. Academic Press, London
7. Philipse AP, Smits C, Vrij A (1989) J Colloid Interf Sci 129:335
8. Hickl P, Ballauff M (1997) Physica A 235:238
9. Kawaguchi T, Hamanaka T (1992) J Appl Crystallogr 25:778
10. Kawaguchi T (1995) J Appl Crystallogr 28:424 and further references given there
11. Bolze J, Ballauff M (1995) Macromolecules 28:7429
12. Bolze J, Hörner KD, Ballauff M (1997) Langmuir 13:2960
13. Bolze J (1997) Dissertation, Karlsruhe; Dingenouts N, Bolze J, Pötschke D, Ballauff M (1998) Adv Polym Sci, in press

Progr Colloid Polym Sci (1998) 110:80–82
© Steinkopff Verlag 1998

Shear induced alignment of kaolinite: Studies using a diffraction technique

A.B.D. Brown
S.M. Clarke
A.R. Rennie

A.B.D. Brown · S.M. Clarke
Dr. A.R. Rennie (✉)*
Cavendish Laboratory
Cambridge University
Madingley Road
Cambridge, CB3 0HE
United Kingdom

Present address
Industrial Materials Group
Department of Crystallography
Birkbeck College, Malet Street
London WC1E 7HX, UK

Abstract Shear alignment in dispersions of kaolinite particles was measured using neutron diffraction. The orientation distribution of particles was determined as a function of flow rate. The plate normals move towards the shear gradient from the compressional direction as shear rate is increased from 0.2 to 20 s^{-1}. The particle alignment also increases with flow. Order parameters are calculated and compared with theoretical models.

Key words Plate-like particles – neutron diffraction – orientation – clay

Introduction

Concentrated dispersions have a complex, non-Newtonian rheological behaviour [1]. This is due to the interactions between, and the resulting mutual ordering of the constituent particles. Much work has been done on structure under shear of spheres, both with long range repulsion (charge stabilisation) [2] and short range or "hard" repulsion (steric stabilisation) [3]. The previous work has shown that the particle interactions play a major role in determining the rheological properties and structure of colloidal materials.

Less information has been reported on anisotropic particles, which would be expected to show orientational as well as positional ordering under shear although small-angle neutron scattering studies have been reported (e.g. [4]). Here the orientational order in kaolinite is investigated. This is readily available and dispersed in water, is widely used in industry, and consists of plate-like particles.

Sample preparation

Kaolinite was purified by ECCI using sedimentation and centrifugation to reduce the polydispersity of the particles and to remove other minerals. The resulting sample had a hydrodynamic radius of 2 μm with a size distribution with half height points at 1 and 4 μm. The aspect ratio of the plates (diameter divided by thickness) was reported by ECCI to be approximately 76. Without stabilizers, kaolinite has negative faces and positive edges to the plates (in water at pH 7). This is not a stable system and the edge to face attraction was observed to cause flocculation and gelation in concentrated systems. To stabilize the particles a sodium salt of a polyacrylate was added (N40 as provided by Allied Colloids). This has a molecular mass of 3200 g mol^{-1} and dissociates almost completely at neutral pH. The resulting anionic polymer is believed to adsorb onto the edges of the plates giving the particles a negative charge over all the surface. This provides charge stabilization and possibly some steric stabilization. The

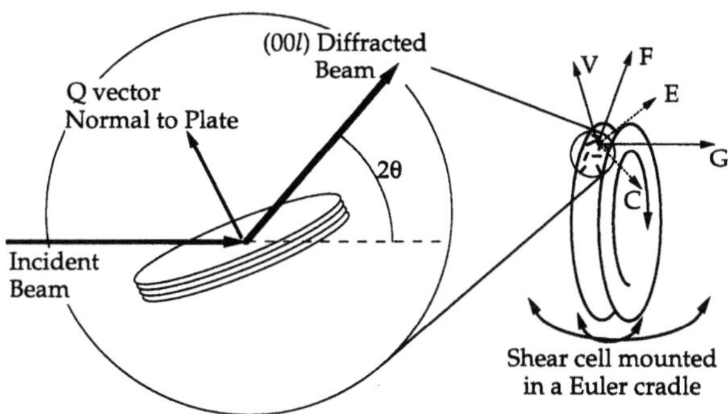

Fig. 1 Schematic diagram of diffraction from plates and the shear cell geometry. The disc/disc cell has a diameter of 200 mm and is mounted in an Euler cradle (not shown) to allow measurements of orientation at all directions with respect to the principal axes of flow. A shear field can be described by three orthogonal axes which are shown as the flow direction, F, the shear gradient direction, G, and the vorticity direction, V. It is sometimes useful to consider the extensional, E, and compressional, C, axes which lie in the plane defined by the flow and gradient directions

sedimentation profile of the sample shows that the stability of the particles is vastly improved upon adding the polymer. Data will be described for a dispersion of 50% w/w.

Diffraction technique

A crystal will diffract if it is oriented to meet the Bragg condition. Thus, the integrated intensity of a diffraction peak from an oriented powder is proportional to the number of crystals in the correct orientation for that peak. If the sample is rotated then all different crystal orientations with respect to the sample can be investigated (see Fig. 1). Kaolinite plates are monocrystals where the crystal axes have a known orientation relative to the particle morphology. The crystal structure and diffraction pattern are well-established [5]. Hence, measuring the variation in intensity of a diffraction peak as the sample is rotated allows the orientation distribution of the plates to be obtained [6].

Neutrons rather than X-rays have been used for their ability to penetrate the sample and sample cells, thus allowing bulk effects to be investigated in a sample a few millimetres thick.

Experiment

Diffraction measurements were made with the new high flux diffractometer, D20 at the ILL, Grenoble, France using a wavelength of 2.41 Å. The large multidetector allowed simultaneous measurement of many Bragg peaks.

The kaolinite was suspended in D_2O to reduce incoherent scattering. The sample was placed in a sealed plate–plate cell made of aluminium and mounted on an Euler cradle. The beam was incident on the cell at distance of 80 mm from the centre of the rotating plate, and the beam diameter was 8 mm, thus the shear is well defined and varies by only 10% across the illuminated volume.

As the sample is rotated the volume of sample illuminated by the beam changes. To allow for this effect the incoherent background from D_2O was used as a calibrant as this scattering comes from the illuminated volume of the sample which varies in the same way as the amount of kaolinite. The peak intensity and background are obtained from least-squares fits of a Gaussian function to an appropriate range of angles for each peak.

The $(0\,0\,l)$ peaks from kaolinite are the most intense, and arise from a Q vector perpendicular to the plate like particle. In this investigation the $(0\,0\,4)$ peak was primarily used as the $(0\,0\,1)$ and $(0\,0\,2)$ peaks were often obscured by the cradle and the shear cell, while the $(0\,0\,3)$ peak was partially obscured by an aluminium diffraction peak. Where possible information from the $(0\,0\,2)$ peak was used, and the intensity/background was scaled appropriately to conform with equivalent information from the $(0\,0\,4)$ peak.

Results and discussion

The data in Fig. 2 show results of a scan of the sample cell that looked at plates with normals in the flow-gradient plane. The different curves show shear rates varying from

Fig. 2 Normalised diffracted intensity for the $(0\,0\,2)$ and $(0\,0\,4)$ Bragg peaks as a function of angle and shear rate. Angle is zero in the gradient direction and positive in the compressional quadrant. The plot shows the variation between the shear gradient and flow axes of $0.22\,\mathrm{s}^{-1}$(○), $1.0\,\mathrm{s}^{-1}$ (●), $1.1\,\mathrm{s}^{-1}$ (■), $3.6\,\mathrm{s}^{-1}$ (♦), $8.4\,\mathrm{s}^{-1}$ (▲) and $23\,\mathrm{s}^{-1}$ (▼)

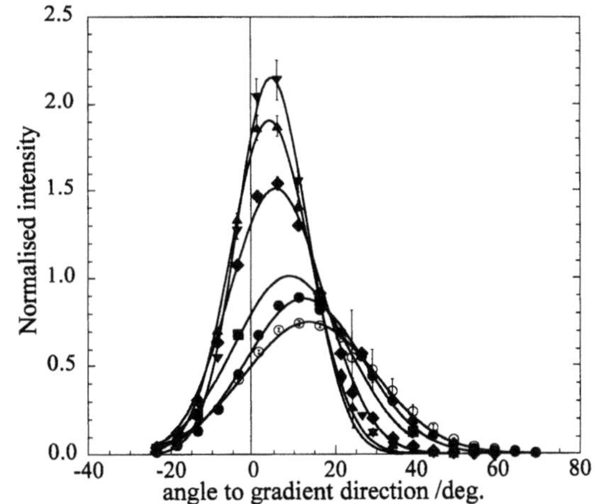

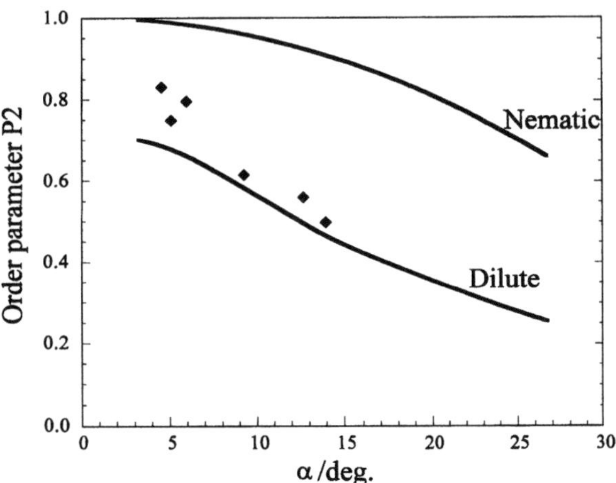

Fig. 3 The order parameter P_2 calculated for each curve in Fig. 2 plotted against the position of the peak in Fig. 2, α, measured away from the shear gradient direction. The two continuous lines below and above the data show the predictions of flow for dilute, non-interacting plates and a nematic fluid [7, 8], respectively

$0.22\,\mathrm{s}^{-1}$ for the flattest curve with peak position at $14.0°$ through to $23\,\mathrm{s}^{-1}$ for the sharpest curve with peak position at $5.1°$. The change in position of the peak in this plot, towards the gradient direction and increase in alignment as the shear rate increases is shown clearly in Figs. 2 and 3. Such behaviour is predicted both for dilute systems [7] and for nematic systems [8] but the magnitude of the effects are different. The angular orientation distribution can be used to calculate order parameters [6] and the data for the first-order parameter P_2 is shown in Fig. 3. A nematic system is expected to show a much higher degree of order for a given direction of alignment than a dilute system. The data collected here for this concentrated but not nematic clay system, lies between the two models suggesting that there is enhanced ordering due to the inter-particle interactions.

Further measurements were also made using time varying flow; the data acquisition was gated to provide series of diffraction patterns as a function of time within the cycle of flow pattern as the sample was aligned and then allowed to relax. This permitted the growth and decay of the orientation as a function of time to be observed. This data, together with information about the variation in order in other directions with respect to the flow, will be described elsewhere [9].

Conclusions

Diffraction of neutrons has been used to study the alignment of crystalline colloidal particles in a well-defined flow field. The use of a well-defined flow geometry has greatly enhanced the ability to understand measurements of the orientation distribution. Order parameters are calculated and apparently lie between the models for dilute dispersions of non-interacting plates and a nematic fluid for the dispersion of clay studied. Further comparisons with theories and computer models can be made.

Acknowledgements We are grateful to the EPSRC for a studentship (ABDB) and to ECC International for the provision and characterisation of clay samples. The ILL provided access to neutron facilities. We particularly thank Pierre Convert for discussions, help with the experiment and construction of the diffractometer.

References

1. Goodwin JW (1990) In: Candau F, Ottewill RH (eds) An Introduction to Polymer Colloids. Kluwer Academic Publishers, Dordrecht, pp 209–223
2. Ashdown S, Markovic I, Ottewill RH, Lindner P, Oberthür RC, Rennie AR (1990) Langmuir 6:303–307
3. Clarke SM, Rennie AR, Ottewill RH (1995) Adv Coll Interf Sci 60:95–118
4. Ramsay JDF, Lindner P (1993) J Chem Soc Farad Trans 89:4207
5. Brindley GW, Brown G (1984) Crystal Structures of Clay Minerals and their X-ray Identification. Mineralogical Society, London
6. Clarke SM, Rennie AR, Convert P (1996) Europhy Letts 35:233–238
7. Weissberger A (ed) (1960) Techniques of Organic Chemistry, Vol I, 3rd ed. Interscience Publishers Ltd, London, pp 2398–2416
8. Carlsson T (1983) J Physique 44:909–911
9. Brown ABD, Clarke SM, Rennie AR, in preparation

Progr Colloid Polym Sci (1998) 110:83–88
© Steinkopff Verlag 1998

Growth and anisotropic ripening in twinned colloidal crystals

M.R. Maaroufi
A. Stipp
T. Palberg

M.R. Maaroufi · A. Stipp (✉) · T. Palberg
Johannes Gutenberg-Universität Mainz
Institut für Physik
Staudinger Weg 7
D-55099 Mainz
Germany
E-mail: STIPP@dipmza.physik.uni-mainz.de

Abstract The heterogeneous crystallization of colloidal suspensions of Yukawa particles on a wall was observed by Bragg microscopy. The growth velocity v_{110} of the fluid–crystal interface can be very well described by a Wilson–Frenkel growth law, while the limiting velocity is found to be lower than in previous investigations. Twinned bcc domains in lateral direction (parallel to the wall) and their ripening behavior are for the first time investigated systematically. A strong anisotropic growth behavior is found: in the direction of the shear flow, which shear melts the suspension prior to the crystallization process, the extension of the domains is much larger than perpendicular to that. The increase in domain size for short and intermediate times is found to follow a Lifshitz–Allen–Cahn ripening law, while unexpectedly for long times the ripening halts and the size of the domains remains constant.

Key words Charged stabilized suspensions – heterogeneous nucleation – crystal growth – ripening

Introduction

Rapid progress in the investigation of the crystallization process and crystallization kinetics in suspensions of colloidal particles has been made in the last two decades [1, 2]. The easily accessible time and length scales as well as the possibility to vary the spherical interaction potential of the particles between the theoretical limits of a hard sphere potential and the one-component plasma make colloidal systems ideal model systems to verify e.g. classical theories of nucleation, growth and ripening during the crystallization process. In addition, crystallization proceeds isothermally since the suspension medium acts as an efficient heat sink. Among the first to observe systematically the crystal growth in colloidal suspensions of charged particles were Aastuen et al. [3, 4]. They measured the growth of homogeneously nucleated crystals by direct microscopic observation and found a linear growth law and an increase of the growth velocity with particle concentration. The best function to describe the concentration dependence of growth velocity was found to be the Wilson–Frenkel law [5]

$$v = v_{\infty}\left[1 - \exp\left(-\frac{\Delta\mu}{k_{\mathrm{B}}T}\right)\right], \tag{1}$$

where v_{∞} is the maximal growth velocity, limited by the elementary step of integrating the particles into the crystal surface, and $\Delta\mu$ is the difference in chemical potential between the fluid and the crystalline phase. A generalized description of the growth process was given by Würth et al. [6], who measured the growth velocity of heterogeneously nucleated crystals from the wall of a shear cell by Bragg microscopy. Shear-induced orientation of the layered melt close to the wall [7], serving as the heterogeneous nucleus for homoepitactic growth results in monolithic crystals with their (1 1 0)-plane parallel to the wall and $\langle 1\,1\,1\rangle$-direction in the former flow direction [8]. To evaluate their data, Würth et al. expressed $\Delta\mu$ in terms of a reduced energy density

$$\Pi^* = \frac{\Pi - \Pi_m}{\Pi_m}, \qquad \Pi = \tfrac{1}{2}\alpha n V(r), \qquad (2)$$

where $n = 3\phi/4\pi a^3$ is the particle number density, a the radius of the particles, ϕ the packing fraction, and α an effective coordination number. The suffix m corresponds to the value at melting. For the pair potential $V(r)$ Würth et al. used the modified DLVO approximation [9]

$$V(r) = V_0 \frac{\exp(-\kappa r)}{r}, \qquad V_0 = \frac{(Z^*_{PBC} e)^2}{4\pi\varepsilon\varepsilon_0} \frac{\exp(\kappa \cdot 2a)}{(1 + \kappa a)^2}, \qquad (3)$$

with the effective screening parameter κ defined via

$$\kappa^2 = \frac{e^2}{\varepsilon_0 k_B T}(2c_s + nZ^*_{PBC}), \qquad (4)$$

where c_s is the salt concentration and Z^*_{PBC} is the effective charge of the particles according to a Poisson–Boltzmann cell model [10]. With

$$\frac{\Delta\mu}{k_B T} = B'\Pi^* \qquad (5)$$

it was possible to describe three different sets of data by the Wilson–Frenkel growth, suggesting a universality of the parameter Π^*. It is an important question whether this behavior can be reproduced by independent measurements on different particles.

Interestingly, the wall crystals grown this way have two possible orientations leading to two possible twins [8]. Therefore, instead of a single crystal a heavily twinned one results. While twinning of face centered cubic crystals has been observed and analyzed before [11–13], no analysis of bcc twinning has been presented up to now. In particular, two aspects are of major relevance. First, what is the shape and relative position of domains and second, what are their ripening kinetics. In this paper we systematically address both questions using a well-characterized model suspension. After a short presentation of experimental procedures first the results on growth are presented, and then the twin domain ripening is discussed.

Experimental

The particles used in this work were carefully characterized polystyrene latex spheres of nominal radius $a = 57.5$ nm and a nominal polydispersity of 4.7% (IDC, Portland, Oregon, USA, Batch-No. 10-95-38.202). These carry an average number of $N = 3600 \pm 100$ sulfate surface groups determined by conductometric titration corre-

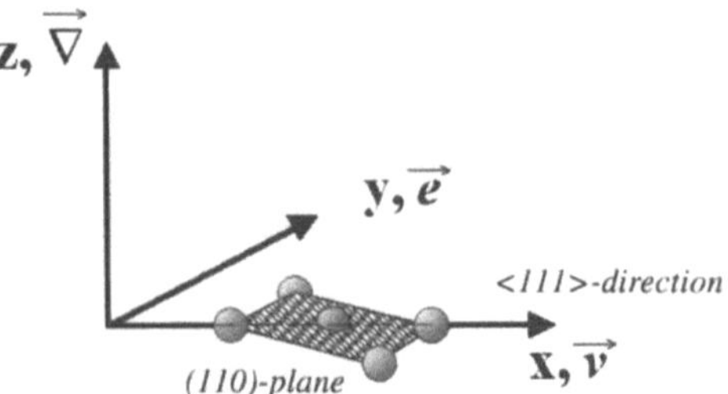

Fig. 1 Sketch of the (1 1 0)-plane, which is parallel to the xy-plane of the cell wall. x corresponds to the direction of former flow $\vec{v}$, z to the gradient $\vec{\nabla}$ and y to the vorticity direction $\vec{e}$

sponding to effective charges $Z^*_\sigma = 730 \pm 20$ (from conductivity measurements) and $Z^*_{PBC} = 805 \pm 50$ (calculated after [10]). Particles were shipped at a packing fraction of $\phi = 0.084$. After filtering by 1 μm filters, mixed bed ion exchange resin (IEX) was added and the suspensions were left to stand for about 1 week. After a second filtering and adding a second batch of IEX, the packing fraction of the resulting stock suspension was determined to be $\phi = 0.063$ by weighing. Further sample preparation and the measurements were performed in a closed tubing system under an Ar-atmosphere including the measuring cells and the preparational units. Details of the continuous deionization procedures are described elsewhere [14]. Leakage of stray ions into the system was estimated from an increase of the conductivity of pure water (55 nS cm^{-1}) of less than 150 nS cm^{-1} h^{-1} to correspond to an NaCl equivalent of 2×10^{-7} mol l^{-1} h^{-1}. The actual samples were prepared from the stock suspension by successive dilution under conductivity control; the uncertainty in the determination of ϕ was $\Delta\phi < 0.03\phi$ at $\phi \approx 10^{-4}$. Under deionized conditions ($c_s \leq 2 \times 10^{-7}$ mol l^{-1}) samples of $\phi > 4.5 \times 10^{-4}$ were body centered cubic (bcc) crystalline at rest. They are shear molten during preparation but readily resolidify once shearing is aborted. Crystallization proceeds either via immediate heterogeneous nucleation at the cell wall, or after some induction time via homogeneous nucleation in the bulk. Using a thin optical cell the first mechanism dominates up to a packing fraction of $\phi = 9 \times 10^{-4}$. A sketch of the coordinate system we use in this paper is given in Fig. 1. The z-direction corresponds to the growth direction of the (1 1 0)-interface into the cell, the x-direction is oriented parallel to $\langle 1 1 1 \rangle$ corresponding to the direction of shear melting flow, and the y-axis (vorticity direction) is perpendicular to both.

Growth kinetics

Aastuen et al. [3, 4] used Bragg microscopy to detect the light scattering of growing crystals by illuminating

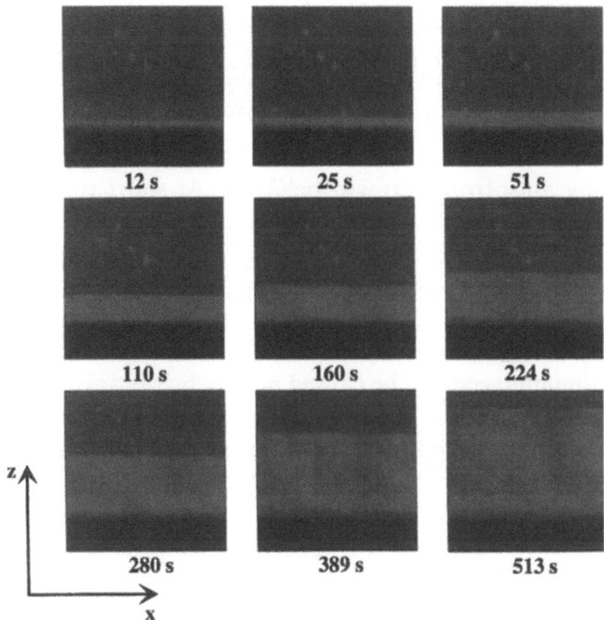

Fig. 2 Growth in z-direction, the shown area corresponds to 3.2×2.4 mm^2, the light region is the wall crystal. Pictures are taken at different times

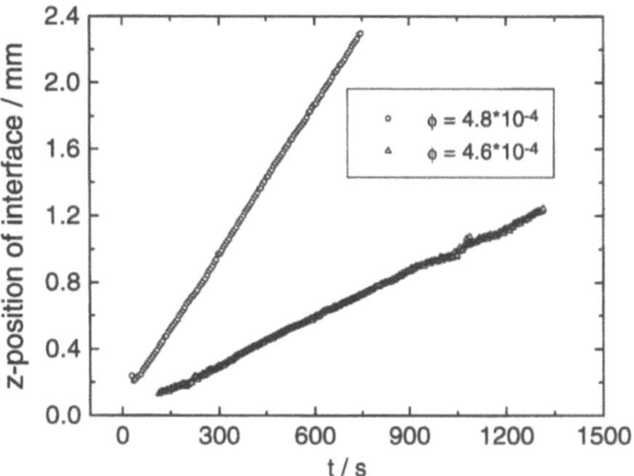

Fig. 3 z-Position of the melt–crystal interface versus time for two different packing fractions. Growth velocities: $\bigcirc$: $v_{110} = 2.98\ \mu\mathrm{m\,s}^{-1}$; $\triangle$: $v_{110} = 0.92\ \mu\mathrm{m\,s}^{-1}$

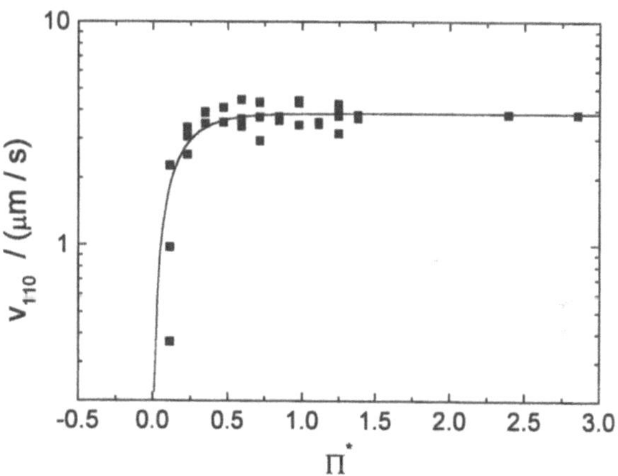

Fig. 4 Growth velocity in z-direction v_{110} versus reduced energy density Π^*. The line is a fit to a Wilson–Frenkel growth law using Eqs. (1)–(5). The best fit parameters are $v_\infty = (3.9 \pm 0.1)\ \mu\mathrm{m\,s}^{-1}$ and $B' = 5.8 \pm 1.9$

the sample with white light and placing a camera under a combination of angles fulfilling the Bragg condition of an individual crystal. Here we use a similar setup and mount the rectangular flow through cell ($30 \times 10 \times 5$ mm^3) on the stage of a microscope, illuminating the cell from below in a manner that the monolithic wall-based crystal scatters into the microscope objective. The cold light source is adjusted to monitor growth in z-direction (off the wall) and pictures are taken by a CCD camera as it is depicted in Fig. 2. The starting time $t = 0$ s is determined by stopping the pumping of the suspension. Contrast is provided by the difference in scattered intensity of the crystal as compared to the fluid. The spatial resolution of the z-position of the moving interface is approximately 20 μm while the growth velocity v_{110} is determined from a plot of z versus time with an accuracy of better than 0.1%. Examples are given in Fig. 3. Growth is observed to be strictly linear for all concentrations.

To evaluate the packing fraction-dependent growth velocities, we follow the procedure of Würth et al. [6] described above using the reduced energy density Π^*. We plot the measured velocities versus Π^* in Fig. 4 and compare them with a Wilson–Frenkel growth law according to Eqs. (1)–(5). The best fit (fixing $v(\Pi^* = 0) = 0\ \mu\mathrm{m\,s}^{-1}$) yields $B' = (5.8 \pm 1.9)$ and $v_\infty = (3.9 \pm 0.1)\ \mu\mathrm{m\,s}^{-1}$. This B'-value compares well with the value of $B' = 6.7 \pm 0.1$ obtained by Würth et al. [6] for particles with a radius $a = 51$ nm and much lower effective charges

of $Z^* \approx 400$. Although in our case the uncertainty of the fit is much larger because at our low number densities even spurious stray ions will significantly change the growth velocity, the universality of the expression for Π^* is confirmed. We note that this is also in line with recent theoretical work, where for a different two parameter potential, B'-values of the order of 4 were obtained [15]. Interestingly, the maximum growth velocity v_∞ is much lower than expected. Assuming the same microscopic growth mechanism in both experiments it should scale with the particle radii. Thus a value of $9.1 \times 115/102 = 10.3$ is expected, i.e. a factor of more than two larger than our measured value. One possible explanation is the lower polydispersity of 1.5% of the smaller particles. A similar observation was reported by Henderson et al. [16], who measured the crystallization kinetics of two colloidal hard sphere suspensions of different polydispersity. The crystallization rates in the suspension with the higher polydispersity were found to be much slower than in the less polydisperse suspension. Although the polydispersities of our particles as well as those of Würth are much lower than for the particles used by Henderson et al., an influence on the growth velocities cannot be excluded.

Twin domain ripening

To investigate the temporal evolution of the twin domains, the cold light source was positioned so that only one twin orientation scatters into the microscope objective. These twins then appear bright, whereas the other orientation appears dark. Images of the twin domains for different times are shown in Fig. 5. The overall increase in intensity corresponds to the continuing growth in z-direction. At early times we observe small domains roundish to irregular seemingly assorted in rows oriented in x-direction. These grow to larger anisotropic objects of less irregular surface. Note that ripening commences already before the sample is fully solidified. To analyze this complex morphology, we restrict ourselves in a first step to the average dimensions of the domains. Mean sizes in x- and in y-direction were determined by computer-aided image processing as follows. A raster of lines in x- and y-direction is

superimposed on the picture, so that every crystal domain is intersected by several lines. The size of the domains is then determined along these lines and the mean value is taken for each picture. Details of the evaluation procedure are given elsewhere [17]. An example for the evolution of domain sizes $x(t)$ and $y(t)$ is given in Fig. 6. We observe a strong anisotropic growth behavior: At short times an increase in size is observed, which is much steeper in $x(t)$, the direction of the former flow, than in $y(t)$. The growth velocity decreases to zero for times larger than $t = 600$ s; at long times, $x(t)$ and $y(t)$ remain constant, but consequently on very different levels. Therefore, the anisotropic shape of twin domains seems to be caused mainly by the anisotropy of growth velocities. The fit functions which described the complete time dependence best for all packing fractions were found to be

$$x(t) = x_0 - x_1 \exp\left(-\frac{t}{\tau_x}\right), \quad y(t) = y_0 - y_1 \exp\left(-\frac{t}{\tau_y}\right)$$

$$\tag{6}$$

and are also plotted in Fig. 6. Here x_0 and y_0 are the domain sizes for $t \to \infty$ in x- and y-direction, respectively; x_1/τ_x and y_1/τ_x are the growth velocities for $t = 0$.

The abortion of the ripening process is a completely unexpected result. Assuming its driving force to be the surface tension, ripening is expected to slow, but not to halt. In fact, Dux et al. [13] report on the annealing of stacking faults in face centered cubic crystals to continue to very large times. To halt the growth either an increasing counter-force or an increasing kinetic barrier has to be assumed. At present we cannot distinguish between these alternatives. It is interesting to note, however, that the time of abortion of ripening decreases with increasing packing fraction even in the range where v_{110} is already constant. This corresponds to a decreasing height z^* of domain

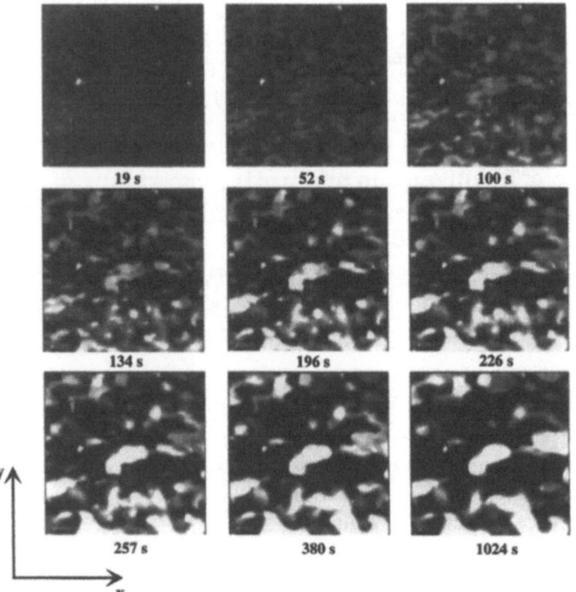

Fig. 5 Development of bcc twin domains in the xy-plane. Pictures are taken at different times

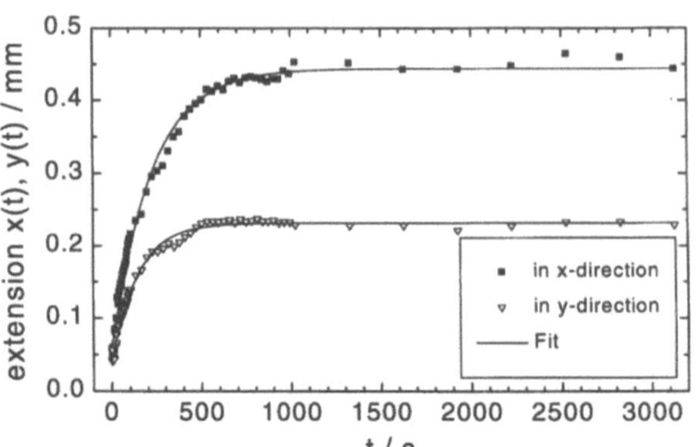

Fig. 6 Domain sizes $x(t)$ and $y(t)$ versus time. The straight line is a fit of Eq. (6)

walls at the time of abortion of ripening from approximately $z^* = 1600\ \mu m$ to $z^* = 400\ \mu m$. Further ripening in x-direction halts at slightly later times than in y-direction.

Ripening mechanism

There are two predictions for the limiting cases of the ripening behavior of crystals. The Lifshitz–Slyozov (or Ostwald) ripening, which is the ripening of a minority phase mediated by a majority phase, predicts a $t^{1/3}$-law for the domain sizes. The Lifshitz–Allen–Cahn (LAC) ripening predicts a $t^{1/2}$-power law; this is e.g. valid for ripening by grain boundary migration. In our case, LAC ripening is therefore expected at short times. Figure 7 shows a double logarithmic plot of the data in Fig. 6. In this example, for more than one order of magnitude in time, a power law t^{α} can be fitted to the data yielding $\alpha = 0.45$ in the x-direction and $\alpha = 0.44$ in the y-direction. Neither a systematic dependence of α on the interaction parameters nor on the direction of ripening is observed. This is shown in Fig. 8, where we plot α_x and α_y for all packing fractions investigated. The data clearly scatter around $\alpha = \frac{1}{2}$, indicating a LAC ripening behavior as expected.

In view of this result, it is tempting to explain the anisotropy in the growth velocity as being due to different local ripening mechanisms. Geometrical considerations show that in the y-direction the domain wall may proceed one lattice spacing if the interfacial layer is collectively moved in the x-direction. In the x-direction much more complex particle motions occur during the advance of the interface. On the other hand, LAC behavior is perfectly compatible with a surface-tension-driven ripening and should continue to infinite times. Instead, the halt of ripening is seen as a clear-cut change in slope in Fig. 7. While in LAC the elementary step of motion is assumed to be self-diffusive, we note that the assumption of a collective motion involved in the ripening mechanism seems to be compatible with an activation barrier increasing in time. Clearly, the microscopic mechanisms present a challenge requiring further experimental and theoretical work.

Conclusions

The presented data, in principle, confirm classical theories of growth and ripening for our system. However, some comments seem to be in place. While the adaptation of the energetic considerations present in these laws to colloidal specific needs seems to work quite well, the kinetic pre-

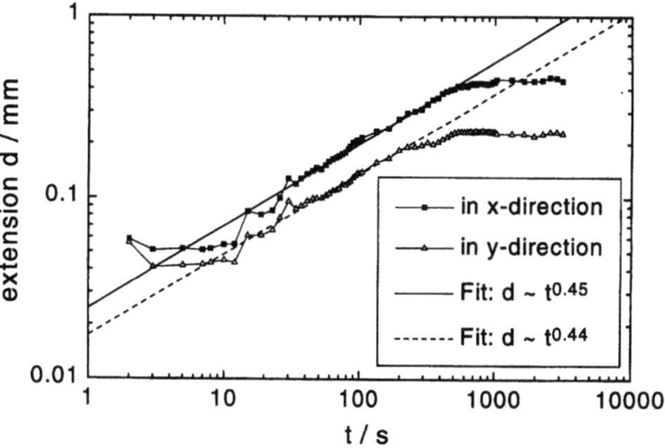

Fig. 7 Double logarithmic plot of Fig. 6 and fit to a power law t^{α} for short and intermediate times

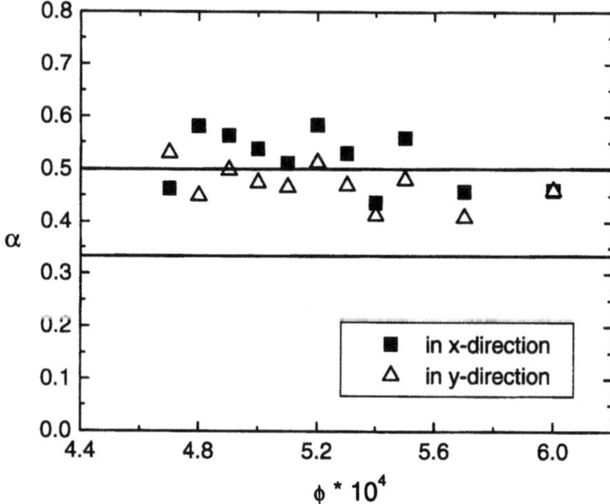

Fig. 8 Density dependence of exponent α. Horizontal lines indicate the values of $\alpha = \frac{1}{2}$ expected for Lifshitz–Allen–Cahn ripening, and $\alpha = \frac{1}{3}$ for Lifshitz–Slyozov ripening

factors deserve further attention. In particular, one should look for possibilities to incorporate collective motions and the effects of polydispersity in the formulation of growth and ripening laws.

Second, the detailed microscopic mechanisms need to be clarified further. This applies to the growth mechanism as well as to the formation of bcc nuclei from the layered melt of hexagonal structure and the particle motions implied in the ripening process which may severely depend on the ripening direction. Significant progress has been reported recently on some of these aspects [7] using high-resolution video microscopy. Also in our case direct light microscopic observations should help to resolve the issues.

References

1. Ackerson B (ed) (1990) Phase Trans 21
2. Palberg T (1997) Curr Op Coll Int Sci 2:607
3. Aastuen DJW, Clark NA, Cotter LK (1986) Phys Rev Lett 57:1733
4. Aastuen DJW, Clark NA, Swindal JC, Muzny CD (1990) Phase Trans 21:139
5. Kelton KF (1993) Solid State Phys 45:75
6. Würth M, Schwarz J, Culis F, Leiderer P, Palberg T (1995) Phys Rev E 52:6415
7. Grier DA, Murray CA (1994) J Chem Phys 100:9088
8. Ackerson BJ, Clark NA (1983) Physica A 118:221
9. Bitzer F, Palberg T, Löwen H, Simon R, Leiderer P (1994) Phys Rev E 50:2821
10. Belloni L (1998) Colloids and Surface A, in press
11. Monovoukas Y, Gast AP (1990) Phase Trans 21:183
12. Elliot MS, Bristol BTF, Poon WCK (1997) Physica A 235:216
13. Dux C, Versmold H (1997) Phys Rev Lett 78:1811
14. Palberg T, Härtl W, Wittig U, Versmold H, Würth M, Simnacher E (1992) J Phys Chem 62:8180
15. Ripoll MS, Tejero CF, Baus M (1996) Physica A 234:311
16. Henderson SI, Mortensen TC, Underwood SM, van Megen W (1996) Physica A 233:102
17. Maaroufi MR, Stipp A, Palberg T (1997) unpublished

Progr Colloid Polym Sci (1998) 110:89–93
© Steinkopff Verlag 1998

R. Piazza
M.J. Eghan
V. Peyre
V. Degiorgio

Light scattering investigation of amphiphile and polymer adsorption on the surface of colloidal particles with low optical contrast

Dr. R. Piazza (✉) · M.J. Eghan*
V. Peyre · V. Degiorgio
Dipartimento di Elettronica
INFM – Dipartimento di Elettronica
Università di Pavia
via Ferrata 1
I-27100 Pavia
Italy

Permanent address
Physics Department
University of Cape Coast
Ghana

Abstract We have exploited the peculiar optical properties of fluorinated polymer colloids, which have a very low refractive index and a partially crystalline internal structure, to obtain accurate adsorption isotherms by combining static and depolarized dynamic light scattering information. Due to the very weak optical contrast between perfluoropolymers and water, the intensity of the light scattered by suspensions of fluorinated colloids strongly depends on the presence of surface adsorbates. At the same time, the depolarized scattering contribution arising from the particle internal optical anisotropy allows to measure the rotational diffusion constant which is strongly dependent on the particle hydrodynamic radius. We have applied such technique to measure the adsorption isotherms of both nonionic and ionic surfactants, and of long-chain block copolymers.

Key words Light scattering – adsorption – amphiphile – polymer

The adsorption properties of polymer and surfactants on the surface of colloidal particles are particularly relevant in industrial applications related to the steric stabilization of emulsions and latices. To be effective, surfactants and polymers should adsorb in appropriate amount and form interfacial layers of sufficient thickness. Additional adsorbates may also be present on the surface of polymer latex particles as residuals of the polymerization process, and compete with the adsorption of the stabilizer. Careful monitoring of adsorption equilibria is therefore required. Traditional adsorption studies are often time-consuming, require a substantial amount of adsorbant, and are intrinsically invasive, since they generally require to separate out the colloidal phase from the supernatant [1]. Moreover, methods which detect variations of the concentration of the adsorbant in solution are not well suited to study weak adsorption effects, where the amount of adsorbate can be only a tiny fraction of the adsorbant remaining in solution.

Light scattering, as a powerful and non-invasive technique which can easily detect changes in the suspended particle morphology, could be a valid substitute to conventional chemicophysical analysis. However, for aqueous solutions of polymer colloids having a refractive index in the range 1.45–1.50, changes of the scattered intensity due to adsorption are very weak, due to the small fraction of particle volume occupied by the adsorbed layer. So far then, the application of light scattering to adsorption studies is confined to a few experiments in which dynamic light scattering was used to determine variations of the particle hydrodynamic radius as a function of the amount of added adsorbant. In the case of surfactant adsorption, changes of the translational diffusion coefficient of the order of a few percent must be detected [2].

We have used static and depolarized dynamic light scattering to probe the adsorption of ionic and nonionic surfactants, and of surfactant block copolymers, on the surface of monodisperse spherical particles made of FEP, a polytetrafluoroethylene copolymer. In the last few years our group has performed extensive studies of the peculiar optical properties of fluorinated polymer colloids [3], and in a preliminary paper we have shown that these properties can be profitably exploited to study surface adsorption processes [4]. Here we simply recall that FEP particles present two very interesting features: they have an average refractive index similar to that of water, and they yield a depolarized contribution to the intensity of the scattered light due to their intrinsic optical anisotropy. As we shall see, the former feature ensures a much stronger sensitivity of the scattered light intensity to the presence of adsorbate layers, while the latter yields a more precise determination of the particle hydrodynamic radius via the measurement of the rotational diffusion coefficient, which is inversely proportional to the particle volume.

Let us first consider a uniform layer of thickness d adsorbed on the surface of a spherical particle of radius R. We will call $\bar{n}$ the average refractive index of the particle, and n_a, n_s, the refractive indices of the adsorbate and of the solvent, so that we will refer to $\bar{n} - n_\mathrm{s}$ and to $n_\mathrm{a} - n_\mathrm{s}$ as to the optical "contrasts" of particle and adsorbate with respect to the solvent. The field scattered by such a core-shell particle can be calculated with a simple trick by summing to the field scattered by a homogeneous particle of radius $R + d$ and refractive index n_a, the field scattered by a homogeneous particle of radius R and refractive index $\bar{n}$ imbedded in a medium of refractive index n_a. The total intensity of the light scattered by N noninteracting particles in the scattering volume V is then given by

$$I = CN \left\{ (n_\mathrm{a}^2 - n_\mathrm{s}^2)\, \tfrac{4}{3}\, \pi (R + d)^3\, F\left[k(R + d)\right] \right.$$
$$\left. + (\bar{n}^2 - n_\mathrm{a}^2)\, \tfrac{4}{3}\, \pi R^3\, F(kR) \right\}^2 , \qquad (1)$$

where k is the scattering wave vector, $F(kx)$ is the form factor of a spherical particle of radius x, and C an instrumental constant. When the particle core is optically anisotropic, an additional depolarized contribution, depending only on the internal anisotropy and not on the properties of the solvent, is also present. The depolarized "incoherent" scattering contributions is very weak compared to the polarized "coherent" component whenever $\bar{n}$ is even slightly different from n_s, and we will take it into account only when dealing with the scattering dynamics. In this work we consider particles which, even in presence of the adsorbed layer, have a radius which is smaller than $1/10$ of the wavelength of the incident light beam. In such a case we can set $F\left[k(R + d)\right] \approx F(kR) \approx 1$, and we can recast Eq. (1) in a simpler and more useful form

$$I = C\, \frac{V^2}{N}\, \left\{ (\bar{n}^2 - n_\mathrm{s}^2)\, \Phi_\mathrm{p} + (n_\mathrm{a}^2 - n_\mathrm{s}^2)\, \Phi_\mathrm{aa} \right\}^2 , \qquad (2)$$

where Φ_p is the particle volume fraction, and Φ_aa the volume fraction of *adsorbed* adsorbant. From Eq. (2) we clearly see that, just because we are summing *fields*, and not *intensities*, we obtain interference terms leading to

a substantial increase of the scattered intensity whenever even a small quantity of high-contrast material is adsorbed on a low-constrast particle. Calling I_0 the intensity scattered by the core particles in absence of added adsorbant, the intensity increase due to adsorption is then given by

$$\frac{I}{I_0} = \left[1 + \frac{(n_\mathrm{a}^2 - n_\mathrm{s}^2)\, \Phi_\mathrm{aa}}{(\bar{n}^2 - n_\mathrm{s}^2)\, \Phi_\mathrm{p}} \right]^2 . \qquad (3)$$

The volume fraction Φ_aa of adsorbate can be simply evaluated from Eq. (3) in terms of experimentally measurable quantities. However, if the adsorption is not complete, as it happens when the amount of added adsorbant exceeds the value required for full surface coverage, the refractive index of the effective solvent in which the particles are dispersed is modified by the fraction of the adsorbant which remains in solution. Assuming additive optical polarizabilities, and calling n_w the refractive index of the original solvent (which in our case is water), and Φ_a the total amount of added adsorbant, the refractive index of the effective solvent in presence of dispersed adsorbant is given by

$$n_\mathrm{s}^2 = n_\mathrm{w}^2 + (n_\mathrm{a}^2 - n_\mathrm{w}^2)\, (\Phi_\mathrm{a} - \Phi_\mathrm{aa}) , \qquad (4)$$

so that Φ_aa can be directly found by combining Eqs. (3) and (4).

The general trend which can be extracted from Eq. (3) is a steady increase of the scattered intensity as a function of the amount of added adsorbant until surface coverage saturation is reached. However, adding further adsorbant ends up in increasing the solvent refractive index, and in reducing then the refractive index contrast between the "dressed" particle and the solvent itself, with a consequent decrease of the scattered intensity. A strong increase of the scattered intensity, compared to the "bare particle" value, can be obtained by using particles which have a refractive index not too different from n_w. In our case, the average refractive index of FEP, as determined by measuring an index-matching curve, is $n_\mathrm{p} = 1.353$, while the refractive indices of the used adsorbants, which are hydrogenated surfactants or polymers, is always in the range 1.45–1.50. The scattering "contrast" between FEP and water is then of the order of $1/10$ of the one between water and adsorbate.

We present now some applications of the previous ideas. Suspensions of fairly monodisperse FEP particles having a radius $R = 41$ nm were extensively dialysed against water, and successively flushed through mixed-bed ionic exchange resins, in order to reduce to a minimum the amount of adsorbed ionic perfluorinated surfactant coming from the polmerization process. Light scattering measurements have been performed by means of our previously described apparatus [4] which includes selective

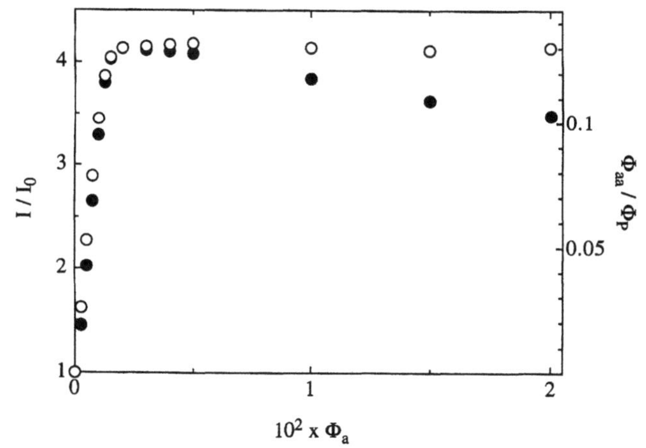

Fig. 1 Adsorption of Triton X-100 on 41 nm FEP particles as a function of the volume fraction of added surfactant Φ_a. Full dots: ratio of the scattered light intensities I/I_0 (left axis). Open dots: ratio of volume fractions Φ_{aa}/Φ_p (right axis)

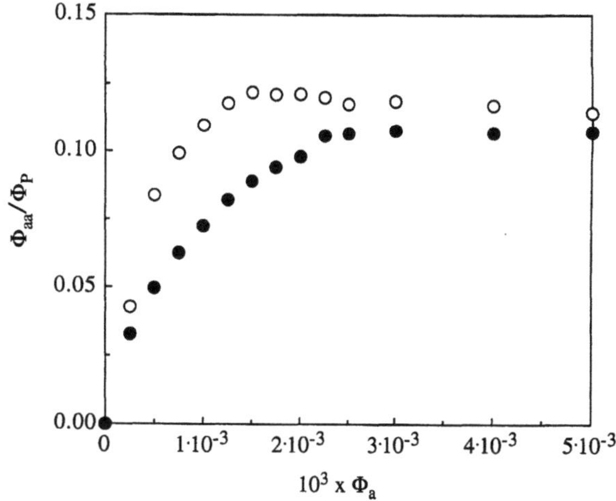

Fig. 2 Volume fraction of adsorbed SDS (relative to the particle volume fraction) as a function of the volume fraction of added surfactant, with no added salt (full dots) and in presence of 50 mM NaCl (open dots)

polarization detection, with the only modification of using a 50 mW frequency-doubled CW Nd–YAG laser source, operating at $\lambda = 532$ nm. All static measurements have been performed at a fixed angle $\theta = 90°$, while depolarized dynamic light scattering data were taken at $\theta = 30°$.

Figure 1 shows the intensity scattering by a suspension of FEP particles at $\Phi_p = 0.01$, as a function of the volume fraction of added nonionic surfactant Triton X-100 (molecular weight $M = 640$, specific volume $v = 0.916$ cm^3/g, refractive index $n_a = 1.492$). The open dots, which give the fractional amount of adsorbate calculated according to Eqs. (3) and (4) (right axis), show that the slow reduction of the scattering intensity which takes place after a much steeper increase is not due to a desorption effect, but rather to the above-mentioned change of the solvent refractive index due to the presence of the free Triton X-100. The adsorption plateau value, $\Phi_{aa} \approx 0.13\Phi_p$, which corresponds to a surface area per Triton X-100 molecule $A = 3Mv\ \Phi_p/(RN_A\ \Phi_{aa}) \approx 0.55$ nm^2, can be reasonably accounted for assuming that Triton X-100 arranges in a monomolecular uniform surface shell of thickness $d \approx 2$ nm [4, 5].

The adsorption curve in Fig. 1 strongly resembles a regular Langmuir isotherm, as we could expect from the adsorption of a noncharged species with strong surface affinity, and can be used to derive the adsorption constant of the nonionic surfactant for fluorinated surfaces [6]. However, things are quite different when we consider the adsorption properties of a charged amphiphile like the anionic surfactant sodium dodecyl sulfate (SDS). Figure 2 shows that, in absence of added salt, the rising part of the adsorption curve is strongly nonlinear, with a progres-

sively decreasing slope (full dots). This can be qualitatively accounted for by considering that, by increasing the surface coverage, the repulsion between the surfactant charged head groups also increases, hindering further adsorption. Screening of the electrostatic interactions by adding salt yields an adsorption curve which resembles more closely with the one found with the nonionic surfactant (open dots). Notice also that, in the presence of 50 mM NaCl, the plateau value is slightly larger, and corresponds to a reduction of something more than 10% of the surface area per head group. It should be mentioned that Ottewill and coworkers [1] have studied some years ago the adsorption of different surfactants on PTFE particles by using standard physico-chemical techniques. A detailed comparison with their data will be published elsewhere.

The effectiveness of our technique for the measurement of slight adsorption changes which may take place due to variations of thermodynamic parameters can be better appreciated by considering the results shown in Fig. 3 which presents the adsorption isotherms of Triton X-100 on FEP at different temperatures. As it is well known, Triton X-100 shows a miscibility gap with water, bounded by an upper critical consolute point at approximately 65 °C. It is interesting to check whether, approaching temperature values which yield phase separation of the surfactant–water mixture, the nonionic surfactant layer adsorbed on the colloidal particle undergoes some modification due to changes in the hydration of the surfactant hydrophilic group. The data in Fig. 3 indicate that, in a wide temperature range extending up to few degrees below the consolute curve of Triton X-100 + water, no

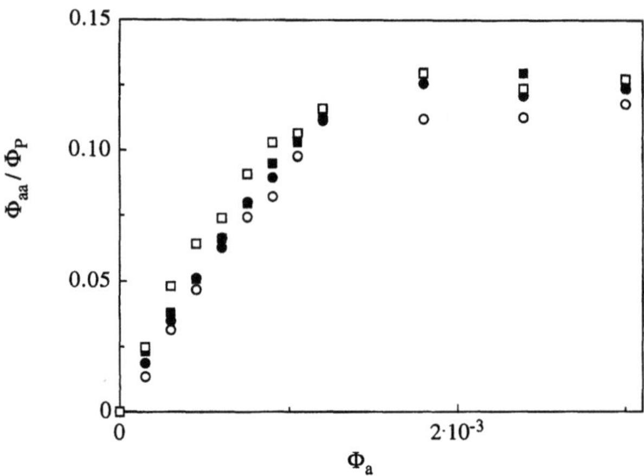

Fig. 3 Temperature dependence of the amount of adsorbed Triton X-100 as a function of the volume fraction of added surfactant. Full dots: $T = 15\,°C$. Open dots: $T = 40\,°C$. Full squares: $T = 50\,°C$. Open squares: $T = 60\,°C$

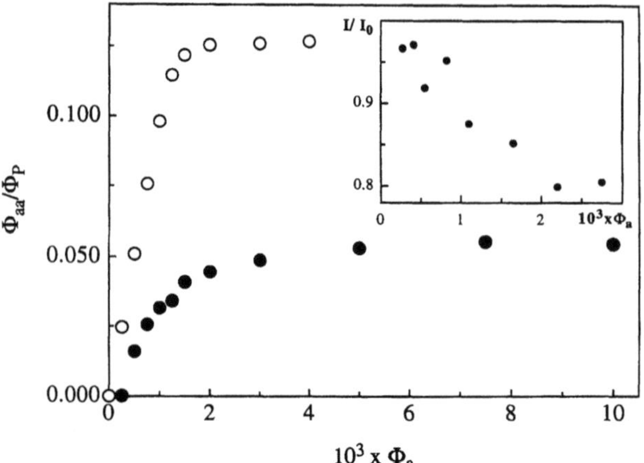

Fig. 4 Relative volume fraction of adsorbed Triton X-100 on FEP particles which have been previously covered with a full layer of the perfluorinated anionic surfactant Galden (full dots). For comparison, the open dots show the adsorption isotherm obtained in absence of added Galden. The inset shows the relative scattered intensity as a function of the volume fraction of added Galden, in absence of Triton X-100

relevant adsorption changes can be detected despite the reduction of the hydration of the hydrophilic part, suggesting that surface adsorption is mainly controlled by the steric hindrance of the hydrophobic backbone.

By playing with adsorbants which present different optical contrasts, it is possible to use static light scattering to investigate the competition between the adsorption of different species. We report in Fig. 4 the results obtained for the adsorption isotherm of Triton X-100 on the surface of FEP particles which have been previously coated with a full layer of Galden, a perfluorinated anionic surfactant. Galden competes strongly with Triton X-100, due to the strong affinity of fluorocarbon chains for PTFE compared to the affinity of hydrocarbons or even of the phenol group which is present in the Triton X-100 molecule. Galden has a refractive index which is quite similar to that of water ($n_G \approx 1.29$). Its adsorption on the surface of FEP leads then only to a slight decrease of the scattering power (see inset of Fig. 4), so that relevant changes of the scattered intensity are essentially due to the adsorption or desorption of Triton X-100. According to Fig. 4, in presence of Galden the maximum amount of adsorbable Triton X-100 is reduced by almost a factor three.

The time dependence of the depolarised scattered intensity from optically anisotropic particles is related to their rotational diffusion coefficient [3]. For suspensions of non-interacting spherical particles, the decay rate Γ of the field correlation function is indeed given by

$$\Gamma = 6D_R + D_T\,k^2$$

where $D_R = k_B T/8\pi\eta R^3$ and $D_T = k_B T/6\pi\eta R$, with η the solvent viscosity, are the particle rotational and translational diffusion coefficients. By performing depolarized dynamic light scattering measurements, D_R can be evaluated by extrapolating the decay rate to zero scattering wave vector (in practice, for particles with a radius smaller than 50 nm, the decay rate is dominated by the rotational contribution, and is almost angular independent). Notice that D_R has a much stronger dependence on the particle radius R than D_T. For instance, full adsorption of Triton X-100 on the surface of 40 nm particles, which yields a surface layer thickness of less than 2 nm, changes D_T by less than 5% (just a couple of times a realistic lower limit of resolution for dynamic light scattering), but decreases D_R by almost 15%. Such enhanced sensitivity can be exploited to detect the morphology of adsorbed polymer layers. The recently developed functional di-block and tri-block copolymers, composed by one or more soluble chains ("buoy") joined to one or more strongly adsorbing blocks ("anchor"), are much more versatile for the stabilization of colloidal lattices than conventional homopolymers or simple surfactants. They are also quite interesting for basic adsorption studies. It is particularly meaningful to consider the case when the "buoy", although still soluble, also has a finite affinity for the surface. By varying the amount of adsorbate, a peculiar structural modification of the surface layer morphology, known as "pancake-to-brush transition" is expected to take place [7]. For low amounts of adsorbate, the polymer chains are evenly spread on the particle surface, and the surface layer is thin. However, when the polymer surface concentration increases, the buoys are suddenly forced to pop out of the surface by their mutual steric repulsion. This kind of transition, which is expected to be first order, has indeed

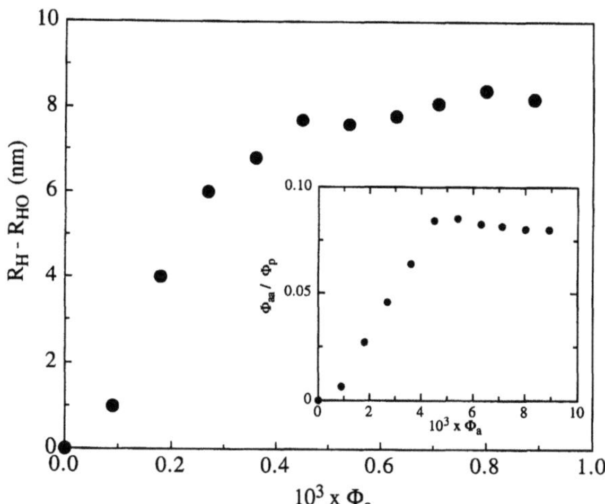

Fig. 5 Dynamic light scattering determination of the thickness of the adsorbed layer as a function of the volume fraction of added JELX block copolymer. R_H–R_{H0} represents the difference between the hydrodynamic radii of covered and bare FEP particles. The inset shows the adsorption isotherm obtained by static light scattering

formula $C_{16}H_{33}$–$(CH_2CH_2O)_{80}$–H, resembling a very long chain nonionic surfactant. For the depolarized dynamic light scattering measurements, FEP suspensions at a volume fraction $\Phi = 0.01$ have been prepared in an index-matched solvent, by adding 16.7% glycerol to water, in order to eliminate completely the coherent part of the polarized scattered intensity which could otherwise leak through the detecting crossed polarizer. The thickness of the adsorbed layer has been determined by measuring the decay rate of the correlation function of the depolarized scattered intensity at an angle $\theta = 30°$, which is sufficiently small to safely neglect any translational contribution.

The data for Fig. 5 show no evidence of abrupt changes either in the thickness of the adsorbed layer or in the amount of adsorbate. This failure in detecting any sign of surface transition can be due to various reasons. First of all, the block copolymers used in this study have a much shorter chain than those used in ref. [8]. This could shift the value of Φ_{aa} for the onset of the transition to quite low values. However, we believe that the main reason could be the presumably very weak affinity of perfluorinated polymers for the hydrophilic hydrogenated group, which could then protrude into the solution well from the start. We are presently trying to confirm these preliminary findings by studying the adsorption of longer chain copolymers.

been observed by Ou-Yang and Gao by adsorbing tri-block copolymers with two hydrophobic groups linked to the ends of a polyethylene oxide buoy, on the surface of polystyrene lattices [8]. We have recently started a similar investigation by using fairly monodisperse di-block copolymers composed by hydrocarbon anchors joined to polyethylene oxide buoys, specially synthesized by Union Carbide for the group of D. Ou-Yang in Lehigh University, and collectively denominated JELX copolymers. Figure 5 shows some preliminary results relative to 25-JELX-6, which has a molecular weight $M_w = 4342$ and chemical

Acknowledgements We thank very much M. Visca and D. Lenti, from Ausimont S.p.A., Milano, Italy, for the kind gift of the FEP lattices and for useful discussions. A special thanks to D. Ou-Yang who kindly gave us the JELX block copolymer used in this study. One of us (M.J. Eghan) benefitted from a fellowship by the International Atomic Energy Agency, Vienna. V. Peyre was a postdoctoral fellow within the frame of the "Colloid Physics" Network (Human Capital & Mobility Program) of the European Commission.

References

1. Bee HE, Ottewill RH, Rance DG, Richardson RA (1983) In: Ottewill RH, Rochester CH, Smith AL (eds) Adsorption from Solution. Academic Press, London, pp 155–171
2. Wu C (1994) Macromolecules 27: 7099–7102
3. Degiorgio V, Piazza R, Bellini T, Visca M (1994) Adv Colloid Polym Sci 48:61–91
4. Piazza R, Degiorgio V (1992) Opt Commun 92:45–49
5. Martin-Rodriguez A, Carbrerizo-Vilchez MA, Hidalgo-Alvarez R (1997) J Colloid Interface Sci 187:139–147
6. Bellini T, Degiorgio V, Mantegazza F, Ajmone Marsan F, Scarnecchia C (1995) J Chem Phys 103:8228–8237
7. Alexander S (1977) J Phys France 38:983
8. Ou-Yang D, Gao Z (1991) J Phys II France 1:1375–1385

Progr Colloid Polym Sci (1998) 110:94–98
© Steinkopff Verlag 1998

Lamellar composite magnetic materials

S. Lefebure
V. Cabuil
D. Ausserré
F. Paris
Y. Gallot
V. Lauter-Pasyuk

S. Lefebure · Dr. V. Cabuil (✉)
Université Pierre et Marie Curie
Laboratoire LI2C., Casier 63
4, Place Jussieu
75252 Paris Cedex 05
France
E-mail: cabuil@ccr.jussieu.fr

D. Ausserré · F. Paris
Université du Maine
Laboratoire LPSI
Le Mans
France

Y. Gallot
Institut Charles Sadron
CNRS
Strasbourg
France

V. Lauter-Pasyuk
Institut Laue Langevin
Grenoble
France
Joint Institute for Nuclear Research
Dubna, Russia

Abstract Synthesis of lamellar auto-assemblies of symetric PS-PBMA copolymers, doped by magnetic nanoparticles is described. As soon as particles are coated by a grafted PS layer, they can be confined in the PS layer of the polymeric smectic. The lamellar order is maintained up to volume fractions of particles of the order of 3% according to Atomic Force Spectroscopy. This volume fraction may be increased by addition of PS homopolymer in the structure.

Key words Smectics – copolymers – nanoparticles – magnetic particles – magnetic fluids – AFM

Introduction

Hybrid materials resulting from the association of lamellar meso-structures and inorganic magnetic nanoparticles are of great interest for applications: original magnetic or optical properties may be determined from such an association. On the theoretical point of view, it is always a tricky problem to mix two colloids, keeping their self-organization and their stability. Magnetic nanoparticles have soon been confined in lamellar lyotropic systems [1, 2]; usual processes to elaborate such hybrids is to mix the lyotropic system with a magnetic fluid, which is a colloidal dispersion of magnetic nanoparticles in water or in an oily solvent [3].

The lamellar system we consider here results from the organization of diblock symmetric copolymers PS-PBMA, where PS lies for polystyrene and PBMA for polybutyl-methacrylate [4]. Self-assembly is driven by the non-miscibility of the two chemical components. When the copolymer is symmetric, thin smectic films may be prepared by spin-coating a solution of the copolymer on a solid substrate and annealing the solid film obtained [5]. Minimization of surface energy at the boundaries dictates which block is present at the air/polymer and at the substrate/polymer interfaces.

We have recently described a procedure which allows to integrate magnetic nanoparticles exclusively inside the PS layers of such a smectic film [6]. The main parameters controlling the structure of the composite we obtain are the nature of the chains coating the particles in order to include them in the PS layers and avoid their aggregation, and the parameters of the particles' size distribution. It has been established that particles have to be coated by polystyrene chains in order to be confined in the polystyrene layers. We shall use in the present work monosized maghemite nanoparticles of different diameters and different length for the polystyrene coating in order to introduce in the composite as many particles as possible.

Experimental

The diblock copolymer is a polystyrene–polybutylmethacrylate symmetric copolymer of molecular weight $M_w = 135\,000$ g/mol, and a polydispersity index $M_w/M_n = 1.05$. The weight fraction of PS is 0.49.

The polymer adsorbed on nanoparticles is an α-lithium polystyrene sulfonate (denoted as PSMS) obtained as described in Ref. [7]. The length of the polystyrene chains are 9000, 4000 or 1500 g/mol.

The magnetic nanoparticles

Particles are maghemite grains (γ-Fe$_2$O$_3$). Colloidal magnetite is synthezised according to Massart's method [8], then oxidized to maghemite, in acidic medium, with ferric nitrate. The particles obtained have cationic surface charges in acidic aqueous solution, which ensure electrostatic repulsions and allow to get a stable dispersion in water, commonly called ferrofluid [9]. Particle size distribution is usually well described by a log-normal law:

$$p(D) = \frac{1}{\sqrt{2\pi}\sigma D} \exp\left[-\frac{1}{2\sigma^2}\left(\ln\frac{D}{D_0}\right)^2 \right],$$

where D_0 is the mean diameter and σ the standard deviation. The synthesis leads to polydisperse samples and a size sorting procedure, based on the thermodynamic properties of aqueous dispersions of nanoparticles, has to be performed in order to obtain from the initial dispersion several fractions of almost monodisperse particles ($\sigma = 0.1$) [10]. We have used particles of diameters 6, 8.5 and 11 nm.

At the end of the size sorting procedure, the nanoparticles are cationic and dispersed in nitric acid solutions. They are coated by polystyrene chains according to the procedure described in [7] and dispersed in toluene, in order to obtain a stable magnetic dispersion. The volume fraction Φ of ferric oxide in the suspension is determined from light absorption at 480 nm. The amount of polystyrene chains adsorbed on the particles or eventually dispersed in the solution is determined from chemical titration of carbon, after elimination of the solvent at the Centre Régional de Microanalyse de l'Université Paris VI.

Synthesis of the composite smectic films

Stable solutions of copolymer in toluene and stable magnetic fluids in toluene are separately synthesized. They are mixed in the required proportions prior to spin coating. These proportions fix the volume fraction of particles inside the structure. Volume fractions of polymer and particles are typically of the order of 10^{-2}. The mixture is deposited on a smooth silicon wafer at a typical rotating speed of 2000 rds/min. The film obtained is annealed 48 h at 150 °C under vacuum.

Characterization of the composite smectic films

The films are firstly characterized by Atomic Force Microscopy (AFM) using a Dimension 3000 nanoscope (Digital) working on the tapping mode.

Neutron Specular Reflexion has been used in order to give details of the films structure. NSR experiments have been performed on the spectrometers D17 (ILL, Grenoble) and SPN (JINR, Dubna). Details are given in Ref [11]. For these experiments fully deuterated PS has been used.

Results

For each sample, incorporation of particles in the PS sequences of the lamellar PS/PBMA smectic was possible as soon as the coating of particles is ensured by a polystyrene chain, whatever its length. Figure 1 is an AFM picture of a smectic film into which we have tried to introduce maghemite particles coated by an usual surfactant: the particles are ejected from the lamellar structure and "volcanos" are observed. Figure 2 corresponds to the case where particles are coated by a polystyrene layer: the surface of the sample does not present any "volcano" anymore, but a regular steps structure is observed, which indicates that the lamellar order is kept. For the non magnetic PS-PBMA smectic, the thickness of the steps allows to deduce the lamellar periodicity (Fig. 3), which is found, for the 135 000 g/mol copolymer, of the order of 30 nm. On Fig. 2, the roughness of the surface and of the steps indicates the presence of particles in the film, and the lamellae thickness cannot be determined anymore.

Whatever the length of the polystyrene layer, the particles have always being dispersed in toluene and included in the film without any difficulty. It is also the case what

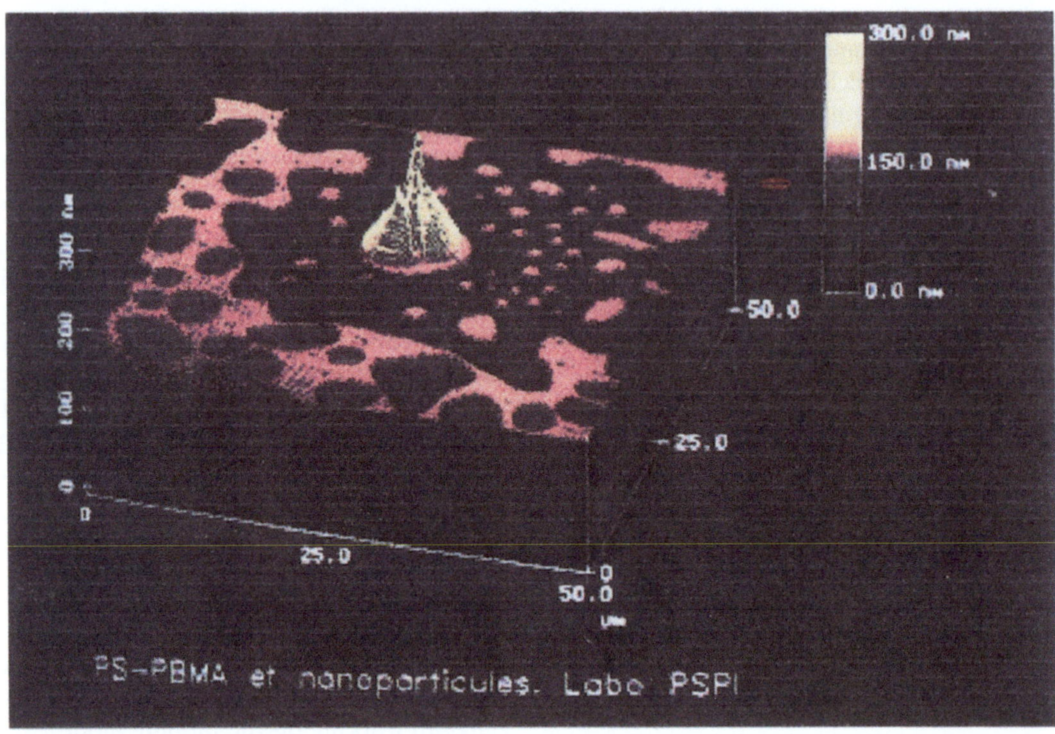

Fig. 1 AFM picture of a PS-PBMA lamellar film, which has been doped by surfactant coated magnetic nanoparticles: the presence of volcanos indicates that particles are ejected from the structure

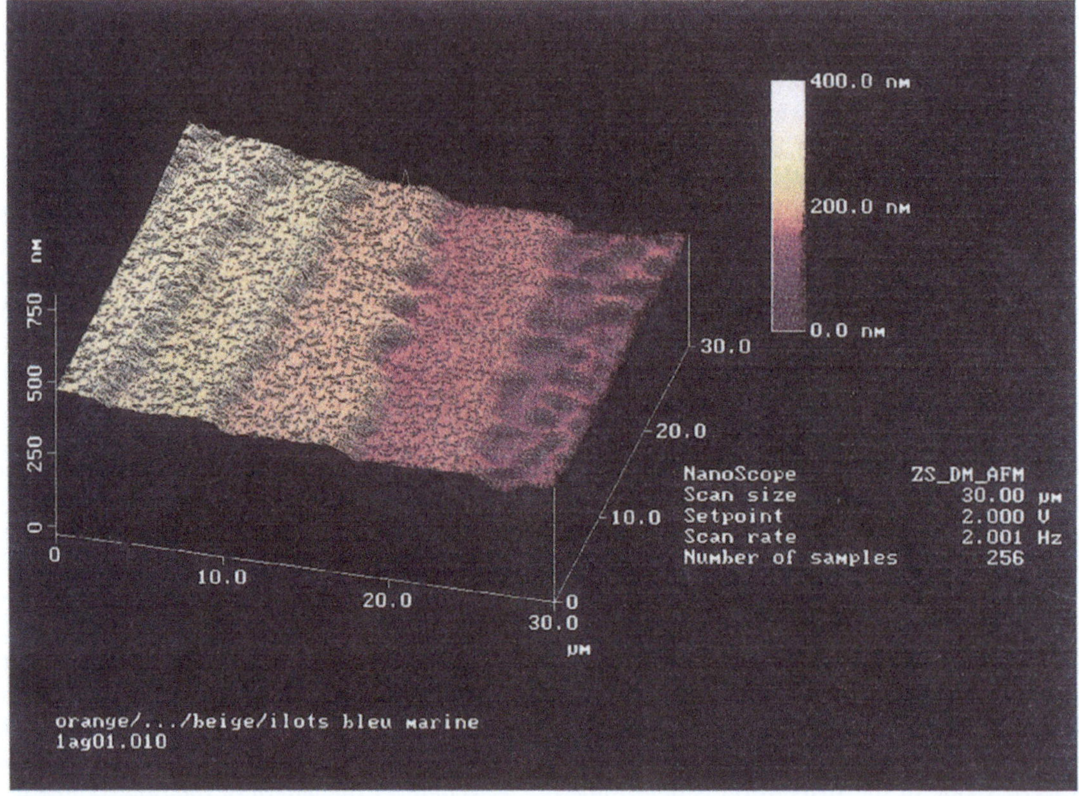

Fig. 2 AFM picture of a PS-PBMA lamellar film, into which PS (9000 g/mol) coated particles (diameter 6 nm) have been introduced. The steps characterize the lamellar order. The roughness of the surface is due to the presence of particles

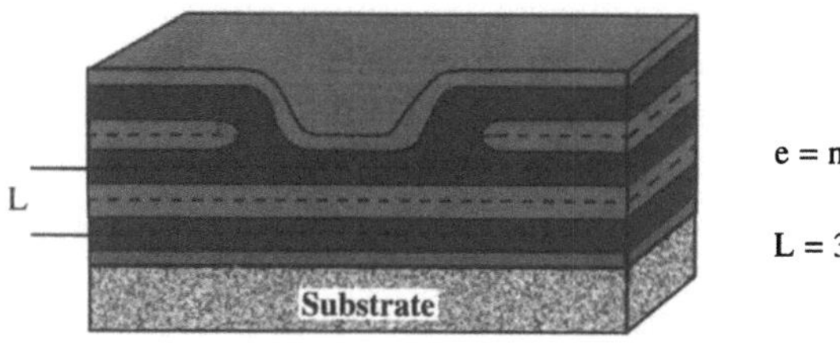

Fig. 3 Schema representing the lamellar organization in a PS-PBMA smectic film. The lamellar period is denoted L

ever the particles diameter (between 6 and 11 nm). The volume fraction of magnetic oxide that may be incuded in the composite without loosing lamellar organization is around 1% for 6 nm particles and may reach 3% for 11 nm particles. For greater volume fractions, particles stay included in the film, but this latter progressively looses its lamellar organization.

Location of particles in the structure has been investigated using Neutrons Specular Reflection. Former studies concerning polydisperse systems of very small particles (average diameter around 4 nm) have shown that particles were located at the boundaries between the PS and the PBMA layers [12]. It is no more the case for greater particles as illustrated by Fig. 4: a best fit is obtained for particles located at the center of PS lamellae.

It appeared that the addition of homopolymer PS of the same weight than the one constituting the copolymer allows to increase the amount of particles that can be introduced in the composite. In the case of 11 nm particles, coated by PS 1500 g/mol, it has been possible to increase the volume fraction of magnetic nanoparticles in the com-

posite up to 7% by adding 12% (volume fraction) of PS 60 000 g/mol.

Discussion

Noticeable magnetic or optical behaviors may be expected from the introduction of magnetic nanoparticles in PS-PBMA lamellar structures provided a sufficient amount of particles may be introduced in the PS lamellae and if particles are big enough. As a matter of fact, each maghemite nanoparticle is a magnetic monodomain which has a permanent magnetic moment of intensity proportional to the particles' volume [10].

Thus, in order to enhance the magnetic properties of the composite, we increased the particle diameter up to 11 nm and reduced the length of the PS chain coating the particles down to a molar mass of 1500 g/mol. Anyway whatever the particle size and the chain length, it was always possible to introduce particles in the structure but the amount of magnetic material that can be confined in the PS lamellae is in fact sterically limited: a simple calculation assuming particles coated by a PS layer of mass 4000 g/mol , arranged on an hexagonal lattice leads to a maximal density of magnetic oxide for a monolayer of 1.42% for 6 nm particles and 4.51% for 11 nm particles. Experimentally, for this coating, we found about 1% for 6 nm particles and 3% for 11 nm, so, as these values are just orders of magnitude, they are not inconsistent with the assumption of particles, surrounded by their PS layer, confined, at the contact, as a monolayer in the center of PS layers.

The presence of particles in PS layers induces local deformations in the film and thus modify the energy

Fig. 4 Experimental reflectivities for d-PS-PBMA films with $M_\mathrm{w} = 135\,000$ g/mol. the open circles corresponds to a pure copolymer film, the filled triangles correspond to the reflectivity from a similar copolymer film but with 6 nm nanoparticles incorporated in the PS lamellae. The solid lines show the fit to the experimental data. The neutron scattering length density profile obtained from the fit to the experimental reflectivity from the sample with nanoparticles is depicted in the insert

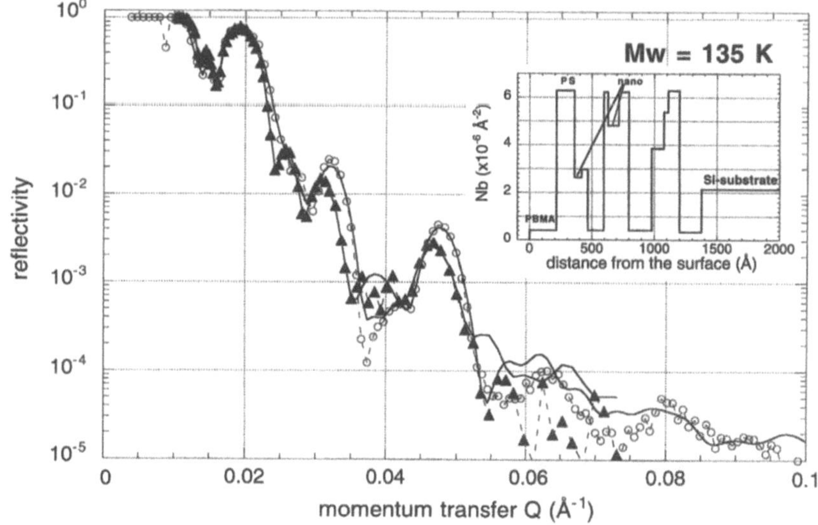

but this latter was inefficient, although addition of PS allowed to increase singnificantly the volume fraction of particles that can be introduced in the PS layers up to 3% for 6 nm particles and up to 7% for 11 nm particles. Our interpretation of this result is illustrated by Fig. 5b. The homopolymer relaxes the stress of the film "filling in" the "holes" between particles. This can explain the fact that the length of PS chains coating particles has little influence on the amount of particles that can be introduced. Long PS chains may play theirselves the role of the homopolymer.

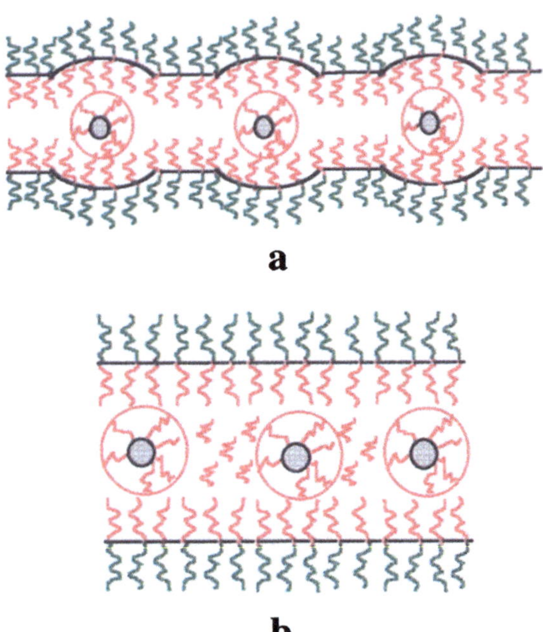

a

b

Fig. 5 Schema representing (a) the stress of the PS-PBMA interface, due to inclusion of nanoparticles and (b) the role of the PS homopolymer to relax these stress

balance creating an excess of interfacial energy. This is very schematically illustrated by Fig. 5a. Introduction of an homopolymer in the system may compensate the film deformations. We thought firstly to addition of PBMA,

Conclusion

These results are obviously preliminary ones. Nevertheless, this work has established that it was possible to introduce inside PS-PBMA lamellar structures monodisperse magnetic particles, coated by a PS layer of different thickness. Particles diameter has been increased up to 11 nm without any problem. Neutron Specular Reflection is a technic well suited for the characterization of the film. These "big" monodisperse particles appear to be located at the center of the PS lamellae, but now swelling laws have to be established and interprated in terms of stress on the interfaces. We propose an "hand-made" interpretation for the effect of an addition of PS homopolymer in the structure. This effect is impressive, but need also to be precised by Neutron Specular Reflection.

Reference

1. Fabre P, Casagrande C, Veyssie M, Cabuil V, Massart R (1990) Phys Rev Lett 64(5):539–542
2. Ménager C, Belloni L, Cabuil V, Dubois M, Gulik-Krzywicki T, Zemb Th (1996) Langmuir 12(14):3516–3522
3. Rosensweig R (1985) Ferrohydrodynamics. Cambridge Univ Press, Cambridge
4. Henkee CS, Thomas EL, Fetters LJ (1988) J Mater Sci 23:1685
5. Coulon G, Russel TP, Deline V, Green PS (1989) Macromol 22:2581
6. Hamdoun B, Ausserré D, Cabuil V, Joly S, Gallot Y, Clissard C (1996) J Phys II France 6:494–501
7. Cabuil V, Hochart N, Perzynski R, Lutz PJ (1994) Progr Colloid Polym Sci 97:71–74
8. Massart R (1981) IEEE Trans Magn 17:131
9. Bacri J-C, Perzynski R, Salin D, Cabuil V, Massart R (1990) J Magn Mater 85:27
10. Dubois E, Cabuil V, Massart R, Hasmonay E, Perzynski R (1995) J Magn Magn Mat 149:1–5
11. Lauter-Pasyuk V, Lauter H J, Ausserré D, Gallot Y, Cabuil V, Kornilov EI, Hamdoum B. Physica B, accepted
12. Lauter-Pasyuk V, Lauter H J, Ausserré D, Gallot Y, Cabuil V, Kornilov EI, Hamdoum B. Physica B, accepted

Progr Colloid Polym Sci (1998) 110:99–104
© Steinkopff Verlag 1998

B. Schlarb
W. Heckmann
E. Schwarzenbach

Multihollow micron-sized latex particles from self-emulsifying polymer blends

Dr. B. Schlarb (✉) · Dr. W. Heckmann
Dr. E. Schwarzenbach
BASF AG
Kunststofflaboratorium
ZKD/B-B 1
D-67056 Ludwigshafen
Germany

Abstract This paper describes a new process for the preparation of micron-sized latex particles with multihollow structures. It is based on blends of solution polymers, which are self-emulsifying in water and form secondary dispersions. Their chemical composition leads to the formation of a multiple W/O/W-emulsion when the highly viscous polymer solution is emulsified in water. The particle size of the multiple emulsion droplets is in the micron range. The mechanism for the formation of the structures as described above was proven by means of fluorescence microscopy: by staining the aqueous phase using a water-soluble fluorescent dye it could be shown that the microvoids are initially formed by inclusions of the aqueous phase and not of organic solvents. These structures are not thermodynamically stable. They can be stabilized by removal of the organic solvents, which leads to "multiple secondary dispersions". Structured polymer particles containing voids as described here could be interesting as white pigments, since they have the ability to scatter visible light.

Key words Multihollow latex particles – multiple emulsion – opacifier – plastic pigment – polymer pigment – secondary dispersion

Introduction: Void containing polymer particles

Hard latex particles with interior cavities that, upon drying, create voids are useful as opacifiers ("polymer or plastic pigments") for paints and coatings [1–3] or paper coatings [4]. Visible light is scattered efficiently by voids having diameters in the range of 200–800 nm [5]. Therefore, the aim of the studies described in this paper was the preparation of micron-sized polymer particles containing many voids of the desired range. Some procedures for the preparation of multihollow polymer particles have been described already. The stepwise treatment of emulsion copolymers with alkali and acid (or subsequent cooling) can lead to submicron-sized multihollow particles [6–10]. Micron-sized cross-linked particles having multihollow structures can be prepared by extracting polystyrene from composite polymer particles with toluene under reflux [11]. A different approach is based on the polymerization of monomers forming the oil phase of a W/O/W-emulsion. These "multiple emulsions" have to be stabilized either by emulsifiers [12] or by block copolymers [13].

In this paper a new process for the preparation of micron-sized latex particles with multiple voids by a simple process using standard building blocks is described.

Experimental details

Materials

Acrylic acid, styrene, *n*-octane, toluene and 25% strength by weight aqueous ammonia solution (BASF AG, technical grade), fluorescein (Riedel-de-Haen) as well as *tert.*-butyl peroctoate (*tert.*-butyl-per-2-ethyl hexanoate, INTEROX TBPEH, Peroxid-Chemie GmbH) were used without further purification.

Preparation of the secondary dispersions with interior cavities

The following feeds were prepared:

Feed 1: 75 g acrylic acid, 425 g styrene,
feed 2: 500 g styrene,
feed 3: 20 g *tert.*-butyl peroctoate, 161 g toluene, 69 g *n*-octane,
feed 4: 71 g aqueous ammonia solution (25 wt%),
feed 5: 1.000 g demineralized water.

A mixture of 119 g toluene and 51 g *n*-octane was charged into a 4 l reactor, which was equipped with a reflux condenser, anchor stirrer and 2 feeds. 182 g of feed 1 was added to the charge and heated under nitrogen to the polymerization temperature of 105 °C. After the temperature was reached, 15 g of feed 3 were added all at once. Thereafter, the remaining amount of feed 1 (318 g) and 15 g of feed 3 were added continuously via separate feeds in the course of 45 min while maintaining the polymerization temperature. The conversion was completed by stirring for an additional 15 min. Then 135 g of feed 2 and 30 g of feed 3 were added over 2 h. After the feed had ended, polymerization was completed by stirring for a further hour. Then the remaining amount of feed 2 (365 g) and a further 130 g of feed 3 were added to the reaction mixture at constant polymerization temperature in the course of 3 h. Thereafter, the remaining amount of feed 3 was added over 2 h and polymerization was continued for 2 h.

The polymer solution was cooled to 65 °C. The following steps were carried out at a constant temperature of the oil bath, which was kept at 70 °C. Feed 4 was added over 30 min. Thereafter, the aqueous dispersion was formed by the addition of feed 5 over a 1 h period. The solvent mixture was then separated off by distillation under reduced pressure to a solvent content of less than 0.5% by weight, the amount of water simultaneously removed being added again (835 g organic solvent and water were removed and 1200 g fresh water were added).

The aqueous multiple secondary dispersion thus obtained had a solids content of 32.9 wt% and pH of 8.1.

For the preparation of the fluorescent dye-containing dispersion, 0.5 mmol of fluorescein were dissolved in feed 4. Under these conditions the fluorescent dye is soluble only in the aqueous and not in the organic phase, as could be shown by a simple visual test.

Scanning electron microscopy

For scanning electron microscopy a drop of dispersion was placed on a small glass slide and mounted on an aluminium sample holder. After room-temperature drying the sample was sputter-coated by a thin gold layer in order to prevent electrostatic charging. The sample was examined with a Hitachi S 400 FE field emission scanning electron microscope.

Transmission electron microscopy

Transmission electron microscopy was performed with a Hitachi H-7100 using an accelerating voltage of 100 kV. Monolayers of particles were prepared by dropping the diluted dispersion on collodion-covered copper grids. No staining was required.

Fluorescence microscopy

The dye-containing dispersion was diluted with water and dropped onto a millicarb filter (PC 0.8 μm). After washing with distilled water the wet filter was mounted between a microscope slide and a coverslip. The fluorescence of the sample was imaged with an oil immersion objective of high numerical aperture (40/1.3) by means of a confocal laser scan microscope (Leica CLSM). Excitation was achieved by irradiation with blue light (488 nm) and on the emission side a longwave pass filter (OG 515) was used.

Gel permeation chromatography

The GPC was carried out at 23 °C, using tetrahydrofuran as solvent. The flow rate was 1 ml/min. For the calibration a poly(styrene) standard (Polymer Laboratories) was used.

Viscosity measurements

The viscosity measurements were carried out using a sample of the organic polymer solution, which was taken before the addition of feed 4. The solids content was 73.8 wt%. For the measurements in the low shear region a shear stress controlled rotational viscometer (Bohlin CS, $R = 20$ mm, angle $= 1°$) was used. At higher shear rates, the measurements were carried out by means of a capillary viscometer (Wolpert KVM, $L/R = 149$, $L = 150$ mm, $R = 1.01$ mm).

Light reflectance

25 g of a pigment dispersion (solids content $= 20$ wt%) was mixed with 25 g of Propiofan® DS 6014 (solids content $= 30$ wt %) as a binder and 5 g of Moviol® 30–92 (solids content $= 10$%) as a thickener. A film (250 μm thick when wet) was applied onto a glass plate and dried under normal conditions. The intensity of reflected visible light was measured by means of an electrical remission photometer (Zeiss ELREPHO).

Results and discussion

The multihollow latex particles described in this paper are prepared by the emulsification of polymer blends in water. The polymer blends consist of two polymers, which differ in their polarity. One is a copolymer containing salt groups and has the function of a protective colloid. The other is a hydrophobic polymer without salt groups. It was shown in an earlier work that such mixtures of polymers are self-emulsifying in water, forming emulsifier-free secondary dispersions [14, 15]. In contrast to these systems, the secondary dispersions described here have a different polymer composition, leading to larger particle diameters. Furthermore, organic solvents with a low solubility in water were used, namely a mixture of toluene (70 wt%) with n-octane (30 wt%).

The process for the preparation of multiple void containing secondary dispersions is depicted schematically in Fig. 1.

In a first step, the polymer blend is prepared by polymerizing two consecutive monomer feeds in organic solution. The first feed contains acidic monomers like acrylic acid together with hydrophobic monomers like styrene. The acidic monomers are needed for the formation of hydrophilic salt groups, which allow the blend to be emulsified in water without the addition of emulsifiers. The glass transition temperature of the whole system should be well above room temperature to allow for the stabilization of the structure, and styrene was therefore chosen as the hydrophobic building block. Pure styrene was used as the second feed. The resulting polymer mixture was neutralized by the addition of an aqueous ammonia solution. After this, water was added slowly to the agitated, highly viscous polymer solution. A solvent-containing multiple emulsion was formed. Thereafter, the organic solvents were removed by distillation in vacuo, which led to an aqueous multiple secondary dispersion.

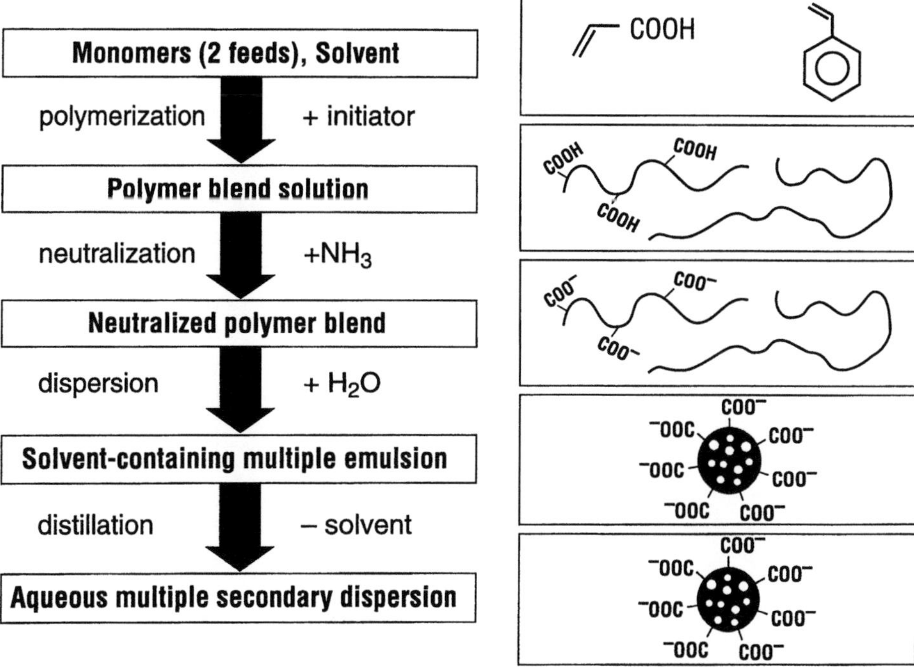

Fig. 1 Schematic representation of the preparation of multihollow micron-sized latex particles from self-emulsifying polymer blends

The formation of the desired multihollow structures occurs during the emulsification step, as shown schematically in Fig. 2.

Since the organic solvents used for the solution polymerization are not miscible with water, a W/O-emulsion is formed during the addition of water to the system. When the phase-inversion point is reached, water becomes the continuous phase. Since the viscosity of the organic phase is very high, the phase separation within the emulsion droplets is not complete and a W/O/W-emulsion is formed. The structure of these multiple emulsions can be observed by means of a light microscope. For a fixation of the multiple emulsion structure the organic solvents have to be removed.

As a typical example, in Fig. 3 a scanning electron micrograph of a dispersion, which had been prepared by this process, is shown. An image of the same sample, which was obtained by transmission electron microscopy, is shown in Fig. 4. The particles have a broad size distribution and most of them are larger than one micrometer. All particles have many voids. The diameters of these voids are in the range of the wavelength of visible light.

Are the interior cavities formed by the aqueous or the organic phase? For the confirmation of the mechanism given in Fig. 2 a simple test, based on fluorescence microscopy, was carried out: the aqueous phase was stained by the addition of the water-soluble fluorescence dye fluorescein. This dye was added to the aqueous ammonia solution which was used for the neutralization of the acidic groups of the polymer. Additional water, which was used for the emulsification, did not contain the fluorescent dye. After removal of the organic solvents, the resulting dispersion was observed by means of a transmission electron

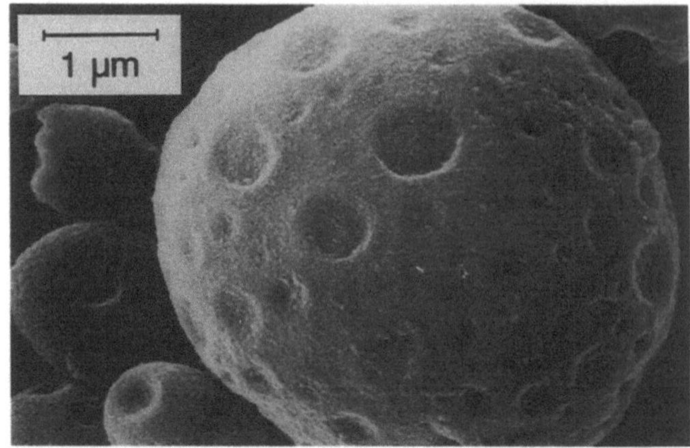

Fig. 3 Scanning electron micrograph of multihollow latex particles (sample A)

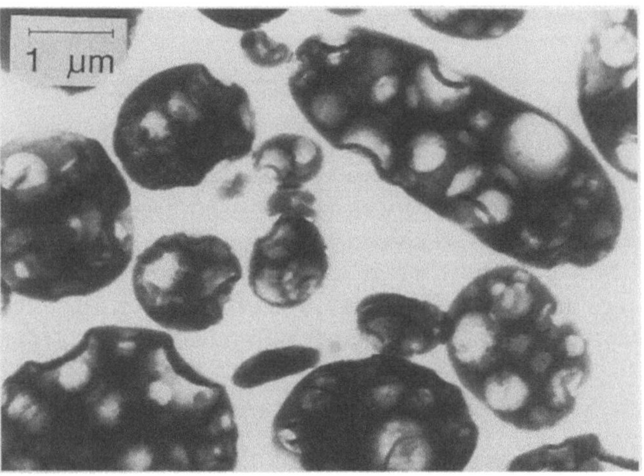

Fig. 4 Transmission electron micrograph of multihollow latex particles (sample A)

Fig. 2 Formation of interior cavities by incomplete phase inversion

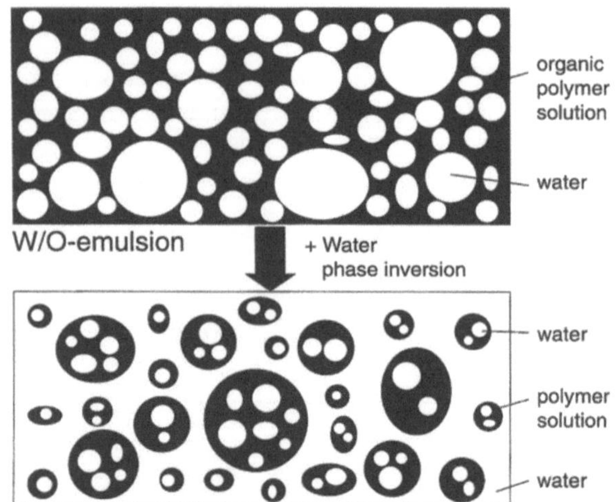

microscope as well as by a light microscope under fluorescence conditions, as depicted in Figs. 5 and 6.

The electron micrograph shown in Fig. 5 makes clear that the fluorescent dye does not disturb the formation of multihollow structures. The fluorescence microscopic investigation of this sample was done by means of a confocal laser-scan microscope. In Fig. 6, one single multihollow particle is shown. Fluorescent light is emitted from the interior cavities with high intensity. This is taken as proof that the cavities contain the aqueous phase, since the fluorescein ammonium salt is not soluble in the organic solvents, as was shown by a simple visual test. This fact supports the mechanism given in Fig. 2.

The formation of a multiple emulsion from the organic polymer solution does not lead to a thermodynamically

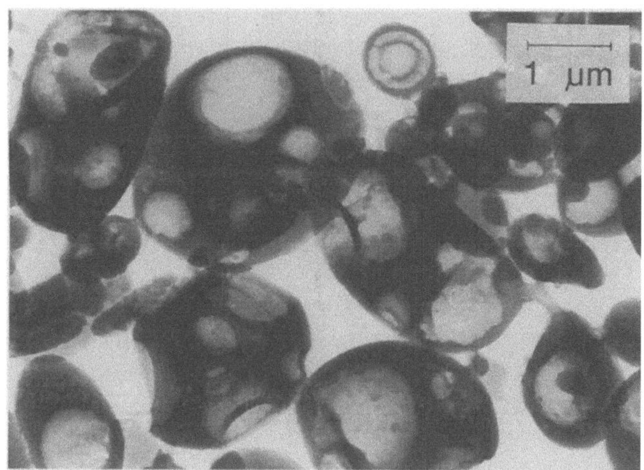

Fig. 5 Transmission electron micrograph of multihollow latex particles (sample B, with fluorescein)

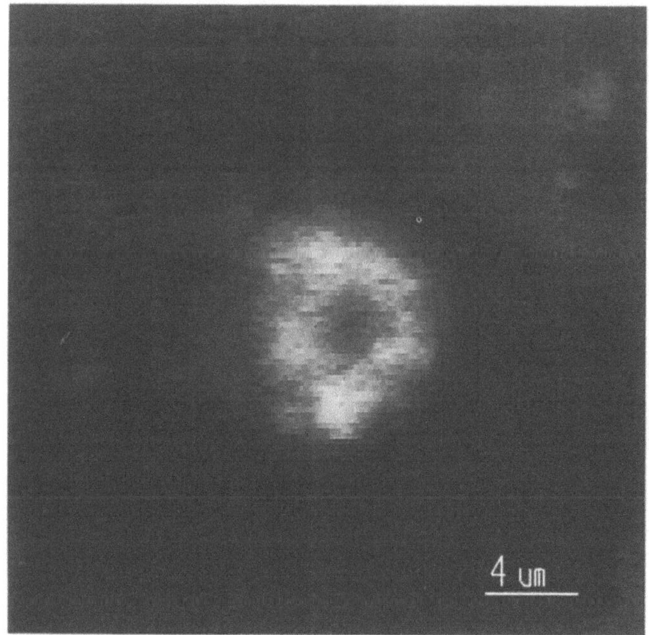

Fig. 6 Optical micrograph (fluorescence conditions) of a latex particle with interior cavities that contain fluorescein

stable state. The high viscosity of the organic (oil) phase must favor the stability of the multiple structure, since the phase separation becomes slower with increasing viscosity. The temperature during the dispersing and distillation steps has therefore also an influence on the structure of the latex particles. Using light microscopy it was found that the solvent-containing multiple emulsions must be cooled down to less than 50 °C to prevent disappearance of the

multiple structure. The fact that the inclusions are visible before the vacuum distillation is carried out demonstrates that the organic solvents are not acting as a blowing agent. The multiple structure is frozen by the separation of the organic solvents, resulting in a "multiple secondary dispersion".

The molecular weight and the viscosity of the organic polymer solution at different temperatures and shear rates was measured, since it seems to be very important for the creation of the multiple secondary dispersions.

The molecular weight of the polymer blend of sample A, as determined by gel permeation chromatography, was found to be 42 000 (M_w) and 11 200 (M_n). The solution viscosity in the range of 60–100 °C was measured at different shear rates. The result is depicted in Fig. 7.

The viscosity of the organic polymer solution is in the range of 10–300 Pa s. Figure 8 shows the relative light scattering ability of the multiple secondary dispersion (sample A) in comparison to calcium carbonate (hydrocarb 90) and titanium dioxide. The value of 74% for the polymer pigment compares well with titanium dioxide.

For the preparation of the samples as shown in Figs. 3–6, the solvent mixture used for the solution polymerization consisted of 70 wt% toluene and 30 wt% n-octane. Further experiments were carried out with different ratios of these solvents. Small changes did not affect the formation of multihollow structures. The use of a solvent mixture consisting of 65 wt% toluene and 35 wt% n-octane also resulted in a multihollow structure. Much higher portions of n-octane are unsuitable, since the polymer precipitates during polymerization. If pure toluene is used, the latex particles have a smaller particle size. These particles were not characterized in more detail.

Fig. 7 Viscosity of the organic polymer solution at different temperatures and shear rates

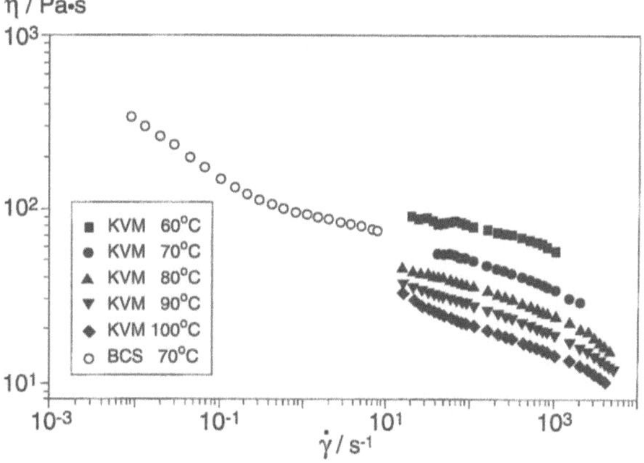

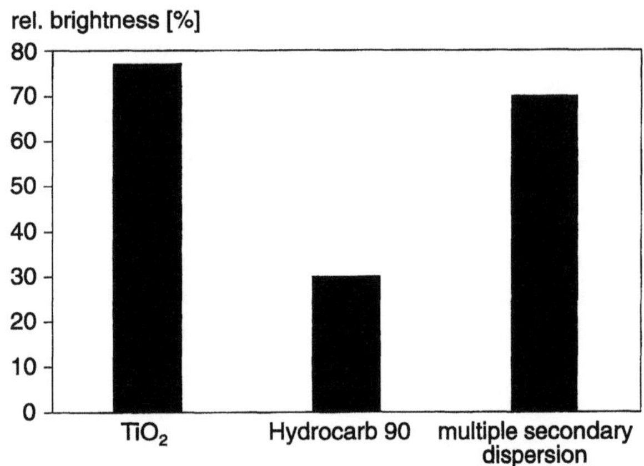

Fig. 8 Comparison of different pigments. Intensity of reflected visible light as determined by means of an electrical reemission photometer

Besides the organic solvents, the composition of the polymer blend was also varied. Originally, the blend consisted of 50 wt% copolymer (15 wt% acrylic acid, 85 wt% styrene) and 50 wt% of the homopolymer (poly(styrene)).

A smaller portion (40 wt%) of the same copolymer also allows the formation of multiple secondary dispersions. If the portion of the copolymer containing salt groups is increased to 60 wt%, a strong foaming during the distillation causes problems.

Conclusions

The emulsification of certain self-emulsifying polymer blends in water leads to the formation of emulsifier-free multiple emulsions. A blend of solution polymers which is well suited for this purpose consists of a copolymer containing salt groups (50 wt%) and poly(styrene) (50 wt%). The copolymer was composed of acrylic acid (15 wt%) and styrene (85 wt%). The best results could be obtained using a mixture of toluene (70 wt%) with n-octane (30 wt%) as a solvent for this process. The multihollow structure can be frozen by the removal of the organic solvents. The polymer (plastic) pigments obtained by this procedure scatter visible light with a high intensity.

References

1. Rennel C, Rigdhal M (1994) Colloid Polym Sci 272:1.111–1.117
2. Fasano DM (1987) J Coat Technol 752:109–116
3. Andrew RW, Dowling DG, Grange B (1986) Polymers Paint Colour J 176:567–572
4. Hemenway CP, Latimer JJ, Young JE (1985) Tappi J 68:102–105
5. Ross WD (1974) Ind Eng Chem Prod Res Develop 13:45–49
6. Okubo M, Ichikawa K, Fujimura M (1991) Colloid Polym Sci 269:1.257–1.262
7. Okubo M, Ito A, Hashiba A (1996) Colloid Polym Sci 274:428–432
8. Okubo M, Ito A, Kanenobu T (1996) Colloid Polym Sci 274:801–804
9. Okubo M, Ito A, Nakamura M (1997) Colloid Polym Sci 275:82–85
10. Okubo M, Nakamura M, Ito A (1997) J Appl Polym Sci 64:1.947–1.951
11. Okubo M, Nakagawa T (1994) Colloid Polym Sci 272:530–535
12. Delgado J (1990) United States Patent 4,968,562
13. Riess G (1993) Makromol Chem Macromol Symp 69:125–131
14. Schlarb B, Rau M G, Haremza S (1995) Prog Org Coat 26:207–215
15. Schlarb B, Haremza S, Heckmann W, Morrison B, Müller-Mall R, Rau MG (1996) Prog Org Coat 29:201–208

Progr Colloid Polym Sci (1998) 110:105–109
© Steinkopff Verlag 1998

A. Schmitt
A. Fernández-Barbero
M.A. Cabrerizo-Vílchez
R. Hidalgo-Álvarez

On the identification of bridging flocculation: An extended collision efficiency model

A. Schmitt (✉) · M.A. Cabrerizo-Vílchez
R. Hidalgo-Álvarez
Grupo de Física de Fluidos y Biocoloides
Departamento de Física Aplicada
Universidad de Granada
Campus de Fuentenueva
E-18071 Granada
Spain

A. Fernández-Barbero
Grupo de Física de Fluidos Complejos
Departamento de Física Aplicada
Universidad de Almería
Cañada de San Urbano s/n
E-04120 Almería
Spain

Abstract A new generalized collision efficiency model for bridging flocculation was developed. It comprises the La Mer, the Moudgil and the Molski models. The extended model allows to separate and quantify the contributions of the different aggregation mechanisms (slow coagulation, weak flocculation and bridging flocculation). The model was tested for polystyrene particles covered with different amounts of bovine serum albumin (BSA). Single-cluster light scattering was employed to study the aggregation kinetics and to obtain the initial aggregation rate constant. It was found that the simple La Mer model cannot explain the experimental results whilst the Molski model gives reasonably good results for fast but fails for slow aggregation processes. Only the extended model works for slow and fast aggregation processes.

Key words Colloidal aggregation – light scattering – bridging flocculation – single-particle detection – collision efficiency

Introduction

Colloidal aggregation may be controlled by a wide variety of different mechanisms. The addition of salt to an electrostatically stabilized colloidal dispersion leads to a compression of the electric double layer and therefore, to salt-induced coagulation. The stability of the colloid may also be affected by macromolecules which irreversibly absorb onto the surface of colloidal particles. At high surface coverage, steric effects usually impede flocculation and ensure an increased colloidal stability. Neverthless, some weak flocculation may occur even for totally covered particles. At lower surface coverage, *bridges* of macromolecules may form between the particles and, thus, give rise to bridging flocculation [1–3].

Bridging flocculation takes place only when a covered part of the surface of one particle collides with the uncovered part of another particle. In this case, the aggregation rate should depend on the degree of surface coverage θ, i.e. the fractional coverage of the particle surface by adsorbed molecules. La Mer assumed the rate k_S of pure bridging flocculation to be proportional to the number of free sites on one particle and the number of occupied sites on the other. He postulated [4–6] that

$$k_S \sim \theta(1 - \theta) \,. \tag{1}$$

This relationship implies that maximum flocculation should occur at half surface coverage ($\theta_{\max} = 50\%$) and no flocculation at all for uncovered and totally covered particles, respectively. This simplified approach to bridging flocculation, however, cannot be entirely true since it assumes that bridging flocculation is the only mechanism of aggregation. It does not take into account that many dispersed systems are significantly unstable in the absence of absorbed molecules (coagulation) as well as when particles are completely covered (weak flocculation) [7]. More detailed models were proposed by Hogg [8], Ash and Clayfield [9], Moudgil [10] and Molski [7]. Nevertheless, all of them show the common feature that bridging flocculation is most efficient at some intermediate surface coverage. This criterion is widely used to identify bridging flocculation. Nevertheless, we will show that bridging flocculation might be present even when no maximum is observed. For this sake we present an extended collision efficiency model which allows to quantify the contribution of the different aggregation mechanisms.

Theoretical background

Colloidal aggregation may be regarded as a process which takes place when two particles collide due to Brownian motion and form a stable bond. If all collisions are effective, the aggregation is said to be entirely diffusion-controlled. This is the fastest possible aggregation mode if no other transport mechanism has to be considered. In this case, the aggregation rate k_S is given by the Smoluchowski formula [11],

$$k_{Smol} = \frac{8k_B T}{3\eta},\tag{2}$$

where k_B is the Boltzmann constant, T the temperature and η the viscosity of the suspension medium. This formula, however, neglects the possible interactions between two approaching particles and therefore often fails to reproduce the experimental values. The discrepancy can usually be explained by the interplay between attractive van-der-Waals forces and repulsive hydrodynamic interactions which typically reduce the theoretical predicted rate constant by a factor one-half [12–16]. Sonntag and Strenge evaluated measurements from different authors and calculated an experimental mean value of $k_S^{diff} = (6 \pm 3) \times 10^{-12}\,cm^3\,s^{-1}$ for diffusion-limited aggregation at room temperature [17].

The sticking probability for a colliding pair of colloidal particles, however, depends not only on the forces acting between the particles but also on whether the colliding part of the surface is covered by macromolecules or not. We can distinguish three possible configurations for two approaching particles:

1. Collision of two uncovered patches of the surface: This process corresponds to the case of aggregation of bare particles and, therefore, will be named coagulation. The sticking probability for this configuration is equal to the coagulation probability α.

2. Collision of two covered patches of the surface: The sticking probability in this case is equal to the weak flocculation coefficient β.

3. An uncovered part of one particle collides with the covered part of another particle: Here the sticking coeffi-

cient γ gives the probability that a *bridge* of macromolecules will form between the particles.

The probability of finding a covered surface patch is given by the fractional surface coverage θ and the probability of finding a bare part of the surface by $(1 - \theta)$. Hence, the fraction of collisions which occur in configurations, 1 and 2 is given $(1 - \theta)^2$ and θ^2, respectively. For the third configuration, the reverse case, i.e. the collision of a covered surface patch with another uncovered one, has also to be considered. Therefore, the probability for collisions to be of type 3 is given by $(1 - \theta)\theta + \theta(1 - \theta) = 2\theta(1 - \theta)$. The three cases are illustrated in Fig. 1.

Multiplying the collision probabilities with the corresponding sticking coefficients yields the overall fraction of effective collisions of each type. Therefore, the total aggregation rate may be written as the sum of the contributions due to the three different aggregation mechanisms:

$$k_S = k_S^{diff}[\alpha(1 - \theta)^2 + \beta\theta^2 + 2\gamma\theta(1 - \theta)]\,.\tag{3}$$

This relationship allows the sticking coefficients α, β and γ to be obtained as fitting parameters for the rate constants k_S measured as a function of surface coverage θ. It should be pointed out that the fitted sticking probabilities, α, β, and γ quantify the contribution of the different aggregation mechanisms (coagulation, bridging and weak flocculation) to the entire aggregation process.

Fig. 1 Schematic view of the three different aggregation mechanisms. The coefficients α, β and γ are the sticking probabilities. θ represents the fractional surface coverage

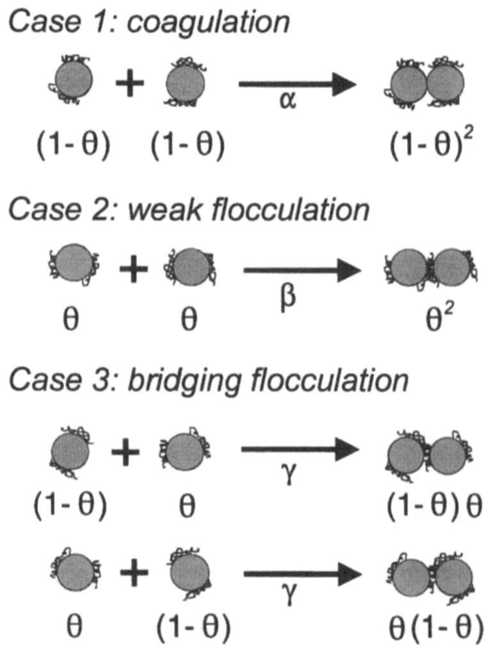

If bridging flocculation were the only aggregation mechanisms, i.e $\alpha = \beta = 0$, and all collisions occuring in bridging configuration lead to the formation of a bond, i.e. $\gamma = 1$, Eq. (3) reduces to

$$k_S = k_S^{\text{diff}} 2\theta(1 - \theta) , \tag{4}$$

which according to Hogg is the correct formulation of the La Mer model [8].

When bridging flocculation is not totally effective, i.e. $\gamma < 1$, but still the only possibility for aggregation Eq. (3) has to be written as

$$k_S = k_S^{\text{diff}} 2\gamma\theta(1 - \theta) , \tag{5}$$

which is exactly the formula given by Moudgil and co-workers [10]. In this case, γ is explained as the square of the fraction of active bridging sites.

If bridging flocculation were the only completely active aggregation mechanism, i.e. $\gamma = 1$, Eq. (3) may be written as

$$k_S = k_S^{\text{diff}} [1 - (1 - \alpha)(1 - \theta)^2 - (1 - \beta)\theta^2] . \tag{6}$$

which is exactly the formula deduced by Molski [7].

Therefore, we conclude that the model, proposed above, may be considered as an extended collision efficiency model which includes the La Mer, Moudgil and Molski models.

Materials and methods

For studying the bridging flocculation mechanism, a monodisperse aqueous suspensions of spherical polystyrene particles, AS2, was chosen as model polymer colloid. The particle had a diameter of (600 ± 17) nm with a polydispersity index of 1.002 and a surface charge density of $-(7.6 \pm 0.4)\,\mu\text{C cm}^{-2}$ due to sulfate groups. Commercially available bovine serum albumin (BSA, Pentex) was employed as a model for adsorbing macromolecules. Particles with a well-known degree of surface coverage were obtained by adding different amounts of BSA to a fixed quantity of colloid. The amount of non-adsorbed protein was then measured photometrically and an adsorption isotherm was determined. For more details see Ref. [18].

Flocculation was induced by adding a small amount of electrolyte to a monodisperse sample with a well-known degree of surface coverage. Mixing was done by injecting equal amounts of sample and buffered electrolyte into a Y-shaped mixing cell. The initial particle concentration in the reaction vessel was $1.0 \times 10^8\,\text{cm}^{-3}$ and the electrolyte concentration 0.1 M KCl. The pH was set to 5 and 9 by low ionic strength acetate and borate buffers, respectively. The temperature was stabilised at $(21 \pm 1)\,°\text{C}$. Immediately after the mixing step, a timer was started.

The aggregation processes were studied by means of single-cluster light scattering. This technique assesses the time evolution of the number of aggregates of different size by simple particle counting. Therefore, no more or less justified assumptions on the detailed cluster-size distribution have to be made for proper interpretation of experimental results and kinetic growth models may be checked almost directly. More detailed descriptions of the experimental set-up and its performance are given in Refs. [14, 19].

The initial flocculation rate constants were obtained from the time evolution of the number of monomeric particles. According to equation

$$\frac{1}{\sqrt{N_1(t)}} = \frac{1}{\sqrt{N_0}}\left(1 + \frac{N_0 k_S}{2} t\right), \tag{7}$$

which corresponds to the constant kernel solution ($k_{ij} = k_{11} = k_S$) of the Smoluchowski equation [20], a straight line was fitted to the onset of the inverse square root of the number of monomeric particles N_1. The initial rate constants were calculated from the slope of the fitted curve using the known initial particle concentration N_0.

Results and discussion

Figure 2a shows the initial rate constant k_S as a function of surface coverage θ for the first series of aggregation measurements carried out at pH 5. The rate constants are almost independent of the surface coverage and no maximum at intermediate surface coverage is observed. Therefore, one intuitively suspects that no bridging flocculation takes place. Obviously, it is impossible to explain the data in terms of the simple La Mer model (dotted curve in Fig. 2a) which takes only bridging flocculation into account. The experimental data, however, lie very close to the value commonly accepted for diffusion-limited aggregation of about $6 \times 10^{-12}\,\text{cm}^3\,\text{s}^{-1}$ which indicates that almost all collisions in coagulation, bridging and weak flocculation configuration must be effective. Therefore, the Molski model should be able to fit the experimental curve since it considers all bridging collisions to be active ($\gamma = 100\%$) and allows for coagulation and weak flocculation at low and high degree of surface coverage, respectively. The best fit of the experimental data to Eq. (6) leads to a coagulation probability of $\alpha_{\text{Molski}} = (73 \pm 13)\%$ and a weak flocculation probability of $\beta_{\text{Molski}} = (55 \pm 13)\%$. Using the extended collision efficiency model (Eq. (3)) the obtained results are very similar yielding a coagulation probability of $\alpha_{\text{ext}} = (81 \pm 12)\%$ and a weak flocculation probability of $\beta_{\text{ext}} = (63 \pm 12)\%$. The probability for bridging flocculation was found to be $\gamma_{\text{ext}} = (66 \pm 21)\%$. This means that two-thirds of all collisions in bridging configuration are effective which contradicts the intuitive

A. Schmitt et al.
Extended collision efficiency model for bridging flocculation

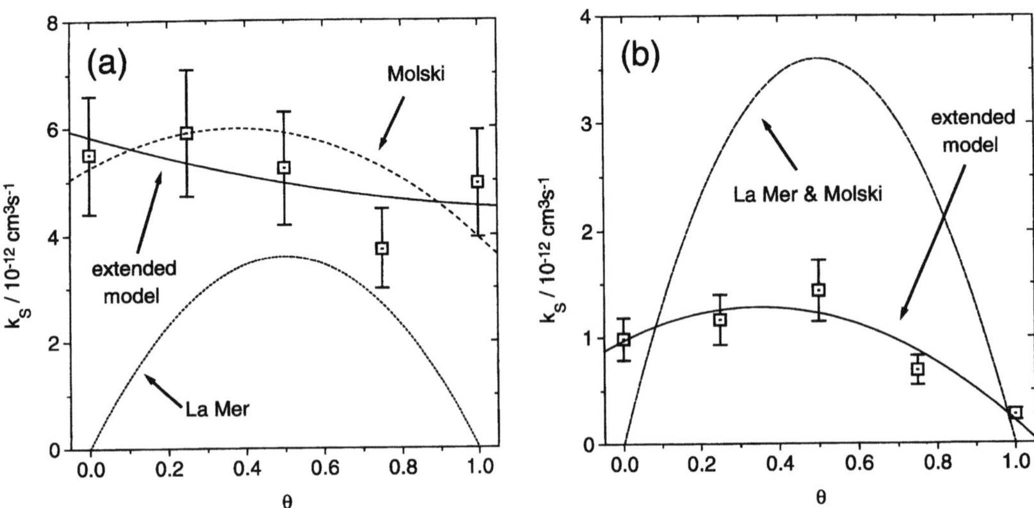

Fig. 2 Initial aggregation rate versus degree of surface coverage for the system AS2/BSA aggregated at 0.1 M KCl. Plot (a) and (b) show the results for pH 5 and 9, respectively

Table 1 Fitting parameters for the different aggregation models

pH	Molski model		Extended model		
	Coagulation probability α	Weak flocculation probability β	Coagulation probability α	Weak flocculation probability β	Bridging probability γ
5.0	0.73 ± 0.13	0.55 ± 0.13	0.81 ± 0.12	0.63 ± 0.12	0.66 ± 0.21
9.0	0.00 ± 0.24	0.00 ± 0.24	0.14 ± 0.03	0.03 ± 0.03	0.26 ± 0.05

statement made earlier that no bridging flocculation is taking place. The Molski model, however, always overestimates the bridging flocculation probability and consequently, tends to underestimate the coefficients for coagulation and weak flocculation.

At pH 5, the BSA molecules are very close to the their isoelectric point and therefore, do not alter the net charge of the colloidal particles. Thus, it is not surprising that probability for bridging flocculation is – within the experimental error – very close to the one for coagulation. The surprising observation, however, is that almost no steric stabilization is observed at high degree of surface coverage for the initial stages of aggregation.

Figure 2b shows the aggregation rates obtained at pH 9 where the BSA molecules are negatively charged and therefore, increase the surface charge of the colloidal particles. In this case, a bell-shaped curve is observed and aggregation is fastest at intermediate surface coverage. This indicates that bridging flocculation should be the predominant aggregation mechanism. Nevertheless, the measured aggregation rates are generally very low and therefore, the sticking probabilities should be smaller than unity for all considered collision configurations; even for

bridging flocculation. This explains why the La Mer and the Molski model do not fit the experimental data correctly since both assume bridging to be 100% effective (see dotted curve in Fig. 2b). Only the generalized collision efficiency model, which allows for slow bridging flocculation, is able to reproduce the experimental data. The best-fit yields a coagulation probability of $\alpha_{ext} = (14 \pm 3)\%$, a weak flocculation probability of $\beta_{ext} = (3 \pm 3)\%$ and a bridging flocculation probability of $\gamma_{ext} = (26 \pm 5)\%$ (see Table 1). This result confirms that bridging is the predominant aggregation mechanism and is almost twice as effective as the coagulation of uncovered particles. Weak flocculation could not be detected which means that, in this case, steric stabilization is totally effective.

Finally, we conclude that the extended model allows to separate and quantify the contributions of the different aggregation mechanisms (slow coagulation, weak flocculation and bridging flocculation) for slow and fast aggregation processes.

Acknowledgements Financial support by the CICYT project MAT97-1024 and the Gottlieb Daimler- und Karl Benz-Stiftung is gratefully acknowledged.

References

1. Dickinson E, Eriksson L (1991) Adv Colloid Interface Sci 34:1–29
2. Gregory J (1989) Crit Rev Envir Control 19(3):185–230
3. Fernández-Barbero A, Cabrerizo M, Martínez R, Hidalgo-Álvarez R (1993) Progr Colloid Polym Sci 93:269–272
4. La Mer VK, Smellie RH (1956) J Colloid Sci 11:704
5. Smellie RH, La Mer VK (1958) J Colloid Sci 23:589
6. La Mer VK (1966) Disk Faraday Soc 42:248
7. Molski A (1989) Colloid Polym Sci 267:371–375
8. Hogg R (1984) J Colloid Interface Sci 102:232
9. Ash SG, Calyfield EJ (1976) J Colloid Interface Sci 55:645
10. Moudgil BM, Shah BD, Soto HS (1987) J Colloid Interface Sci 119:466
11. v Smoluchowski M (1917) Z Phys Chem 92:129
12. Spielman LA (1970) J Colloid Interface Sci 33:562
13. Honig EP, Roebersen GJ, Wiersema PH (1971) J Colloid Interface Sci 36:97
14. Fernández-Barbero A, Schmitt A, Cabrerizo-Vílchez MA, Martínez-García R (1996) Physica A 230:53–74
15. Fernández-Barbero A, Marín-Rodríguez A, Callejas-Fernández J, Hidalgo-Álvarez R (1994) J Colloid Interface Sci 162:257
16. Holthoff H, Schmitt A, Fernández-Barbero A, Borkovec M, Cabrerizo-Vílchez MA, Schurtenberger P, Hidalgo-Álvarez R (1997) J Colloid Interface Sci 192:463–470
17. Sonntag H, Strenge K (1987) Coagulation Kinetics and Structure Formation, VEB Deutscher Verlag der Wissenschaften, Berlin
18. Schmitt A, Fernández-Barbero A, Cabrerizo-Vílchez MA, Hidalgo-Álvarez R (1997) Progr Colloid Polym Sci 104:144–147
19. Fernández-Barbero A, Cabrerizo-Vílchez MA, Martínez-García R, Hidalgo-Álvarez R (1996) Phys Rev E 53(2):4981
20. Drake RL (1972) In: Hidy GM, Brock JR (eds) Topics in Current Aerosol Research. Vol 3. Pergamon Press, New York, p 201

Progr Colloid Polym Sci (1998) 110:110–113
© Steinkopff Verlag 1998

M. Tirado-Miranda
A. Schmitt
J. Callejas-Fernández
A. Fernández-Barbero

Experimental evidence of rearrangement in fractal clusters

M. Tirado-Miranda (✉) · A. Schmitt
J. Callejas-Fernández
Grupo de Física de Fluidos y Biocoloides
Departamento de Física Aplicada
Universidad de Granada
Campus de Fuentenueva
E-18071 Granada
Spain

A. Fernández-Barbero
Grupo de Física de Fluidos Complejos
Departamento de Física Aplicada
Universidad de Almería
Cañada de San Urbano s/n
E-04120 Almería
Spain

Abstract Dynamic and static light scattering experiments were performed to measure the time evolution of the cluster-size distribution and cluster morphology for aggregating polystyrene particles. Systems with different particle size were studied. Dynamic scaling was observed in every case. The homogeneity parameter, λ, of the Van Dongen model, the fractal dimension of the clusters and the dimer formation rate constant, k_{11}, were determined. These parameters allowed the aggregation mechanism and its relation with the cluster structure to be determined. A rearrangement process is proposed to explain our results.

Key words Colloidal aggregation – dynamic scaling – cluster size distribution – light scattering – fractal structure – mesoscopic systems

Introduction

Considerable interest has been focused on the understanding of the dynamic nature of processes leading to cluster formation. The aggregation of colloidal particles is a good model for studying the two basic aspects of this phenomenon: aggregation kinetics and cluster morphology. The former is described using the Smoluchowski coagulation equation [1] and the Van Dongen and Ernst model [2]. The structure is described taking into account that clusters are fractals [3, 4]. An universal behaviour, independent of particle nature, is predicted for the cluster-size distribution and structure of the aggregates [5, 6].

In this paper, we report light-scattering measurements on the aggregation of monodisperse polystyrene microspheres induced at high salt concentration. We used dynamic light scattering (DLS) to measure the mean hydrodynamic radius of the aggregates and static light scattering (SLS) for studying the fractal structure of the clusters. The combined evaluation of the experimental results allows the number average cluster size to be determined. Dynamic scaling solutions [2] are used to determine the aggregation mechanism at long times. Finally, the rate constant for dimer formation, k_{11}, was determined from the first cumulant of the intensity autocorrelation function.

Theory

We use the Smoluchowski coagulation equation [1] for the cluster-size distribution, $n_l(t)$,

$$\frac{\mathrm{d}n_l}{\mathrm{d}t} = \frac{1}{2} \sum_{i+j=l} k_{ij} n_i n_j - n_l \sum_{i=1}^{\infty} k_{li} n_i , \tag{1}$$

where n_l is the number of clusters formed by l monomers. This equation is only useful for dilute systems, where only binary collisions are relevant. The kernels k_{ij} are the rate at which i-mers bind to j-mers to form $(i + j)$-mers and contain all the physical information about the systems.

Most coagulation kernels used in literature are homogeneous functions of i and j, at least for large i and j.

Van Dongen and Ernst [2] introduced a classification scheme for this type of kernels, where the dependence of the kernel on i and j at large i and/or j, is given by

$$k_{ai,aj} \propto a^{\lambda} k_{ij} , \tag{2}$$

where a is a positive constant. The homogeneity parameter λ describes the tendency of a large cluster to bind to another large cluster and governs the overall rate of aggregation. It should take the value 0 for diffusion limited cluster aggregation (DLCA), and 1 for reaction limited cluster aggregation (RLCA). λ will be used to characterise the aggregation mechanism.

Assuming the scaling limit, i.e. long times and large clusters, the cluster-size distribution has the form [7]

$$n_l(t) \propto s^{-2} \phi\left(\frac{l}{s}\right) , \tag{3}$$

where $s = s(t)$ is related to the average cluster size and $\phi(l/s)$ is a time-independent cluster-size distribution. This relationship implies that the relative shape of the cluster-size distribution is constant. It should be pointed out that $\phi(x)$ is an universal function of the variable $x = l/s$ and does not depend on initial conditions. It characterises the aggregation mechanism.

Accepting fractal structure for cluster, the dependency of the hydrodynamic radius $\langle R_h \rangle$ with the number average mean cluster size $\langle n \rangle$ is [8, 9]

$$\langle R_h \rangle = \langle R_0 \rangle \langle n \rangle^{-d_f} , \tag{4}$$

where $\langle R_0 \rangle$ is the monomer radius and d_f is the fractal dimension.

Using the dynamic scaling solution (3) and the fractal morphology of clusters, $\langle R_h \rangle$ is given by [10]

$$\langle R_h \rangle \propto t^{1/d_f(1-\lambda)} \tag{5}$$

for $\lambda < 1$. So, the time evolution of the number average mean cluster size may be expressed by

$$\langle n \rangle = \left(\frac{\langle R_h \rangle}{\langle R_0 \rangle}\right)^{d_f} \propto t^{1/(1-\lambda)} . \tag{6}$$

This formula will be used to determine λ from the number average mean cluster size.

Using the scaling solution (3) and considering the α-th moment of the cluster-size distribution, Olivier and Sorensen [11] obtained the following equation for the first cumulant of the intensity autocorrelation function

$$\mu_1(t) = \mu_1(0)\left(1 + \frac{t}{t_c}\right)^{-1/d_f(1-\lambda)} , \tag{7}$$

where $t_c = 2/N_0 k_{11}$ is the characteristic aggregation time when a constant kernel is considered. It is expressed as a function of the initial particle concentration, N_0, and the

rate constant for dimer formation, k_{11}. Equation (7) allows the characteristic time aggregation, t_c, to be obtained once d_f and λ are known. After that, k_{11} will be obtained from t_c using the initial particle concentration, N_0.

Materials and methods

Two monodisperse suspensions of spherical polystyrene particles, AS5 and SS40, with different size were used as experimental systems. They were cleaned by serum replacement. Table 1 shows the particle size $\langle R_0 \rangle$, polydispersity (p.d.i.) and critical coagulation concentration (c.c.c.) for both samples determined by DLS. The smaller particles were stabilised by surfactant molecules (SDS) adhered on the particle surface. All samples were prepared with water purified by inverse osmosis using Millipore equipment. They were sonicated for 15 min prior to aggregation in order to break up any initial clusters. Aggregation was initiated by adding a KBr solution to the stable colloid. We used a Y-shaped device for mixing equal amounts of monodisperse particles and electrolyte. In every case, the final electrolyte concentration was 1.0 M, above the c.c.c. The temperature was set at $(25 \pm 1)\,^\circ$C. The experimental light scattering set-up was a 4700 System from Malvern Instruments (UK) working with a 488 nm wavelength argon laser. The scattering angle range was 10° to 140°. This instrument allowed to follow the aggregation processes by DLS and SLS. Both techniques present the advantage that they do not perturb the samples during aggregation. SLS was used to assess the cluster structure and DLS for measuring the mean hydrodynamic radius from the scattered intensity autocorrelation function.

Results and discussion

Long time behaviour

We first obtained the fractal dimension by SLS using the long time asymptotic relationship $I(q) \propto q^{-d_f}$, valid in the range $\langle R_h \rangle^{-1} < q < \langle R_0 \rangle^{-1}$, where $q = (2\pi n/\lambda)\sin(\theta/2)$ is the scattering vector [3]. We found 1.7 and 2.4 for system AS5 and SS40 [12], respectively (see Table 2). After that the mean hydrodynamic radius $\langle R_h \rangle$ was measured by

Table 1 Characteristics of the polystyrene particles

Samples	$\langle R_0 \rangle$ [nm]	p.d.i.	c.c.c. [mM]
AS5	120 ± 20	0.03 ± 0.02	~ 400
SS40	20 ± 2	0.03 ± 0.03	~ 1000

Table 2 Characteristic parameters of the aggregation processes for the latixes

Samples	d_f	λ	k_{11} [cm^3 s^{-1}]
AS5	1.7 ± 0.1	0.03 ± 0.01	$(3.8 \pm 0.1) \times 10^{-12}$
SS40	2.4 ± 0.1	0.06 ± 0.03	$(3.9 \pm 0.1) \times 10^{-12}$

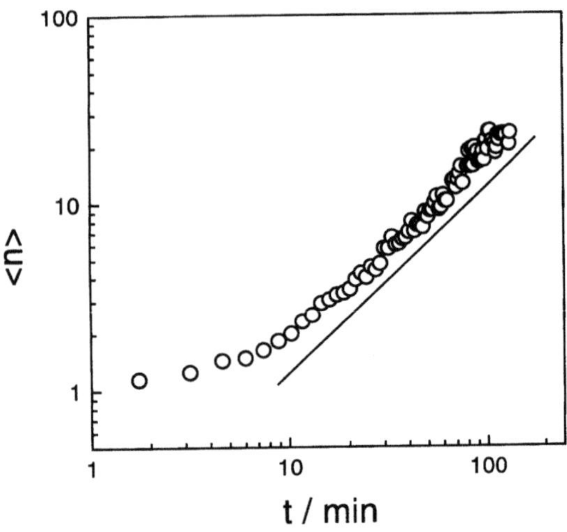

Fig. 1 Average mean clustersize as a function of time t for system AS5

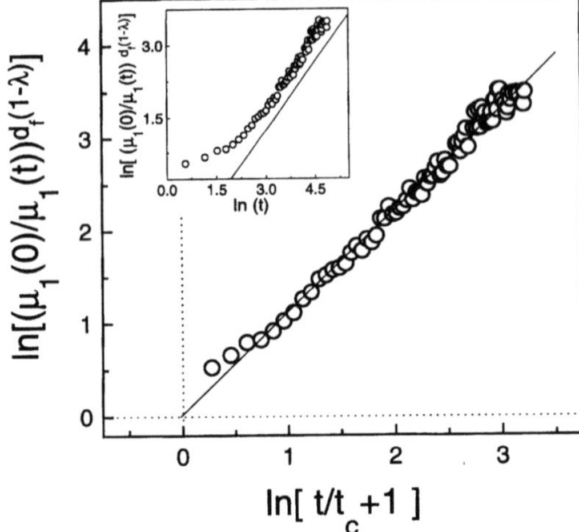

Fig. 2 Time evolution of the normalized first cumulant scaled in time, for particles AS5. From t_c, the rate of aggregation for dimer formation is obtained

DLS as a function of time and using Eq. (6), we assessed the number average mean cluster size $\langle n \rangle$. An asymptotic power law at long time is observed in Fig. 1. The homogeneity parameter, λ, was obtained from the slope according to Eq. (6). The results are also summarised in Table 2. In each case, λ is close to 0 which indicates that the aggregation processes are totally controlled by the diffusion of the clusters. So, the probability for two large clusters to form a stable bond is the same as for two small clusters. This result is in line with other DLCA experiments [5, 6, 12–14].

Short time behaviour

Now, information about the aggregation at short aggregation times will be obtained using Eq. (7) for the evolution of the first cumulant. Figure 2 shows $[(\mu_1(0)/\mu_1(t))^{d_\mathrm{f}(1-\lambda)}]$ as a function of time (upper left-hand corner) and as a function of the scaled time $[t/t_\mathrm{c} + 1]$, both for the system AS5 aggregating under DLCA conditions. t_c was chosen considering that after time scaling, the new curve should be a straight line with a slope equal to unity. From this plot t_c was estimated and k_{11} measured using the known value for the initial particle concentration (see Table 2). The same plot was repeated for system SS40 in order to check for the influence of the particle size and the different

cluster compactness reflected by the higher fractal dimension. The experimental rate constants k_{11} are similar, and independent of particles size and initial particle concentration which is in agreement with the theoretical predictions [15]. Furthermore, k_{11} are in agreement with the value commonly accepted for DLCA conditions, considering the effect of hydrodynamic interactions [6, 16, 17].

We can observe that the fractal dimensions for both systems are not related to the aggregation mechanism at short and long times (DLCA). Only AS5 shows the characteristic fractal dimension for DLCA. This indicates that the final cluster morphology in SS40 is reached as a consequence of another mechanism added to the diffusion of the clusters. An internal rearrangement of the cluster is proposed as such.

Conclusions

The time evolution of the average mean cluster size and the angular dependence of the scattered light intensity were determined for two systems with different particle sizes. Therefrom, the homogeneity parameter λ, the rate constant, k_{11} and the fractal dimension d_f were determined. The experimental results for λ and k_{11} indicate that the aggregation processes are controlled by diffusion in both cases. Nevertheless, the final cluster compactness for SS40 is higher than for AS5, probably as a consequence of an internal rearrangement of the clusters.

Acknowledgements This work was supported by CICYT (Spain), Project MAT 97-1024. A.S. is grateful for the fellowship granted by the Gottlieb Daimler- und Karl Benz-Stiftung.

Progr Colloid Polym Sci (1998) 110:110–113
© Steinkopff Verlag 1998

References

1. Smoluchowski MV (1917) Z Phys Chem 92:129
2. Van Dongen PGJ, Ernst MH (1985) Phys Rev Lett 54:1396–1399
3. Vicsek T (1992) Fractal Growth Phenomena. World Scientific, Singapore
4. Mandelbrot BB (1982) The Fractal Geometry of Nature. Freeman (ed) San Francisco
5. Fernández-Barbero A, Cabrerizo-Vílchez MA, Martínez-García R, Hidalgo-Álvarez R (1996) Phys Rev E 53(5): 4981–4989
6. Fernández-Barbero A, Schmitt A, Cabrerizo-Vílchez MA, Martínez-García R (1996) Physica A 230:53–74
7. Jullien R, Botet R (1987) Aggregation and Fractal Aggregates. World Scientific, Singapore
8. Fernández-Barbero A, Cabrerizo-Vílchez MA, Martínez-García R, Hidalgo-Álvarez R (1997) Phys Rev E 56(4):4337–4344
9. Berne B, Pecora R (1990) Dynamic Light Scattering. Wiley, New York
10. Ziff RM (1984) Kinetics of Aggregation and Gelation. North-Holland, Amsterdam
11. Olivier BJ, Sorensen CM (1990) J Colloid Interface Sci 134:139
12. Tirado-Miranda M, Schmitt A, Callejas-Fernández J, Fernández-Barbero A (1997) Progr Colloid Polym Sci 104:138–140
13. Lin HY, Lindsay HM, Weitz DA, Ball RC, Klein R, Meakin P (1989) Nature 339:360; (1990) J Phys Condens Matter 2:3093
14. Asnaghi D, Carpineti M, Giglio M, Sozzi M (1992) Phys Rev A 45: 1018–1023
15. Sonntag H, Strenge (1987) Coagulation Kinetics and Structure Formation. Plenum Press, New York
16. Fernández-Barbero A, Martín-Rodríguez J, Callejas-Fernández J, Hidalgo-Álvarez R (1994) J Colloid Interface Sci 162:257
17. Spielman LA (1970) J Colloid Interface Sci 33:562

Progr Colloid Polym Sci (1998) 110:114–118
© Steinkopff Verlag 1998

A.C. Nieuwkerk
A.T.M. Marcelis
E.J.R. Sudhölter

Interactions between hydrophobically modified poly(maleic acid-*co*-alkyl vinyl ether)s and dodecyltrimethylammonium bromide

A.C. Nieuwkerk · A.T.M. Marcelis
Prof. Dr. E.J.R. Sudhölter (✉)
Wageningen Agricultural University
Laboratory of Organic Chemistry
Dreijenplein 8
6703 HB Wageningen
The Netherlands

Abstract Binding of dodecyltrimethyl-ammonium bromide (DTAB) to poly(maleic acid-*co*-*n*-[(4-cyano-4'-biphenylyl)oxy]hexyl vinyl ether) and poly(maleic acid-*co*-*n*-[(4-cyano-4'-biphenylyl)oxy]dodecyl vinyl ether) has been investigated by potentio-metry using a surfactant-selective electrode and surface tension measurements. The binding between DTAB and the polyelectrolytes already starts at very low DTAB concentrations and is noncooperative. The strong and noncooperative binding of DTAB is thought to result from the formation of microdomains by the polyelectrolytes. Although the polyelectrolytes are only weakly surface active when solubilized in water at pH 10, highly surface active complexes are formed in the presence of small amounts of DTAB. These complexes are more hydrophobic than the individual components, resulting in an accumulation of the complexes at the air–solution interface.

Key words Potentiometry – surface tensiometry – microdomains – noncooperative interaction

Introduction

Complex water-based fluids containing polymers and sur-factants find important practical applications in various areas such as detergency, paints and coatings and tertiary oil recovery. They also play a key role in many biological systems [1]. For example, they control the functionality and the stability of cell membranes. The interactions be-tween model synthetic or natural polymers and surfactants have been studied extensively over the past 30 years [2–4]. It has been generally recognized that the details of interac-tion between a particular polymer and a surfactant is very important to the application of such a polymer–surfactant system. For systems containing oppositely charged linear polyelectrolytes and surfactants, it has been found that the interaction strength increases with surfactant chain length and polyelectrolyte hydrophobicity and that binding is a cooperative process [5, 6]. This cooperativity indicates that the surfactants bind to the polymer in the form of micelle-like aggregates [7]. However, for hydrophobically modified polyelectrolytes the interaction increases in strength upon increasing polyelectrolyte or surfactant hy-drophobicity, whereas the cooperativity of interaction de-creases [8, 9]. These polyelectrolytes can form micelle-like aggregates themselves, so-called microdomains, to which individual surfactant molecules bind. The change in co-operativity has been explained in terms of the free energy differences between binding of an oncoming surfactant next to an already bound surfactant molecule and binding to a site where it interacts with the copolymers side chains. At low polyelectrolyte hydrophobicity binding next to an already bound surfactant molecule has a more negative free energy than separate binding, whereas at high poly-electrolyte hydrophobicity the binding to polyelectrolyte microdomains has a more favourable free energy [10, 11].

Poly(maleic acid-*co*-alkyl vinyl ether)s carrying (cyano-biphenylyl)oxy chromophores at the ends of the alkyl chains have previously been studied by us [12]. The cyano-biphenylyl)oxy units have been incorporated to modify the polyelectrolyte's hydrophobicity and to act as intrinsic probes to monitor changes in polyelectrolyte conformation upon addition of surfactants [13]. By use of UV spectroscopy, the conformation of the polyelectrolytes in aqueous solution was seen to change from a more compact globular coil at low pH to the more extended, swollen conformation at high pH [12]. An effect of the side chain hydrophobicity was seen on both the compactness of the microdomains and on the degree of aggregation between the chromophores.

In the present paper we report on the interaction between DTAB and the chromophore labelled polyelectrolytes as studied by potentiometry using a DTAB-selective electrode and by surface tension measurements.

Experimental

Materials

The synthesis of poly(maleic acid-*co-n*-[(4-cyano-4′-biphenylyl)oxy]alkyl vinyl ether)s **I-n** ($n = 6, 12$) has been described previously [12]. *N-n*-dodecyl-*N,N,N*-trimethyl-ammonium bromide, DTAB (Aldrich), was recrystallized from ethanol before use.

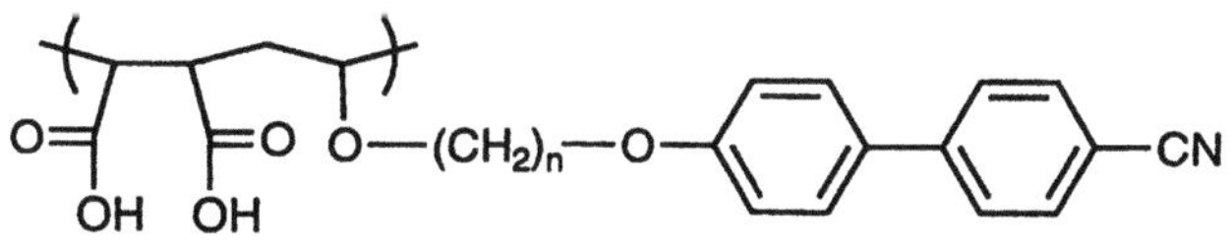

I-n, n = 6, 12

Potentiometry

The DTAB-selective electrode was prepared as described by Fluka Chemika [14]. The membrane was prepared from poly(vinyl chloride) plasticized by *bis*(2-ethylhexyl)-phthalate (20:80 w/w). As electrode body a Philips IS561 (Unicam) was used. The inner filling solution contained 4.4 mM DTAB in 5.4 mM KBr [11]. The reference electrode was a calomel reference electrode (Corning-Eel Scientific Instruments). The experiments were performed in a thermostatted vessel at $25 \pm 5\,°C$, with the following electrode setup:

Ag/AgCl electrode	DTAB reference solution	membrane	investigated solution	calomel electrode

The experiments were performed according to Zana et al. [11] using a computer-controlled titration system (Autolab®, ECO Chemie). The method was dynamic titra-tion, meaning that aliquots of surfactant solution were added only when the emf had reached a constant value. The calibration curves were linear from well below $10^{-5}\,M$ to the cmc of DTAB. The slope was 55.4 mV/decade, where the theoretical slope is 59.1 mV/decade at 25 °C.

Aqueous solutions of **I-n** (10^{-3} mol binding site l^{-1}) were prepared by dissolving the appropriate amount of dry polymer into 5 ml of THF. This solution was then added to an aqueous 5 mM KBr solution at pH 12 (final volume 25 ml). The THF was removed by stirring under a nitrogen flow.

Surface tensions were measured with a ring tensiometer (Sigma 70, KSV Instruments) using a Metrohm Dosimat 665 titrator to add the DTAB solution. The temperature was 20 ± 0.1 °C. The reproducibility between measurements on different samples of the same polymer was $\pm 0.5\,mN\,m^{-1}$. Polymer solutions were prepared in pure water as described for the potentiometric measurements.

Results and discussion

Potentiometry

The interaction between polymers and surfactants are usually presented in the form of binding isotherms which give a quantitative measure of the extent of interaction. The binding isotherms of the interaction between DTAB and polymers **I-n** were obtained by potentiometry using a surfactant selective electrode. The concentration of free, non bonded surfactant is measured by this electrode. In Figure 1 the potentiometric curve for binding of DTAB to **I-6** is presented. The potentiometric curve deviates from the calibration curve because of surfactant adsorption to the polyelectrolyte. It can be seen that binding of DTAB already starts at very low concentrations. At any C_s, corresponding to an emf value E, the concentration of free surfactant, C_s^f, is obtained from the difference with the calibration curve, at the same E value. Using these values the binding isotherm can be calculated using

$$\beta = C_s^b/C_p = (C_s - C_s^f)/C_p, \tag{1}$$

where C_s^b is the concentration of surfactant bound to the polymer, C_s is the total surfactant concentration, C_s^f is the free surfactant concentration, and C_p is the concentration of polymeric binding sites [15]. With this definition $\beta = 1$ corresponds to one bound DTAB molecule per carboxylic acid group.

Figure 2 displays the binding isotherms of DTAB to **I-6** and **I-12**. At a total DTAB concentration of about 0.5 mM, corresponding to a ratio of one DTAB molecule

A.C. Nieuwkerk et al.
Polyelectrolyte–surfactant interactions

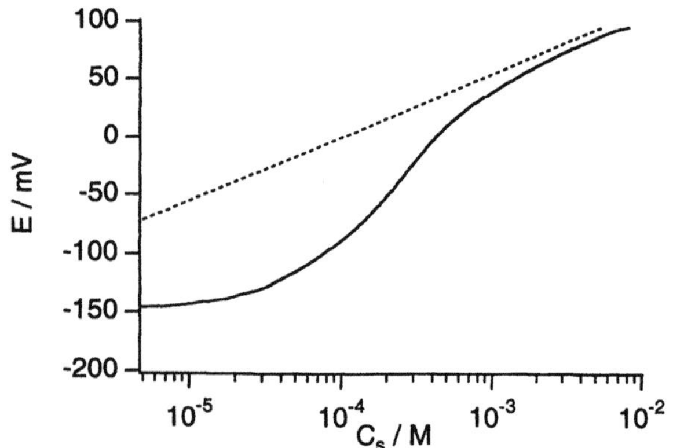

Fig. 1 Electrode response to the DTAB concentration C_s in a 1 mM I-6 solution (binding sites l^{-1}) at pH 8.3 in 5 mM KBr at 25 °C. The calibration plot (in the absence of I-6) is the dotted line

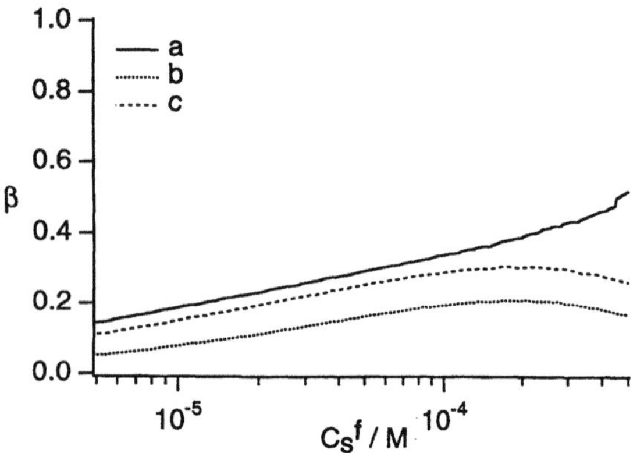

Fig. 2 Binding isotherms of DTAB to I-6 at pH 8.3 (a) and 11.9 (b), and to I-12 at pH 11.8 (c)

per repeating unit, precipitation of the formed polyelectrolyte-DTAB complexes started. This precipitation might interfere with the electrode response. Furthermore, the error in β at high C_s can be large because the calibration plot and the E vs. log C_s plot in the presence of polyelectrolyte are fairly close. For these reasons the results at high C_s are not discussed.

A slow and continuous increase in β with C_s^f is observed, indicating noncooperative binding. Cooperative surfactant binding is indicated by a sudden increase of β and completion of binding in a very narrow concentration range, and may be compared to the formation of micelles at their cmc. For the I-n-DTAB systems, the surfactant molecules partition between the aqueous phase and the hydrophobic microdomains formed by the polyelectrolytes. The noncooperative behavior can thus be ascribed to the hydrophobicity of the polyelectrolyte.

The polyelectrolyte charge density, which is modified by changing the pH, and the spacer length have small effects on the binding isotherms. These small effects are

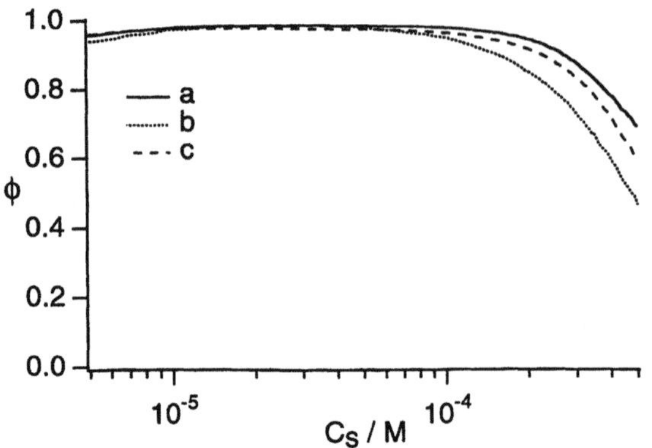

Fig. 3 Variation of the fraction of bound DTAB (ϕ) with the total DTAB concentration for I-6 at pH 8.3 (a) and 11.9 (b), and for I-12 at pH 11.8 (c)

thought to result from differencess in compactness of the microdomains.

In the low surfactant concentration range investigated most of the surfactant molecules are bound to the polyelectrolytes. For instance, for I-12 at $C_s^f = 5 \times 10^{-6}$ M, $\beta = 0.11$, indicating that $C_s^b = 1.1 \times 10^{-4}$ M and that 96% of the surfactants is bound to I-12. Likewise, at $C_s^f = 1 \times 10^{-4}$ M ($\beta = 0.29$), $C_s^b = 2.9 \times 10^{-4}$ M and 74% of the surfactants is bound. Similar data hold for I-6, and under the experimental conditions used C_s^b is very close to C_s.

Figure 3 displays the variation in the fraction of bound surfactant $\phi = C_s^b/C_s$ with C_s for I-12 at pH 11.8 and for I-6 at pH 8.3 and 11.9. Only when C_s approaches C_p, ϕ starts to deviate from 1, illustrating the very strong binding of DTAB to the hydrophobic polymers.

For poly(maleic acid-co-alkyl vinyl ether)s PSX, Zana et al. observed cooperative surfactant binding for the short side chains ($X \leq 4$), whereras noncooperative binding is observed for side chains containing more than 4 carbon atoms ($X > 4$) [8, 9, 11]. Similar results are reported by Kwak and co-workers for various hydrophobically modified maleic acid polymers with cationic surfactants [16, 17]. The interaction with DTAB depends strongly on the conformation of the polyelectrolyte. For PS4 at low charge density, when microdomains are present in aqueous solution, the interaction with DTAB is noncooperative, whereas at high charge density, when the PS4 adapts the extended coil conformation, cooperative interaction is observed [8, 9, 11]. When microdomains are present, surfactant molecules will preferentially bind to these microdomains, releasing their counterions into solution. Such binding is highly favoured from the energetic point of view because it allows both hydrophobic interactions between bound surfactant alkyl chains and polyelectrolyte side chains constituting the microdomains, and electrostatic interactions between the polyelectrolyte and the oppositely charged surfactant headgroups. The absence of cooperativity in binding comes from the fact that once a first surfactant molecule is incorporated into a microdomain, a next surfactant molecule will preferably incorporate into another microdomain for entropic reasons. Furthermore, binding of a surfactant molecule will reduce the overall charge of a microdomain resulting in a weakening of the attraction between the microdomain and surfactant molecules.

The strong and noncooperative binding of DTAB to **I-6** and **I-12** can be ascribed to the hydrophobicity of these polyelectrolytes which causes the formation of microdomains in water [12]. Surfactant molecules bind to these microdomains with their charged headgroups in close proximity to the charged groups of the polyelectrolyte. Mixed micelles are formed in which the counterions of the surfactant are expelled from the aggregate surface and the surfactant alkyl chains swell the microdomains [9].

Surface tensiometry

Surface tension measurements afford a simple and informative method to study mixtures of two components, of which one is highly surface active and the other relatively inactive. Many researchers have investigated the interaction between polymers and surfactants using surface tensiometry [2, 18–20]. Generally, the addition of surfactant to a polymer containing aqueous solution results in a decrease in surface tension as compared to the addition of the surfactant to pure water. If the polymer itself is not surface active the lowering in surface tension can only be ascribed to interactions between the polymer and surfactant.

In Fig. 4 the surface tensions of pure water and aqueous solutions of **I-12** are displayed as a function of the DTAB concentration. A DuNoüy ring was used to measure the surface tensions. In pure water a gradual decrease in surface tension is observed until the cmc of DTAB, 15.4 mM at 20 °C, is reached. At this point micellar aggregates are formed and no further decrease in surface tension is anticipated because all extra DTAB will form micelles and the free DTAB concentration remains constant. The slight increase in surface tension, observed upon increasing DTAB concentration above the cmc, is due to impurity of the surfactant [21]. Repetitive recrystallization of DTAB did not yield better results.

When DTAB is added to a solution containing polyelectrolyte a synergistic lowering of the surface tension is observed, even after the first addition of DTAB (concentration 1×10^{-5} mol l^{-1}, corresponding to a surfactant:binding site ratio of 1:100). Polyelectrolytes **I-6** and **I-12** are relatively inactive at the air–water interface at this low polyelectrolyte concentration.

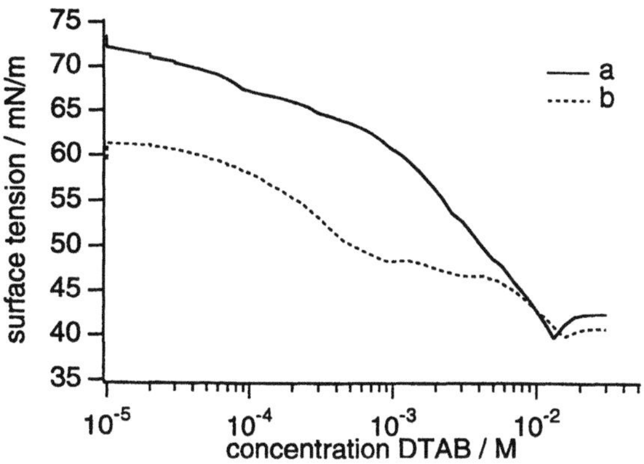

Fig. 4 Surface tension of DTAB in pure water (a) and in the presence of **I-12** (b). Polymer concentration 1×10^{-3} mol binding sites l^{-1}, pH 9.9, 20 °C

Table 1 Surface tension (γ in mN m^{-1}) of 1×10^{-3} M solutions of **I-n** (binding sites l^{-1}) in the absence and presence of 1×10^{-5} M DTAB at 20 °C and pH 9.9

Polymer	γ in pure water	γ in DTAB	$\Delta\gamma$
I-6	66.4	58.3	8.1
I-12	69.8	61.3	8.5

Upon increasing the DTAB concentration the surface tension decreases until a plateau is reached. At this point the DTAB concentration roughly equals half the concentration of binding sites. In this plateau the formation of a precipitate is observed due to charge neutralization of the polyelectrolyte and resulting decreased water solubility. The final drop in surface tension at high DTAB concentration results from the formation of free micelles.

Table 1 shows the surface tension of aqueous solutions of **I-n** in pure water and in the presence of 1×10^{-5} M DTAB (ratio surfactant:binding sites 1:100). The surface tension in pure water decreases upon decreasing spacer length. UV absorption spectroscopy shows stronger interactions between the chromophores of **I-12** as compared to **I-6** at pH 10 [12]. This indicates that the microdomains formed by **I-6** are more open than the microdomains formed by **I-12**. For **I-6** the side chains are expected to be exposed to water to a larger extent than the side chains of **I-12**, resulting in some accumulation at the air–solution interface reducing the surface tension.

The reduction in surface tension upon addition of DTAB implies coadsorption of the polyelectrolytes and DTAB molecules at the air–solution interface and the formation of a highly surface active polyelectrolyte–surfactant complex. For both polyelectrolytes a similar lowering in surface tension is observed. This might be explained by the balance between the compactness of the microdomains and the hydrophobicity of the polyelectrolytes. The micro-domains of **I-6** have a more open structure at pH 10 as compared to **I-12**, which facilitates the interaction with DTAB molecules. On the other hand, **I-12** is more hydrophobic due to its longer spacer as compared to **I-6**, which results in stronger interactions with DTAB. Overall, the hydrophobicity of the polyelectrolyte–DTAB complex has increased as compared to the individual components. To stabilize the solution, this increased hydrophobicity is counteracted by some accumulation of the formed complexes at the air–solution interface, resulting in a lowering of the surface tension upon complexation.

Conclusions

Potentiometric measurements using a DTAB-selective electrode have shown the noncooperative binding of DTAB to **I-6** and **I-12**. Only when the total surfactant concentration approaches the polyelectrolyte binding site concentration the fraction of bound DTAB ($\phi = C_s^b/C_s$) starts to deviate from 1 showing the very strong binding of DTAB to these polyelectrolytes. Surface tension measurements also show the very strong binding between the polyelectrolytes and DTAB which results in a synergistic lowering of the surface tension. In the presence of DTAB the hydrophobicity of the polyelectrolyte aggregates is reduced. Therefore, the solubility of the formed complex in water decreases and accumulation of the complex at the interface increases.

Both techniques used to study the interactions between **I-n** and DTAB show that binding is influenced by the charge density and side chain length of the polyelectrolytes. Besides the direct influence of these two factors on the electrostatic and hydrophobic attraction of DTAB to **I-n**, they also determine the compactness of the microdomains formed by the polyelectrolytes. This compactness seems to be very important for the interaction.

References

1. Ringsdorf H, Schlarb B, Venzmer J (1988) Angew Chem 100:117–162
2. Goddard ED (1986) Colloids Surf 19:301–329
3. Robb ID (1981) In: Lucassen-Reynders E (ed) Anionic Surfactants in Physical Chemistry of Surfactant Action, Marcel Dekker, New York, pp 109 and references therein
4. Hansson P, Lindman B (1996) Current Opinion Colloid Interface Sci 1:604–613
5. Hayakawa K, Kwak JCT (1983) J Phys Chem 87:506–509
6. Thalberg K, van Stam J, Lindblad C, Almgren M, Lindman B (1991) J Phys Chem 95:8975–8982
7. Malovikova A, Hayakawa K, Kwak JCT (1984) J Phys Chem 88:1930–1933
8. Anthony O, Zana R (1996) Langmuir 12:1967–1975
9. Anthony O, Zana R (1996) Langmuir 12:3590–3597
10. Low polyelectrolyte hydrophobicity indicates the presence of short alkyl side chains and/or high charge density, e.g. poly(maleic acid-*co*-alkyl vinyl ether)s PSX with $X \leq 4$ [11]. Polyelectrolytes of high hydrophobicity have long alkyl side chains and/or low charge density, e.g. PSX with $X > 4$
11. Benrraou M, Zana R, Varoqui R, Pefferkorn E (1992) J Phys Chem 96:1468–1475
12. Nieuwkerk AC, Marcelis ATM, Sudhölter EJR (1995) Macromolecules 28:4986–4990
13. Nieuwkerk AC, Marcelis ATM, Sudhölter EJR (1997) Langmuir 13:3325–3330
14. Ionophores for Ion-Selective Electrodes (1988) Fluka Chemika AG, Switzerland, p 9
15. Moroi Y (1992) In: Moroi Y (ed) Micelles: Theoretical and Applied Aspects. Plenum Press, New York, pp 233–239 and references therein
16. Shimizu T, Seki M, Kwak JCT (1986) Colloids Surf 20:289–301
17. Shimizu T, Kwak JCT (1994) Colloids Surf A: Physicochem Eng Aspects 82:163–171
18. Asnacios A, Langevin D, Argillier J-F (1996) Macromolecules 29:7412–7417
19. Chang Y, Lochhead RY, McCormick CL (1994) Macromolecules 27:2145–2150
20. Merta J, Stenius P (1995) Colloid Polym Sci 273:974–983
21. Clint JH (1992) Surfactant Aggregation, Blackie, Glasgow and London, pp 110–111

Progr Colloid Polym Sci (1998) 110:119–124
© Steinkopff Verlag 1998

F. Le Berre
E. Pefferkorn

Structural characteristics of sheared suspensions of hydrated colloids

F. Le Berre · Dr. E. Pefferkorn (✉)
Institut Charles Sadron
6, rue Boussingault
F-67083 Strasbourg Cedex
France
E-mail: peffer@ics.u-strasbg.fr

Abstract The mass and size frequencies of aggregates formed under low shear by encounters between hydrated colloids of micrometric size was investigated under marginal stability conditions. The mass $c(n)$ and size $c(d)$ frequency curves were determined by particle counting and laser diffractometry, respectively. Correlation between reduced mass and size distributions unequivocally led to the fractal dimension of the aggregates. Aggregation/fragmentation under low shear rate gave rise to aggregates of relatively low fractal dimension when compared to situations of high shear rates. Under marginal stability conditions, the particle concentration and ionic strength differently induced the aggregate formation. In 0.15 M NaCl suspension, aging effects were determined above a threshold concentration while in 0.075 M NaCl suspension, aging effects were observed at all concentrations. Aggregation/fragmentation in 0.15 M NaCl suspensions led to aggregates of fractal dimension 1.4 while in 0.075 M NaCl suspensions, aging and restructuration led to aggregates of fractal dimension 1.8. These relatively low values of the fractal dimension were interpreted by the existence of a preferential orientation of the aggregates in the low shear flow.

Key words Orthokinetic aggregation – aggregate mass distribution – aggregate size distribution – hydrated colloids – aggregation in a Couette system – aggregate fractal dimension

Introduction

Suspended colloids in aquatic environments may undergo various migration processes depending on hydrodynamic conditions and hydrosphere nature [1]. Brownian diffusion may exert in very quite waters such as lakes and inland seas. Hydrodynamic forces may exert in rivers, streams and groundwaters. From very elementary considerations, flow in groundwater may be modelled by flow through porous media, where shear rates vary in a great domain and the rate of groundwater transport depends on the texture, porosity and permeability of the water zones [2]. Similarly, stream flow may resemble turbulent flow. The situation in quiet rivers may be simpler when suspended colloids migrate under laminar flow. Actually, soil erosion is the primary mechanism by which microcolloids are transferred from soils to the aquatic environment. The microcolloids initially collected at the river surface sink into bulk water and flow with the current. The fate of the immersed colloidal particles depends on the nature of the colloid/water interface and hydrodynamic characteristics [2]. Actually, since usual hydrophilic colloids may rapidly aggregate and settle, it is of greater interest to determine the fate of hydrated colloids. Such particles should undergo very slow aggregation processes in the aquatic environment where aggregate growth and breakup processes may develop successively or concomitantly.

In order to model this complex situation, we employed hydrated latex particles characterized by a high surface density of hydrophilic groups (hydroxyls, carboxylate and sulfate groups) which revealed an unexpected stability in electrolyte medium in spite of a very low surface potential and we fixed the hydrogen ion and electrolyte concentrations to induce a perceptible colloid aggregation in capillary flow [3]. Aggregates of nanometric particles are of lesser importance in environmental pollution problems than aggregates of particles of micrometric sizes because the first ones migrate with the flux without settling while the second ones may contribute to sediment pollution after aggregation [4]. To this end, monosized polystyrene latex spheres of 1.09 μm diameter were employed in this study. They were sheared in a Couette flow system and, periodically small samples of the suspension were carefully collected and analyzed for particle mass and size distribution determinations.

From a fundamental point of view, slow aggregation processes have been found to develop with features of reaction-limited processes whose characteristics are well known. Suspended aggregates may be described by their mass and size distributions and characterized by their average masses and sizes. The present work addressed the fractal nature of aggregates and the self-similarity of the mass and size distribution curves. The aggregates mass corresponds to the number of single particles (of mass equal to 1) contained in the aggregate and is derived from the volume of the solid phase. The aggregate size corresponds to the total volume occupied by the aggregate, including the total volume of single particles and that of the liquid phase contained in the porous structure. Numerical simulations of slow reaction-limited aggregation processes and related experiments have demonstrated the validity of scaling laws to characterize the self-similarity of the mass frequency curves and the temporal variation of the average masses [5–8]. We started with this usual approach of perikinetic processes to describe the stability of the colloid suspension under very low shear. Finally, we determined the aggregate size distribution curves and established their self-similarity in order to derive the fractal dimension of the aggregates from comparison with the corresponding reduced mass distribution curves.

Material and methods

Colloidal particles. Spherical polystyrene latex particles of diameter 1.09 μm were prepared by emulsifier free polymerization using potassium persulfate as initiator and kindly provided by G. Graillat (Lyon, France). Electrochemical and stability characteristics of the latex particles have been described elsewhere [3]. These particles were found to present usual characteristics of electrocratic sols and, as a result of their unexpected stability in 0.15 M NaCl aqueous suspension, may be considered as being hydrated. The hydrated shell could be attributed to the combined effect of hydroxyl and carboxylic acid groups present at a concentration of about 12 μC/cm^2 surrounding the dissociated sulfate groups present at a concentration of 6 μC/cm^2.

Perikinetic aggregation. All experiments were performed at 25 °C. Following usual procedures, the latex suspension was added to the electrolyte solution and gentle tumbling of the test-tube was performed twice before flocculation started. In order to eliminate particle sedimentation, experiments were systematically carried out in a mixture of deuterium oxide/water/eletrolyte having a density of 1.045.

Orthokinetic aggregation. The Plexiglas® home-made Couette system composed of an external fixed cylinder of 5.2 cm internal diameter and a concentric central cylinder of 5.0 cm diameter and a height of 15 cm may contains about 35 ml of suspension into the gap. The inner cylinder was fitted with a pulley driven by a strap system connected to a stepper motor. Immediately after preparation the suspension was slowly introduced by gravity flow into the nearly horizontal slanting Couette system. After filling, the Couette system is slowly set upright. Samples of the suspension are collected near the bottom with the aid of a Teflon® tubing. The sample collection is induced by lowering the tubing aperture of about 1 cm in relation to the upper limit of the fluid in the gap. After withdrawal of the suspension remained in the tubing, the collected sample is immediately analyzed for mass or size distributions. In order to verify that the sample collection procedure does not modify the suspension characteristics, perikinetic aggregation was performed in the Couette system and the results were found to be in agreement with those obtained employing the usual procedure. In some instances, to exclude any artifact due to a possible slow settling of very large aggregates, water was replaced by the deuterium oxide/water mixture. No sensible difference could be detected between experiments carried out in water and those in the isodense medium. All experiments were carried out at 5 rpm corresponding to a shear rate of 14 s^{-1}.

Particle counting. Counting by the Coulter counter Multisizer II (Coultronics) is performed on samples taken from the suspension or collected at the outlet of the Teflon® tubing after dilution by an electrolyte solution having the same composition than the liquid phase suspending the latex suspension. The technique and methodology for obtaining the aggregate mass distribution ($c(n)$ vs n) have been described elsewhere [9, 10].

The suspension characteristics are determined by the number $N(t)$ and weight $S(t)$ average masses of the aggregates. These values defined by Eq. (1) were calculated from the aggregate mass distribution $c(n, t)$ derived from the histogram given by the particle counter:

$$N(t) = \frac{\sum_n nc(n, t)}{\sum_n c(n, t)}, \qquad S(t) = \frac{\sum_n n^2 c(n, t)}{\sum_n nc(n, t)}, \tag{1}$$

where $\sum_n nc(n, t) = N_1$ corresponding to the number of particles initially contained in the suspension. Due to self-similarity of the aggregate mass distributions throughout the process [11, 12], we represent the reduced concentration of particle of mass n, $S^2(t)c(n, t)/N_1$ as a function of the reduced mass $n/S(t)$ which usually provides a unique slope τ defined by

$$c(n) \propto n^{-\tau}. \tag{2}$$

Particle sizing. The particle size distribution was measured with the Coulter LS100 (Coultronics). The scattered light is focused using a Fourier optic. The intensity in the focal plane at certain distances from the optical axis corresponds to certain scattering angles. The instrument applies a software routine assuming a size distribution to calculate a radial intensity distribution [13]. Due to the large size of the particles and aggregates, Fraunhofer theory is applied so that the complex refractive index is not required. All parameters measured by the instrument are related to the volume of the aggregates.

Self-similarity of the size distribution is evidenced by the variation of reduced concentration of particles of diameter d, $S^2(t) \times c(d, t)/N_1$ as a function of the reduced size $d/S(t)$ and is expressed by

$$c(d) \propto d^{-\alpha} \tag{3}$$

and the correlation between mass and size leads to the following equation for the aggregate fractal dimension f;

$$f = \alpha/\tau. \tag{4}$$

Results and discussion

Aggregation/fragmentation kinetics

Figures 1 and 2 show the number average mass of aggregates as a function of time for experiments carried out in 0.15 and 0.075 M NaCl media, respectively. The variable parameter is the initial particle concentration. In each figure, the values of $N(t)$ obtained under perikinetic conditions are indicated by the dashed line (black symbols). The low shear exerts an obvious influence on the rate and

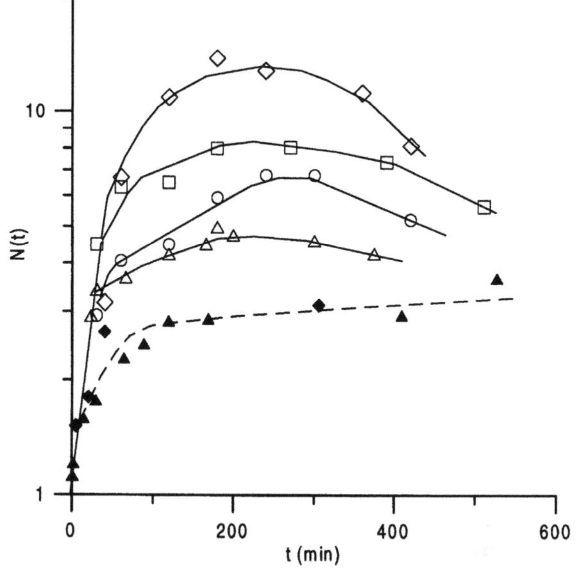

Fig. 1 Aggregation in 0.15 M Nacl suspension. Representation of the number average mass $N(t)$ of aggregates as a function of the aggregation time (min) for different initial concentration of latex particles (g/l) for orthokinetic: (○), 0.0875; (□), 0.1325; (◇), 0.175; (△), 0.2625 and perikinetic processes: (◆), 0.175; (▲), 0.2625

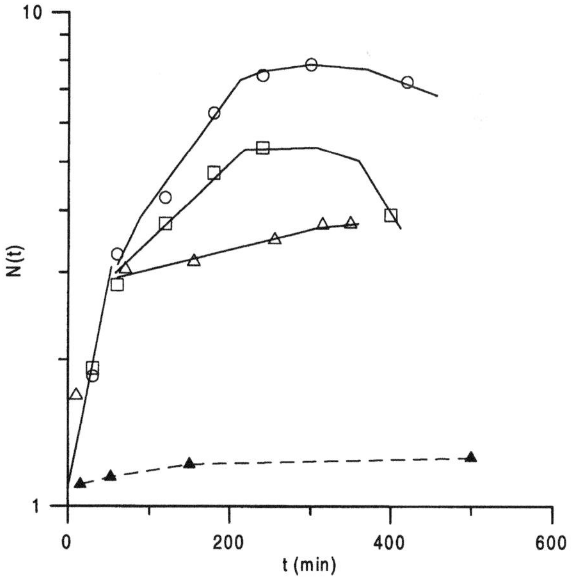

Fig. 2 Aggregation in 0.075 M NaCl suspension. Representation of the number average mass $N(t)$ of aggregates as a function of the aggregation time (min) for different initial concentration of latex particles (g/l) for orthokinetic: (○), 0.0875; (□), 0.1325; (△), 0.2625 and perikinetic processes: (▲), 0.2625

development of aggregation and fragmentation processes in 0.15 M NaCl media while it exerts a strong influence in 0.075 M NaCl media. In the perikinetic process (in 0.15 M NaCl suspension) and during the period corresponding to the relatively fast initial increase of the average mass $N(t)$ in the shear-induced process, the aggregation rate does not depend on the particle concentration. However, the particle concentration exerts a role when aggregates grow more slowly, during the period of equal importance of aggregation and fragmentation processes and when the fragmentation process become predominant after about 250–300 min. The aggregate internal cohesion may be quantified by the value of the average mass corresponding to the period were aggregation and fragmentation cancel each other. In 0.15 M NaCl, $N(t)$ increases with the initial particle concentration below a concentration close to 0.2 g/l while above this value $N(t)$ values are strongly reduced. The internal cohesion in 0.075 M NaCl decreases with the particle concentration in all situations. The cumulative effects of hydrodynamic forces and variable low energy of interparticle bonds are responsible of the present concentration dependent aggregate cohesion.

Furthermore, the correlation $S(t) = N^2(t)$, usually obtained for particles sustaining the reaction-limited aggregation process is determined to apply in the present situation during the first period where aggregation prevails over fragmentation, the dynamic equilibrium and the last period of prevailing fragmentation. Conclusively, shear-induced fragmentation progressively restores the previous masses and mass polydispersity to the colloid suspension.

Reduced mass distribution curves

Figures 3a and b show the reduced mass distribution of aggregates formed in 0.15 and 0.075 M NaCl media, respectively, for the different latex concentrations. A unique slope describes the variation of the reduced aggregate concentration as a function of the reduced mass, in the domain of small values of the reduced masses. The value 1.5 of the slope is determined for both predominant aggregation, dynamic equilibrium and fragmentation processes.

One may fully compare the situation at small and large times where equal average masses and mass distribution curves were obtained. The unique slope means that (i) fragmentation progressively breaks interaggregate links at the same position they were progressively formed and, (ii) breakup of aged links requires longer periods than that of links being established immediately before fragmentation became predominant.

Fractal dimension of aggregates

Figure 4 represents the reduced size distribution curves corresponding to experiments carried out in 0.15 M NaCl

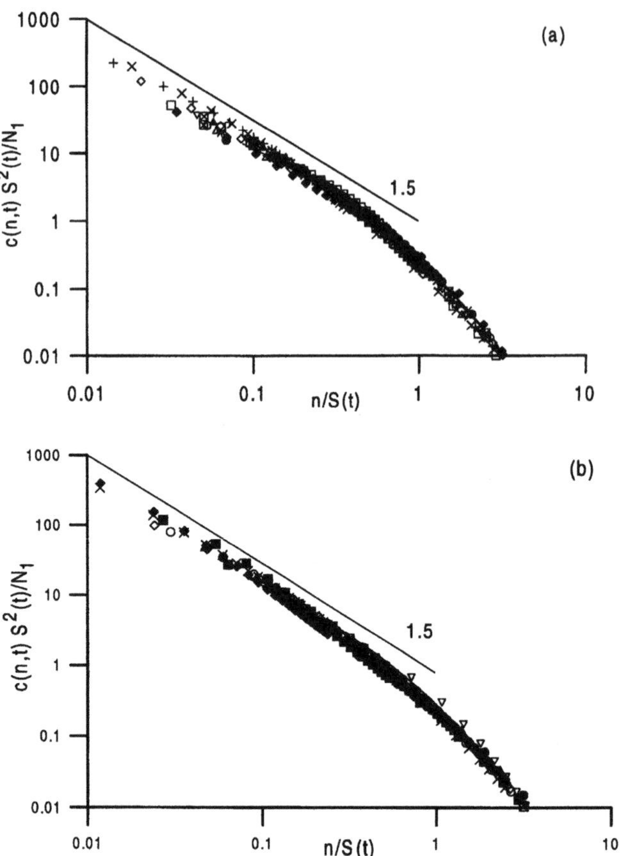

Fig. 3 Orthokinetic aggregation in 0.15 M (a) and 0.075 M (b) NaCl suspensions. Representation of the reduced mass distribution $c(n, t)S^2(t)/N_1$ as a function of the reduced mass $n/S(t)$. The straight lines correspond to a slope of -1.5. The different symbols correspond to experiments reported in Figs. 1 and 2

suspension at the concentrations of 0.1725 g/l in (a), 0.1325 g/l in (b) and 0.0875 g/l in Fig. 4c. Figure 4d shows the situation for experiments carried out in 0.075 M NaCl at the two concentrations of 0.1325 and 0.0875 g/l. According to Eq. (3), we determine α to be close to 2.1 and 2.7 for aggregates formed in 0.15 and 0.075 M NaCl suspension, respectively. From Eq. (4) the fractal dimension of aggregates is

$$f = \begin{cases} 1.4 & \text{for } 0.15 \text{ M NaCl suspensions}, \\ 1.8 & \text{for } 0.075 \text{ M NaCl suspensions}. \end{cases} \tag{5}$$

This result may be interpreted as follows.

In 0.15 M NaCl suspensions and for concentrations between 0.0875 and 0.175 g/l, the average aggregate mass attains higher levels in concentrated suspensions, showing that the aggregate structure do not require long aging times to transform the favorable situation of closest approach into sticking. Nevertheless, it should be noted that the process is different from that induced by diffusion-limited aggregation, where each collision leads to sticking.

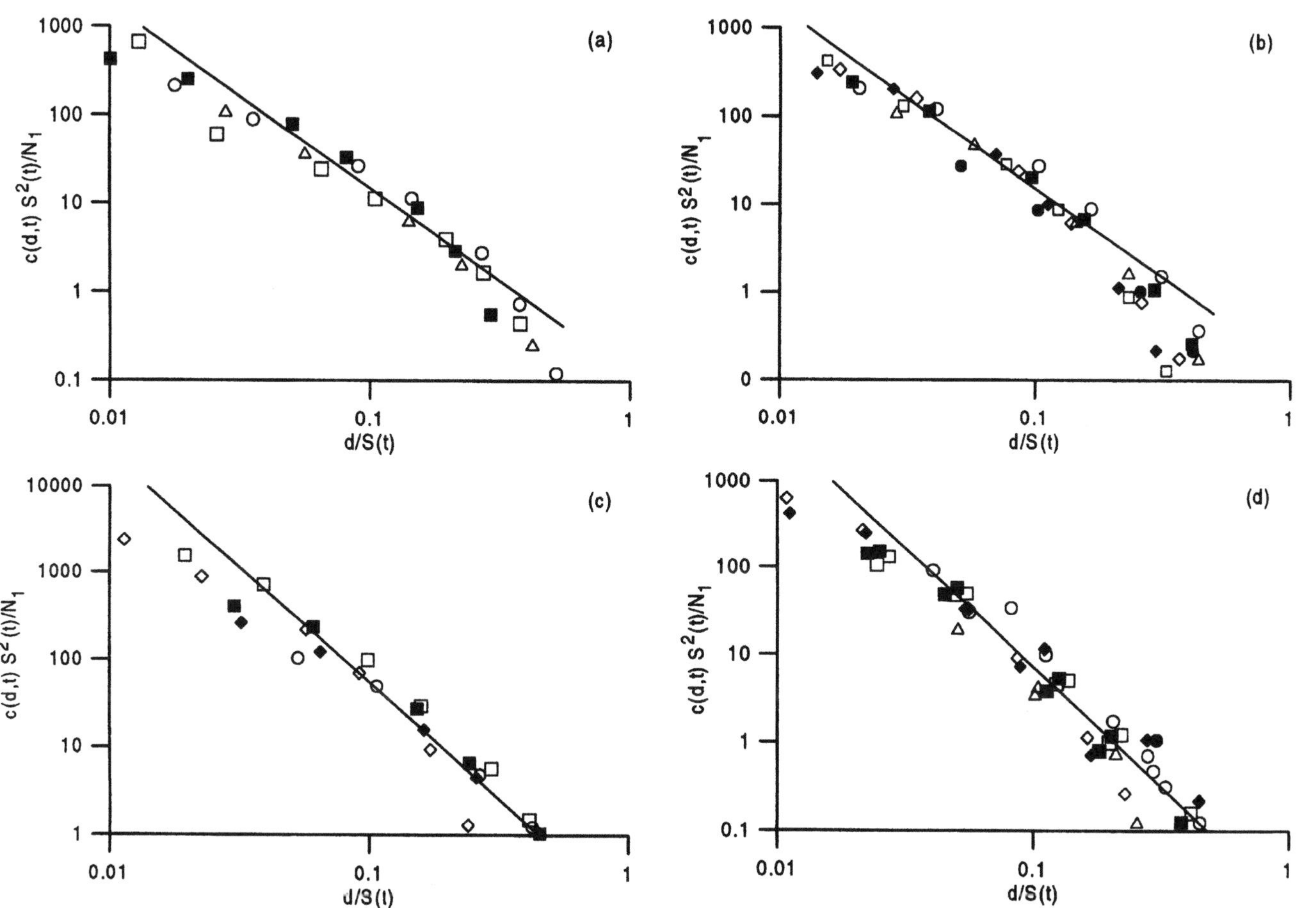

Fig. 4 Representation of the reduced size distribution $c(d, t)S^2(t)/N_1$ as a function of the reduced size $d/S(t)$ corresponding to experiments carried out in 0.15 M NaCl suspensions at the concentraitons of 0.1725 g/l in (a), 0.1325 g/l in (b) and 0.0875 g/l in (c). (d) draws the reduced size distributions for two separate experiments carried out in 0.075 M NaCl at the concentrations of 0.1325 and 0.0875 g/l. The symbols correspond to different sizes d of the aggregates (μm) present at the concentration $c(d, t)$: ($\diamond$), 2; ($\blacklozenge$), 5; ($\square$), 8; ($\blacksquare$), 15; ($\circ$), 21; ($\bullet$), 29; ($\triangle$), 40

In the present situation of reaction-limited aggregation, many connections may be experienced before establishment of the favorable situation. Since the interparticle connections are rapidly frozen under the conformation existing at the collision moment, the absence of significant inter-penetration of colliding aggregates (and particles) and restructuration may explain the unexpectedly small fractal dimension of 1.4. On the other hand, when branched aggregates are formed they may preferentially breakup and recombine to form elongated structures.

In 0.075 M NaCl suspensions, the limiting average aggregate masses are smaller in concentrated suspensions. This means that the collision efficiency decreases and that the formation of stable aggregates requires long aging in addition to the required favorable positioning of colliding aggregates [14]. Therefore, aggregates may sustain multiple restructurations prior establishment of the stable conformation. The value 1.8 of the fractal dimension is too small to really correspond to a compact arrangement of particles. Our supplementary assumption is that shear induced rotation may oppose the formation of shapes of spherical symmetry to the benefit of that of elongated ones. Actually, the more dense restructured aggregates may rotate without sustaining a great fragmentation and this phenomenon may give rise to aggregates characterized by a fractal dimension appreciably lower than 2.0. Similar low values of the fractal dimensions were determined experimentally and interpreted by a shear induced preferential orientation of the aggregates [15, 16]. Similar results were obtained numerically and provoked by the existence of sites of attractive and repulsive forces (polarization effects) [17–20].

Conclusion

We determined that the aggregate mass and size growths were limited to values which depended of the ionic strength and particle concentration of the suspension. This limitation was provoked by a typical breakup mechanism involving the shear induced aggregate restructuration. In fact, during the two regimes of prevailing aggregation and fragmentation and for the intermediary moment where the two processes developed at similar rates, a good correlation was observed between the mass and size characteristics $N(t)$, $S(t)$, $c(n, t)$ and $c(d, t)$. Prolonged low shear restored the suspension in early existing situations with conservation of the reduced mass distributions ($\tau = 1.5$ at the two ionic strengths) and the reduced size distributions ($\alpha = 2.1$ and 2.7 at 0.15, and 0.075, respectively). The fractal dimensions of 1.4 and 1.8 characterized the aggregate shapes in 0.15 M and 0.075 M NaCl media. The elongated shape of aggregates formed in 0.15 M NaCl medium might result from the faster rate of aggregate aging and restructuring which were found to developed more slowly under the less energetic attractive forces in the 0.075 M NaCl medium.

Finally, since the mass polydispersity $S(t)/N(t)$ of the system was determined by $N(t)$ in all situations, the low shear induced the aggregates breakup hierarchically and not randomly. The clear cut information was that fragmentation of old interaggregate links established at the beginning of the prevailing aggregation phase required a long time whereas fragmentation of newly formed links rapidly succeeded their establishment.

Acknowledgements We acknowledge support from the Institut Français du Pétrole (IFP) in the form of a studentship (to F.L.B.). G. Chauveteau id acknowledged for helpful discussions and G. Graillat (Laboratoire de Chimie et Procédés de Polymérisation, Lyon, France) for kindly providing the latex particles. The authors are grateful to J. Widmaier for experimental assistance.

References

1. Valioulis IA (1986) Adv Colloid Interface Sci 24:81
2. Pierzynski GM, Sims JT, Vance GF (eds) (1993) Soils and Environmental Quality. Lewis Publishers, Boca Raton
3. Le Berre F, Chauveteau G, Pefferkorn EJ (1998) Colloid Interface Sci 199:1
4. Buffle J, van Leuven HP (1992) Environmental Particles. IUPAC Environmental Analytical and Physical Chemistry Series. IUPAC, Chelsea, MI
5. Family F, Landau DP (eds) (1984) Kinetics of Aggregation and Gelation. North-Holland, Amsterdam
6. Stanley HE, Ostrowsky N (eds) (1986) On Growth and Form. Nijhoff, Dordrecht
7. Jullien R, Botet R (1989) Aggregation and Fractal Aggregates. World Scientific, Singapore
8. Vicsek T (1989) Fractal Growth Phenomena. World Scientific, Singapore
9. Walker PH, Hutka J (1971) In Division if Soils. Tech Paper 1, 3
10. Pefferkorn E (1995) Adv Colloid Interface Sci 56:33
11. Swift DL, Friedlander SK (1964) J Colloid Sci 19:621
12. Lushnikov AA (1973) J Colloid Interface Sci 45:549
13. Singer JK, Anderson JB, Ledbetter MT, McCave LR, Jones KPN, Wright R (1988) J Sediment Petrol 58:534
14. Bos MTA, van Opheusden JHJ (1996) Phys Rev E 53:5044
15. Hurd A, Schaefer D (1985) Phys Rev Lett 54:1043
16. Hurd AJ, Flower WL (1988) J Colloid Interface Sci 122:178
17. Meakin P, Chen Z-Y, Evesque P (1987) J Phys Chem 87:630
18. Jullien R (1985) Phys Rev Lett 55:1697
19. Meakin P, Muthukumar M (1989) J Phys Chem 91:3212
20. Meakin P (1990) J Colloid Interface Sci 134:235

Progr Colloid Polym Sci (1998) 110:125–128
© Steinkopff Verlag 1998

R.C. van Duijvenbode
M. Borkovec
G.J.M. Koper

Odd-even shell ionization of Astramol™ dendrimers

R.C. van Duijvenbode (✉) · G.J.M. Koper
Leiden Institute of Chemistry
Leiden University
Gorlaeus Laboratories
PO Box 9502
2300 RA Leiden
The Netherlands

M. Borkovec
Department of Chemistry
Clarkson University
Potsdam, NY, 13699-5814
USA

Abstract Potentiometric titrations of five generations of poly(propylene imine) dendrimers were performed at 0.1, 0.5, and 1.0 M KCl and NaCl. The titration curves reveal two-step protonation behavior similar to linear polyelectrolytes. The difference to linear polyelectrolytes is that the position of the intermediate plateau lies at 2/3 of the total ionizable groups. The intermediate plateau results from the stability of an onion-like structure where all odd shells of the dendrimer are protonated, while the even shells remain deprotonated. The titration curve features two distinct steps around pH 6 and 10. The Ising model permits a quantitative analysis of this protonation behavior.

Key words Dendrimer – potentio-metric titration – Ising model

Introduction

The ultimate goal of studies on acid–base properties of polyelectrolytes is the prediction of the ionization behavior of arbitrary molecules from structural information. For linear polyelectrolytes, the linear chain Ising model turns out to be rather successful, and allows to rationalize acid–base properties of linear polyelectrolytes and their oligomeric analogs quantitatively [1–5].

The Ising model with nearest-neighbor interactions was recently extended to branched polyelectrolytes [6]. In contrast to linear molecules, experimental studies on the acid–base properties of a branched polyelectrolyte, where an ionized site has more than two charged neighbors, are scarce. To investigate the extended Ising model more thoroughly, further experimental data are needed.

In this paper we demonstrate that poly(propylene imine) dendrimers (recently introduced as Astramol™ dendrimers) are a system from which detailed information about the ionization of branched structures can be extracted.

Highly charged linear polyelectrolytes protonate in two steps. The intermediate plateau in the titration curve is related to a stable intermediate protonation state, where protonated and deprotonated groups alternate along the chain. This two-step behavior is caused by the short-range character of the interaction potential between ionizable sites. The splitting between the two protonation steps is related to the interaction energy of forming a pair of nearest-neighbor ionized sites.

As discussed below, two main differences between ionization behavior of dendrimers and the corresponding linear polyelectrolytes emerge [6]: (i) The intermediate plateau in the titration curve of a dendrimer arises at a degree of protonation 2/3, whereas for their linear analogues this occurs at half protonation. (ii) Since every dendrimer site has three nearest-neighbors, the overall effect of interactions is much stronger than for the linear polyelectrolyte with only two nearest-neighbors. In the latter case the splitting between the two protonation steps is therefore of the order of two times the splitting between the two protonation steps in a corresponding diprotic acid or base, whereas for the dendrimers this splitting is three times as large splitting.

Potentiometric titrations

The titration curves of five generations of 1,4-diaminobutane poly(propylene imine) dendrimers, obtained from DSM, the Netherlands, are shown in Fig. 1. A schematic picture of this sequence is represented on the right-hand side of Fig. 1, all structures emanating from a 1,4-diaminobutane core with propylene imine monomers. In the following these dendrimers will be abbreviated with DAB-*dendr*-(NH$_2$)$_x$ with $x = 4, 8, 16, 32$, and 64. Packing

Fig. 1 Experimental titration curves of poly(propylene imine) dendrimers DAB-*dendr*-(NH$_2$)$_x$ (schematic representations on right-hand side) plotted as the degree of protonation θ of as a function of pH in KCl as background electrolyte. The ionic strengths are 0.1 M (▲), 0.5 M (■) and 1.0 M (●). The curves are translated along the ordinate for clarity. On the left-hand side (pH 2) the degree of protonation corresponds to one, while on the right-hand side (pH 13) the degree of protonation is zero. For DAB-*dendr*-(NH$_2$)$_{64}$ the most important protonation microstates are also shown (top); a closed circle indicates a protonated site and an open circle a deprotonated state. The two-step protonation curve results from the stability of the intermediate onion-like protonation state, where 2/3 of the sites are protonated. Solid lines are best fits with the Ising model with parameters shown in Table 1

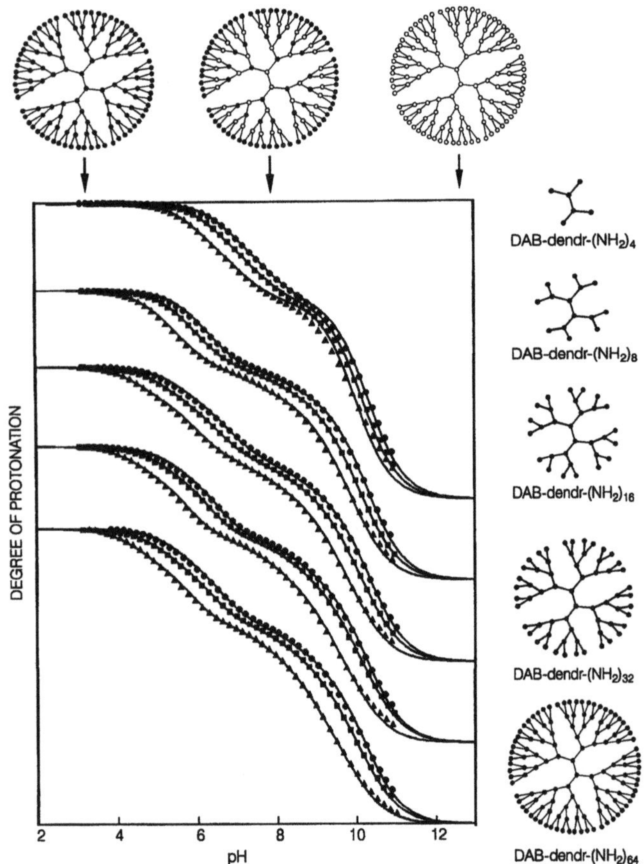

constraints with the outermost shells make the complete synthesis of higher generations difficult.

All potentiometric titrations were carried out at a temperature of $25 \pm 1\,°C$, with HCl and carbonate-free NaOH/KOH. NaCl/KCl was used as the supporting electrolyte to keep the ionic strength constant at 0.1, 0.5 and 1.0 M during the experiment. The titration curves have an error of ± 0.05 on the pH scale and an error of $\pm 10^{-2}$ in the degree of protonation. More detailed information about the experimental conditions and analysis methods is described in Ref. [7].

Ising model

The protonation behavior of polyelectrolytes can be rationalized in terms of an Ising model [4–6, 8, 9]. Each of the N-ionizable sites can be characterized by a state variable s_i such that $s_i = 0$ corresponds to a deprotonated site and $s_i = 1$ to a protonated site. Including pair interactions only, the free energy of a given protonation state F relative to the completely deprotonated state can be written as

$$\frac{F(s_1, \ldots, s_N)}{kT \ln 10} = \sum_i (\text{pH} - p\hat{K}_i)s_i + \sum_{i<j} \varepsilon_{ij}s_i s_j , \qquad (1)$$

where kT denotes thermal energy. In this equation we have introduced the following abbreviations: pH is the negative common logarithm of the proton activity (in M), $p\hat{K}_i$ the negative common logarithm of the microscopic dissociation constant of site i given all other groups are deprotonated, and $\varepsilon_{ij} > 0$ characterizes the strength of the pair interactions between sites i and j. These pair interaction parameters are assumed to be nonzero for nearest neighbors only. A more extended explanation of the evaluation of the Ising model to the titration data of a branched polyelectrolyte is described in Ref. [6].

The simplest Ising model which describes the experimental titration data of the 1,4-diaminobutane poly(propylene imine) dendrimers in a satisfactory fashion involves four different microscopic pK values and three nearest-neighbor pair interaction parameters, see Fig. 2.

The microscopic pK values are denoted by $p\hat{K}^{(1)}$, $p\hat{K}^{(3,o)}$, $p\hat{K}^{(3)}$ and $p\hat{K}^{(3,i)}$. The outermost primary amines have the values $p\hat{K}^{(1)}$. We have three groups for the tertiary amines. The outermost shell of tertiary amine groups has $p\hat{K}^{(3,o)}$, the two innermost tertiary amines $p\hat{K}^{(3,i)}$, while all the remaining ones $p\hat{K}^{(3)}$.

The three pair interaction parameters are abbreviated as ε_1, ε_2 and ε_3. The outermost bonds between primary and tertiary amines have the value ε_3. The innermost bond, which consists of four carbons (1,4-diaminobutane), has the value ε_1. All remaining bonds have the value ε_2. The

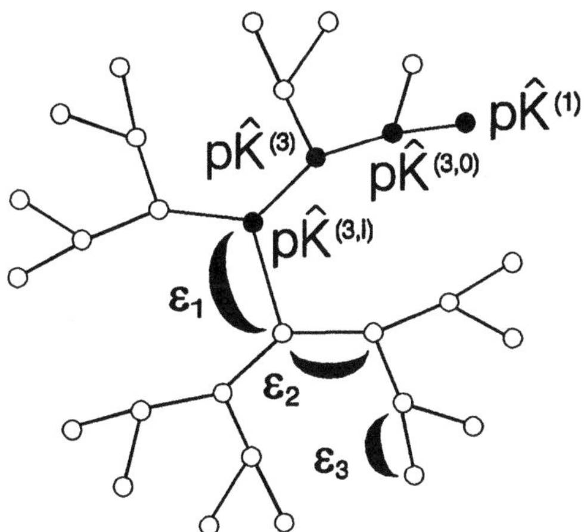

Fig. 2 Schematic representation of a third generation dendrimer molecule, showing the various parameters used to fit the Ising model prediction to the experimental curves. Additional shells in higher generations are described with $p\hat{K}^{(3)}$

pair interaction parameters ε_1 and ε_2 were estimated from macroscopic pK values of linear amines [5] and were not adjusted further.

The titration curves of DAB-*dendr*-$(NH_2)_4$ were fitted by adjusting two parameters $p\hat{K}^{(1)}$ and $p\hat{K}^{(3,i)}$. The titration curve of DAB-*dendr*-$(NH_2)_8$ was rationalized by introducing two additional parameters; $p\hat{K}^{(3,o)}$ in order to account for a smaller proton affinity of the middle shell and by assuming a larger interaction parameter ε_3 for the outermost shell. The parameters $p\hat{K}^{(3,i)}$ and ε_3 determined from DAB-*dendr*-$(NH_2)_8$ were used for all higher dendrimers. The titration curves of DAB-*dendr*-$(NH_2)_{16}$, DAB-*dendr*-$(NH_2)_{32}$, and DAB-*dendr*-$(NH_2)_{64}$ were fitted by introducing one additional parameter $p\hat{K}^{(3)}$ which reflects the different proton affinity of the additional shells.

Results

The solid lines in Fig. 1 are least-squares fits of the Ising model to the titration data of poly(propylene imine) dendrimers at salt concentrations of 0.1, 0.5, and 1.0 M, the resulting parameter values are given in Table 1.

One observes that the tertiary group is slightly more acidic than the primary one, which is in line with the observation that primary amine groups are more basic than tertiary ones [10]. The innermost amine groups are more acidic because of their different coordination which involves a butyl chain instead of a propyl chain. The fact that the pair interaction parameter is larger for the outermost shell than for the remaining ones is probably related to a more effective solvation of these bonds.

In spite of the fact that at maximum four parameters are determined from a single-titration curve (DAB-*dendr*-$(NH_2)_8$), these parameters remain remarkably constant throughout all generations investigated. Thus, the present model captures the intrinsic features of the dendrimer protonation sequence. The only exception is $p\hat{K}^{(1)}$ at 0.1 M, which shows a systematic variation with generation number. This parameter decreases with increasing size of the dendrimer. Such pK shifts are induced most likely by neighboring amine groups, which are situated on different chains. Such effects have been observed for other amines at lower ionic strengths as well [11]. For higher ionic

Table 1 Ising model parameters for different dendrimers of generation k and at ionic strength I

I [M]	k	x	N	$p\hat{K}^{(3,i)}$	$p\hat{K}^{(3)}$	$p\hat{K}^{(3,o)}$	$p\hat{K}^{(1)}$	ε_1 [a]	ε_2 [a]	ε_3
0.1	1	4	6	9.02	–	–	9.97	0.61	1.05	–
	2	8	14	8.19	–	9.66	9.85	0.61	1.05	1.57
	3	16	30	8.19	7.99	9.72	9.79	0.61	1.05	1.57
	4	32	62	8.19	8.02	9.71	9.70	0.61	1.05	1.57
	5	64	126	8.19	7.95	9.66	9.56	0.61	1.05	1.57
0.5	1	4	6	9.41	–	–	10.14	0.59	1.03	–
	2	8	14	8.82	–	9.91	10.12	0.59	1.03	1.46
	3	16	30	8.82	8.54	9.96	10.13	0.59	1.03	1.46
	4	32	62	8.82	8.59	10.00	10.15	0.59	1.03	1.46
	5	64	126	8.82	8.47	9.93	10.05	0.59	1.03	1.46
1.0	1	4	6	9.57	–	–	10.29	0.56	0.97	–
	2	8	14	9.07	–	9.64	10.33	0.56	0.97	1.21
	3	16	30	9.07	8.75	9.72	10.34	0.56	0.97	1.21
	4	32	62	9.07	8.74	9.67	10.29	0.56	0.97	1.21
	5	64	126	9.07	8.72	9.67	10.25	0.56	0.97	1.21

[a] Taken from Ref. [5] and not adjusted further.

strengths, the screening is more effective and these effects disappear.

Note that only DAB-*dendr*-$(NH_2)_4$ does not fit into the scheme equally well as the other molecules. The present scheme is not sufficently flexible and does not allow a better fit of its titration curve than the one shown in Fig. 1. But since this molecule represents the first generation in the whole series of dendrimers, we are not particularly concerned with minor deficiencies of the model for this particular molecule.

Onion-like shell protonation behavior

The titration experiments reveal a characteristic protonation curve with two protonation steps around pH values of 6 and 10 with an intermediate plateau where 2/3 of the ionizable sites are protonated. In Fig. 3 the individual shell protonation behavior of the four shells of DAB-*dendr*-$(NH_2)_{16}$ at 1.0 M KCl is shown. Therefore parameters from Table 1 were used. Figure 3 shows that in the stabilized intermediate structure at pH 8 (Fig. 1, DAB-*dendr*-$(NH_2)_{16}$ at 1.0 M) mainly the shell of primary amines and the second shell of tertiary amines are protonated. This effect leads to the intermediate plateau in the titration curve of a dendrimer. For a large dendrimer the stable intermediate protonation state is an onion-like structure where all odd shells are protonated. Ref. [12] discusses experimental measurements of such titration curves of the individual shells by NMR-N[15], which confirm this picture quantitatively.

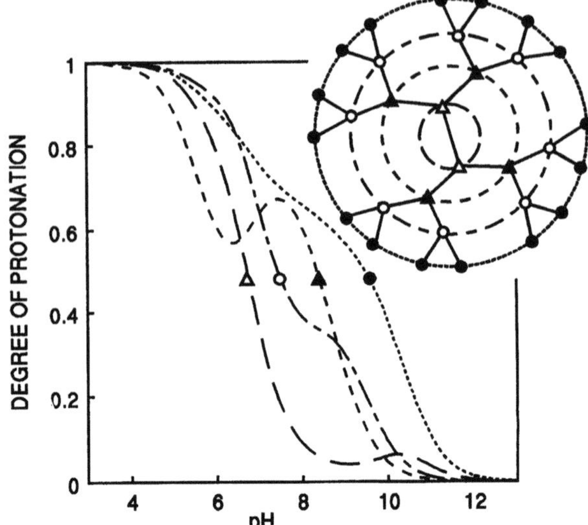

Fig. 3 Calculated protonation behavior of the four individual shells for DAB-*dendr*-$(NH_2)_{16}$ at 1.0 M KCl, based on the fit parameters as provided in Table 1. A schematic representation of the dendrimer is shown, in which each of the four shells is specified with a symbol and a dashed circle, corresponding to the titration curve of the individual shell. The fraction of protonated amines in the different shells at pH 8 corresponds to the stabilized intermediate structure in the general titration curve

We have shown that the Ising model is able to describe the protonation behavior of the Astramol™ dendrimers quantitatively. The concepts developed here for poly(propylene imine) dendrimers are expected to be applicable more generally. For other types of ionizable dendrimers similar odd-even shell protonation behavior is expected.

References

1. Marcus RA (1954) J Phys Chem 58: 621
2. Harris FE, Rice SA (1954) J Phys Chem 58:725
3. Katchalsky A, Mazur J, Spitnik P (1957) J Polym Sci 23:513
4. Smits RG, Koper GJM, Mandel M (1993) J Phys Chem 97:5745
5. Borkovec M, Koper GJM (1994) J Phys Chem 98:6038
6. Borkovec M, Koper GJM (1997) Macromolecules 30:2151
7. van Duijvenbode RC, Borkovec M, Koper GJM (1997) Polymer 39:2657
8. Borkovec M, Koper GJM (1996) Ber Bunsenges Phys Chem 100:764
9. Koper GJM, Borkovec M (1996) J Phys Chem 104:4204
10. Perrin DD, Dempsey B, Serjeant EP (1981) pK_a Prediction for Organic Acids and Bases. Chapman & Hall, London
11. Smith RM, Martell AE (1989) Critical Stability Constants. Plenum Press, New York
12. Koper GJM, van Genderen MHP, Elissen-Román C, Baars MWPL, Meijer EW, Borkovec M (1997) J Am Chem Soc 119:6512–6521

Progr Colloid Polym Sci (1998) 110:129–133
© Steinkopff Verlag 1998

T. Preis
R. Biehl
T. Palberg

Phase transitions in a colloidal dispersion flowing through a cylindrical capillary

T. Preis (✉) · R. Biehl · T. Palberg
Johannes Gutenberg Universität Mainz
Institut für Physik, KoMa 336
Staudingerweg 7
D-55099 Mainz
Germany

Abstract The flow of a charged-stabilized colloidal dispersion of crystalline equilibrium structure through a cylindrical capillary is investigated. The simultaneous existence of up to three differently ordered states is observed under conditions of stationary flow. The evolution of their concentric arrangement along the capillary is determined by Bragg microscopy. For sufficient low fluxes, stationary phase distributions are obtained. While the apparent viscosity is non-monotonous and non-Newtonian, the corresponding velocity profiles are found to be piecewise parabolic and are compatible to numerical calculations. Furthermore, we test the assumption of a constant yield stress determining the radial position of the interface between the polycrystalline core and the surrounding layer phase.

Key words Colloidal dispersion – non-equilibrium phase transitions – non-Newtonian rheology

Introduction

Monodisperse colloidal dispersions of spherical particles are valuable model systems to study the structure and dynamics of complex fluids. [1, 2] Due to the colloid-specific time and length scales, their properties are easily accessible by light scattering and/or microscopy. The pair interaction is experimentally variable between the theoretical limits of hard spheres and the one-component plasma. Throughout the last years there has been a tremendous increase in our knowledge of their equilibrium behavior. In the particular case of highly charged spheres the phase diagram comprises regions of fluid, body centred cubic (bcc), face centred cubic and glass-like structure. Also other properties like elastic behavior and equilibrium transport behavior have been experimentally determined and successfully modelled by theory and simulation.

In addition, a large number of experimental and theoretical investigations on the structural and rheological properties in simple shear flows have been performed [3–11]. One main result was the establishment of sequences of non-equilibrium phase transitions with increasing shear rate: shear melting of a bcc crystal, e.g. proceeds via an intermediate state with hexagonally closed packed layers (hex), which may move relative to each other in several fashions [3–6]. The phase transitions are accompanied by pronounced changes in the system viscosity [4, 7, 8]. In two recent papers these phases were observed to coexist in a capillary viscosimeter which in turn led to a pronounced non-linear, non-monotonous rheology [9, 10]. Such coupling between phase transition and flow behavior is also known for other systems such as worm-like micelles [11] and is of great importance to the processing of technical dispersions.

In particular, the authors of [9, 10] observed the shear melt entering the capillary to partially solidify along it. For low fluxes dV/dt stationary concentric distributions of phases were observed. At the position of the phase boundaries the velocity profiles $v(r)$ showed a jump in the shear rate dv/dr. Finally the apparent viscosity showed a general decrease with increasing flux, but a step-like increase, when the number of coexisting phases changed. This jump was most pronounced when a fluid phase appeared in addition to the coexisting polycrystalline core and hexagonal close packed layers.

We here explore possibilities to model these data in two different respects: We first show that the measured velocity profiles can be qualitatively understood on the basis of the Navier–Stokes equation (NS). A quantitative evaluation, however, will require a detailed knowledge of the pressure drop along the tube. Secondly we give an estimate of the critical stress at the position of the phase boundary between crystalline core and layer phase. From both discussions, suggestions for advanced rheo-optic experiments are derived.

Experimental

Details of the experimental set-up [7, 8], the preparation of the colloidal melt [9], the phase behavior and other equilibrium properties [10] have been given in previous papers. We here only note that polystyrene latex spheres of nominal diameter $2a = 102$ nm and charge $Z \approx 6 \times 10^2$ (Lot #2011 M9R, Seradyn, USA) were used at deionized conditions ($c_{\mathrm{salt}} \leq 2 \times 10^{-7}$ mol l^{-1}) and at a packing fraction of $\Phi = 0.0035$. At rest the suspension shows a bcc structure.

The suspension is subjected to flow through a horizontally mounted capillary of 4 mm inner diameter and 500 mm length. It flows under solely hydrostatic pressure difference $\Delta p = \rho g \Delta h$ between two reservoirs of fixed height difference Δh, with ρ being the density and g the gravitational acceleration. The reservoirs are connected by separate tubings containing an inert gas atmosphere. The suspension enters the capillary in a completely shear molten state and partially solidifies along it. The phase distribution strongly depends on the overall flux which was varied between 5.9 mm^3 s^{-1} ≤ dV/dt ≤ 152.1 mm^3 s^{-1} by adjusting Δh. A peristaltic pump refills the upper reservoir through a closed tubing system containing an ion exchange chamber and a conductivity measurement to maintain and control conditions of complete deionization.

Phase distribution and flow profiles

The phase distributions along the capillary for different fluxes were determined by Bragg microscopy [10]. Three representative examples are given in Fig. 1, where the intensity of light scattered off a laser beam crossing the cell and observed by a CCD-camera under an angle corresponding to the fulfilment of the Bragg condition for the layer phase is shown. Being presented like this the

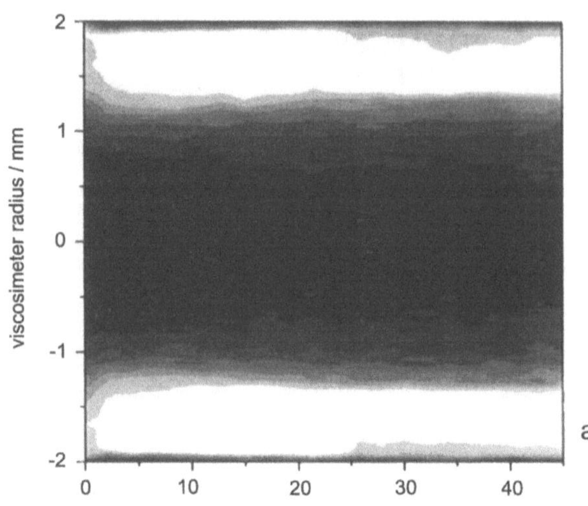

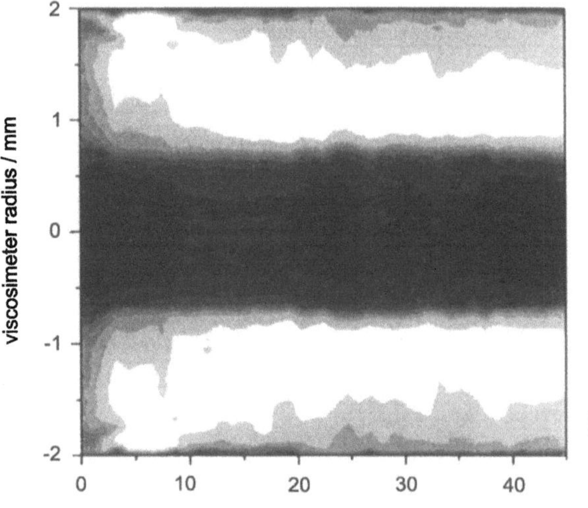

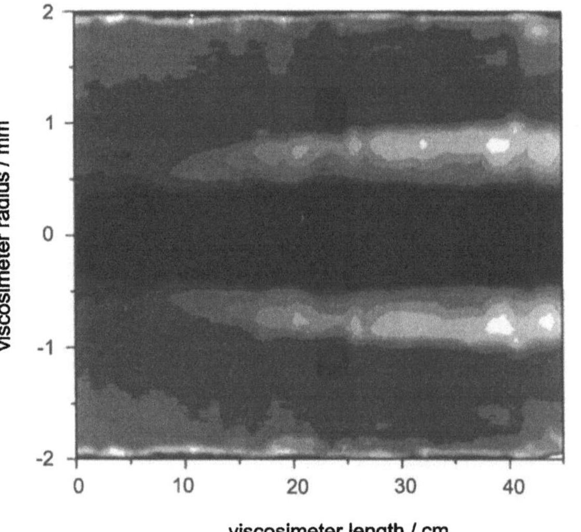

polycrystalline core appears black to dark grey, the melt medium grey, while the layers scatter with highest intensity and appear light grey to white.

Simultaneously the velocity profiles were determined using Laser-Doppler Anemometry with fiber optic detection providing a spatial resolution of 10 μm. In Fig. 2 three examples are shown. Figure 2a shows the data for $dV/dt = 58.6\,\mathrm{mm}^3\,\mathrm{s}^{-1}$, where the hexagonal closed packed layers are found to have their largest extension. The data can be well fitted by a parabolic velocity profile, while the local shear rate varies over more than an order of magnitude. This result corresponds to a shear rate independent viscosity of the layer phase, i.e. Newtonian flow behavior. The same holds for Fig. 2b, where the velocity profile is shown for the extended fluid phase at $dV/dt = 152.1\,\mathrm{mm}^3\,\mathrm{s}^{-1}$. We are thus able to confirm the assumption of Newtonian behavior within the individual phases made in our previous paper.

In Fig. 2c corresponding to the phase distribution of Fig. 1c we show that a good fit is obtained also for a flux of $dV/dt = 105.3\,\mathrm{mm}^3\,\mathrm{s}^{-1}$, where four regions of differently ordered phases coexist. We here use a piece-wise parabolic flow profile:

$$v(r) = v_0 - b(r - a)^2 , \tag{1}$$

where v_0 is the velocity at the center of each parabola and a and b are fit parameters corresponding to the opening and the position of the parabola, respectively. The variable r covers the values, where the respective phases are observed. We explicitly note that within experimental uncertainty b as obtained from best fits of Eq. (1) to the data is identical for both layer phase regions, while it is significantly smaller for the fluid. The core in all cases flows as a solid plug and we thus conclude that $\eta_{\mathrm{bcc}} \gg \eta_{\mathrm{fluid}} > \eta_{\mathrm{hex}}$.

Modelling of flow properties

To qualitatively reproduce the flow profiles we employ a simple model calculation based on several assumptions: First we assume the NS to apply to our problem of flow with phase transitions. Second, we assume Newtonian flow behavior for all individual phases. Third, pressure drop is approximately to be linear in capillary length l. This corre-

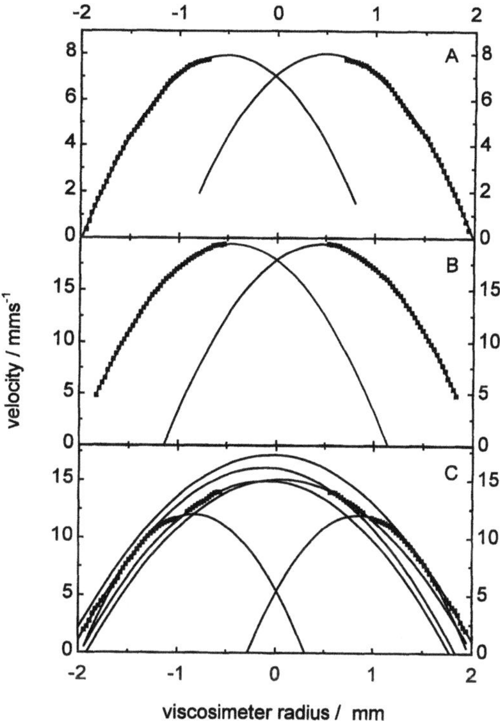

Fig. 2 Flow profiles measured shortly before the outlet of the capillary by Laser Doppler Anemometry. (A) $dV/dt = 58.6\,\mathrm{mm}^3\,\mathrm{s}^{-1}$; (B) $dV/dt = 152.1\,\mathrm{mm}^3\,\mathrm{s}^{-1}$; (C) $dV/dt = 105.3\,\mathrm{mm}^3\,\mathrm{s}^{-1}$. Solid lines are best fits to the data of the individual phases

sponds to a l-independent phase distribution as is experimentally observed only for fluxes $dV/dt < 66\,\mathrm{mm}^3\,\mathrm{s}^{-1}$. The NS is then numerically integrated under the boundary conditions of $v(r = 2\,\mathrm{mm}) = 0$ for cylindrical geometry. As an additional parameter we vary the number of phases and the position of the phase boundaries.

In Fig. 3a–c we show three representative profiles. Curve (A) is for a highly viscous core surrounded by a low viscous region. For $\eta_{\mathrm{bcc}} = \infty$ this corresponds to the experiments at $dV/dt < 66\,\mathrm{mm}^3\,\mathrm{s}^{-1}$, where the hexagonal closed packed layers lubricate the solid core. Curve (B) is a demonstration of the flow profile expected for a low viscosity core. Curve (C) approximates the experimental situation in Fig. 2c with the melt separating two regions of hexagonal closed packed layers and the ratio of viscosities taken as $\eta_{\mathrm{fluid}}/\eta_{\mathrm{hex}} = 1/0.4$, while $\eta_{\mathrm{bcc}} \equiv \infty$. Our modelling thus qualitatively reproduces the unexpected flow profiles observed in the experiment, including the case of large fluxes, where the melt is lubricated by the hexagonal closed-packed phase.

All further analysis by fits of the modelled profiles to the experimental data to quantify the viscosities of the respective phases requires additional detailed knowledge of the pressure drop along the complete instrument. While

Fig. 1 Intensity scattered off a laser beam crossing the capillary as detected by a CCD camera positioned under an observation angle corresponding to the Bragg condition for the hexagonal layer phase. Thus, different intensities correspond to different phases. Dark, low-intensity: polycrystalline body centered cubic; medium grey, medium intensity: isotropic shear melt; white, high intensity: hexagonal layer phase. (A) $dV/dt = 22.5\,\mathrm{mm}^3\,\mathrm{s}^{-1}$; (B) $dV/dt = 58.6\,\mathrm{mm}^3\,\mathrm{s}^{-1}$; (C) $dV/dt = 105.3\,\mathrm{mm}^3\,\mathrm{s}^{-1}$

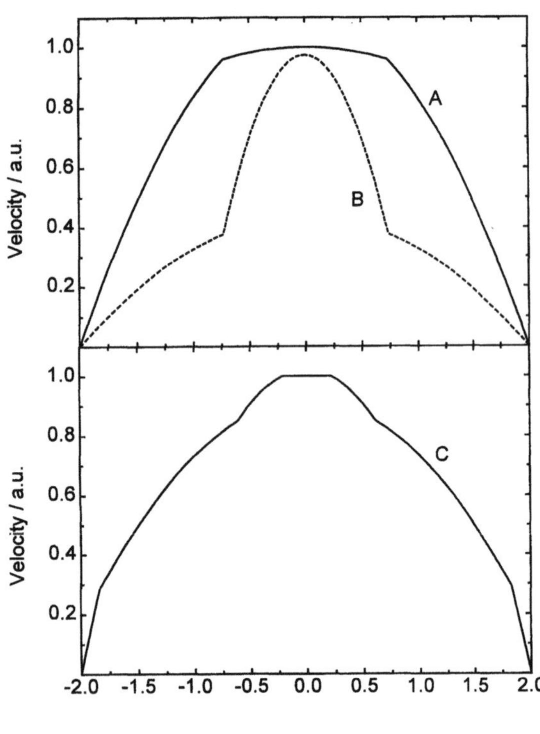

Fig. 3 Flow profiles calculated from the Navier–Stokes equation. (A) a low-viscosity phase surrounds a high viscosity phase; (B) a high-viscosity phase surrounds a low viscosity phase. (C) Number of phases and viscosity ratio $\eta_{\text{bcc}} \gg \eta_{\text{fluid}} > \eta_{\text{hex}}$ are chosen to approximate the situation given in Figs. 1c and 2c

the pressure drop over the connections between reservoir and capillary (1 m flexible tubing of $R = 1.2$ mm inner radius) is determined by the R^4 – dependence in the Hagen–Poiseuille law with a pure fluid phase, the pressure drop over the capillary varies in a similar manner but with an additional dependence on the phase distribution in it. Therefore, the overall pressure drop $\Delta p = \Delta p_{\text{con}} + \Delta p_{\text{cap}}$ is a linear function of the height difference Δh but the ratio between the partial pressure differences is a strong function of the phase distribution in the capillary. Further the radius of the connections may not be determined to sufficient accuracy, these should be omitted in future experiments.

Even then it would remain difficult to determine the viscosities of the individual phases. While the positions of the phase boundaries in principle may be known to 2% accuracy, fluctuations of these positions may be on the order of 10–15% of the capillary radius. The latter are most pronounced for fluxes corresponding to a change in the number of coexisting phases where the flow is observed to show pronounced time dependence. Taking the uncertainty in the phase boundary position as 10% yields a residual uncertainty in η of at least 40%.

The model presented above assumed a linear pressure drop along the capillary. In the experiments, however, a stationary phase distribution is reached only after some distance past its inlet, and for the largest fluxes the hexagonal layer phases continue to grow over the whole capillary length. We therefore modelled the pressure drop along the tube by fixing the individual viscosities as in Fig. 3c but allowing for shifts in the boundary position. The pressure drops approximately linearly for fluxes $\mathrm{d}V/\mathrm{d}t < 66$ mm^3 s^{-1}, where solid core and layer phase coexist. At larger fluxes a pronounced non-linearity evolves, i.e. since except for the solid core the suspension is shear molten closely past the inlet, the growth of the low viscosity layer phase lead to a decrease in $\mathrm{d}p/\mathrm{d}l$.

The latter situation may in principle be modelled, but in addition to known viscosities η_i requires the determination of the l-dependence of the flow profiles. Finally we mention that also the influence of the capillary wall on the kinetics of phase transitions remains to be clarified.

Positions of the phase boundaries

For fluxes $\mathrm{d}V/\mathrm{d}t < 66$ mm^3 s^{-1} the interface between solid core and hexagonal closed packed layers is found to be at radial positions constant along the capillary. If we assume the pressure drop along the capillary and the overall flux $\mathrm{d}V/\mathrm{d}t$ to be proportional to Δh (i.e. we neglect the connections) we can calculate the corresponding stress profile:

$$\sigma(r) = \frac{\rho g \Delta h}{2l} r. \tag{2}$$

In Fig. 4 we plot the stress at the positions of the phase boundary as a function of the overall flux. This critical stress is found to be approximately constant at a value of 600 ± 100 mPa for $\mathrm{d}V/\mathrm{d}t < 66$ mm^3 s^{-1}. The corresponding ratio of $\sigma_{\text{crit}}/G = 1.5$ is significantly larger than the

Fig. 4 Critical shear stresses from a comparison of calculated stress profiles to the position of the phase boundary between crystalline core and layer phase

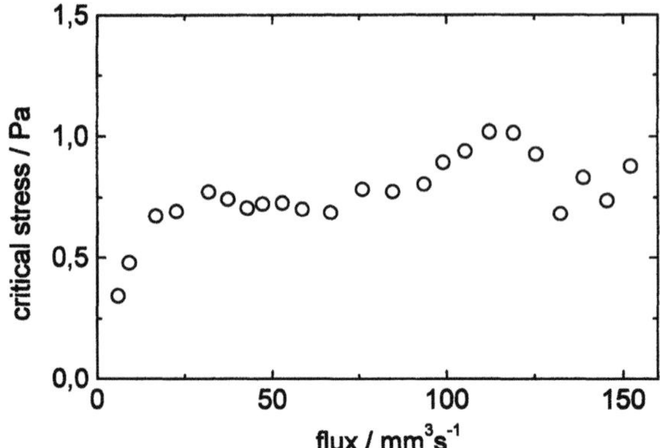

value of 0.04 obtained by Chen and Zukoski for a similar system [7].

Conclusions

While the set-up was optimized for structural analysis, an equally well defined access to rheologic boundary conditions is of uttermost importance. The exact knowledge of the stationary phase distribution may then allow the determination of the rheological properties of the individual phases as well as of the whole system. Our analysis clearly shows that due to the complex feedback mechanisms presented, only the combination of optical and rheological experiments will lead to a comprehensive understanding of flows with shear-induced phase transitions.

References

1. Pusey PN (1989) In: Hansen JP, Levesque D, Zinn-Justin J (eds) Liquids, Freezing and Glass Transition. 51st Summer School in Theoretical Physics. Les Houches (F), Elsevier, Amsterdam, 1991, p 763
2. Ackerson BJ (1990) (ed) Phase Transitions 21:2–4
3. Ackerson BJ (1983) Physica A 128:221
4. Laun HM, Bung R, Hess S, Loose W, Hess O, Hahn K, Hädicke E, Hingmann R, Schmidt F, Lindner P (1993) J Rheol 36:1057
5. Loose W, Ackerson BJ (1994) J Chem Phys 101:7211
6. Stevens MJ, Robbins MO (1993) Phys Rev E 48:3778
7. Chen LB, Zukoski CF (1990) Phys Rev Lett 65:44
8. Chen LB, Chow MK, Ackerson BJ, Zukoski CF (1994) Langmuir 10:2817
9. Palberg T, Würth M (1996) J Phys I (France) 6:237–244
10. v Hünerbein S, Würth M, Palberg T (1996) Progr Coll Polym Sci 100:241–245
11. Berret JF, Roux DC, Porte G (1994) J Phys II (France) 4:1261

Progr Colloid Polym Sci (1998) 110:134–138
© Steinkopff Verlag 1998

M.J. García-Salinas
F.J. de las Nieves

A study of the primary electroviscous effect in monodisperse carboxyl polystyrene latex suspensions

M.J. García-Salinas
Prof. F.J. de las Nieves (✉)
Complex Fluids Physics Group
Department of Applied Physics
Faculty of Experimental Sciences
University of Almería
E-04120 Almería
Spain

Abstract In the present work, an experimental research of the primary electroviscous effect has been carried out with suspensions of carboxyl polystyrene latexes. The latex was previously characterized from a surface and electrokinetic point of view. The surface charge density and its variation with the pH were determined by conductimetric and potentiometric titration. The electrophoretic mobility of this latex particles was measured against the NaCl concentration. The mobility data were translated to zeta-potential values by using different theories (Smoluchowski, O'Brien–White and Dukhin–Semenikhin). The primary electroviscous coefficient (p) has been obtained by measuring viscosities of suspensions as a function of the volume fraction of solid. The viscosities were measured at several pH and electrolyte (NaCl) concentrations. The experimental values of the p coefficient were compared to those obtained theoretically by several authors (Smoluchowski, Krasny–Ergen, Booth …). Because zeta-potential has to be used in these theories, the p coefficient was calculated by including different zeta potential values. The p values calculated by using Booth's theory and the O'Brien–White zeta potential were very similar to the experimental p coefficients.

Key words Electroviscous effect – polystyrene latex – zeta potential

Introduction

Rheological properties of model colloidal particles dispersions have been widely studied. The hydrodynamic behaviour of the fluid is considerably affected by the presence of the electrical double layer (e.d.l.) around the spheres. These systems exhibit characteristic viscosity behaviour as a result of electroviscous effects, usually classified into first-, second- and third-order effects [1–3].

The primary electroviscous effect occurs when a suspension of charged particles is sheared. When the system is undergoing flow, the double layer around the particles is distorted by the shear field, leading to an extra-disipation of energy and an increased viscosity. It manifests itself as a correction factor "p" to the Einstein equation [3, 4]

$$\eta_r = 1 + 2.5(1 + p)\phi \ . \tag{1}$$

We briefly summarize here some of the theoretical and experimental contributions. With regard to the former, many researchers have studied the subject and developed different theories. Smoluchowski [5] presented the following value for the "p" coefficient in 1916:

$$p_S = \frac{(2\varepsilon_0\varepsilon_r\zeta)^2}{\lambda_0\eta_0 a^2} \ , \tag{2}$$

where ε_0 is the permittivity of free space, ε_r the relative permittivity, ζ the zeta potential, λ_0 the specific conductivity of the continuous phase and η_0 its viscosity. This was expected to be valid for large values of κa. Later, Krasny–Ersgen [6] derived Smoluchowski's value multiplied by 3/2.

A more general analysis of the primary electroviscous effect was carried out by Booth [7]. He calculated the viscosity of a colloidal suspension as a series expansion in Q (the charge of the particles) or alternatively in the zeta potential. A dependence with ζ^2 was found for the primary electroviscous effect coefficient:

$$p_{BO} = q^* \left(\frac{e\zeta}{KT}\right)^2 Z(\kappa a)(1 + \kappa a)^2 , \tag{3}$$

where q^* is a dimensionless coefficient and $Z(\kappa a)$ is a complicated function which expresses the deformability of the counterion cloud and increases with decreasing κa. A simpler form of Eq. (3) given by Honig et al. [8] is used in this work.

Finally, Watterson and White [9] obtained the solution of the problem by solving a system of coupled differential equations. They also discussed Booth's theory, finding that it was valid for small values of the zeta potential (where the higher order terms in Booth's expansion were less important).

However, the experimental results do not always agree with the theoretical estimation of the p coefficient. McDonogh and Hunter [10] obtained experimental data higher than those theoretically expected. In their conclusions, the necessity of more experimentation was pointed out. Similar discrepancies were found by Delgado et al. [11]. In these papers, the calculations of the "p" coefficient using Booth's theory were similar to those obtained using Watterson and White's theory, although some differences could be found for low values of κa and high values of zeta potential (obtained using the O'Brien and White [12] conversion theory). This was also concluded in [13], where an experimental examination of the function $Z(\kappa a)$ of Eq. (3) was carried out. But anyway, the theory always seemed to underestimate the experimental effect and no good agreement is found in the literature except some very specific cases. In ref. [8], for example, the fitting of the experimental relative viscosity data to the Booth's equation was done adapting two parameters: particle charge and residual electrolyte concentration. A similar procedure was done in [14] where good fits were obtained adapting the surface charge density.

Added to all that, in the theoretical calculations of the electroviscous coefficient, the zeta potential has always been used given by the O'Brien and White [12] conversion theory. It would be interesting to use other theories, as they can provide very different values for the zeta potential [15] and could change the theoretical estimation of the electroviscous effect.

This brief account of the subject raises the necessity of taking it up again. So, in the present work, an experimental

research of the primary electroviscous effect has been carried out with suspensions of carboxyl polystyrene latex. The p coefficient of this effect has been obtained by measuring viscosities of suspensions as a function of the volume fraction of solid. The viscosities were measured at several pH and electrolyte (NaCl) concentrations. The idea is to vary the surface charge density of the particles by changing the pH of the solution and to study the influence of these changes on the primary electroviscous coefficient. By changing the ionic strength we can modify the electric double layer thickness, and therefore vary the energy dissipated during its distortion.

The experimental values of electroviscous effect have been compared to those obtained theoretically by several authors (Smoluchowski, Krasny-Ergen, Booth, and Watterson and White). Besides, as the "p" coefficient is a function of κa and the ζ potential, its influence on the different mobility to ζ potential conversion theories has also been studied.

Materials and methods

All the chemicals in this study were of analytical grade and were used without further purification. In all experiments ultrapure water with electrical conductivity less than $1\ \mu S/cm$ was used.

The carboxylated latex used in this work was synthetized by our group using the emulsifier-free polymerization method in a discontinuous reaction [16]. The carboxyl groups were provided by the 4,4'-Azo-bis(4-Cyanovaleric acid) initiator, with the advantage that the surface characteristics are those of the polystyrene and there is no emulsifying agent present. Styrene (Merck) was previously distilled under low pressure (10 mmHg and 40 °C). The latex was cleaned by serum replacement until the conductivity of the supernatant was similar to that of the water.

The diameter of the particles, as obtained by TEM, was 187 ± 7 nm. The particle size was also measured by Photoncorrelation Spectroscopy (PCS) measurements.

The surface charge density was obtained by conductimetric and potentiometric titrations, in order to detect the change of the surface charge with the pH [15, 16]. The maximum surface charge density (for basic pH) was $\sigma_0 = 22.2 \pm 1.2\ \mu C/cm^2$ and Fig. 1 shows the dependence of σ_0 with pH.

The electrophoretic mobility measurements were carried out with a Malvern Zetamaster S device. The mobilities were obtained by taking the average of at least six measures considering their standard deviation as the experimental error for different salt (NaCl) concentrations and pH conditions. The zeta potentials were calculated

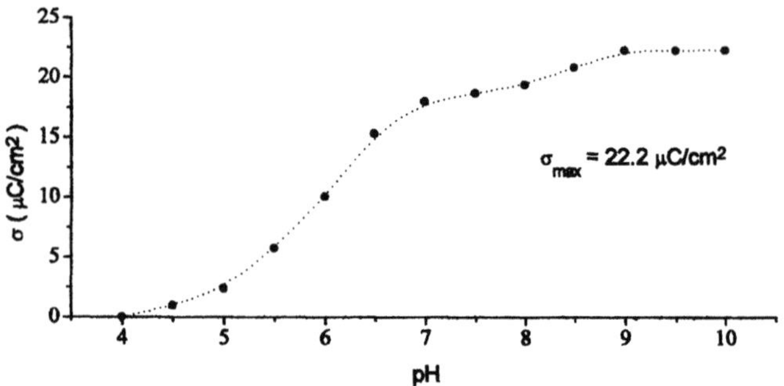

Fig. 1 Dependence of the surface charge density with pH

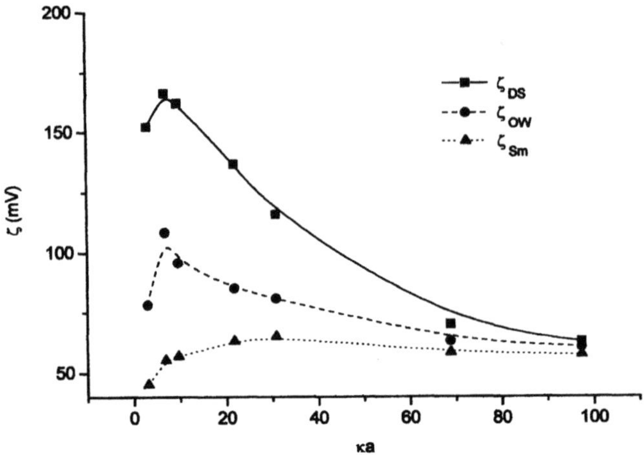

Fig. 2 Zeta potential (ζ) for the carboxyl latex with no buffer solution against the electrokinetic radius using Dukhin–Semenikhin (DS), O'Brien–White (OW) and Smoluchowski (Sm) conversion theories

using the Dukhin and Semenikhin [17], O'Brien and White [12], and Smoluchowski [18] theories.

The concentration of the stock latex solution was determined by evaporating to dryness at about 90 °C. Using the dried weight data obtained, the volume fractions of the suspensions were carefully calculated and prepared. The density of the samples was calculated considering the amount of latex and salt present according to [19].

The viscosity of the samples was measured with a Schott-Geräte equipment using Ubbelohde capillary viscometers in a thermostatic bath (with refrigeration and agitation) keeping a constant temperature of 25.0 ± 0.1 °C. Measurement takes place after an equilibration time of 15 min, and the flow time is approximately 320 s.

Before the viscosity measurements, the suspensions were placed in an ultrasonic bath for at least 10 min. Also, before and after these measurements the presence of aggregates was tested with PCS measurements. Conductivity and pH were also checked in order to detect any possible anomaly.

The capillary was cleaned after each measurement with HCl, acetone and water, and then tested with water (whose viscosity is known) to reassure that the cleaning process was right.

Results and discussion

First, the mobility measurements were translated into zeta potential data by using several theories. Figure 2 shows the results of the conversion for the mobilities with no buffer solution (approximately pH 6). Higher zeta-potential values but with similar trend were found for pH 7 (results not shown). The clear maximum in the mobility, which appears at a concentration of 10^{-2} M of NaCl, almost disappears and is less pronounced when the zeta values are calculated by the Dukhin and Semenikhin theory [17].

For fixed pH and salt concentration, the viscosities of several suspensions with different volume fractions were determined. From the slope of the η vs. ϕ plot the value of the p coefficient is obtained. Then, the same is done for other salt concentrations. We have to point out that the experimental p values are very sensitive to any slight change in the slope of η vs. ϕ plot, which can produce anomalous results. Figure 3 shows the experimental p values versus the electrokinetic radius (κa, being κ the reverse of the Debye length and a the particle radius) and a wrong point when $\kappa a \approx 20$ can be observed. In principle, a nearly constant value of p can be observed. To compare the experimental electroviscous coefficients with the theoretical ones, Fig. 3 shows the Smoluchowski's coefficients (Eq. (2)) as calculated by using three zeta potential values (Smoluchowski, O'Brien and White and Dukhin and Semenikhin). The best agreement is found for the one which uses Smoluchowski's potential, because all the theories predict a decrease in p at low κa, although the Smoluchowski theory predicts the lower one.

Figure 4 shows the same results as in Fig. 3 but the theoretical p values are calculated by the Booth [7] theory (Eq. (3)). In this case a better agreement is found between the Booth's coefficient and the experimental one, when the O'Brien and White theory is used to estimate the zeta-potential. Thus, the use of Eq. (2) gives good results with the Smoluchowski theory of mobility/zeta-potential conversion, while a more advanced theory for p (as that from Booth) gives better results with a more advanced theory of conversion in zeta.

Finally, the p coefficient provided by the Watterson and White theory [9] is less sensitive to changes in zeta potential, and its estimation does not improve the agreement found for the experimental coefficient with the Booth one, when O'Brien and White zeta potential coversion is used.

The same study was done for pH 7 and similar results were found in relation with the theoretical and experimental comparison at low κa values. Figure 5 shows the

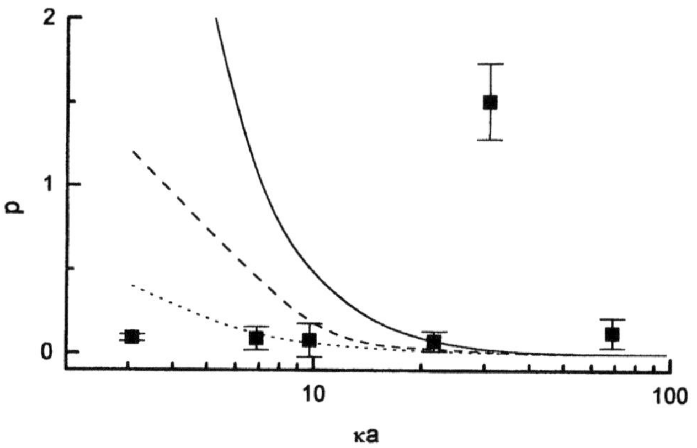

Fig. 3 Experimental and Smoluchowski's theoretical values of the primary electroviscous effect coefficient (p) versus the electrokinetic radius for no buffer solution. Three zeta potentials have been used: Dukhin–Semenikhin (——); O'Brien–White (– – – –); and Smoluchowski (· · · · ·)

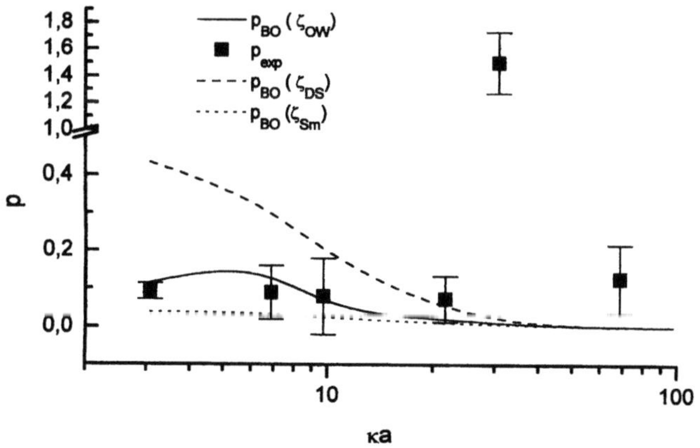

Fig. 4 Experimental and Booth's theoretical values (using the three zeta potentials) of the primary electroviscous effect coefficient (p) versus the electrokinetic radius for no buffer solution

experimental p values at pH 6 (without the wrong point) and 7, together with the best fitting as obtained by using Eq. (3) and the O'Brien and White theory of conversion. The differences in surface charge and mobility when the pH changes from 6 to 7, are reflected in the electroviscous effect with higher values at pH 7. For the experimental data at pH 7, however, the best theoretical fit cannot be chosen using zeta potential as a parameter: the differences between the theoretical (Booth) values using O'Brien and White or Dukhin and Semenikhin zeta potentials are

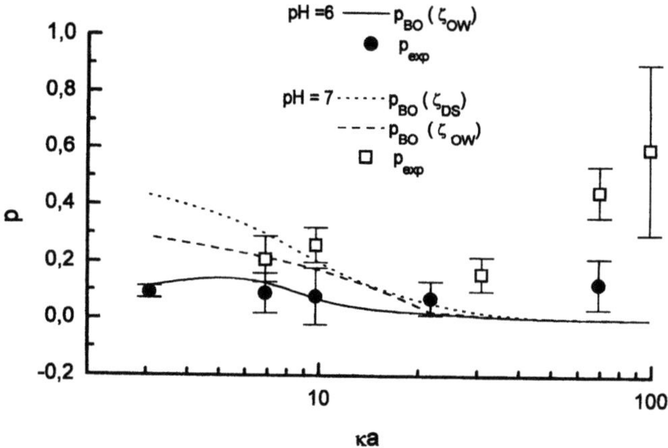

Fig. 5 Experimental values of the primary electroviscous effect coefficient (p) versus the electrokinetic radius for pH 7 and 6 (symbols) and theoretical fits (lines). The values for no buffer solution are also shown for comparison

found mostly for the low κa values, where there are less experimental data.

In spite of the discrepancies found in the previous figures, the results of this work show a better agreement between theory and experiment than those found in the literature. There is still a rather significant discrepancy at high values of κa. This apparently increase in viscosity has been discussed previously [10] looking for some possible causes. The presence of a small hairy layer was found to be a reasonable factor. However, in Fig. 5 it can be seen that the higher experimental values are only found for high values of κa, and low zeta potentials, so it would be reasonable to assume the presence of aggregates in the suspension as an important factor. For the samples with no buffer solution PCS measurements were done before and after each viscosity measurement but no significant differences were found so as to prove the presence of aggregates. At this moment, no clear explanation has been found for the increase in viscosity at higher electrolyte concentration and more experiments have to be conducted to confirm this results at higher pH, when the surface charge of the carboxyl latex is higher.

Acknowledgements The financial support provided by the Comisión Interministerial de Ciencia y Tecnología (CICYT), under project MAT96-1035-C03-03, is greatly appreciated. The authors express their gratitude to Mr. A. Puertas and Mr. M.S. Romero for synthetizing the carboxyl latex.

References

1. Saunders FL (1990) In: Advances in Emulsion Polimerization and Latex Technology, Vol. 2. Lehigh University Press, Bethlehem

2. Hunter RJ (1981) Zeta Potential in Colloid Science. Principles and Applications. Academic Press, London

3. Hiemenz PC, Rajagopalan R (1997) In: Principles of Colloid and Surface Chemistry. Marcel Dekker, New York

4. Einstein A (1906) Ann Phys (Leipzig) 19:298
5. Smoluchowski M (1916) Kolloid Z 18:190
6. Krasny-Ergen W (1936) Kolloid Z 74:172
7. Booth F (1950) Proc R Soc London A 203:533
8. Honig EP, Pünt WFJ, Offemans PHG (1989) J Colloid Interface Sci 134:169
9. Watterson IG, White LR (1980) J Chem Soc Faraday Trans II 77:1115
10. McDonogh RW, Hunter RJ (1982) J Rheol 27(3):189
11. Delgado A, González-Caballero F, Cabrerizo MA, Alados I (1987) Acta Polymerica 38:66
12. O'Brien RW, White LR (1978) J Chem Soc Faraday Trans II 74:1607
13. Yamanaka J, Ise N, Miyoshi H, Yamaguchi T (1995) Phys Rev E 51:1276
14. Chabalgoity-Rodríguez A, Martín-Rodríguez A, Galisteo-González F, Hidalgo-Alvarez R (1991) Progr Colloid Polym Sci 84:416
15. Moleón-Baca JA, Rubio-Hernández FJ, de las Nieves FJ, Hidalgo-Alvarez R (1991) J Non-Equilib Thermodyn 16:187
16. Bastos D, Ortega JL, de las Nieves FJ, Hidalgo-Alvarez R (1996) J Colloid Interface Sci 176–232
17. Semenikhin NM, Dukhin SS (1975) Kolloid Zh 137:1127
18. Smoluchowski M (1918) Z Phys Chem 92:129
19. Millero FJ (1971) Chem Rev 71(2):147

Progr Colloid Polym Sci (1998) 110:139–143
© Steinkopff Verlag 1998

J. Zipfel
P. Lindner
W. Richtering

Shear induced order and disorder in lyotropic lamellar phases

J. Zipfel* · Dr. W. Richtering (✉)
Universität Freiburg
Institut für Makromolekulare Chemie
Stefan-Meier-Str. 31
D-79104 Freiburg
Germany
E-mail: rich@ruf.uni-freiburg.de

P. Lindner
*Institut Laue-Langevin BP 156
F-38042 Genoble
France

Abstract Small angle light scattering (SALS) and small angle neutron scattering (SANS) has been used to study the influence of shear on two different lyotropic lamellar phases: (1) a defective lamellar L_α^h-phase of the ternary system sodium dodecyl-sulfate/decanol/water and (2) the lamellar L_α-phase of a semi-dilute solution of the nonionic surfactant tetra ethylenglycole dodecylether (C12E4) in water. Three shear induced states were observed with the defective L_α^h-phase. (1) At low shear rates, the lamellae were aligned parallel to the walls of the shear cell. (2) Multilamellar vesicles were obtained at intermediate shear rates. (3) At high shear rates, the vesicles were destroyed and lamellae aligned with the layer normal along the neutral (vorticity) direction were observed. The C12E4 L_α-phase revealed the formation of shear induced vesicles already at low shear rates and at higher shear rates, a butterfly pattern was recently found in light scattering. Here we present extensive SANS measurements at a very broad range of momentum transfers q. At high shear rates a characteristic butterfly pattern with a scattering peak revealing both the structure and the supramolecular structure of the system could be detected at very low q. After cessation of flow the butterfly pattern disappeared and the diffraction pattern showed a ring at the same q. The scattering data can be interpreted with a model in real space which can explain both the butterfly and the peak.

Key words Lamellar phase – neutron scattering – vesicles – shear flow – butterfly

Introduction

While static properties of colloidal systems have been examined in great detail, the investigation of dynamics of complex fluids like lyotropic and thermotropic liquid crystals, polymer and colloidal solutions is of great relevance today. Especially liquid crystalline systems exhibit a complex rheological behavior which cannot be described with simple models. There is a fundamental interest in understanding the microscopic structure and dynamics of such complex fluids as the macroscopic material properties might change with the application of an external perturbation like shear. Shear flow is known to have a profound influence on the orientation and structure of complex fluids. During the last years a lot of work has been done to study the influence of shear on micellar liquid crystalline systems [1–8].

Surfactants are known to form lyotropic mesophases in concentrated aqueous solutions. The lyotropic lamellar phase (L_α) occurs in the phase diagram of many amphiphilic molecules within a wide range of stability. When the lamellar bilayer curve and close, the resulting aggregates are called vesicles. Both unilamellar (ULV) and multilamellar vesicles (MLV) are known. Roux and coworkers reported on the formation of shear induced multilamellar vesicles out of lyotropic lamellar phases at intermediate shear rates [2]. At low shear rates the lamellae were oriented parallel to the wall of the shear cell but with many defects moving with the flow. At high shear rates the orientation of the lamellae was similar but defects were suppressed. Alternatively to the highly ordered state, called region III in the orientation diagram of Roux and coworkers, a state of hexagonally ordered assemblies of multilamellar vesicles – similar to the shear induced ordering in colloids – has been observed in other lamellar systems at high shear rates [6]. Another feature which could be observed at high shear rates is the reorientation from the orientation with the lamellae parallel to the wall of the shear cell (with the director of the lamellae parallel to the velocity gradient direction) to an orientation with the lamellae normal parallel to the neutral direction [3, 5].

Obviously the influence of shear flow on the orientation in lamellar systems can be manifold and investigations of different systems are necessary in order to understand the complex behavior of lamellar phases under shear.

Here we report on small angle neutron scattering results obtained from two different systems. (1) The L_α^h-phase in the ternary system containing sodium dodecyl sulfate (SDS), decanol and water. This system is characterized by a defective structure of water holes in the surfactant lamellae [9]. (2) The L_α-phase of the nonionic surfactant C12E4. Results from small angle light scattering from these lamellar phases under shear have been reported before [10–12].

Experimental

The sodium dodecyl sulfate/decanol/water sample investigated in this study corresponded to the previously studied sample which had a water mass fraction of 0.65 [10]. The molar fraction of decanol in the surfactant/cosurfactant mixture was 0.36. Synthesis and purification of the surfactant C12E4 were performed with standard techniques. The concentration was 33.6% (w/w) in D_2O. The systems undergoes a phase transition from an isotropic to a lamellar phase at ca. 16 °C. To avoid preshearing effects this sample was always loaded in the isotropic phase.

Neutron scattering experiments were performed at the instrument D11 of the Institut Max von Laue-Paul Langevin in Grenoble. The Couette type shear cell used for this study consisted of two concentric quartz cylinders with a gap of 1 mm. Two scattering configurations were used: (i) in the "radial" beam, the primary neutron beam passed the sample along the velocity gradient direction. (ii) in the "tangential" beam, the neutron beam passed the sample along the flow direction (i.e. through the gap of the shear cell). All experiments were performed at 25 °C.

Results and discussion

First we wish to discuss results obtained from the defective L_α^h-phase. Figure 1 shows the SANS results obtained from the SDS/decanol/water system. The Bragg peak at high momentum transfer q ($q = (4\pi/\lambda)\sin(\theta/2)$, λ: wavelength, θ: scattering angle) corresponding to the lamellar spacing could be observed with radial and tangential beam, respectively.

At a low shear rate of $4.5\ \mathrm{s}^{-1}$ a small peak along the neutral direction was observed with the radial beam. With the tangential beam, however, a higher scattering intensity was found along the direction of the velocity gradient. The latter beam configuration thus clearly showed that the lamellae were aligned parallel to the walls, i.e. with the layer normal along the direction of the velocity gradient. The spacing of these layers cannot be detected with the radial beam, and therefore the small peak along the neutral direction observed with the radial beam is caused by a small fraction of lamellae oriented with their layer normal along the neutral direction.

At a shear rate of $26\ \mathrm{s}^{-1}$ the Bragg peak was observed in all three direction indicating that the surfactant double layers were not aligned. Rheo-SALS experiments demonstrated the formation of vesicles in this range of shear rates [10]. Vesicles consist of closed surfactant double layers and consequently the Bragg peak can be observed in all directions with both beam configurations.

With increasing shear rate the scattering became strongly anisotropic and finally the peak was only observed along the neutral direction, with both radial and tangential beams. This shear rate corresponds to region III in the notation of Roux and coworkers, i.e. the vesicles were destroyed by the shear flow and aligned lamellae were obtained. The orientation of the lamellae, however, is different as compared to the system reported by Roux and coworkers. The lamellae are not aligned parallel to the walls of the shear cell, but the layer normal was parallel to the neutral direction (i.e. perpendicular to both the directions of flow and velocity gradient). The transitions between the three states were fully reversible.

Next we want to discuss the behavior of the lamellar L_α-phase of the nonionic surfactant C12E4 in water. First, we performed some SANS measurements at high q ($\geq 0.01\ \text{Å}^{-1}$). According to previous results we found

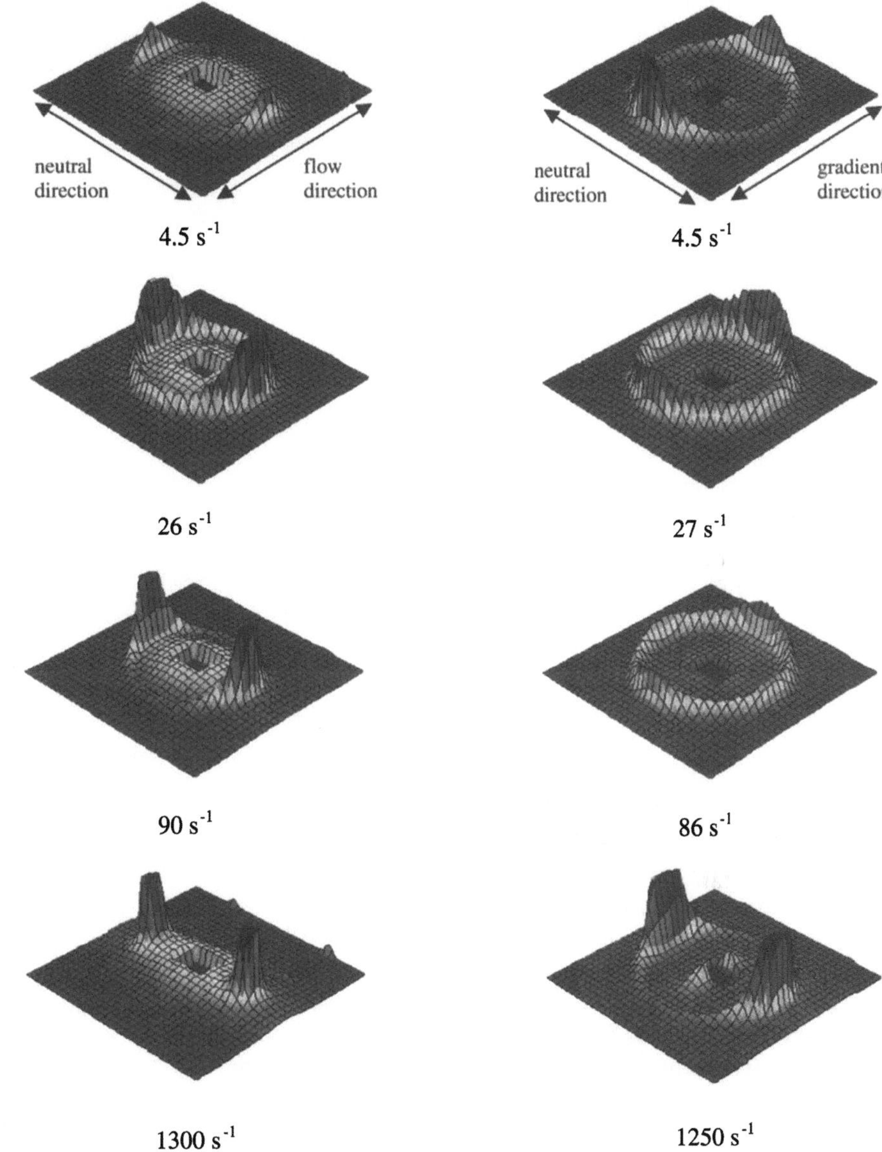

Fig. 1 SANS spectra from the defective L_α^h-phase of the SDS/decanol/water system at various shear rates, q-range: $0.28\,\text{nm}^{-1} < q < 2\,\text{nm}^{-1}$ left: radial beam, right: tangential beam

a peak at $q_{max} = 0.071\,\text{Å}^{-1}$, revealing a lamellar spacing of 9 nm. In the quiescent state we found a radially isotropic scattering on the two-dimensional detector. Under shear the intensity distribution on the two-dimensional detector became anisotropic: (i) with the radial beam, the scattering intensity was higher parallel to the neutral direction as compared to along the flow direction, (ii) in the tangential beam, however, an isotropic scattering ring was found. Figure 2 displays 3D-surface plots of the scattering patterns with the tangential beam at two shear rates. The intensity maximum in the neutral direction became weaker at higher shear rates and the pattern became less anisotropic. At small angles, however, strong scattering appeared. The high-q results thus show that elongated vesicles were present, which were not destroyed within the

accessible range of shear rates. Scaling relations were found in rheological measurements. With increasing shear rate, the size of the vesicles, determined by means of SALS, and the viscosity decreased, while the storage modulus increased. The shear thinning behavior could be characterized by $\eta \propto \dot{\gamma}^{-0.7}$, the size of the vesicles by $d \propto \dot{\gamma}^{-0.3}$ and for the storage modulus G' we found the relation $G' \propto \dot{\gamma}^{0.40}$.

Additional SANS experiments were performed in the radial beam configuration at very low q ($0.0009\,\text{Å}^{-1} < q < 0.0063\,\text{Å}^{-1}$) in order to investigate the structure under shear on a greater length scale. A striking butterfly pattern was observed at shear rates of about $1200\,\text{s}^{-1}$ (see Fig. 3). It disappeared immediately after cessation of flow. This butterfly pattern was accompanied by a scattering peak at

Fig. 2 SANS spectra with the tangential beam at high q from the C12E4 L_α-phase at two shear rates

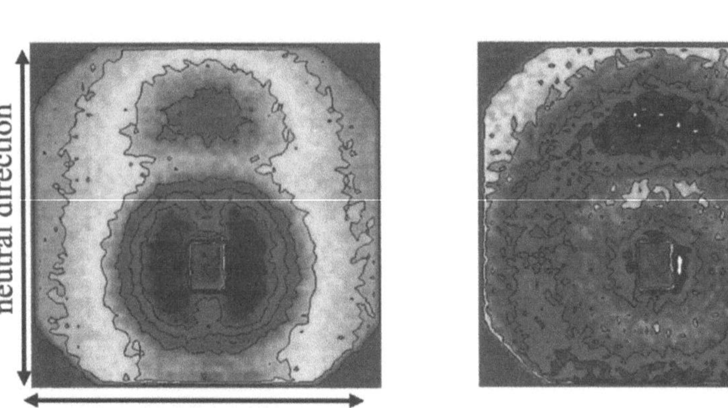

Fig. 3 Left: butterfly scattering pattern form the C12E4 L_α-phase at a shear rate of $1200\ \text{s}^{-1}$; right: scattering pattern after cessation of shear

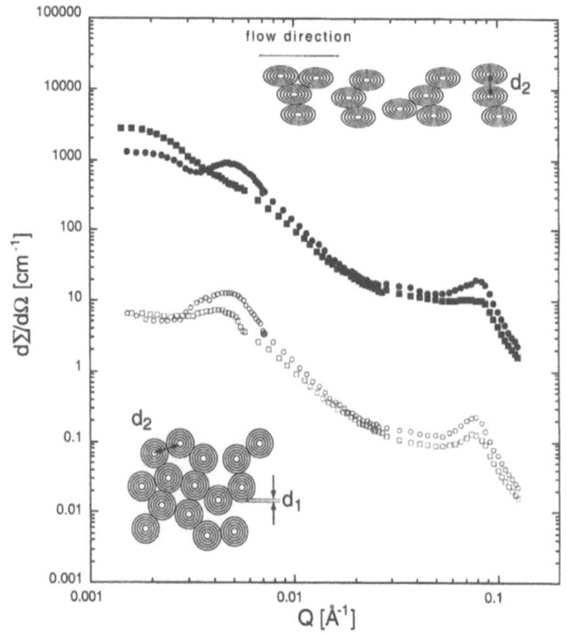

0.0048 Å^{-1} along the neutral direction. After cessation of flow the diffraction pattern showed a ring at the same q but with an intensity maximum still perpendicular to the flow direction. After relaxation the sample was sheared at lower shear rates. Reducing the shear rate to $80\ \text{s}^{-1}$, a butterfly pattern could be observed again and after cessation of shear again a ring was found.

To combine the scattering intensity for both higher and very low q we performed additional measurements in a intermediate q-range at the same shear conditions. The data were brought to an absolute scale according to standard procedures and averaged over 30°-sectors parallel and perpendicular to the flow direction. Figure 4 displays the angular dependence of the absolute scattering intensity over the total q-range for a shear rate of $1200\ \text{s}^{-1}$ and the corresponding relaxation pattern.

Two features can be seen in Fig. 4 at low q: (i) the enhanced scattering intensity along the direction of flow, (ii) the scattering peak along the neutral direction in the scattering intensity after shear. At high q the scattering peak due to the distance between surfactant double layers can be observed both perpendicular and along the flow direction.

Figure 4 also shows a model for the solution structure in real space. The butterfly pattern indicates the formation of flow-enhanced concentration fluctuations. It is known from semi-dilute polymer solutions and polymer networks

Fig. 4 Absolute scattering intensity under shear (top, full symbols) and at rest (bottom, open symbols, intensity values divided by 100). Circles represent the neutral direction, squares the flow direction. Model for structure under shear and at rest is shown in the top and bottom, respectively

and is usually linked to elastic properties of the material [13, 14]. Under shear the vesicles were elongated and the dense packing was perturbed along the flow direction. However, a distinct average distance between vesicles was still present along the neutral direction giving rise to the scattering peak. At rest, the vesicles have a random packing and the peak can be observed in all directions on the two-dimensional detector. The butterfly corresponds to a new structure of a lyotropic lamellar phase at high shear rates, in addition to the previously reported defect free lamellar phase and the hexagonal assembly of vesicles.

Conclusions

The two examples discussed above demonstrate the strong influence of shear flow on the stucture of lamellar mesophases. Three regions were observed with the defective lamellar phase: (i) at low shear rates, the lamellae were aligned parallel to the walls, i.e. the layer normal was parallel to the direction of the velocity gradient, (ii) at intermediate shear rates multilamellar vesicles were formed, (iii) at high shear rates, the vesicles were destroyed and the lamellae were aligned with the their layer normal parallel to the neutral (vorticity) direction. The transitions between these three states were fully reversible.

The lamellar phase of the nonionic surfactant showed different behavior. Vesicles were formed already at the lowest accessible shear rate. The vesicles became smaller with increasing shear rate, but a transition to oriented lamellae was not found. Instead a butterfly pattern was obtained by SANS at very small q, indicating the formation of flow-enhanced concentration fluctuations along the flow direction.

Acknowledgments This work was supported by the Deutsche Forschungsgemeinschaft. J.Z. acknowledges a Marie-Curie-Research Fellowship from the European Community.

References

1. Diat O, Roux D, Nallet F (1993) J Phys II France 3:1427
2. Safinya CR, Sirota EB, Bruinsma RF, Jeppesen C, Plano R, Wenzel L (1993) Science 261:588
3. Mang JT, Kumar S, Hammouda B (1994) Europhys Lett 28:489
4. Bergenholtz J, Wagner NJ (1996) Langmuir 12:3122
5. Penfold J, Staples E, Khan Lodi A, Tucker I, Tiddy GJT (1997) J Phys Chem B 101:66
6. Diat O, Roux D, Nallet F (1995) Phys Rev E 51:3296
7. Lukaschek M, Müller S, Hasenhindl A, Grabowski DA, Schmidt C (1996) Colloid Polym Sci 274:1
8. Schmidt G, Richtering W, Lindner P, Alexandridis P (1998) Macromolecules 31:2293
9. Berger K, Hiltrop K (1996) Colloid Polym Sci 274:269
10. Läuger J, Weigel R, Berger K, Hiltrop K, Richtering W (1996) J Colloid Interface Sci 181:52
11. Weigel R, Läuger J, Richtering W, Lindner P (1996) J Phys II France 6:529
12. Richtering W (1997) Progr Colloid Polym Sci 104:90
13. Boué F, Lindner P (1994) Europhys Lett 25:421
14. Ramzi A, Mendes E, Zielinski F, Rouf C, Hakiki A, Herz J, Oeser R, Boué F, Bastide J (1993) J Phys IV France 3:91

Progr Colloid Polym Sci (1998) 110:144–149
© Steinkopff Verlag 1998

P.A. Nommensen
M.H.G. Duits
J.S. Lopulissa
D. van den Ende
J. Mellema

Rheology of suspensions stabilized by long grafted polymers

P.A. Nommensen · M.H.G. Duits (✉)
J.S. Lopulissa · D. van den Ende · J. Mellema
University of Twente
Department of Applied Physics
Rheology Group
P. O. Box 217
7500 AE Enschede
The Netherlands

Abstract The synthesis of colloidal silica spheres grafted with long poly(dimethyl siloxane) chains as described in literature has been successfully reproduced. The particles have been elaborately characterized, using different techniques and taking advantage of the availability of the unreacted bare particles and free polymer. Rheological measurements were performed on dispersions in heptane, all made from a single stock. An effective volume fraction was defined, based on the viscosity at low concentrations. Low shear viscosities and linear viscoelastic properties were measured at effective volume fractions up to 0.81. A transition from liquid-like to solid-like behavior is observed at $\phi \cong 0.60$. Up to this volume fraction, the particles behave much like Brownian hard spheres. At higher concentrations, softness effects become noticeable. Here, also differences from otherwise comparable soft spheres become pronounced.

Key words Dispersions–silica particle–polymer brush – characterization–rheology – hard spheres

Introduction

Since the first experimental results on the rheological behavior of sterically stabilized colloidal hard spheres were published about a decade ago, there has been a growing interest in the dispersion rheology of particles stabilized by longer chains. D'Haene et al. [1] performed an interesting set of experiments with poly(methyl methacrylate) (PMMA) particles of different sizes, coated with (end-grafted) poly(hydroxy stearic acid) (PHS) chains of the same length. In this way, the deviations from hard sphere behavior were systematically varied. Another example is the work of Buitenhuis and Förster [2], who investigated a series of block-copolymer micellar systems, with the molar mass of the outside block as a variable.

In most systems, the (grafted) chains have a "brush-like" character, i.e. the lateral dimensions are (much) smaller than the layer thickness. With this characteristic, a hairy particle system can still be anything between "nearly hard spheres" and "nearly star polymers". To classify a particle between these two extremes, a softness parameter can be invoked. However, different softness parameters are used in literature. For instance, D'Haene et al. use δ/a with δ the thickness of the outer layer and a the particle radius, whereas Buitenhuis and Förster also suggest the use of ε/a with ε based on the blob size of the polymer. For monodisperse systems, the volume fraction where the low shear viscosity diverges has also been used as a measure for the softness. Comparisons between the rheology of different polymerically stabilized systems have hardly been made yet.

In this paper we present a rheological study of hairy spheres consisting of silica cores coated with end-grafted poly(dimethyl siloxane) (PDMS) chains. This particle system is relatively new and was selected for a number of reasons. The size distribution for the silica cores can be kept narrow with an appropriate synthesis route. For the grafting of PDMS chains a promising recipe has appeared in the literature. In addition, since the grafting is performed in a separate procedure, a more elaborate characterization of the particle is possible via separate analysis of the core and the free polymer. The mentioned characteristics of the silica–PDMS system should allow for meaningful comparisons with other (model) systems, as well as with theory. In this paper, the emphasis will be on the characterization as well as on the first rheological results. A comparison with other systems will be made.

Preparation

Silica seed particles having a radius of approximately 50 nm were synthesized according to the method of Stöber et al. [3] and grown out to a final radius of about 80 nm by the addition of TEOS (TetraEthylOrthoSilicate, Merck) in four steps. Next, H_2O and NH_3 were removed from the suspending medium by repeated careful rotatory evaporation at 50 °C and slightly reduced pressure. Subsequently, ethanol was replaced by DEK (DiEthylKetone, Merck) using the same technique. GLC-analysis was used to confirm that all H_2O, NH_3 and ethanol had been removed. The PDMS (ABCR, code PS347.5) used had a M_w of 110.000 g/mol with $M_w/M_n = 1.4$, and was silanol terminated at both ends.

For the grafting of PDMS onto silica, different methods have been presented in the literature [4–8]. However, we and others [6] have observed that reproducing these recipes can be difficult. We have essentially followed the method of Auroy [4] in which the grafting is effected by a condensation reaction between the silanol groups of the PDMS and the silica surface. (Differences in the) details will be outlined below.

The silica dispersion was diluted with DEK plus an amount of freshly distilled DIEA (DiIsopropylEthyl Amine, Merck) to obtain weight fractions of 5% for the silica and 4% for DIEA. The function of the amine is to increase the nucleophility of the silica surface [5]: 1500 g of this solution were mixed with an equal amount of a 20% (by weight) PDMS-solution in DEK in a 4 L round-bottom flask placed in an oil bath. A first grafting was carried out by maintaining the mixture temperature at 100 °C for 16 h while gently stirring. In a second step, more PDMS was added (now in pure form) to increase its weight percentage in the mixture from 10 to 30%, followed by equilibration overnight, after which the reaction was continued for another 16 h at 100 °C.

The resulting particles were purified from unreacted PDMS by repeated centrifugation/redispersion in heptane. Here the speed was set not higher than 3000 rpm (corresponding to 8500 m/s^2) since otherwise the sedi-

mented particles could not be redispersed anymore. The presence of free PDMS in the product was monitored by drying the supernatant, redissolving it in toluene and shaking to see if any foam could be produced. Four centrifugation steps were necessary to obtain a supernatant "free of PDMS" (containing less than 0.005%). The final yield amounted 60 g of PDMS-grafted silica particles.

All sample dispersions were made from a single stock (in heptane). Concentrated samples were made by centrifuging a weighed amount of stock in a glass tube at 8500 m/s^2 (3000 rpm), pipetting off the calculated weight of heptane and subsequently vigorous shaking of the tube for 5–10 min in a whirl mixer. From the weight fractions w, weight concentrations c were calculated according to $1/c = (v_p - v_s) + v_s/w$ with v_p and v_s the specific volumes of the particles and the solvent, respectively. This equation implies a zero excess mixing volume. For v_s, 1.472 ml/g was used, whereas the value of v_p was determined from mass-density measurements of a concentration series with a density meter (Mettler KEM DA-200) and was found to be 0.526 ± 0.006 ml/g. From the weight concentrations, particle volume fractions can be calculated using the relation $\phi = q^*c$ where q is the appropriate specific volume.

Particle characterization

The PDMS content was determined in two ways, after drying the particles overnight at 100 °C. Using solid-state ^{29}Si-NMR, the number of Si-atoms present in PDMS was found to be 0.10 ± 0.02 times the total number of silicons belonging to the silica (as either Q^4 or Q^3); this corresponds to a PDMS content of 0.11 ± 0.02 by weight. Elemental analysis (EA) was also used to estimate the PDMS content of the particles. Here the weight percentages C, H and N of the bare and the grafted particles were compared. Assuming that the compositional changes are due to PDMS only, the weight fraction of the PDMS was estimated to be 0.12 ± 0.02. Infrared spectroscopy was also done, but for our particles the PDMS absorption bands at 1259 or 814 cm^{-1} were rather difficult to distinguish from the strong signal caused by the silica.

Transmission Electron Microscopy was used to determine the size distribution of the silica cores. The average radius amounted 80 nm with a standard deviation of 8%. Static and dynamic lightscattering (SLS and DLS) measurements were performed on dilute dispersions. The optical radius found with SLS was 82 ± 5 nm, in agreement with the TEM result. The hydrodynamic radius of the grafted particle in heptane was found to be 140 ± 5 nm, indicating a layer thickness of 58 ± 7 nm.

The hydrodynamic specific volume q_h was determined via intrinsic viscosity measurements using an automated

(Schott) Ubbelohde capillary viscometer. Eleven data points at relative viscosities up to 1.08 were plotted as $(\eta_r - 1)/c$ versus c, and fitted with a straight line. Using Einstein's equation, a q_h of 2.23 ml/g was found from the intercept. Comparing this number with the specific volume of 0.449 ml/g found for the bare particles, the ratio a_H/a_c between the hydrodynamic and the core radius can be calculated. We obtained a value of 1.77 ± 0.1, in good agreement with the 1.7 ± 0.1 found with light scattering.

From the above results, a number of derived quantities can be calculated. Using a radius of 82 nm and a mass density of 2.2 g/ml for the core, and assuming for the polymer a (monodisperse) molar mass of 110.000 g/mol, a layer thickness of 58 nm (i.e. $a_H/a_c = 1.7$) and a weight fraction of 0.11, the average area per polymer chain is found to be 25 nm^2 at the silica surface and 72 nm^2 at the periphery. These calculated estimates show that the lateral dimension of the polymer chains is much smaller than their extension normal to the silica surface, and hence the polymers can be considered as "brush-like".

We have also calculated some other derived quantities, in order to allow a comparison with the work of others. The grafting density expressed in mg/m^2 amounts 8 ± 1 in our case, which is of the same order of magnitude as the values found for other stable silica–PDMS systems: 4.65 [4], 6.2, 7.7 [5] and 9.8, 14.3 [8]. The number of arms per particle amounts about 3000, which is well above the predicted critical value that allows crystallization of the particles [9]. Indeed crystallization under gravity has been observed.

Rheological experiments

Flow curves were measured with a Contraves Low Shear 40 rheometer equipped with a homemade vapourlock, using a Couette geometry with inner and outer radii of, respectively, 5.5 and 6.0 mm. The particle concentration was varied between 0.076 and 0.40 (weight fractions). The corresponding range for the hydrodynamic volume fraction ϕ_h ($= q_h^*c$) is 0.12 to 0.81. Temperature was 25.0 °C for all measurements. The shear rate was varied between 2×10^{-4} and 80 s^{-1}. Reproducibility was checked by probing the same set of shear rates at least twice.

Linear viscoelastic properties were measured with a Bohlin VOR rheometer also equipped with a homemade vapourlock, using a cone-plate geometry. Ten samples with hydrodynamic volume fractions ϕ_h from 0.43 to 0.81 were measured, all at 25.0 °C. Prior to the measurements, all samples were presheared at 81 s^{-1}, with short test-measurements in between to monitor the time dependence of G' at 1 Hz. Measurements were started as soon as this quantity became stable: this could take 30 min for concen-

trated samples. Frequency sweeps were carried out after the critical shear amplitude marking the end of the linear regime had been determined at 1 Hz.

Results and discussion

An overview of our results is given in Fig. 1 (low shear plateau viscosity) and Fig. 2 (frequency-dependent linear viscoelasticity). From both figures it can be seen that a drastic change in the rheological behavior occurs around ϕ_h ($= q_h^*c) = 0.6$. The low-shear viscosity diverges at $\phi_h = 0.60$. Above this volume fraction, a low-shear Newtonian plateau is no longer observed. In the linear viscoelastic behavior, a transition from a fluid-like to a mainly elastic behavior is observed at $\phi_h = 0.61$. The data at this volume fraction showed a poor reproducibility which appeared to be inherent to the transition.

It is interesting to note that our rheological data show a remarkable resemblance to those obtained by D'Haene et al. [1] for a particular colloidal soft sphere system: PMMA particles with a diameter of 129 nm coated with a PHS-layer of 9 nm thickness and dispersed in decalin. The polydispersity of this system is 12%.

Since the two systems are chemically rather different, the resemblance of their rheological behavior poses the question, under which conditions the specific properties of the polymer layer become important to the rheology. For this reason, a comparison with the mentioned system of D'Haene will be made in the following. First, a comparison with the rheology of hard spheres will be made for both systems.

Rheological behavior for $\phi_h < 0.6$

Fitting the data in Fig. 1 to Quemada's expression [10],

$$\eta_r = (1 - \phi/\phi_m)^{-2} \,, \tag{1}$$

a "maximum" volume fraction ϕ_m ($= q_h c_m$) of 0.60 ± 0.02 was obtained. This value is within the range of values obtained for other systems of spherical particles with a very short-ranged repulsive ("hard sphere") potential. We have also compared the low shear viscosities themselves with the various "hard spheres" data compiled by Phan et al. [11]: the viscosities for the silica–PDMS system are slightly higher than the compiled data, but still agree within their error ranges. For D'Haene's 129 nm particles, ϕ_m was found to be 0.60 as well.

Next, we will compare the linear viscoelastic behavior of our particles with that of the hard spheres of Van der Werff et al. [12], and D'Haene's 129 nm particles. Assuming that hairy particles behave like hard spheres, it should

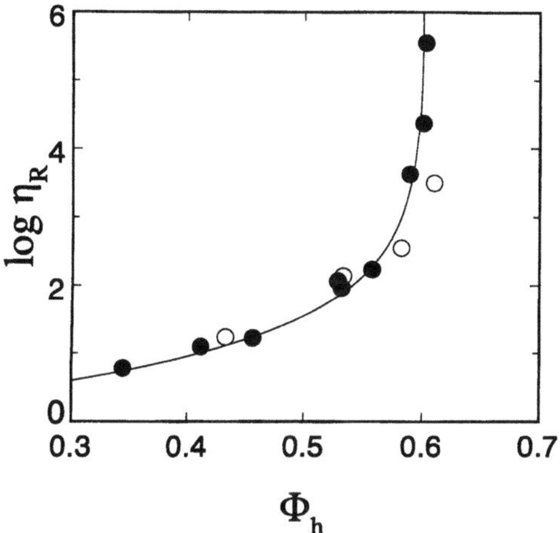

Fig. 1 Relative low shear (●) and low-frquency (○) limiting viscosity versus volume fraction. The drawn line represents the fit to Eq. (1)

be possible to scale out the hydrodynamic particle radius a_h and solvent viscosity η_s. This scaling should lead to a collapse of all data onto master curves, though it should be mentioned that some differences between presumed hard sphere systems have been observed [13]. Below we will analyze the three quantities most characteristic for hard sphere behavior: the low-frequency limit of the real part of the complex viscosity η' ($\omega \to 0$), and the dominant relaxation time (τ) and strength (G) for low frequencies.

The $\eta'(\omega \to 0)$ data are incorporated in Fig. 1 since on theoretical grounds, $\eta'(\omega \to 0)$ should be equal to $\eta(\dot{\gamma} \to 0)$. Obviously, the agreement with the low-shear data is good. The dominant relaxation time and strength for low frequencies were obtained from a fit to $G'(\omega)$ and $G''(\omega)$ using

the relaxation spectrum description [14]:

$$G'(\omega) = \sum_k G_k \frac{\omega^2}{\omega^2 + 1/\tau_k^2} (+ G_0) , \qquad (2)$$

$$G''(\omega) = \sum_k G_k \frac{\omega/\tau_k}{\omega^2 + 1/\tau_k^2} (+ \omega\eta'_\infty) , \qquad (3)$$

where k designates a relaxation, characterized with a strength G_k and a time τ_k. Like for the other two systems, we have not observed any G_0. The contribution of η'_∞ has been neglected because of the low frequencies involved.

Some differences exist between the types of spectra that have been used by the different authors to describe $G'(\omega)$ and $G''(\omega)$. This is partly due to the different (normalized) frequency ranges used in the different studies. In addition, there are some differences in the precise shapes of $G'(\omega)$ and $G''(\omega)$, but these are not too important for the estimation of the typical relaxation time and strength.

D'Haene reported a "nearly Maxwellian" behavior (only one relaxation): hence G and τ can be directly obtained from the cross-over point between $G'(\omega)$ and $G''(\omega)$. Van der Werff et al. used a *spectrum* of coupled relaxations to describe the behavior at very high frequencies. To characterize the behavior at low frequencies, it is justified to use only the slowest relaxation, characterized by G_1 and τ_1.

For our particle system, a fit with two relaxations gave the best description of $G'(\omega)$ and $G''(\omega)$. In that case, the mean relaxation time is the most appropriate characteristic time:

$$\tau_M = \sum_k G_k \tau_k^2 / \sum_k G_k \tau_k . \qquad (4)$$

Figure 3 shows the volume fraction dependence of the relaxation time after multiplication with $k_B T/(6\pi\eta_s a_h^3)$.

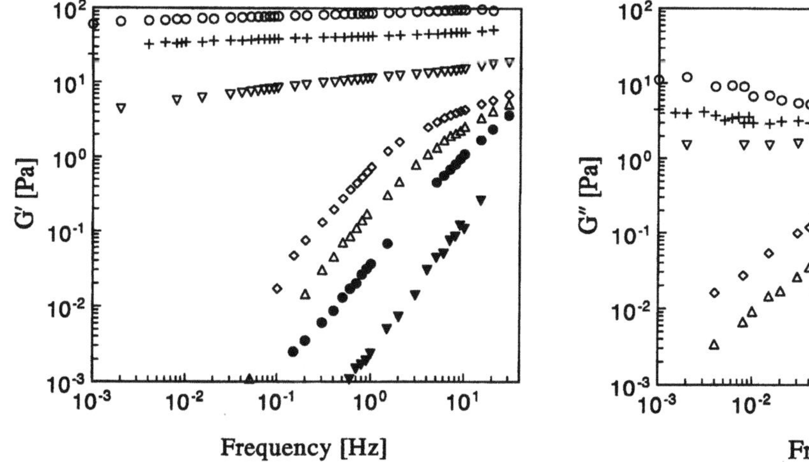

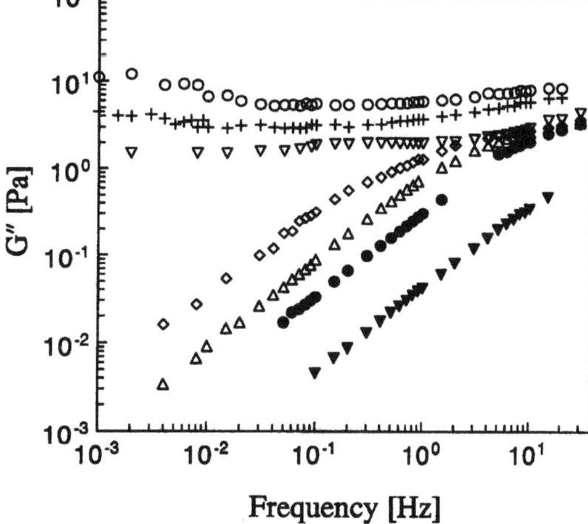

Fig. 2 Linear viscoelastic moduli versus the oscillation frequency. Volume fractions are: ▽: 0.430, ●: 0.533, △: 0.582, ◇: 0.584, ▼: 0.615, +: 0.746, ○: 0.813

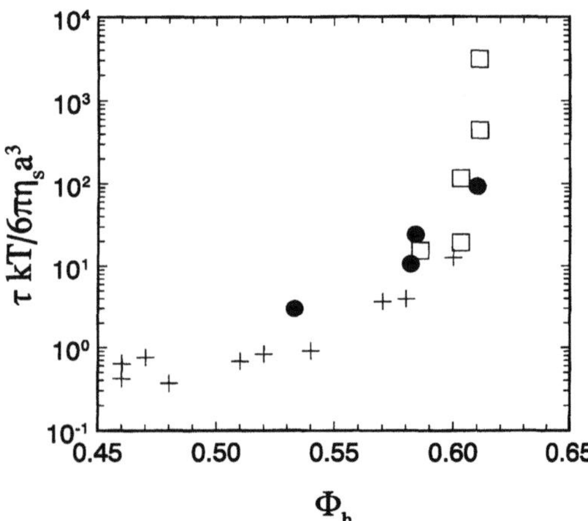

Fig. 3 Normalized relaxation time versus volume fraction: $+$: hard spheres, ●: silica–PDMS, □: PMMA–PHS

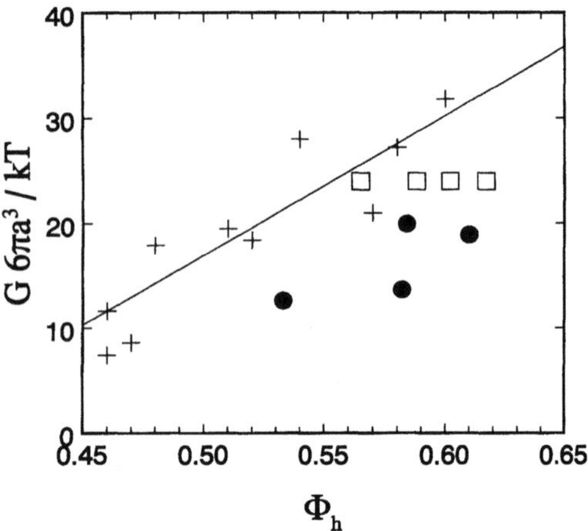

Fig. 4 Normalized relaxation strength versus volume fraction: $+$: hard spheres, ●: silica–PDMS, □: PMMA–PHS

Here η_s is the solvent viscosity. The background of this scaling is that the relaxation is related to a restorement of the equilibrium pair distribution function via Brownian motion. Considering that rather different systems are compared in the figure, our normalized data are in reasonable agreement with those of the other two systems.

To obtain the typical relaxation strength for the silica–PDMS particles, we have fitted our data with the spectrum as used by Van der Werff, using τ_M for τ_1 and fitting for G_1. Figure 4 shows the volume fraction dependence of the relaxation strength after multiplication with $6\pi a_h^3 / k_B T$. Clearly, the volume fraction dependence of the relaxation strength is much weaker than that of the relaxation time, for all three systems. In addition, the magnitudes of the normalized G values are roughly comparable. As evidenced both by our own data and by those of

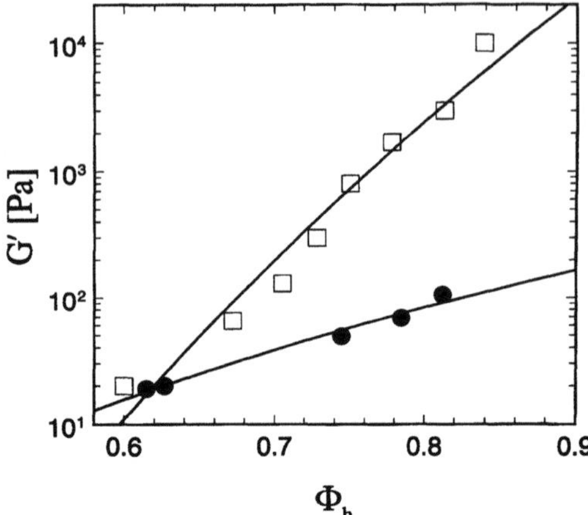

Fig. 5 Plateau value of G' versus volume fraction. ●: silica–PDMS, □: PMMA–PHS

Van der Werff, the uncertainties in the G values are large: this appears to be inherent to their extraction from G' and G'' functions that vary over three or more orders of magnitude.

Considering all foregoing observations, it appears that the different soft-sphere systems can be considered to behave much like hard spheres up to high volume fractions. Apparently, the hairy spheres have to approach each other very closely (or even be compressed slightly) before the deviations from hard sphere behavior (i.e. the detailed interactions between the polymer chains belonging to different particles) become important.

Rheological behavior for $\phi_h > 0.6$

In Fig. 5 the plateau values for G' in the elastic regime are presented as a function of ϕ_h. Fitting the data with a power law yields a power of 6. For D'Haene's particles [15], a much stronger concentration dependence of the G' plateau was found. Fitting their data we found a power of 18. Bearing in mind the relation between G' (plateau) and the particle pair potential [16], this finding indicates that our particles show more softness, when compressed.

Conclusions

The synthesis of monodisperse silica spheres coated with end-grafted PDMS, as developed by Auroy, was successfully reproduced. Working on a 4 L scale, a yield of 60 g particles was obtained. The grafting densities as determined with Si-NMR and Elemental Analysis are in good agreement with each other, as well as comparable to the results obtained by others. The thicknesses of the PDMS layer as measured with lightscattering and as calculated from the intrinsic viscosity compare favourably with each other. The microscopic picture that emerges from these results is a layer of stretched (i.e. brush-like) polymers with a thickness of 0.7 times the particle radius.

When dispersed in heptane, the particles repel each other via a soft potential as evidenced by ordering phenomena, and long-term colloidal stability. Using a volume fraction definition based on the intrinsic viscosity, volume fractions of 0.8 (and possibly higher) can be obtained. At $\cong 0.60$ on this scale, a transition from a liquid-like to a solid-like behaviour is observed. Comparison with the rheological behaviour of hard spheres reveals that up to a (hydrodynamic) volume fraction of 0.60, our particles behave much like hard spheres. A comparison with data for a **PMMA–PHS** soft-sphere reference system shows that the rheological behaviour is very comparable up to $\phi = 0.60$, but above this volume fraction our particles turn out to be softer.

Acknowledgements We like to thank Jean-Christophe Castaing for a fruitful discussion about the particle synthesis. The work described in this paper was financially supported by the Netherlands Foundation for Chemical Research (SON).

References

1. D'Haene P (1992) PhD Thesis, Katholieke Universiteit Leuven, Belgium
2. Buitenhuis J, Förster S (1997) J Chem Phys 107:262–272
3. Stöber W, Fink A, Bohn E (1968) J Colloid Interface Sci 26:62–69
4. Auroy P, Auvray L, Léger L (1992) J Colloid Interface Sci 150:187–194
5. Castaing J-C, Allain C, Auroy P, Auvray L, Pouchelon A (1996) Europhys Lett 36:153–158
6. Ketelson HA, Brook MA, Pelton RH (1995) Polym Adv Technol 6:335–344
7. Ketelson HA, Brook MA, Pelton RH (1995) Chem Mater 7:1376–1383
8. Edwards J, Lenon S, Toussaint AF, Vincent B (1984) Am Chem Soc Symp Ser 240:281–296
9. Witten T, Pincus P, Cates M (1986) Europhys Lett 2:137–140
10. Quemada D (1977) Rheol Acta 16:82–94
11. Phan SE, Russel WB, Cheng Z, Zhu J, Chain PM, Dunsmuir JH, Ottewill RH (1996) Phys Rev E 54:6633–6645
12. VanderWerff JC, deKruif CG, Blom C, Mellema J (1989) Phys Rev A 39: 795–807
13. Mellema J (1997) Curr Opin Colloid Interface Sci 2:411–419
14. Tschoegl NW (1990) In: The Phenomenological Theory of Linear Viscoelastic Behaviour. Springer, Berlin
15. D'Haene P, Mewis J (1994) Rheol Acta 33:165–174
16. Zwanzig R, Mountain RD (1965) J Chem Phys 43: 4464–4471

Progr Colloid Polym Sci (1998) 110:150–155
© Steinkopff Verlag 1998

The wavelength dependence of the high-frequency shear viscosity in a colloidal suspension of hard spheres

M.W. Heemels
C.P. Lowe
A.F. Bakker

M.W. Heemels (✉) · C.P. Lowe
A.F. Bakker
Computational Physics
Delft University of Technology
Lorentzweg 1
2628 CJ Delft
The Netherlands

Abstract Using a lattice-Boltzmann method to simulate the dynamics of a colloidal suspension of hard spheres, we have calculated the wavelength-dependent viscosity. By extrapolating to small wave vectors we have calculated values for the high-frequency shear viscosity, v, up to volume fractions of 59%. Where comparison is possible, our values are in good agreement with existing numerical, theoretical and experimental values. Examining the wave-vector-dependent viscosity, $v(k)$, we find very similar behavior to that found in a much simpler system – the hard sphere fluid. At small k we find that $(v(k) - v)$ scales proportionally to $k^{3/2}$ and only for values of the wave-vector an order of magnitude smaller than the reciprocal particle radius can the viscosity be considered essentially independent of wave vector.

Key words Suspension – viscosity – computer simulation – hydrodynamics – rheology

Introduction

A colloidal suspension consists of particles which are by molecular standards very large (the colloidal particles) dispersed in a solvent. While from a static or equilibrium point of view these systems can be treated by classical statistical mechanics in much the same way as simple fluids, the large size difference between the colloidal particles and the solvent can lead to quite complex dynamical behavior. From a practical point of view, the most important dynamical property is the viscosity because it determines the "flow" characteristics of the suspension. Notably, the viscosity of a suspension v is greater than the viscosity of the solvent v_s and this enhancement of the viscosity depends strongly on the volume fraction ϕ occupied by the colloidal particles. Einstein derived the relation between v and v_s to first order in the volume fraction [1],

$$\frac{v}{v_s} = 1 + \frac{5}{2}\phi + O(\phi^2) \, . \tag{1}$$

By including higher-order effects, more recent theoretical work [2] has succeeded in predicting the suspension viscosity up too volume fractions where Eq. (1) is inadequate.

In practice, the viscosity of a colloidal suspension is a complex quantity which will generally depend on the frequency (or time scale) [3] and the shear rates [4] one is considering. It is therefore important to be clear about which regime one is addressing. In this article we describe results obtained in the limit of low shear rate and high frequency. Taking the limit of low shear rates means we can assume that the suspension has its equilibrium structure. By taking the limit of high frequencies we can neglect the "collisions" between suspended particles, driven by Brownian motion, which are responsible for their osmotic pressure. Thus, the only effect we consider is the transfer of momentum through the fluid and between the colloidal particles – the so-called hydrodynamic interactions. To see why it is ever sensible to study this regime it is necessary to consider some relative time scales involved. There is a very large difference between the time it typically takes for momentum to be transported through the system (which is short) and the time it takes for Brownian motion to displace the particles a significant distance (which is long). To illustrate this it is useful to consider the decay of one Fourier component of the instantaneous transverse velocity field

$$\Delta v_x(\mathbf{r}, t = 0) = a(\mathbf{k}_T)\exp(i\mathbf{k}_T\cdot\mathbf{r}) , \tag{2}$$

where the wavevector $\mathbf{k}_T$ is orthogonal to the x direction and $a(\mathbf{k}_T)$ is the Fourier transform of the velocity field. In a simple Newtonian fluid one expects such a perturbation to decay as

$$\Delta v_x(\mathbf{r}, t) = a(\mathbf{k}_T)\exp(- k_T^2 vt) , \tag{3}$$

where v is the kinematic viscosity of the fluid. For a given wavevector the perturbation therefore decays on a time scale $t_v \sim 1/k_T^2 v$. On the other hand, the time it takes a particle to diffuse a distance of the order of its own radius is of the order $t_D \sim d^2/D$, where d is the particle diameter and D the diffusion coefficient. The diffusion coefficient can be estimated from the Stokes–Einstein equation

$$D = \frac{kT}{6\pi\rho va} , \tag{4}$$

where T is the temperature, k the Boltzmann's constant, ρ the solvent density and a the particle radius. If we now consider, for illustrative purposes, the case of a micronsized particle dispersed in water at room temperature then we find that $t_D \sim 1$ s and $t_v \sim 100/k_T^2$ s. The "high-frequency" condition that $t_v \ll t_D$ is therefore satisfied so long as $k_T \gg 10$ cm^{-1}. Thus, there exists a large range of values of k_T, corresponding to real lengths much greater than the dimensions of the colloidal particles, where we can safely neglect the effects of Brownian motion.

In this article we describe numerical results for the high-frequency viscosity calculated using the lattice-Boltzmann method [5, 6]. Ladd reported values calculated using this approach in ref. [6]. Two different methods were used to extract the viscosity from the simulations; evaluating the steady-state response to an externally imposed shear and integrating the stress–stress correlation function. The former reportedly required some considerable time to reach a homogeneous steady state, whilst the latter was complicated by the lack of true equipartition in the lattice Boltzmann model. Our aims here are twofold. One is simply to find an easier way of extracting the viscosity. To do so we apply a method based on examining the time-dependent response of the suspension to an initial velocity perturbation of the form given in Eq. (2). The method also allows us to address our second question; on what length scales might a suspension behave like a simple Newtonian fluid? We also note that extensive calculations of suspension viscosities as a function of shear

rate have been reported using Stokesian Dynamics [4]. This method is probably more suitable than the lattice-Boltzmann method for studying long time (or low-frequency behavior).

Description of the model

The method we used involved simulating configurations of hard spheres embedded in a simple model fluid (the solvent). The initial configurations of colloidal spheres were generated using a standard Monte Carlo method. The solvent was modelled using a lattice Boltzmann equation which reproduces the time-dependent hydrodynamic interactions between the hard spheres. Solvent particles are only allowed to have a discrete set of velocities $\mathbf{c}_i$ so that they are constrained to move along the vertices of a lattice and be located at lattice sites at integer times (see e.g. [7]). The state of the fluid is then characterized by the single-particle distribution function $n_i(\mathbf{r}, t)$. This describes the average number of particles at a particular site of the lattice $\mathbf{r}$, at a time t, with the discrete velocity $\mathbf{c}_i$. The hydrodynamic fields, mass density ρ, momentum density $\mathbf{j}$, and the momentum flux density Π are simply moments of this velocity distribution:

$$\rho = \sum_i n_i , \quad \mathbf{j} = \sum_i n_i \mathbf{c}_i , \quad \Pi = \sum_i n_i \mathbf{c}_i \mathbf{c}_i . \tag{5}$$

The lattice model used in this work is the 4D face-centered hyper cubic (FCHC) lattice. A two- or three-dimensional model can then be obtained by projection in the number of required dimensions, here we work in three dimensions. This FCHC model is used because three-dimensional cubic lattices do not have a high enough symmetry to ensure that the hydrodynamic transport coefficients are isotropic. The time evolution of the distribution functions n_i is described by the discretized analogue of the Boltzmann equation [8]

$$n_i(\mathbf{r} + \mathbf{c}_i, t + 1) = n_i(\mathbf{r}, t) + \Delta_i(\mathbf{r}, t) , \tag{6}$$

where Δ_i is the change in n_i due to instantaneous "molecular" collisions at the lattice sites. The post-collision distribution $n_i + \Delta_i$ is propagated in the direction of the velocity vector C_i. The collision operator, $\Delta_i(\mathbf{r}, t)$, is constructed in such a way that it conserves mass and momentum and generates the correct equilibrium distribution. Shear stresses are partially relaxed at every lattice site, the rate of stress relaxation being related to the kinematic viscosity of the fluid. A complete description of the collision process is given in ref. [5].

The motion of the colloidal particles is determined by the forces and torques exerted on them by the fluid. These in turn depend on the boundary conditions applied at the solid/fluid interface. Here we follow the usual practice of assuming that, at the interface, the velocity of the solvent and of the solid are equal – the stick boundary condition. Stick boundary conditions with a stationary boundary can be implement in a very simple manner.

All fluid particles which attempt to cross the solid/fluid interface are simply reflected at half a time step and return to their original site. The solid/fluid interface is thus considered to be midway between a lattice site located inside the objects and a lattice site outside. For a moving boundary a similar procedure is followed but some of the particles moving in the same direction as the solid object are allowed to "leak" through the interface and thus match the fluid velocity to the object velocity at the boundary. The equations of motion of the colloidal particles are then integrated according to a rule [9] whereby the force and torque which act on an object give the same new velocities for both particle and fluid. The algorithm was implemented in parallel using the reduced storage method described in ref. [10]. This makes it possible to study systems containing several thousand particles.

In going to a Boltzmann level of approximation (for the solvent) spontaneous fluctuations in the state of the fluid are averaged out. Thus, in the absence of any externally imposed fluctuations, a colloidal particle in a Boltzmann fluid does nothing. Fluctuations can be "reinstated" in the lattice-Boltzmann model by adding a suitable random noise term to the stress tensor [5], but we have chosen a different approach. According to Onsager's regression hypothesis the decay of a fluctuation which we impose on the (otherwise) purely dissipative system should be the same as the decay of a spontaneous fluctuation in the "real" fluctuating system. The advantage of following the decay of an imposed, rather than spontaneous, fluctuation is that it does not involve adding any additional noise to the system and so the results are relatively free from statistical error. A philosophically similar procedure was applied in ref. [11] to study diffusion in suspensions. In this work, however, the fluctuation we impose corresponds to one Fourier component of the transverse velocity field, i.e. we apply an initial velocity perturbation of the form

$$v_x(\mathbf{r}, t = 0) = C \exp(i\mathbf{k}_T \cdot r) \,, \tag{7}$$

Where C is a constant to both the fluid and the colloidal particles. We then calculate the correlation function

$$C(t) = \sum_{i=1}^{N_l} v_x(\mathbf{r}_l, 0)\, v_x(\mathbf{r}_l, t) + \sum_{j=1}^{N_p} v_x(\mathbf{r}_p, 0)\, v_x(\mathbf{r}_p, t) \tag{8}$$

where N_l is the number of lattice sites outside the spheres, N_p the number of colloidal spheres and $\mathbf{r}$ the position

vector of either the lattice site (subscript l) or the centre of the particle (subscript p). In keeping with our "high-frequency" (or short time) assumption the positions of the particles do not change during the course of the simulation. One expects a priori that the function $C(t)$ will decay as [12]

$$\frac{C(t)}{C(0)} = \exp(-k_T^2 v(k) t) \tag{9}$$

where $v(k)$ is a wavevector-dependent viscosity. Hence, in principle, one can extract $v(k)$ by calculating $C(t)$. This approach is of course phenomenological in that it assumes $C(t)$ decays as predicted by Eq. (9). This remains to be shown. If, however, Eq. (9) does hold then this is a convenient means of extracting the viscosity because one only has to established the rate of decay of $C(t)$. This should require a relativley small number of timesteps and, furthermore, one would not expect this number to depend on system size. In contrast, if one takes the alternative approach of looking for the steady state response to an external perturbation, then the time required for the system to reach a steady state increases dramatically as the system size increases. System size is an important consideration for two reasons. Firstly, hydrodynamic interactions are long-ranged and finite size effects on transport coefficients can be large [13]. Secondly, the system size determines the minimum value of k_T which can be studied and only in the limit of k_T approaching zero can one expects $v(k_T)$ to approach the limiting value v_0 (the true, in our case high-frequency, viscosity). If we are merely interested in extracting the viscosity then it is obviously important that we can simulate wavevectors small enough such that, to a good approximation, $v(k_T) = v_0$ or, alternatively, that one can extrapolate reliably to this limit. An analogous linear response technique for calculating the viscosity can be applied in molecular dynamics simulations and extracting a reliable estimate of the zero wavevector value can indeed be problematic [14]. On the other hand, the dependence of the viscosity on the wavevector is of some interest in its own right. For a simple Newtonian fluid, $v(k_T)$ is a constant independent of k_T so, for wavelengths sufficiently small that in essence $v(k_T) = v_0$, the suspension behaves like an "effective" Newtonian fluid. By this we mean a Newtonian fluid with the viscosity of the suspension replacing the viscosity of the solvent. Conversely, for values of k_T where $v(k) \neq v(0)$ the suspension cannot be considered to behave in this simplified manner. Thus, by examining the wave vector dependence we can hope to both calculate v_0 and also glean information about the length scales on which a modified continuum view of the suspension might suffice.

Results

A typical example of the results we obtain for the correlation function $C(t)$ is plotted in linear-log form in Fig. 1. These particular results are for a suspension consisting of 2738 spheres at a volume fraction of 35% and the function has been averaged over 10 configurations. The dimensionless wavevector k^* (defined as $k_T a$, where a is the particle radius) was equal to 0.2. In the simulation, the radius of one sphere was nominally equal to 2.5 lattice spacings. The particle radius in these lattice units is an important parameter because it basically determines the spacial resolution of the simulation. Increasing this radius leads to a better resolution of the hydrodynamics interactions between spheres which are almost touching (the lubrication forces), at the expense of bigger system sizes. One sees from the figure that, after the decay of some short-lived initial transients, an exponential decay is observed (the plot becomes linear). Thus, we find the decay which we expected and we can use Eq. (9) to calculate a wavevector dependent viscosity. We also note that the number of time-steps required before we can establish the form of the exponential decay is relatively few (~ 100), whereas for steady-state calculations on much smaller systems [6] the number of steps required was hundreds of thousands. To illustrate the dependence of the viscosity on k^* we focus on one particular intermediate volume fraction, $\phi = 0.35$. In Fig. 2 the wavevector-dependent viscosity is plotted as a function of k^*. The results were again averaged over ten independent configurations and these results are for a sphere of radius of 4.5 lattice units. This simulations were repeated with

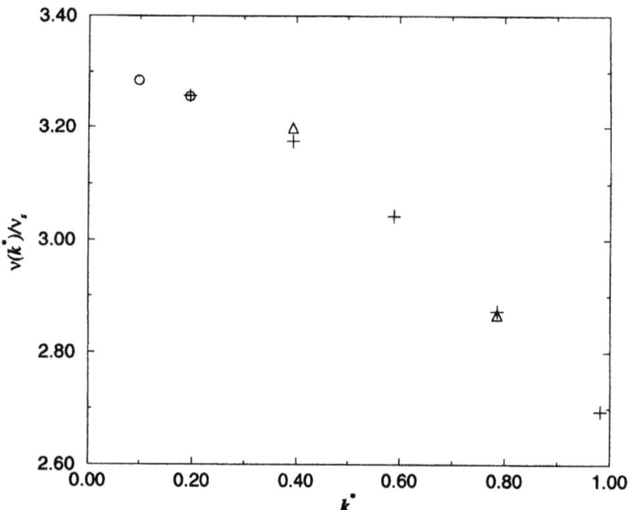

Fig. 2 The wavevector-dependent viscosity for a simulation of a suspension of colloidal spheres which occupy a volume fraction $\phi = 0.35$. The symbols correspond to systems of 21 904 spheres (circles), 2738 spheres (pluses) and 342 spheres (triangles). The statistical errors are of the order of, or less than, the symbol size

a larger value for the radius ($a = 6.5$), but, at this particular volume fraction, the values obtained for $v(k^*)$ were not significantly different, indicating that a radius of 4.5 gave adequate resolution. The conclusion one can draw from Fig. 2 is that the limiting ($k^* = 0$) value has almost been reached for the smallest wavevector we have studied ($k^* = 0.098$). With the aid of a small extrapolation, based on a polynomial fit to the data shown in Fig. 2, we can therefore obtain an accurate value for v_0. The uncertainty in the extrapolation is in fact of the same order as the statistical error associated with averaging over a number of configurations.

By following this procedure we have calculated the high-frequency viscosity over the range of volume fractions $0 < \phi < 0.59$. The highest value is very close to the maximum possible for a random packing of hard spheres $\phi = 0.605$. The values we obtained are plotted in Fig. 3. In the figure we also show the effect of changing the spacial resolution (the radius, in lattice units, of the sphere). Clearly, this needs to be increased as the volume fraction increases but not dramatically. At a volume fraction of 50% spheres of radii 6.5 and 8.5 give results which differ only by a few percent, a difference which is not statistically significant. However, at these high-volume fractions smaller spheres are clearly not adequate. In the figure we have also plotted accurate numerical results [13], the theoretical prediction due to Beenakker [2] and the results of experimental measurements by van der Werff et al. [3]. The agreement between all three is very good. However, at a volume fraction of 45% our results are in somewhat better agreement with the theory than those of Ladd. In

Fig. 1 Log-linear plot of the correlation function $C(t)$ (defined in the text) as a function of time. The data refer to a simulation of a system of 2738 spheres of nominal radius 2.5 lattice units, the spheres occupying a volume fraction of 35%. The reduced wavevector ($k_T a$) was equal to 0.2

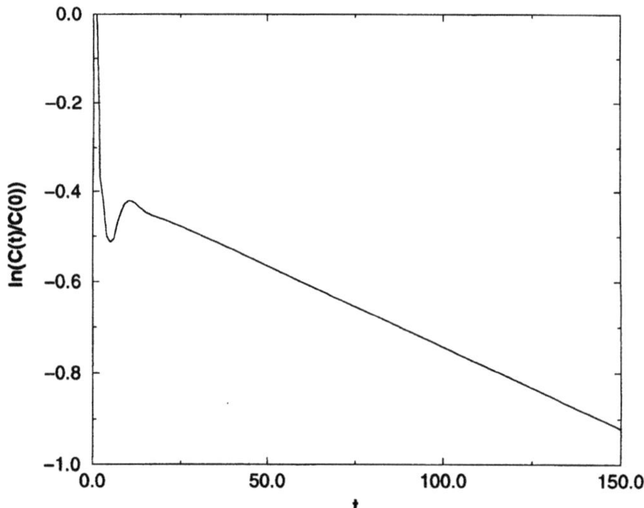

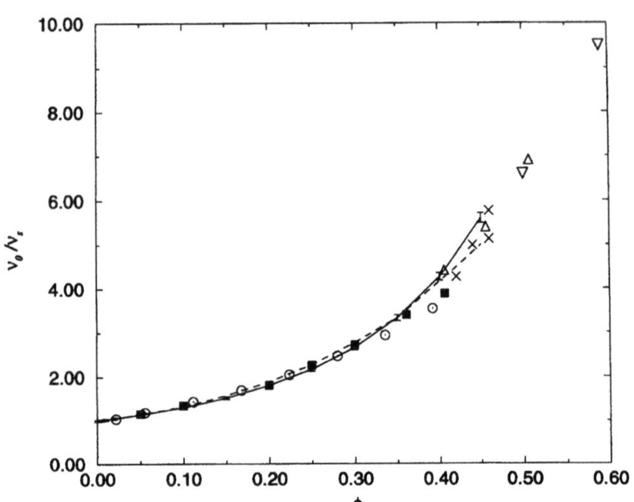

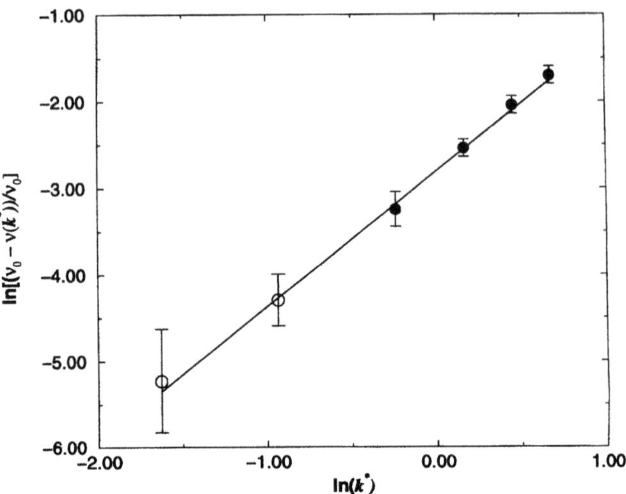

Fig. 3 The high-frequency suspension viscosity, v_0 as a function of volume fraction ϕ. The solid line is a spline through the numerical values calculated by Ladd [13]. The dashed line is the theoretical result due to Beenakker [2]. The crosses are the experimental results of van der Werff et al. [3]. The remaining symbols are the numerical results described in this work using spheres of nominal radius 2.5 (circles), 4.5 (squares), 6.5 (upward triangles) and 8.5 (downward triangles). The statistical errors are of the order of, or smaller than, the symbol size

Fig. 4 Log–log plot of the difference between the wavevector dependent viscosity, $v(k^*)$, and the suspension viscosity v_0 as a function of the reduced wavevector k^*. The data refer to simulations of a suspension of 21 904 (open symbols) and 2738 (filled symbols) spheres, occupying a volume fraction of 35%. The solid line is the result of a linear least-squares fit to the data

addition, our results for higher volume fraction do not appear to increase as rapidly with increasing volume fraction as the trend suggested by Ladd's results (which only extends (to $\phi = 0.45$) would indicate. It is therefore possible that the theoretical values are quite good even at volume fractions higher than 45%. Unfortunately, the values quoted in ref. [2] only extend up to $\phi = 45\%$ so this we can only summize.

If we now return to the wave vector dependence of the viscosity (Fig. 2) we remarked that at the smallest wavevector we studied ($k^* = 0.098$), $v(k^*)$ was almost, but not quite, equal to v_o. There is still a statistically significant difference. What is somewhat surprising is that there is any measurable discrepancy at all. This value of k^* corresponds to dimensionless wavelengths, $\lambda^*(= l/a)$, of about 60, i.e. real space lengths which are very much greater than the typical separation between colloidal particles. Indeed, the theory of Beenakker [2] predicts that we should see no wavevector dependence for $k^* < 1$. It also apears that v_o is aproached relatively slowly with decreasing k^*. In order to analyse this more quantitatively we have also analysed the data shown in Fig. 2 by plotting, in log–log form $(v_o - v(k^*))$ as a function of k^* (Fig. 4). The plot appears to be linear, a linear fit to the data yielding a slope of 1.55 ± 0.05. This, interestingly, is very similar to observations made by Evans [15], who performed a similar analysis for a soft sphere fluid and Alley and Alder [16], who performed a similar calculation for a hard sphere

fluid. They found small k^* behavior compatible with the form

$$v(k^*) = v_o - ck^{*3/2} \tag{10}$$

with c a positive constant. In fact, if one compares Fig. 4 with the equivalent plot for a hard sphere fluid then the similarity is not just qualitative but quantitative. For example, at $k^* = 1$ we find $v(k^*)/v_o = 0.8$ whereas for the hard sphere fluid it is about 0.75. Given that the two systems are quite different (in a hard sphere fluid there is no solvent, the particles undergo ballistic motion between direct collisions) the similarity is surprising.

Discussion

We have described a method for calculating, with the lattice Boltzmann framework, the high-frequency, low shear rate viscosity of a colloidal suspension. The method seems to have some advantages over those previously employed in that it only requires the dynamics of the suspension be simulated for a very short period of time. The values for the viscosity that we have calculated using this approach are, up to volume fractions of 45%, in good agreement with other simulations, experiment and theory. We have also extended the calculations up to volume fractions approaching the maximum possible for random (that is "glass-like" as opposed to "solid-like") configurations of spheres. A comparison with theory at these higher volume fractions would clearly be desirable. This would allow one to establish to what extent the theory brakes down at higher volume fractions – at 45% it still works remarkably well. A comparison with experiment would also be desirable as it would establish if, at sufficiently high-volume fractions, our approximation of colloidal particles as perfect hard spheres breaks down. In reality, the colloidal particles studied experimentally are only approximately hard spheres (they have polymer coats and interact via a short-ranged electrostatic interaction). One might expect that eventually, at high enough volume fractions, the hard sphere approximation would break down and some difference would become apparent.

Finally, we examined the k^* dependence of the viscosity. We found that the low k^* limit was approached relatively slowly with decreasing k^*, in a manner very similar to that observed for simple model "atomic" fluids. The k^* dependence was apparent at wavelengths much smaller than the theory contained in ref. [2] would suggest. This means that the suspension can only be expected to behave like a simple (continuum) Newtonian fluid on spacial lengths scales very much greater than the size of the colloidal particles. This may be relevant if one considers a suspension flowing in some constricted geometry with dimensions not much larger than the colloidal particles themselves. In this case the suspension may well behave as if it had a somewhat lower viscosity than in a bulk sample.

This remains to be examined. Finally, it is worth commenting on the similarity between the k^* dependence we found for a suspension and that found in a hard sphere fluid. A hard sphere fluid is most fundamentally different from a suspension in that, on the length scales of the particles making up the fluid, the fluid cannot be treated as a continuum. One might therefore expect that any deviations between the dynamics of a hard sphere fluid and the dynamics of a continuum Newtonian fluid (for instance a k^* dependent viscosity) would originate from this discrete nature of the fluid. However, this is not true for a suspension because the solvent, on the length scale of the colloidal particles, does appear as a continuum. Deviations from simple Newtonian fluid behaviour in the suspension can only originate from the complex nature of the hydrodynamic interactions between the particles. The observation that the (small) k^* dependence of the viscosity in a suspension is very similar to that of a hard sphere fluid suggests that, in a simple "atomic-like" fluid, the k^* dependence of the viscosity has its origins more in the hydrodynamics of the fluid than in its discrete nature.

Acknowledgement The research of Dr. Lowe has been made possible by a fellowship of the Royal Netherlands Academy of Arts and Sciences. We would like to acknowledge HPαC (High Performance Applied Computing centre) for making the Cray T3E available for our calculations.

References

1. Einstein A (1906) Annln Phys 19:289
2. Beenakker CWJ (1984) Physica 128A:48
3. Van Der Werff JC, de Kruif CG, Blom C, Mellema J (1989) Phys Rev A39:795
4. Bossis G, Brady JF (1989) J Chem Phys 91:1866
5. Ladd AJC (1994) J Fluid Mech 271:285
6. Ladd AJC (1994) J Fluid Mech 271:311
7. McNamara GR Alder BJ (1992) In: Mareschal M, Holian BL (ed) Microscopic Simulation of Complex Hydrodynamic Phenomena. Plenum, New York
8. Frisch U, d'Humières D, Hasslacher B, Lallemand P, Pomeau Y, Rivet J-P (1987) Complex Systems 1:649
9. Lowe CP, Frenkel D, Masters AJ (1995) J Chem Phys 103:1582
10. Bakker AF, Heemels MW (1996) Proceedings of the second annual conference of the Advanced School for Computing and Imaging. p 21
11. Lowe CP, Frenkel D (1996) Phys Rev E 54:2704
12. Hansen J-P, McDonald IR (1986) The theory of simple liquids. Academic Press, London
13. Ladd AJC (1990) J Chem Phys 93:3483
14. Allen MP, Tildesley DJ (1987) Computer simulation of liquids. Oxford University Press, Oxford
15. Evans DJ (1982) Mol Phys 47:1165
16. Alley WE, Alder BJ (1983) Phys Rev A 27:3158

Progr Colloid Polym Sci (1998) 110:156–162
© Steinkopff Verlag 1998

Effect of temperature and dispersed phase content on the behavior of the dielectric constant of ionic microemulsions below the percolation onset

Y. Alexandrov
N. Kozlovich
A. Puzenko
Y. Feldman

Y. Alexandrov · N. Kozlovich
A. Puzenko · Dr. Y. Feldman (✉)
Graduate School of Applied Science
The Herbrew University of Jerusalem
91904 Jerusalem
Israel

Abstract A statistical model is developed to describe the dielectric polarization of ionic microemulsions at a region far below percolation in which the microemulsions consist of spherical single droplets with water in the central core surrounded by a layer of surfactant molecules. The model describes the effect of temperature and dispersed phase content on the behavior of the dielectric polarization of ionic water-in-oil microemulsions and explains the experimentally observed increase of the static dielectric permittivity as a function of temperature. The microemulsions formed with the surfactant sodium bis(2-ethylhexyl) sulfosuccinate (AOT) have been analyzed with the help of this model. It is shown that the droplet polarizability is proportional to the mean-square fluctuation dipole moment of the droplet. The mean-square dipole moment and the corresponding value of the dielectric increment depend on the equilibrium distribution of counterions within a diffuse double layer. The density distribution of ions is determined by the degree of the dissociation of the ionic surfactant. The relationship between the dielectric permittivity, the constant of dissociation, the content of the dispersed phase and the temperature has been ascertained.

Key words Ionic microemulsions – dielectric constant

Introduction

The microemulsions formed with the surfactant, sodium bis(2-ethylhexyl) sulfosuccinate (AOT), reside in the L_2 phase over a wide temperature range, i.e., the microemulsions consist of nanometer-sized spherical droplets with water in the central core surrounded by a layer of surfactant molecules [1–3]. Molecules of AOT can dissociate into anions containing negatively charged head groups SO_3^{2-}, staying at the interface and positive counterions Na^+, distributed in the droplet interior.

In our prior research [4, 5], it was shown that the ionic microemulsions start to exhibit percolation behavior that is manifested by a rapid increase in the static dielectric permittivity ε and electrical conductivity σ when the temperature reaches the percolation onset T_{on}. The appearance of the percolation reveals that in the region $T > T_{on}$ the droplets form transient clusters.

Below the percolation onset, both the conductivity σ and static dielectric permittivity ε of the microemulsions increase as a function of the volume fraction of droplets φ and/or temperature T. However, this increase is not particularly significant, as it is within the percolation region [4]. On the one hand, the increase of the conductivity versus temperature and volume fraction of droplets below the percolation onset ($T < T_{on}$) can be described by the charge fluctuation model [6, 7]. In this model, the conductivity is explained by the migration of charged aqueous non-interacting droplets in the electric field. The droplets acquire charges owing to the fluctuating exchange of charged surfactant heads at the droplet interface and the oppositely-charged counterions in the droplet interior. The conductivity is then proportional to the temperature and volume fraction, $\sigma \sim \varphi T$. On the other hand, the temperature behavior of static dielectric permittivity of water-in-oil microemulsions with ionic surfactant below percolation has been recently explained in our paper [8] in terms of the fluctuation model of dielectric polarization. The mechanism described in [8] is related to the temperature dependence of the fluctuation dipole moment of non-interacting and therefore non-aggregating droplets dispersed in oil. The temperature behavior of the fluctuation dipole moment is modulated by the temperature dependence of the dissociation of the ionic surfactant. However, the practical utility of the considered model was restricted since the results were obtained only numerically. The purpose of this paper is further to develop the model of fluctuation mechanism of dielectric polarization. The analytical solution of the Poisson–Boltzmann equation will be used for a calculation of the mean square dipole moment of a droplet. This approach enables us to present the dielectric parameters of ionic microemulsions below the percolation region as functions of both the temperature and composition of the system.

Theoretical part

Static dielectric constant of ionic microemulsions

We have shown [8] that the dielectric polarization of ionic microemulsions can be represented as a sum of the two contributions associated with the term, due to displacements of mobile ions in the diffuse double layer, and a term due to all other displacements in the mixture related to the contributions from various processes of polarization in the microemulsion that is treated as a heterogeneous system composed of various components. The dielectric permittivity ε can be described by the relationship

$$\frac{(\varepsilon - \varepsilon_{\mathrm{mix}})(2\varepsilon + \varepsilon_{\mathrm{mix}})(2\varepsilon + \varepsilon_{\mathrm{w}})}{\varepsilon^2} = \frac{9\varphi \langle \mu^2 \rangle}{R_{\mathrm{d}}^3 k_{\mathrm{B}} T} \tag{1}$$

where $\langle \mu^2 \rangle$ is the mean-square dipole moment of a droplet, φ is the volume fraction of the droplets, R_{d} is the radius of the surfactant-coated water droplet, T is the temperature, and k_{B} is the Boltzmann constant and $\varepsilon_{\mathrm{mix}}$ is the dielectric permittivity due to the polarization of heterogeneous system. Since each droplet consists of a water core surrounded with a surfactant layer in a continuous phase made of oil, this polarization can be accurately regarded

by using a Maxwell–Wagner mixture formula (one-shell model [9]).

Equation (1) establishes a dependence of the dielectric permittivity of a microemulsion on the temperature T, volume fraction of droplets φ, and apparent dipole moment of a droplet $\mu_{\mathrm{a}} = (\langle \mu^2 \rangle)^{1/2}$.

Fluctuating dipole moment of a droplet

The calculation of the mean-square dipole moment of droplet, $\langle \mu^2 \rangle$, are carried out within the framework of the following assumptions:

(a) The droplets are considered as identical and the interaction of the droplets is neglected.

(b) A nanodroplet contains N_{a} surfactant molecules, N_{s} of which are dissociated. For electroneutrality the numbers of the negatively charged surfactant molecules, N^-, and the number of positively charged counterions, N^+, are equal, i.e., $N^+ = N^- = N_{\mathrm{s}}$.

(c) The ions are treated as point charges.

(d) The average spatial distribution of counterions inside the droplets is continuous and governed by the Boltzmann distribution law.

(e) All the negatively charged surfactant molecules are assumed to be located in the interface at the spherical plane of radius R_{w}, corresponding to the radius of the droplet water pool.

In the model a single droplet is described by the spherical coordinate system shown in Fig. 1. The dipole moment of the single droplet is given by

$$\mu = e \sum_{i=1}^{N_s} (\mathbf{r}_i^+ - \mathbf{r}_i^-) \, , \tag{2}$$

where $\mathbf{r}_i^+$ and $\mathbf{r}_i^-$ are the radius-vectors of the positively charged counterion and negatively charged surfactant head, respectively; and e is the magnitude of the ion charge.

The quantity of interest is the mean-square dipole moment $\langle \mu^2 \rangle$ of a droplet. It can be expressed in terms of the mean-squared fluctuations of the dipole moment μ by

$$\langle \mu^2 \rangle = \langle (\varDelta \mu)^2 \rangle + \langle \mu \rangle^2 \, , \tag{3}$$

where $\langle (\varDelta \mu)^2 \rangle = \langle (\mu - \langle \mu \rangle)^2 \rangle$.

The apparent dipole moment of the droplet has a fluctuation nature [8], i.e.

$$\langle \mu^2 \rangle \approx \langle (\varDelta \mu)^2 \rangle \, . \tag{4}$$

In order to calculate the value of $\langle \mu^2 \rangle$, we square the left- and right-hand sides of Eq. (2) and average the result

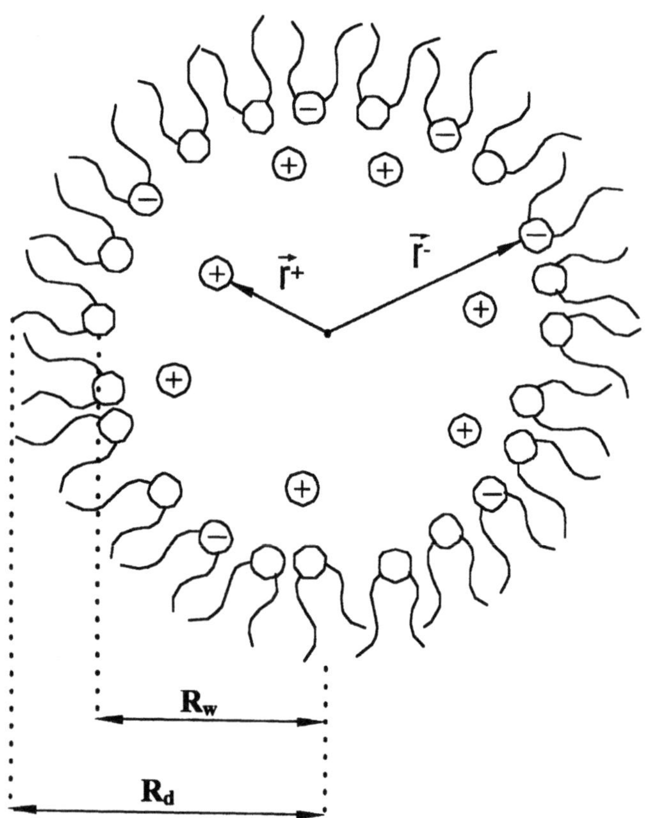

Fig. 1 Schematic picture of the spherical water–surfactant droplet. The reference point is chosen at the center of the droplet. The ith non-counterion pair is represented by radius-vectors of ion and counterion, r_i^+ and r_i^-

by the ensemble of the realizations of random positions of ions. Then, retaining the main terms in the quadratic form, we obtain

$$\langle(\Delta\mu)^2\rangle \approx e^2 \sum_{i=1}^{N_s} [\langle(\mathbf{r}_i^+)^2\rangle + \langle(\mathbf{r}_i^-)^2\rangle] \tag{5}$$

It is easy to show that a condition of applicability of approximation (5) is fulfilled when the following correlation coefficients are small:

$$Q_{i,j}^{\pm,\,\pm} = \frac{\langle\mathbf{r}_i^\pm\mathbf{r}_j^\pm\rangle}{\sqrt{\langle(\mathbf{r}_i^\pm)^2\rangle\langle(\mathbf{r}_j^\pm)^2\rangle}} \ll 1 \quad \text{for } i \neq j \tag{6a}$$

$$Q_{i,i}^{+,\,-} = \frac{\langle\mathbf{r}_i^+\mathbf{r}_i^-\rangle}{\sqrt{\langle(\mathbf{r}_i^+)^2\rangle\langle(\mathbf{r}_j^-)^2\rangle}} \ll 1 \;. \tag{6b}$$

We note that inequalities (6a) and (6b) cannot be examined directly because no relationships were found in the literature regarding the two-particle coordinate distribution function for non-uniform distribution of ions in the restricted volume. However, the two-particle distribution function for continuous uniform medium are well known from the Debye–Hückel theory [10, 11]. Using this function for estimations, one can show that inequalities (6a) and (6b) are fulfilled for the considered range of droplet sizes and small electrolyte concentration.

A calculation of the terms $\langle(\mathbf{r}_i^+)^2\rangle$ and $\langle(\mathbf{r}_i^-)^2\rangle$ entering Eq. (5) can be performed using the one-particle probability distribution functions $W_1^+(\mathbf{r}^+)$ and $W_1^-(\mathbf{r}^-)$ that are proportional to the ion density

$$W_1^+(\mathbf{r}^+) = W_1^+(|\mathbf{r}^+|) = \frac{1}{N_s}c(r)\;, \tag{7a}$$

$$W_1^-(\mathbf{r}^-) = \frac{1}{4\pi R_w^2}\delta(|\mathbf{r}_i^-| - R_w)\;, \tag{7b}$$

where R_w is the radius of the water core, $c(r)$ is the density of the counterions at the distance $r = |\mathbf{r}^+|$ from the center of the droplet, and N_s is the total number of the counterions in the droplet interior:

$$N_s = 4\pi \int_0^{R_w} r^2 c(r)\,\mathrm{d}r\;. \tag{8}$$

By taking into account Eqs. (7a) and (7b), Eq. (5) reads

$$\langle(\Delta\mu)^2\rangle = e^2\left\{4\pi \int_0^{R_w} r^4 c(r)\,\mathrm{d}r + N_s R_w^2\right\}\;. \tag{9}$$

According to Eq. (4), this relation allows us to calculate the apparent dipole moment $\mu_a = (\langle(\Delta\mu)^2\rangle)^{1/2}$ of a droplet.

Counterion density distribution

The distribution of the counterions in the droplet interior is assumed [12–17] to be governed by the Poisson–Boltzmann equation

$$\Delta\Psi = -\frac{4\pi e c(r)}{\varepsilon_w} = -\frac{4\pi e c_0}{\varepsilon_w}\exp\left(-\frac{e(\Psi(r) - \Psi(0))}{k_B T}\right),$$
$$\tag{10}$$

where Ψ is the electrostatic potential and ε_w is the dielectric permittivity of the water core. Here the reference point is chosen at the center ($r = 0$) of the spherical droplet, where the counterion density is c_0 and the electronic potential is $\Psi(0)$.

Equation (10) reads in the dimensionless form as

$$\psi'' + \frac{2}{x}\psi' = -\mathrm{e}^{-\psi} \tag{11}$$

with the boundary conditions

$$\psi(0) = 0 \tag{12}$$

and

$$\psi'(0) = 0 \,, \tag{13}$$

where ψ and x are the dimensionless potential $\psi = e[\Psi - \Psi(0)]/k_B T$ with respect to the center and the dimensionless distance $x = r/l_D$, respectively. Here the characteristic thickness of the counterion layer near the surface of the water core

$$l_D = \left(\frac{\varepsilon_w k_B T}{4\pi e^2 c_0}\right)^{1/2} \tag{14}$$

is the Debye screening length.

The Poisson–Boltzmann equation (11) can be solved by numerical integration [8, 12–14, 17] or by the expansion of the potential ψ in the radial coordinate x [16]. We look for the solution of Eq. (11) in the form of a logarithm of a power series:

$$\psi(x) = -\ln \sum_{j=0}^{\infty} a_j x^{2j} \,, \tag{15}$$

where the coefficients a_j are to be found. By substituting Eq. (15) into Eq. (11), we obtain a simple relationship for the charge density

$$c(x) = c_0 \sum_{j=0}^{\infty} a_j x^{2j} \,. \tag{16}$$

Differentiating Eq. (15) and substituting the corresponding series for ψ and the derivatives ψ' and ψ'' in Eq. (11), we then obtain the following recurrence formula for a_j:

$$2 \sum_{m,n}^{\infty} a_n a_m m[2(n - m) - 1]\delta_{m,j-1}$$

$$= - \sum_{q,p,k}^{\infty} a_q a_p a_k \delta_{k,j-1-p-q} \,, \tag{17}$$

In order to satisfy the boundary condition (13) we set $a_0 = 1$, then, for the next few coefficients we obtain: $a_1 = \frac{1}{6}$, $a_2 = \frac{1}{45}$, ... We note here that, as a matter of fact, the number of terms of the series in the recurrence formula (17) is finite due to the Kronecker symbol. The convergence of the power series in Eqs. (15) and (16) is dependent on the value of x being considered. Applying the standard ratio test [18], we found that the interval of the convergence of the series (15) and (16) is $0 < x < 3.27$.

Calculation of the fluctuation dipole moment of a droplet and the dielectric permittivity of ionic microemulsions

Equation (16) describes the distribution of the counterions in the droplet interior and can be substituted in Eq. (9) for the calculation of the mean-square dipole moment of a droplet. The total number of the counterions in the droplet interior N_s entering Eq. (9) can be obtained by substituting Eq. (16) into Eq. (8) and integrating it term by term, namely,

$$N_s = 4\pi l_D^3 c_0 \sum_{j=0}^{\infty} \frac{a_j}{2j + 3} x_R^{2j+3} \tag{18}$$

where $x_R = R_w/l_D$. Furthermore, using Eqs. (9), (16) and (18), we obtain the relationship for the mean-square dipole moment $\langle \mu^2 \rangle$ of a droplet as

$$\langle \mu^2 \rangle = 4\varepsilon_w k_B T R_w^3 \sum_{j=0}^{\infty} \frac{a_j(j + 2)}{(2j + 3)(2j + 5)} x_R^{2(j+1)} \,. \tag{19}$$

In order to find the counterion density at the center of a droplet c_0, entering Eq. (14) for the Debye length l_D, the counterion concentration $c(r)$ must be related to the dissociation of the surfactant molecules in the water core of the droplet. The dissociation of the surfactant molecules is described by the equilibrium relation [12, 16, 17]

$$K_s(T) = c_0 e^{-\psi(x_R)} \frac{N_s}{N_a - N_s} \,, \tag{20}$$

where N_a is the micelle aggregation number, K_s is the equilibrium dissociation constant of the surfactant, and $\psi(x_R)$ is the dimensionless electrical potential near the surface of the water core (i.e. at $r = R_w$). Substituting $\psi(x_R)$ in Eq. (20) with the power series given by Eq. (15), we obtain

$$K_S(T) = \frac{c_0 N_S}{N_a - N_S} \sum_{j=0}^{\infty} a_j x_R^{2j} \,. \tag{21}$$

The system of coupled equations (18), (19) and (21), along with the recurrence formula (17) for a_j and Eq. (14) for l_D, constitute the model describing the temperature and geometry dependence of the mean-square dipole moment of the droplet $\langle \mu^2 \rangle$. Furthermore, by inserting the calculated values of the mean square dipole moment into Eq. (1), we can obtain the equation

$$\frac{(\varepsilon - \varepsilon_{mix})(2\varepsilon + \varepsilon_{mix})(2\varepsilon + \varepsilon_w)}{\varepsilon^2} = 36\varphi\varepsilon_w \left(\frac{R_w}{R_d}\right)^3$$

$$\times \sum_{j=0}^{\infty} \frac{a_j(j + 2)}{(2j + 3)(2j + 5)} x_R^{2(j+1)} \,. \tag{22}$$

This enables us to calculate the values of the dielectric permittivity ε of the system.

Approximate relationships for the fluctuation dipole moment of a droplet and the dielectric permittivity of ionic microemulsions

An adequate approximate relationship for the calculation of the mean-square dipole moment $\langle \mu^2 \rangle$ in the case of a small droplet and/or the small dissociation of surfactant $(R_w \leq l_D)$ can be obtained in the first approximation, by taking into account the first term $(j = 0)$ only, in the series of the expressions (18)–(20). In this approximation, by using the relationship $x_R = R_w/l_D$, and taking into account Eq. (14), we obtain for the mean-square dipole moment,

$$\langle \mu^2 \rangle \approx \frac{8\varepsilon_w k_B T R_w^5}{15 l_D} = \frac{32}{15}\pi e^2 c^0 R_w^5 . \tag{23}$$

An approximate relationship for the counterion density at the droplet center c_0 can be obtained by using Eqs. (18) and (21) in the first approximation, which reads

$$c_0 \approx \left(\frac{3K_s}{R_w A_s}\right)^{1/2} , \tag{24}$$

where $A_s = 4\pi R_w^2/N_a$ is the average area (cm^2/molecule) on the surface of a water core associated with one surfactant molecule. After combining Eqs. (23) and (24), the mean-square fluctuation dipole moment of a droplet becomes

$$\langle \mu^2 \rangle \approx \frac{32\sqrt{3}}{15}\pi e^2 \left(\frac{R_w^9 K_s}{A_s}\right)^{1/2} . \tag{25}$$

In order to get a tractable relationship for the dielectric permittivity ε, we can further simplify Eq. (1), by taking into account the relative magnitudes of ε, ε_{mix} and ε_w. For $\varepsilon_w \approx 78$ and $\varepsilon \sim \varepsilon_{\text{mix}} \ll \varepsilon_w$, we approximate in Eq. (1) the term $(2\varepsilon + \varepsilon_w)$ by ε_w and $(2\varepsilon + \varepsilon_{\text{mix}})$ by 3ε. Then by substituting Eq. (25) in Eq. (1), we obtain

$$\varepsilon \approx \frac{\varepsilon_{\text{mix}}}{1 - X} , \tag{26}$$

where

$$X = \frac{32\sqrt{3}\pi\varphi e^2}{5R_d^3 k_B T \varepsilon_w} \left(\frac{R_w^9 K_s}{A_s}\right)^{1/2} . \tag{27}$$

It is easy to show that in the case of small droplet concentrations, $X \ll 1$, thus an approximate relationship for ε is

$$\varepsilon \approx \varepsilon_{\text{mix}}(1 + X) . \tag{28}$$

In order to explain the experimental temperature behavior of the dielectric permittivity of the system we have to consider the temperature behavior of the dissociation constant of the surfactant K_s, which has an Arrhenius behavior [12]:

$$K_s(T) = A \exp\left(-\frac{\Delta H}{k_B T}\right) , \tag{29}$$

where ΔH is the apparent activation energy of the dissociation of the surfactant in the water pool of a droplet, and A is the pre-exponential factor. It is easy to show that the dielectric permittivity of microemulsions obtained from Eqs. (26) or (28) for the Arrhenius behavior of the dissociation constant is a growing function of temperature in the temperature range of $T < \Delta H/2k_B$, which is always fulfilled in the measured temperature interval for any reasonable value of the activation energy.

Experiment

We have studied ternary sodium bis(2-ethylhexyl) sulfosuccinate (AOT)–water–oil (decane) microemulsions, with compositions of (1.9:2.4:93.7), (5.9:7.1:87.0), (11.7:14.2:74.1), and (17.5:21.3:61.2), respectively. All compositions are represented as percent by volume. In order to keep the radius of all the droplets fixed at 48 Å the molar ratio of water to surfactant has the value of $W = [\text{water}]/[\text{AOT}] = 26.3$ for all the microemulsions. AOT was purchased from American Cyannamid, USA. Decane and hexane were purchased from Sigma. The materials were used without further purification. Deionized and bidistilled water was used throughout the experiments.

Dielectric measurements were carried out by means of the Dipole TDS Ltd. time domain dielectric spectroscopy system TDS-2. The general principles of Time Domain Dielectric Spectroscopy and a detailed description of the set-up and a procedure of our measurements has been described elsewhere [19]. All samples were measured in a temperature range much inferior to that of the percolation region. Hence, for each microemulsion the measurements were started at 2 °C and continued until the temperature reaches the onset of the percolation region T_{on}.

Results and discussion

For numerical evaluations of the model we have to set the values of the parameters matching the studied systems.

The value of the dielectric permittivity of water was assumed to be equal to that of the bulk water at the corresponding temperature throughout all the calculations, as $\varepsilon_w = 87.74 - 0.40008t + 9.398 \times 10^{-4}t^2 - 1.41 \times 10^{-6}t^3$ [20], where t is the temperature in Celsius. The value of 2 for the dielectric permittivity of decane was adopted in the present calculations. The effective value of 8.5 was used [21] for the dielectric permittivity of AOT. The value of ε_{mix} was calculated by using the one-shell model [9]. The aggregation number N_a was estimated to be 244 molecules per droplet. The value of $A_s = 65\ \text{Å}^2$ for the average area on the surface of the water core associated with one AOT molecule was adopted [22].

We have to bear in mind that the model developed can be applied within the special ranges of the droplet radius and ionic dissociation of the surfactant only. On the one hand, the droplets cannot be too small. The radius of the water core must be larger than 15 Å (the water to surfactant ratio $W > 10$) to ensure that a core of "free" water exists [23]. On the other hand, the droplets cannot be too large, since the applicability range of the solution (15) is restricted by condition $R_w < 3.27l_D$. This condition also may be expressed in terms of the strength of the electrolyte in the droplet interior pK_s ($pK_s \equiv -\log K_s$) and/or by the degree of dissociation of surfactant $\alpha = N_s/N_a$. The correspondent $pK_s - (R_w)_{max}$ and $\alpha - (R_w)_{max}$ diagrams are depicted in Fig. 2.

Regarding K_s of AOT surfactant, little is known and we did not find any reliable experimental data for it in the literature. Thus, K_s can be considered as an adjustable

parameter of the theory which can be calculated from the inverse problem, i.e. we can determine K_s from the knowledge of experimentally measured dielectric permittivity of the studied microemulsions. The equilibrium dissociation constant K_s can be calculated by using Eqs. (14), (21), (22) or, in the case of small droplets and/or small dissociation of surfactant, by the approximate Eq. (26).

Figure 3 compares the values of the experimental apparent dipole moment μ_a of the studied microemulsions, obtained from Eq. (1), together with the theoretical values obtained on the basis of Eq. (22). One can see that the apparent dipole moment $\mu_a = (\langle \mu^2 \rangle)^{1/2}$ of the studied microemulsions increases versus temperature. For all the

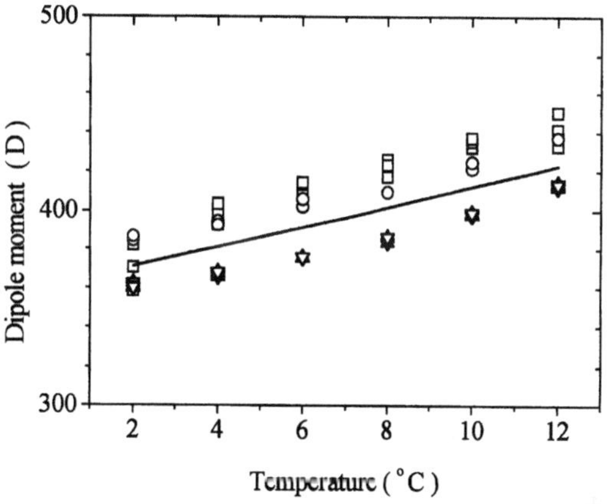

Fig. 3 Temperature dependence of the experimental and calculated on the basis of Eq. (21) macroscopic apparent dipole moments of a droplet of the AOT/water/decane microemulsions. The experimental values of the dipole moment are shown for various volume fractions φ of the dispersed phase: 0.043 (□); 0.13 (○); 0.26(△) and 0.39 (▽). The calculated values are shown by the solid line

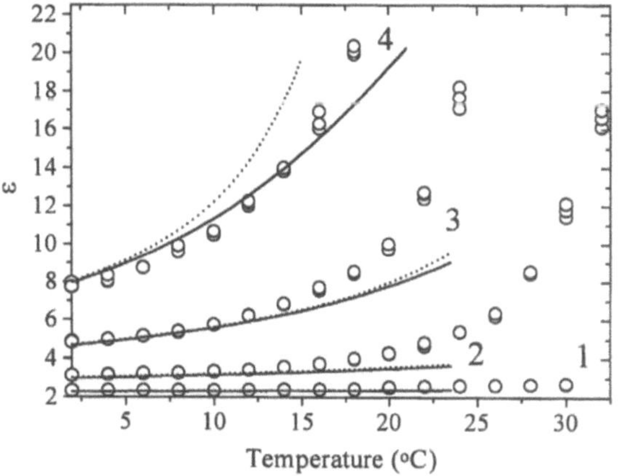

Fig. 4 Experimental (○) and calculated static dielectric permittivity versus temperature for the AOT-water-decane microemulsions for various volume fractions φ of the dispersed phase: 0.39 (curve 4); 0.26 (curve 3); 0.13 (curve 2); 0.043 (curve 1). The calculations were performed by using the formulae (22) and (26) (dashed line)

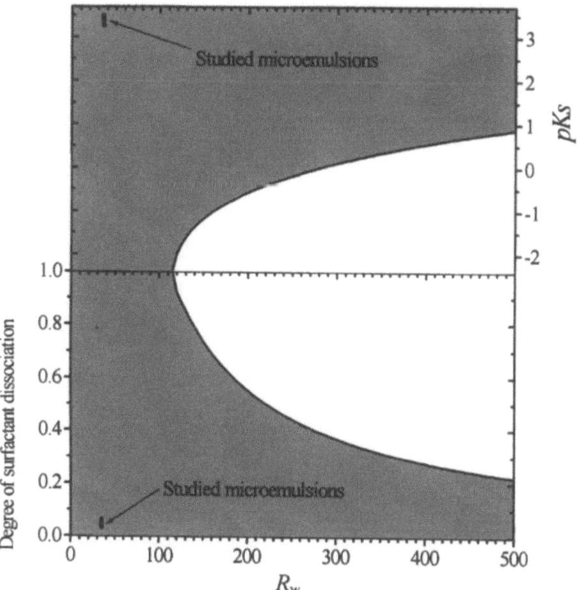

Fig. 2 The $pK_s - (R_w)_{max}$ and $\alpha - (R_w)_{max}$ diagrams of the validity of solution (15)

studied microemulsions the magnitude of the dipole moments for various volume fractions φ of the dispersed phase does not depend on φ within a degree of accuracy better than 10% which confirms the assumption of the model that droplets in the system can be considered as non-interacting for such concentrations of droplets.

As a final comment, let us briefly discuss the dielectric permittivity of the microemulsions studied. Figure 4 shows the temperature dependencies of the experimental dielectric permittivity and the results of the calculations on the basis of the developed model performed by using Eqs. (26) and (22). The difference between the values of ε obtained from these formulae can only be observed at high φ. The calculated values of ε concur well with the experimental data in the region far below the onset of percolation ($T < T_{on}$), where the assumptions of the model are fulfilled. At the temperatures close to the percolation onset T_{on} and beyond it, deviations of the theoretical values from the experimental data are observed. These deviations indicate the structural changes in the system that appear at percolation.

Acknowledgements We would like to express our appreciation to Dr. Ido Nir for the help with the experimental work. Yu. A. acknowledges the support of the Levy Eshkol Foundation of the Israel Ministry of Science and Technology. Yu. F. is grateful to the donors of the Bruno Goldberg Endowment Fund for partial support of this work.

References

1. Kotlarchyk M, Chen S-H, Huang JS, Kim MW (1983) Phys Rev A 28:508
2. Kotlarchyk M, Chen S-H, Huang JS, Kim MW (1984) Phys Rev A 29:2054
3. Langevin D (1992) Annu Rev Phys Chem 43:341
4. Feldman Y, Kozlovich N, Nir I, Garti N (1995) Phys Rev E 51:478
5. Feldman Y, Kozlovich N, Alexandro Y, Nigmatullin R, Ryabov Y (1996) Phys Rev E 54:5420
6. Eicke HF, Bercovec M, Das-Gupta B (1989) J Phys Chem 93:314
7. Hall DG (1990) J Phys Chem 94:429
8. Kozlovich N, Puzenko A, Alexandrov Y, Feldman Y (1998) Colloids and Surfaces A 104:N2–3
9. Takashima S (1989) Electric Properties of Biopolymers and Membranes. Adam Hilger, Bristol and Philadelphia, p 155
10. Watt RO, McGee IJ (1976) Liquid State Chemical Physics. Wiley, New York, p 272
11. Fisher IZ (1964) Statistical theory of liquids. The University of Chicago Press, Chicago, p 113
12. Beunen JA, Ruckenstein E (1983) J Colloid Interface Sci 96:469
13. Ruckenstein E, Beunen JA (1988) Langmuir 4:77
14. Sjöblom J, Jönsson B, Nylander C, Lundström I (1983) J Colloid Interface Sci 96:504
15. De Rozieres J, Middleton MA, Schechter R (1988) J Colloid Interface Sci 124:407
16. Tomic M, Kallay N (1992) J Phys Chem 96:3874
17. Kapre P, Ruckenstein E (1990) J Colloid Interface Sci 137:408
18. Boas M (1983) In: Mathematical Methods in Physical Sciences. Wiley New York, p 12
19. Feldman Y, Andrianov A, Polygalov E, Romanychev G, Ermolina I, Zuev Y (1996) Rev Sci Instrum 67:3208
20. Hasted JB (1973) Aqueous Dielectrics. Chapman and Hall, London
21. Belletete M, Lachapelle M, Durocher G (1990) J Phys Chem 94:5337
22. Dijk MA, Joosten JGH, Levine YK, Bedeaux D (1989) J Phys Chem 93:2506
23. D'Angelo M, Fioretto D, Onori G, Palmieri L, Santucci A (1995) Phys Rev E 52:R4620

Progr Colloid Polym Sci (1998) 110:163–170
© Steinkopff Verlag 1998

M.H.G.M. Penders
D.P. Jones
D. Needham
E.G. Pelan
D.R. Underwood
D.J.P. Scollard
A.P. Davies

Kinetics and thermodynamics of tea cream formation: A colloidal approach

M.H.G.M. Penders (✉) · D.P. Jones
D. Needham · E.G. Pelan · D.R. Underwood
D.J.P. Scollard · A.P. Davies
Unilever Research Colworth
Colworth House
Sharnbrook, Bedford MK44 1LQ
United Kingdom

Abstract We report on the effect of decaffeination and pH on the kinetics and thermodynamics of tea cream formation using turbidimetry and time-resolved (static and dynamic) light scattering.

Decaffeination enhances the solubility of tea solids (polyphenols) in black tea infusions, resulting in a shift of the location of the phase diagram to lower temperatures in comparison to "standard" black tea material. The phase diagrams for the studied tea samples displays similar trends to those of (classical) simple mixtures which dissolve at higher temperatures but separate into immiscible phases below the (upper) critical solution temperature. In the case of decaffeinated tea it did prove possible to access the metastable region of the miscibility gap with nucleation and growth being the important mechanism in the dilute part of the phase diagram. More often than not, however, the mechanism responsible for the formaion of tea cream is governed by demixing through spinodal decomposition caused by the increased insolubility of polyphenols.

On lowering the pH, the location of the phase diagram is shifted to higher temperatures and larger particles of associated (polyphenol and polyphenol/caffeine) structures are identified, reflecting the decrease in solubility of tea solids in black tea infusions as a result of decreased electrostatic interactions. At natural pH of 4.8 the electrostatic repulsion between the charged droplet surfaces protects the droplets from coagulation, whereas at pH = 2.0 close to the iso-electric point coagulation of tea cream particles takes place resulting in a rapid settling out of tea solids, as observed previously by Harbron. Results at pH = 3.0 and $c = 0.3$ wt%, however, do not indicate aggregation/coagulation between different particles, which means that under these conditions there is apparently enough electrostatic repulsion preventing cream particles from coagulation.

Key words Tea cream – colloids – light scattering – coagulation – caffeine

Introduction

It is commonly known that hot aqueous black tea infusions on cooling can produce noticeably turbid colloidal coacervates referred to as "tea cream". Tea cream is a complex mixture of many components with mainly polyphenols (thearubigins and theaflavins) and caffeine [1–3]. Rutter [4, 5] and Collier et al. [6] concluded that the formation of tea cream is governed by various molecular types of interaction including polyphenol–caffeine complexation and polyphenol–polyphenol interactions. According to these authors [4–6] polyphenol–caffeine complexation is influenced by the number of gallate ester and hydroxy groups of the polyphenols.

More recently, Dickinson [7] reviewed our current state of knowledge of physical and colloidal aspects of beverage stability and clarity. This review clearly displays the importance of thermodynamic and kinetic aspects of tea cream formation which affect the clarity of tea as a beverage. A detailed study of these aspects using turbidimetry and time-resolved (static and dynamic) light scattering is summarised in a previous paper from our laboratory [8] in which the mechanistic pathways responsible for tea cream formation are elucidated. The application of light scattering to black tea colloids has not received much attention until recently [8, 9] but has proven to be very successful for identifying possible mechanisms responsible for the formation of tea cream like demixing through spinodal decomposition and nucleation and growth.

As demonstrated by several authors [10–13] the formation of tea cream is governed by various parameters i.e. temperature, total tea solids concentration, chemical composition and pH. The effect of tea solids concentration has been discussed in more detail by Bee et al. [10]. They showed that tea cream consists of spherical droplets a few microns in diameter at solid contents ≥ 5 wt% using (electron and optical) microscopy methods. They also demonstrated that at high tea cream phase volumes the phenomenon of phase inversion will take place with a dilute phase dispersed into a continuous cream phase. Harbron [12] and Smith [13] showed that a change in pH will drastically alter the physico-chemical properties of tea cream formation. Smith [13] found that depending on the total tea solids concentration and temperature the maximum yield of tea cream formed is situated around pH = 3.0. Obviously, the pH-influence is of crucial interest in black tea infusions from a fundamental scientific point of view.

This paper contains a overview of recent developments in our laboratory with regards to a conceptual mechanistic understanding of tea cream formation using time-resolved (static and dynamic) light scattering and turbidimetry. The effect of decaffeination and pH on the kinetics and thermodynamics of tea cream formation will be discussed in more detail.

Experimental

Sample preparation

Black tea infusions were prepared from a bulked black tea material of dried extract, dissolved in boiling Millipore water. The extract was prepared by hot water countercurrent extraction of black Sri-Lankan tea made at 3% tea solids and freeze dried to a fine powder. A decaffeinated form of this black tea was also studied with decaffeination achieved by chloroform extraction according to the method of Rutter and Stainsby [5]. Henceforth we will refer to the two materials as "standard" and "decaffeinated". pH-adjustments were made on adding a few drops of (semi-concentrated) aqueous solutions of hydrochloric acid HCl obtained from BDH Laboratory Supplies or adding a few drops of (semi-concentrated) aqueous solutions of sodium hydroxide NaOH obtained from Sherman Chemicals to raise the pH.

In the case of light scattering experiments removal of dust and large impurities is essential, as they largely dominate the light scattering data, masking the scattering from the majority of smaller scatterers of interest. This was effectively achieved by ultracentrifugation of tea infusions for 30 min at temperatures above the "cream points" at a centrifugal field of 184×10^3 g using a Beckman L8-M Ultracentrifuge with a Type 70 Ti Fixed Angle Rotor (average radius $r = 65.7$ mm). As a result of centrifugation, the insoluble grey-white pellet of material, which was centrifuged to the bottom, is of unknown composition but represented less than 1% of the total tea solids. The initially clear liquid supernatants were used for the scattering studies.

The method of sample purification by ultracentrifugation proved unnecessary for turbidimetric (or light transmittance) measurements. The latter technique is not very sensitive to the presence of dust and/or multiple scattering [14, 15] and proved well suited to the determination of "tea cream" phase boundaries. In particular we ascertained that the temperature at which the onset of cream formation took place (the so-called cream point) was not appreciably sensitive to the purification of tea infusions by centrifugation.

Location of phase boundaries

Light transmittance (turbidity) measurements at $\lambda = 800$ nm were carried out using a UV-2101 PC Shimadzu spectrophotometer to locate the phase boundaries of black tea infusions at different concentrations and/or pH. Quartz cells ranging in path length from 2 to 10 mm enabled a wide range of tea solid concentrations (0.5–15 wt%) to be studied. The temperature was controlled by the circulation of thermostatically regulated water through a large black metal block, in which the cells were housed. The transmittance was recorded as a function of temperature on cooling with a rate of 0.25 K/min. The onset-temperature at which a (relatively) sharp decrease in transmittance is observed due to the formation of tea cream was taken as the "cream"-point T_{cream}. The procedure of the determination of T_{cream} (within 1–2 °C) is described in more detail elsewhere [8]. It has to be admitted that the phase boundaries so determined are not strictly thermodynamic (in the sense e.g. of a sharp binodal) but rather kinetically controlled and to some extent sensitive to the choice of cooling rates.

Time-resolved light scattering measurements

Time-resolved measurements of scattering intensities and particle sizes and scattering intensities for various tea infusions were performed using the ALV-3000 (Langen, Germany) dynamic light scattering (DLS) set-up equipped with a Spectra-Physics Series 2000 high power (up to about 500 mW) Krypton ion red laser (wavelength $\lambda = 647.1$ nm). Temperatures were chosen in the vicinity of the cream-point of the black tea infusions with very small quench depths (in the range of 1–2 K). We prefer the use of red lasers ($\lambda > 550$ nm) over green and/or blue lasers ($\lambda < 550$ nm) as in the latter case the reduction of transmitted intensity for black tea infusions is caused by both scattering and absorption. We found that for $\lambda < 550$ nm tea infusions strongly absorb light, in accordance with earlier results by Harbron [12].

Both static (SLS) and dynamic (DLS) light scattering are feasible option for the determination of particle sizes.

In principle, SLS provides the radius of gyration R_g from the angular dependence of the scattered intensity $I(\theta)$. For not too large particles (radii typically <100 nm) one makes use of the Guinier approximation (e.g. [16]):

$$I(\theta) \sim \exp\left(- K^2 R_g^2/3\right) \tag{1}$$

with

$$K = (4\pi n/\lambda)\sin(\theta/2) . \tag{2}$$

Here K represents the wave vector, n the refractive index of the sample and θ the scattering angle. For spherical particles R_g equals $(3/5)^{0.5}a$ with a the radius of the spheres.

DLS provides the diffusion coefficient D of the particles, from which the hydrodynamic radius R_H is obtained via the Stokes–Einstein equation $R_H = kT/(6\pi\eta D)$. Here k represents Boltzmann's constant, T the absolute temperature and η the viscosity of the continuous phase (in this case water). The standard deviation in the diffusion coefficient (or hydrodynamic radius) can be obtained from the so-called cumulant-analysis [17], with the second cumulant Q representing a measure for the variance in D. The size distribution for the tea cream particles was in most cases fairly/moderately monodisperse.

Comparison of SLS and DLS particle size measurements yielded a value for the ratio R_g/R_H of 0.80 ± 0.05. For non-porous spheres this ratio is expected to be $(3/5)^{0.5} \simeq 0.77$, and for hollow spheres it is 1.0 [18]. Electron microscopy studies (not shown here) indicate that the tea cream particles are spherical in general. The measured value for the ratio R_g/R_H of 0.80 ± 0.05 would not be inconsistent with substantially non-porous spherical particles. The second cumulant Q, inferred by DLS, was usually <0.10 indicating relatively narrow, monomodal size distributions.

According to the Rayleigh–Gans Debye (RGD) theory the intensity at 90°, I_{90}, of monodisperse spherically symmetric particles in dilute dispersions, neglecting interparticle interactions, is given by

$$I_{90} \sim \rho(n_p - n_s)^2 V^2 P_{90} \tag{3}$$

with ρ the number of tea particles per unit volume (number density), n_p the refractive index of the particle, n_s the refractive index of the solvent, $V \sim a^3$ the volume of the particles and $P(\theta)$ the so-called form factor. The term $|n_p - n_s|$ is often called *contrast*. Equation (3) demonstrates that a constant contrast and number density, I_{90}/a^6 or I_{90}/R_H^6 is time-invariant. This scaling principle will be applied to scattering properties of "standard" and decaffeinated tea material.

Results and discussion

Phase diagrams

Effect of decaffeination

In Fig. 1 phase diagrams (T_{cream} vs c_{solids}) for "standard" and decaffeinated black tea materials are displayed. These phase diagrams show trends similar to those of (classical) simple mixtures which dissolve at high temperatures but separate into immiscible phases below the (upper) critical solution temperature. The cream-points of the tea infusions were determined from light-transmittance and light-scattering measurements. In general the cream-points for decaffeinated tea are situated at lower temperatures compared to those for "standard" black tea, which reflects the enhancement of tea solids in black tea infusions in the absence of caffeine. This is in agreement with earlier observations by Rutter [4] and Smith [13].

Figure 1 contains a "schematic" picture of the (possible) mechanisms causing tea cream formation for both tea systems. Here the spinodal-line demarcates the metastable from the unstable region. In the latter region phase separation takes place spontaneously, whereas in the former region the process of nucleation and growth prevails

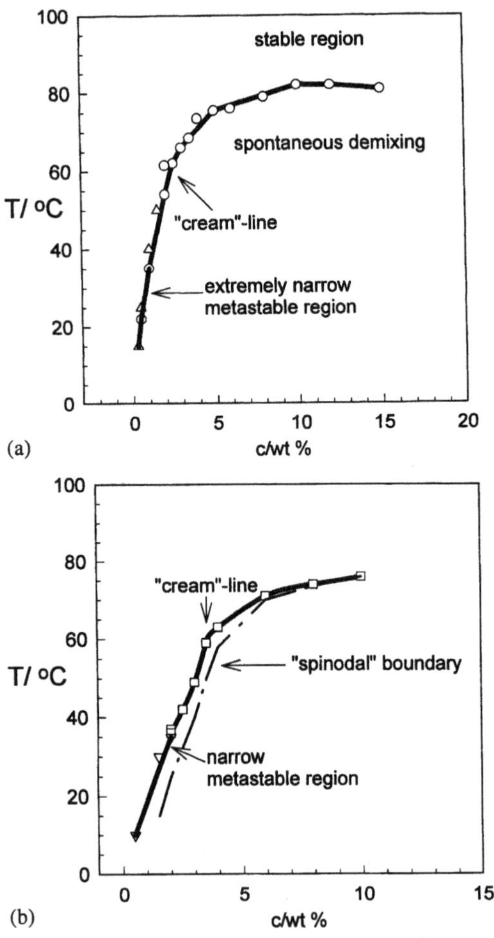

Fig. 1 "Schematic" phase diagrams at natural pH of 4.8 based on turbidimetric and time-resolved light (static and dynamic) scattering results for (a) "standard" black tea material and (b) decaffeinated black tea material. The "cream"-line demarcates the stable region from the metastable region, which is extremely narrow for "standard" black tea. The "spinodal" boundary forms the separation between the unstable and the metastable region as explained in the text. (○) "standard" black tea, turbidimetry results, (△) "standard" black tea, light scattering results, (□) decaffeinated black tea, turbidimetry results and (▽) decaffeinated black tea, light scattering results

and an energy barrier has to be overcome to induce phase separation. As will be discussed later only for decaffeinated tea (see Fig. 1b) under special conditions (relatively low concentrations and small temperature quenches) did it prove possible to access the narrow metastable region of the phase diagram, permitting in principle study of (nucleation and/or) growth of tea particles. In the more concentrated region of the phase diagram ($c > 2.0$ wt%), the metastable region becomes extremely narrow. This implies that with a small temperature quench one again directly accesses the unstable region of the phase diagram. In this region, tea infusions usually separate into two or more phases due to insolubility of polyphenols.

pH effects

In Fig. 2a the pH-dependence of the cream-point for "standard" black tea infusions is displayed for various tea solids concentrations. The cream-points of the tea-infusions were determined from light-transmittance and light-scattering measurements as discussed in more detail in the

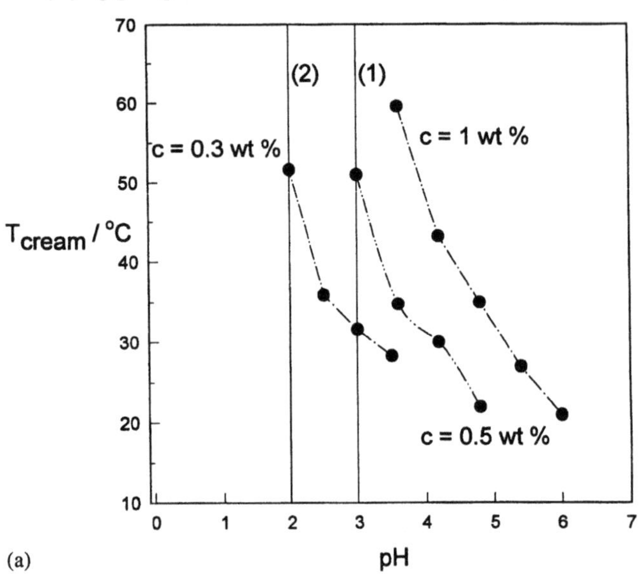

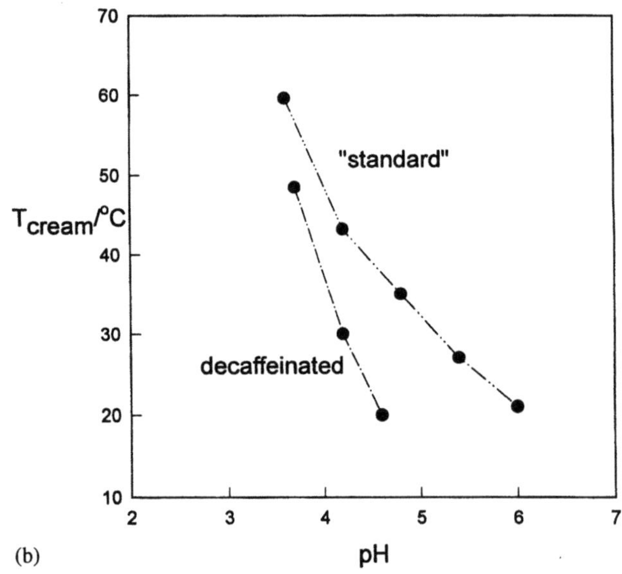

Fig. 2 Cream-point vs pH for "standard" and decaffeinated black tea. (a) Phase boundaries at various tea solids concentrations for "standard" black tea. As explained later in the text aggregation between different particles takes place at pH = 2 whereas at pH = 3 results suggest associated structures within the particles. (b) Comparison "standard" with decaffeinated black tea at 1.0 wt% tea solids concentration

experimental section. Figure 2a shows that changing pH from 4.8 to 3.6 at a total tea solids concentration of $c = 1.0$ wt% causes an increase in cream-point from 35 to 60 °C, showing the large significance of pH. These results are in accordance with earlier results from e.g. Harbron and Smith [12, 13] showing an increase in amount of tea cream at decreasing pHs as a result of a decreased solubility of tea solids. Smith [13] also observed that the mass of tea cream goes through a maximum on decreasing the pH. The location of the maximum depends on the total solids concentration. Figure 2b shows the influence of decaffeination on the location of the cream-point vs pH. The decrease in cream-point as a result of decaffeination demonstrates that in the absence of caffeine, the solubility of the tea solids in black tea infusions is enhanced. The increase in solubility is in agreement with earlier observations by Rutter [4] and Smith [13].

Time-resolved light scattering experiments

Effect of decaffeination

Figure 3a and b show time-resolved measurements of hydrodynamic radius at 90° for "standard" and decaffeinated black tea material at various concentrations. The measurements were carried out in the dilute region of the phase diagram (≤ 1.5 wt%) at temperatures T located in the vicinity of the cream-points with relatively small quench-depths $|T - T_{\mathrm{cream}}|$ in the order of 1–2 K. From Fig. 3b it can be inferred that for decaffeinated tea initially there is a rapid growth in hydrodynamic radius, and at later time, the particle size more or less approaches a plateau. The polydispersity, however, stays relatively constant. For "standard" black tea changes in particle size (Fig. 3a) and polydispersity with time are relatively minor.

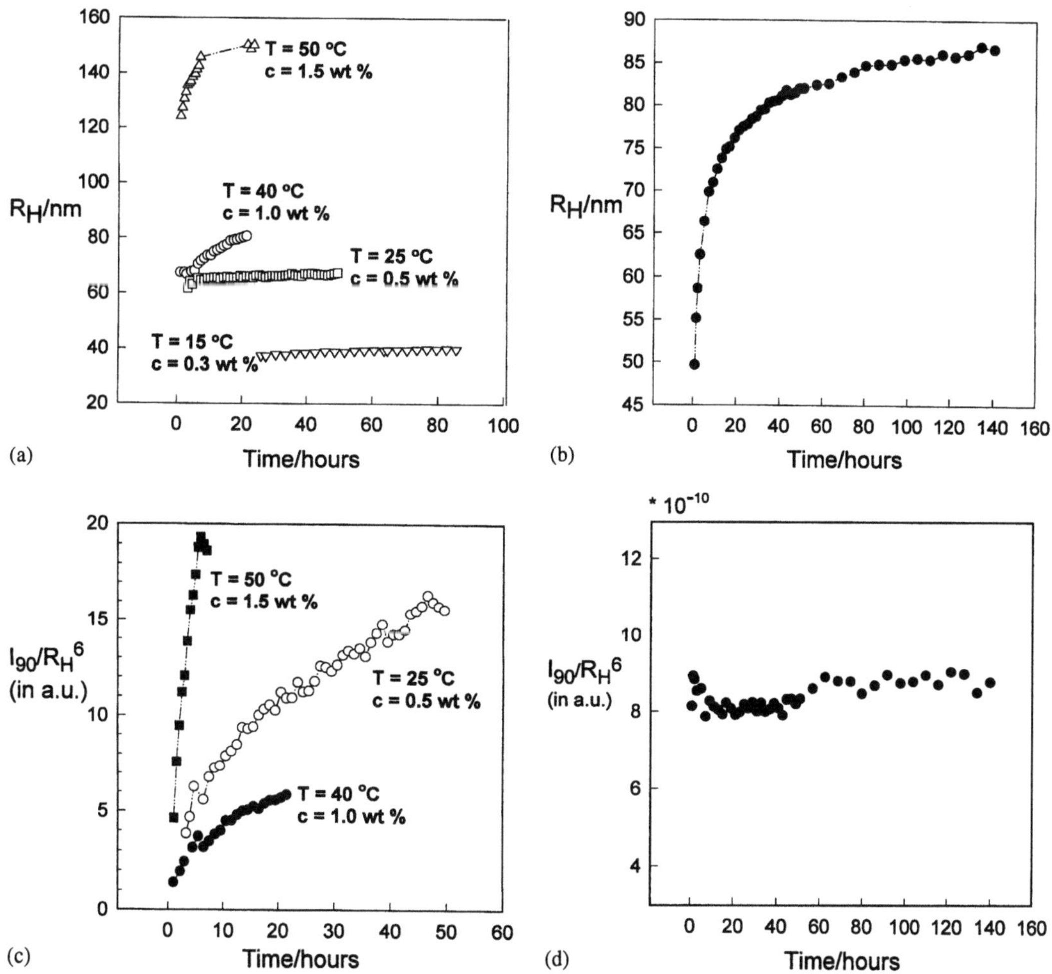

Fig. 3 Time-resolved light scattering for "standard" and decaffeinated black tea material at small quench depths. (a) Hydrodynamic radius vs time for "standard" black tea at different concentrations, (b) hydrodynamic radius vs time for decaffeinated black tea at $c = 1.5$ wt% and $T = 30$°C, (c) $(I_{90}/I_{\mathrm{ref}})/R_{\mathrm{H}}^6$ vs. time for "standard" black tea at different concentrations and (d) $(I_{90}/I_{\mathrm{ref}})/R_{\mathrm{H}}^6$ vs. time for decaffeinated black tea at $c = 1.5$ wt% and $T = 30$°C

Apparently, the particle size is determined mainly by the total tea solids concentration and to much lesser extent by time.

In Fig. 3c and d the scaling parameter I_{90}/R_H^6 is plotted versus time for "standard" and decaffeinated tea to test the time-invariance of this parameter. Figure 3d demonstrates that over a large time-interval the quantity I_{90}/R_H^6 is time-invariant. This proves, indeed, that in the case of the *decaffeinated* tea material particles grow with time at fixed number density ρ and contrast $|n_p - n_s|$ (see Eq. (3)). Apparently, there is a difference in scattering behaviour in comparison to "standard" black tea material, where this scaling does not apply and I_{90}/R_H^6 increases with time (see Fig. 3c). In the latter case a strong growth in scattering intensity with time was observed at relatively constant particle size, especially at concentrations ≥ 0.5 wt%. This appears to conform well to some of the predictions of the Cahn–Hilliard theory considering the (relatively) early stages of spinodal decomposition assuming our measured hydrodynamic radius gives a reasonable representation of the dominant wavelength of the spinodal decomposition [19, 20]. Lack of appropriate K-range prevents us for testing this hypothesis further using more conventional procedure of studying $I(K)$ vs K with time.

For "standard" black tea two possible explanations can be considered: either more particles are formed or particles become more compact. The lack of time-dependence in both particle size and polydispersity argues against the formation of new particles as the origin of the observed increase in scattering intensity. Instead, on the basis of Eq. (1) this implies an increase in contrast $|n_p - n_s|$ (i.e. density of tea within the particles) with time at relatively constant particle volume V and number density ρ. The fact that the particle size and polydispersity in size remain relatively constant with time indicates that late stage coarsening mechanisms appear not to be very important, i.e. particles exhibit a high level of stability with respect to late stage coarsening effects like flocculation (aggregation), coalescence and Ostwald ripening. The stability against flocculation and coalescence is probably due to electrostatic repulsion between the charged surfaces of the particles, as has been suggested by Harbron et al. [11].

pH effects

The effect of pH on the scattered intensity and on the particle size with time for "standard" black tea infusions is presented in Figs. 4 and 5. All the light scattering measurements took place at temperatures in the vicinity of the cream-point. As expected, on increasing the pH lower scattering intensities and smaller particle sizes were observed (Fig. 4) reflecting the clearer (black) appearance of the tea infusions. Lowering the pH, however, results in a markedly increase in scattering intensity and particle size reflecting the more opaque appearance of the tea infusions. This is demonstrated in Fig. 5, showing an increase in hydrodynamic radius from about 35 to about 100 nm for black tea at drinking strength ($c = 0.3$ wt%) on decreasing the pH from 4.8 to 3.0.

In the case of pH $= 3.0$ a large increase in scattering intensity with time was observed, whereas the particle size remained relatively unchanged. As discussed before, we

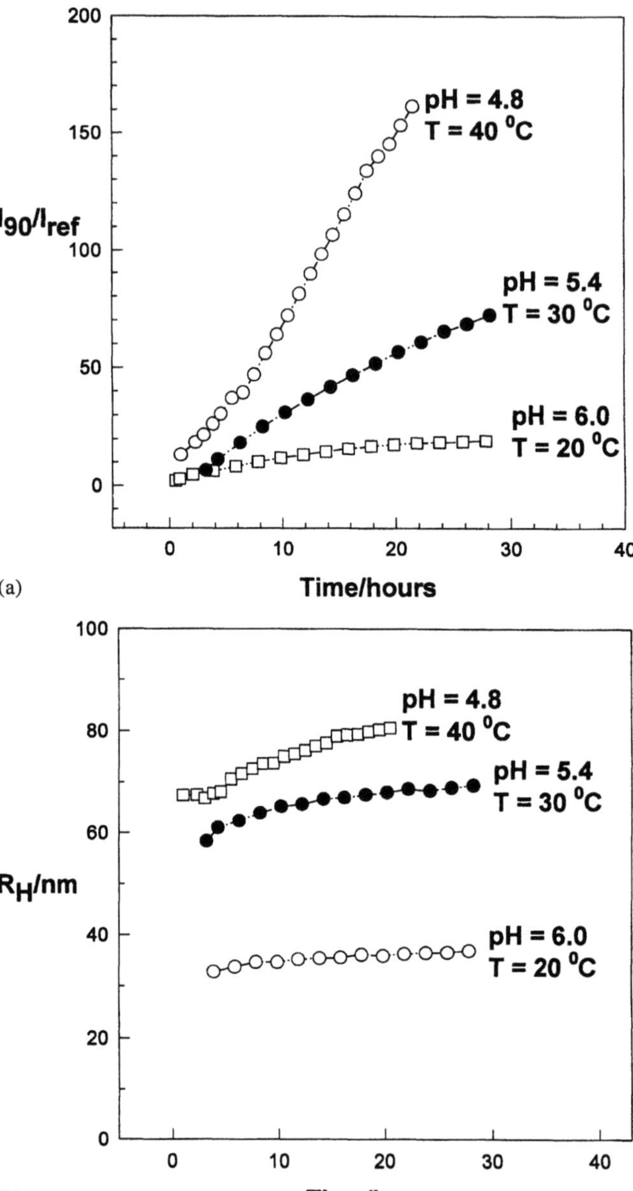

Fig. 4 (a, b) Intensity at 90° and R_H vs. time at various pHs for 1 wt% "standard" black tea material. All temperatures were chosen in the vicinity of the cream-points

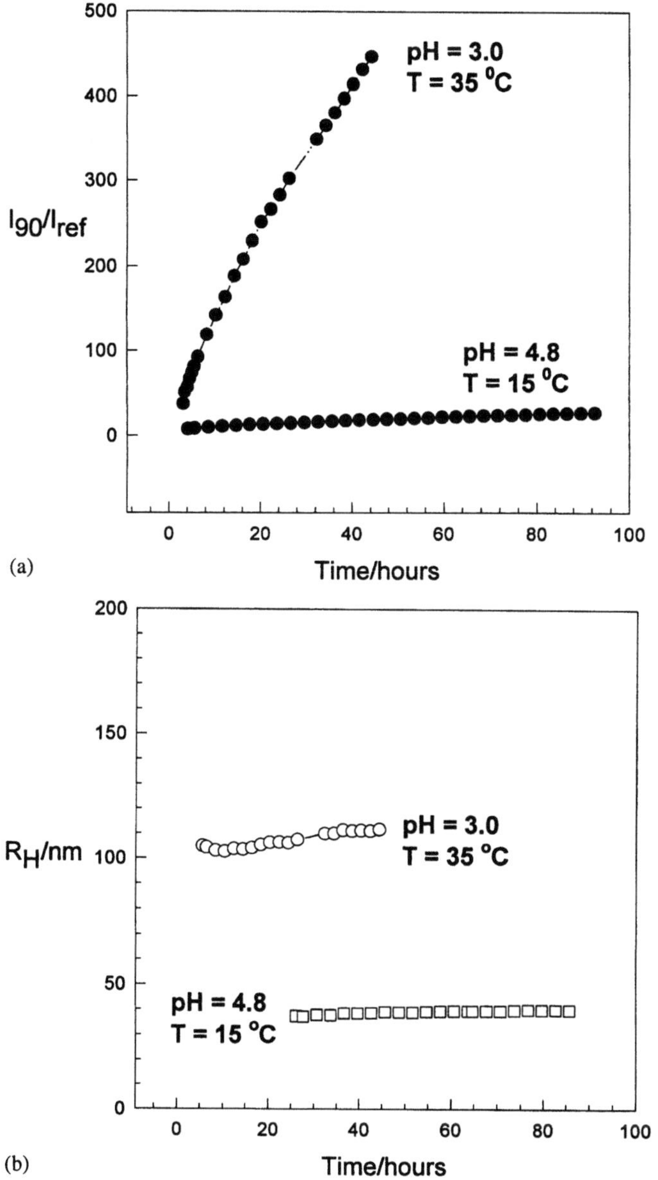

Fig. 5 (a, b) Intensity at 90° and R_H vs. time at various pHs for 0.3 wt% "standard" black tea material

ascribe this behaviour to the fact that at increasing times the particles become denser at constant number density. Results do not indicate coagulation/aggregation between different particles, but associated (polyphenol and polyphenol/caffeine) structures within the particles as a result of decreased electrostatic repulsions at this pH compared those at the natural pH of 4.8. Static and dynamic light scattering results did not show any evidence of the type of floc/fractal structures observed in other colloidal systems i.e. gold sols [21] and dispersions of silica particles coated with octa-decyl chains in linear alkanes [22]. Higher tea

solids concentrations and/or lower pHs might very likely cause coagulation between particles.

At pH = 2.0 coagulation of tea cream particles was observed by Harbron in the microscope of a micro-electrophoresis apparatus [12]. At this pH, a rapid settling out of tea solids was seen [12] which was consistent with the reported iso-electric point at pH $\simeq$ 2 as determined with micro-electrophoresis. The findings at pH = 2.0 and 3.0 are summarised in Fig. 2a.

All reported pH-results can be interpreted in terms of electrostatic stabilisation. At natural pH of 4.8 particles carry a net negative charge due to the presence of ionized acidic groups ($pK_a \sim 3$). These particles appear to exhibit a high level of stability at this pH with respect to flocculation and coalescence [7, 11], as was inferred from electrophoretic mobility and colloid coagulation studies. In accordance with the classical theory of lyophobic colloids [23], the electrostatic repulsion between the charged droplet surfaces protects the droplets from coagulation.

As was discussed before at pH = 3 and $c = 0.3$ wt% an increase in particle size was observed compared to the size at pH = 4.8 as result of decreased electrostatic interactions. No coagulation/flocculation was observed. Instead, results indicate associated (polyphenol and polyphenol/caffeine) structures within the particles. Apparently, even at pH = 3.0 there appears to be enough electrostatic repulsion preventing cream particles from further coagulation.

The large coagulation of tea cream particles resulting in a rapid settling out of tea solids at pH = 2.0 as observed by Harbron can be explained by a large decrease in electrostatic repulsion between the ("uncharged") droplets. This seems consistent with the reported iso-electric point at pH $\simeq$ 2 as determined with micro-electrophoresis.

Conclusions

We have investigated the effect of decaffeination of pH on the kinetics and thermodynamics of tea cream formation using turbidimetry and time-resolved static and dynamic light scattering measurements.

Decaffeination enhances the solubility of tea solids (polyphenols) in black tea infusions, resulting in a shift of the location of the phase diagram to lower temperatures in comparison to "standard" black tea material. The phase diagrams for the studied tea samples display similar trends to those of (classical) simple mixtures which dissolve at higher temperatures but separate into immiscible phases below the (upper) critical solution temperature. In the case of decaffeinated tea did it prove possible to access the metastable region of the miscibility gap with nucleation and growth being the important mechanism in the dilute part of the phase diagram. More often than not, however, the mechanism responsible for the formation of tea cream is governed by demixing through spinodal decomposition caused by the increased insolubility of polyphenols.

On lowering the pH, the location of the phase diagram is shifted to higher temperatures, reflecting the decrease in solubility of tea solids in black tea infusions as a result of decreased electrostatic interactions. At natural pH of 4.8 the electrostatic repulsion between the charges droplet surfaces protects the droplets from coagulation. Larger particles composed of associated (polyphenol and polyphenol/caffeine) structures were identified at pH = 3.0 and $c = 0.3$ wt% compared to substantially smaller particle sizes at natural pHs. Results do not indicate aggregation/ coagulation between different particles, which means that at pH = 3.0 there is apparently enough electrostatic repulsion preventing cream particles from coagulation. Coagulation of tea cream particles takes place at pH = 2.0 resulting in a rapid settling out of tea solids, as observed by Harbron. This can be explained by a large decrease in electrostatic repulsion between the ("uncharged") droplets at this pH, which seems consistent with the reported isoelectric point at pH $\simeq$ 2 as determined with micro-electrophoresis [12].

Acknowledgment Dr. Alex Lips and Dr. Ian T. Norton are thanked for critical reading of the manuscript and for stimulating discussions.

References

1. Roberts EAH (1963) J Sci Food Agric 14:700–705
2. Wickremasinghe RL, Perera KPWC (1966) Tea Quart 37:131–133
3. Smith RF (1968) J Sci Food Agric 19:530–534
4. Rutter P (1971) A physico-chemical study of tea cream. PhD thesis, Leeds University
5. Rutter P, Stainsby G (1975) J Sci Food Agric 26:455–463
6. Collier PD, Mallows R, Thomas PE (1972) Phytochemistry 11:867
7. Dickinson E (1994) Food Chem 51: 343–347
8. Penders MHGM, Jones DP, Needham D, Pelan EG (1998) Food Hydrocolloids, in press
9. Groening R, Baroth V, Breitkreuz J (1995) Pharm Pharmacol Lett 2:77–79
10. Bee RD, Izzard MJ, Harbron RS, Stubbs JM (1987) Food Microstructure 6:47–56
11. Harbron RS, Ottewill RH, Bee RD (1989) In: Food Colloids. R Soc Chem 75:283–294
12. Harbron RS (1986) A colloid-chemical study of tea cream from instant black tea infusion. PhD thesis, University of Bristol
13. Smith MA (1989) A physico-chemical study of the processes involved in the formation of tea cream. PhD thesis, University of Bristol
14. Penders MHGM, Vrij A (1990) J Chem Phys 93:3704–3711
15. Jansen JW, Vrij A, De Kruif CG (1986) J Colloid Interface Sci 114:492–500
16. Guinier A, Fournet G (1955) Small Angle Scattering of X-rays. Chapman and Hall, London
17. Koppel DE (1972) J Chem Phys 57: 4814–4820
18. Buchard W (1983) Adv Polym Sci 48: 1–124
19. Cahn JW (1965) J Chem Phys 42:93–99
20. Cahn JW, Hilliard JE (1958) J Chem Phys 28:258–267
21. Weitz DA, Huang, JS (1984) In: Landau DP (ed) Kinetics of Aggregation and Gelation. Elsevier, Amsterdam, pp 19–28
22. Rouw PW, de Kruif CG (1980) Phys Rev A 39:5399–5408
23. Verwey EJW, Overbeek JThG (1948) In: Theory of the Stability of Lyophobic Colloids. Elsevier, Amsterdam

Progr Colloid Polym Sci (1998) 110:171–174
© Steinkopff Verlag 1998

Phase behavior of polyoxyethylene dodecyl ether–water systems

K.-L. Huang
K. Shigeta
H. Kunieda

K.-L. Huang · K. Shigeta · H. Kunieda (✉)
Graduate School of Engineering
Yokohama National University
Tokiwadai 79-5
Hodogaya-ku, Yokohama 240
Japan

Abstract Phase diagram of poly-oxyethylene dodecyl ethers–water system was constructed as a function of the polyoxyethylene (EO) chain length at 25 °C to figure out the correlation between HLB (hydrophile–lipophile balance) of surfactant and the types of self-organizing structures. The EO chain length is varied from 1 to 9. Various self-organizing structures such as discontinuous-type cubic phase (I_1), hexagonal liquid crystal (H_1), normal bicontinuous-type cubic phase (V_1), lamellar liquid crystal (L_α), and reverse bicontinuous-type cubic phase (V_2) are formed depending on the HLB of the surfactant and/or the surfactant concentration. The effect of the EO-chain length on the structure of L_α phase was also investigated by means of small-angle X-ray scattering. The new phase diagram of water–$C_{12}EO_9$ system is also presented.

Key words Phase behavior – polyoxyethylene dodecyl ether – liquid crystal

Introduction

Phase behavior and self-organization of polyoxyethylene-type nonionic surfactant have been extensively investigated [1–8]. When phase diagram of the binary water–surfactant system is constructed, temperature and surfactant concentration are generally chosen as variables as is shown in Fig. 1.

However, in order to avoid the temperature effect on the self-organization of surfactant molecules, the polyoxyethylene (EO) chain length should be successively changed by mixing surfactants with different EO chains [8]. It is considered that the mixing of surfactants with different polyoxyethylene chain lengths does not cause serious changes in phase behavior within water system. The monomeric solubility or cmc of ordinary nonionic surfactant is extremely small and the homologues are very soluble in each other in aggregates. Hence, it can be regarded that all the surfactants practically form aggregates except in a extremely dilute region.

In this context, we constructed the phase diagrams of water–polyoxyethylene dodecyl ethers ($C_{12}EO_n$) system as a function of the EO chain length at 25 °C. The range of the EO chain length is between 1 and 9. The effect of the EO chain length on the structure of lamellar liquid crystal was also investigated by means of small-angle X-ray scattering.

Experimental section

Materials

Homogeneous polyoxyethylene dodecyl ethers were obtained from Nikko Chemicals Co. They are abbreviated as $C_{12}EO_n$ where $n = 1$–9 is the number of EO units. Two surfactants were mixed to obtain $C_{12}EO_n$, where the EO number, n is not an integer. For example, if n is between 5 and 6, $C_{12}EO_5$ and $C_{12}EO_6$ are mixed.

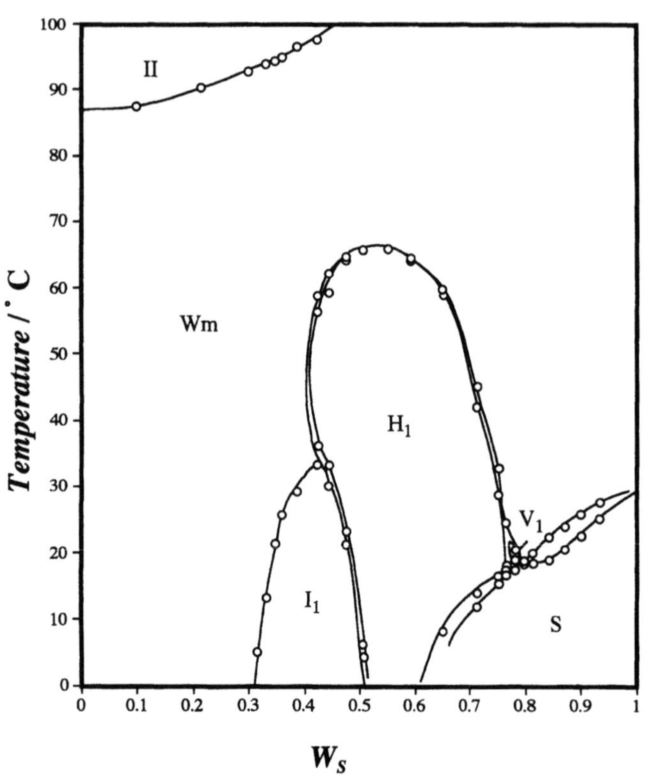

Fig. 1 Phase diagram of water–$C_{12}EO_9$ system as a function of temperature. The weight fraction of surfactant in the system, W_s, is plotted horizontally. II indicates a two-phase region

Small-angle X-ray scattering

Interlayer spacing of lamellar liquid crystal was measured using small-angle X-ray scattering (SAXS), performed on a small-angle scattering goniometer with an 18 kW Rigaku Denki rotating anode generator (RINT-2500) at ∼25 °C. The samples were covered by plastic films for the SAXS experiment (Mylar seal method). Lamellar and hexagonal liquid crystalline phases were distinguished by the SAXS peaks [9, 10]. The ratio of interlayer spacing from first and second peaks is $1 : 1/2$ for the lamellar type and $1 : 1/\sqrt{3}$ for the (reverse) hexagonal type, respectively. The type of liquid crystal was also identified by a polarizing microscope.

Calculation of the volume fraction of dodecyl group

The densities of surfactants were measured by the digital density meter (Anton Paar 40). We assume that the density of surfactant in a liquid state is unchanged even in solutions or liquid crystals at a constant temperature. The molar volume of surfactant is calculated by the following equation:

$$V_S = \frac{M_S}{\rho_S} \tag{1}$$

where M_S and V_S are the molecular and the molar volume of surfactant, respectively. It is also assumed that arithmetic additivity is held concerning the molar volume of each functional group in the surfactant [8, 11]. Then, the molar volume of $C_{12}EO_n$ is the sum of molar volume of each group in the surfactant and the following relation is held.

$$V_S = V_L + nV_{EO} + V_{OH}, \tag{2}$$

where V_L, V_{EO} and V_{OH} are the molar volumes of the lipophilic chain (dodecyl group), the oxyethylene unit and the hydroxyl group, respectively. n is the number of oxyethylene units. V_S, V_L, V_{EO} and V_{OH} were determined from the density data for homogeneous polyoxyethylene dodecyl ethers ($C_{12}EO_{1-6}$) and pure dodecanol because they are in a liquid state at 25 °C. The obtained values of V_L, V_{EO} and V_{OH} are 215, 38.8 and 8.8 cm³ mol⁻¹, respectively.

Notation

We use the following notations to distinguish each phase. The subscript 1 denotes "hydrophilic" or "normal"-type self-organizing structure or phase, in which the average curvature of surfactant layer is positive and convex toward water. The subscript 2 indicates "lipophilic" or "reverse"-type assemblies.

H_1: hexagonal liquid crystal
I_1: discontinuous-type cubic phase (water-continuous)
V_1: bicontinuous cubic phase
V_2: reverse bicontinuous cubic phase
L_α: lamellar liquid crystal
D_2: isotropic bicontinuous surfactant phase (reverse type) This phase is denoted by D′ or L_3 phase in the previous papers [6, 12].
Wm: aqueous phase containing surfactant aggregates
Om: oily phase like reverse micellar solution phase or surfactant liquid.
W: excess water phase
S: solid-present region

Results and discussion

Phase diagram of water–$C_{12}EO_n$ system at 25 °C

The phase diagram of water–polyoxyethylene dodecyl ethers constructed as a function of the EO-chain length at 25 °C is shown in Fig. 2.

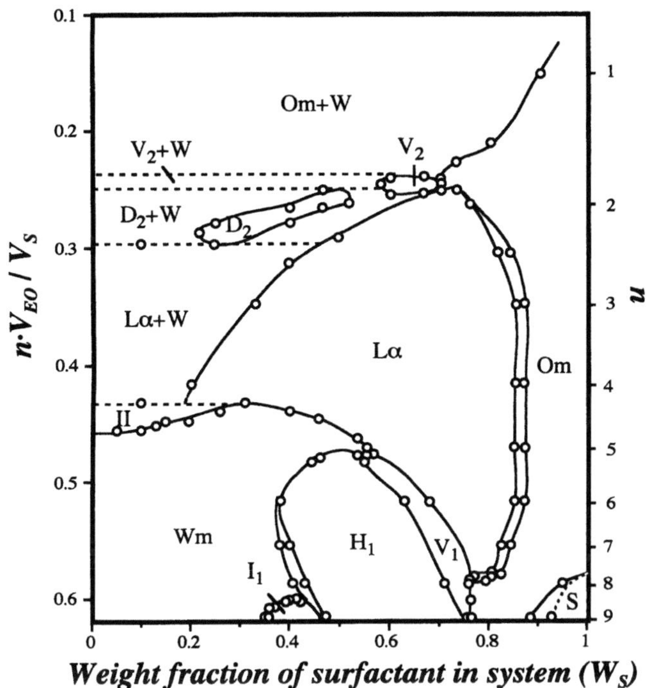

Weight fraction of surfactant in system (Wₛ)

Fig. 2 Phase diagram of water–$C_{12}EO_n$ system as a function of the EO chain length at 25 °C. II indicates a two phase region

The weight fraction of $C_{12}EO_n$ in the system, W_S, is plotted horizontally whereas nV_{EO}/V_S is plotted vertically. nV_{EO} means the molar volume of polyoxyethylene chain of the surfactant and V_S is the molar volume of the surfactant. This ratio is related to the Griffin's HLB number [13],

$$\text{HLB number} = 20(\rho_{EO_n}/\rho_s)(nV_{EO}/V_S),\qquad (4)$$

where ρ_S and ρ_{EO_n} are the densities of surfactant and polyoxyethylene chain, respectively. The average number of EO unit, n, is also plotted on the right-hand axis in Fig. 2. The $nV_{EO}/V_S = 0$ indicates dodecanol whereas the $nV_{EO}/V_S = 1$ means the dodecyl surfactant having infinitely long EO chain. The I_1 phase is considered to be a discontinuous-type cubic phase as is shown in Fig. 1. Since the maximum temperature for the I_1 phase is 18 °C in the $C_{12}EO_8$ system [5], this phase is considered to dominate the phase diagram in a longer EO-chain surfactant system. It is reported that both face- and body-centered cubic structures are present in the I_1 phase [14]. However, it was difficult to distinguish them in the phase diagram by SAXS because the temperature is close to its melting temperature and the numbers of measured SAXS peaks are not enough to figure out the detailed structure.

It is clear from Fig. 2 that the types of self-organizing structures are highly dependent on the EO-chain length or

the HLB number of surfactant, because the effective cross-sectional area of each surfactant in the aggregates depends on the EO-chain length [15]. By decreasing the EO-chain number of surfactants, the effective cross-sectional area of surfactant at the interface continuously decrease, and the curvature of surfactant layer would also decrease. In the dilute region, aqueous micellar phase, Wm, is changed to surfactant liquid, Om, via $L_α$, D_2, and V_2 phases with decreasing the EO-chain, because the effective cross sectional area of surfactant at the interface continuously decreases. Also in the high concentration region, along the line as the weight fraction (W_S) of surfactants is 0.7, when the EO chain decreases, the continuous phase transitions as the order of hexagonal (H_1), bicontinuous cubic phase (V_1), lamellar ($L_α$) and reversed-bicontinuous-cubic phase (V_2) can be observed. The change can be interpreted by the same reason.

By increasing the concentration of surfactants, steric hindrance is considered to affect the formation of the self-organizing structure. Phase transition from cubic (I_1) to hexagonal phases (H_1) or hexagonal (H_1) to bicontinuous cubic phases (V_1) can be explained by the above reason.

The phase diagram resembles that of an ordinary water–polyoxyethylene alkyl ether system (ref. Fig. 1) as a function of temperature. In the present phase diagram, however, almost all the phases appear because the increase in thermal motion is eliminated, though one important liquid crystal, reverse hexagonal liquid phase (H_2) is not formed. We recently found the H_2 phase in water–polyoxyethylene oleyl ether system and the melting temperature is low compared with other liquid crystals [8]. Therefore, the stability of H_2 phase is highly dependent on the lipophilic chain length of surfactant. Note that the H_2 phase is formed in polyoxyethylene dodecyl ether system by the addition of oil [16].

Effect of EO-chain on the $L_α$ phase

The interlayer spacing for the $L_α$ phase in water–$C_{12}EO_3$ and water–$C_{12}EO_4$ systems were measured by means of SAXS and the results are shown in Fig. 3.

The effective cross-sectional area for one surfactant molecule, $a_s(= 2v_S/d\phi_S)$ and the half thickness of lipophilic layer, d_L, are also plotted in Fig. 3, where v_S is the volume of one surfactant molecule and ϕ_S is the volume fraction of surfactant in the system. d is the measured interlayer spacing. The cross-sectional area, a_s, of the $C_{12}EO_4$ system is almost constant and is larger than that in the $C_{12}EO_3$ system, because a_s depends on the EO chain

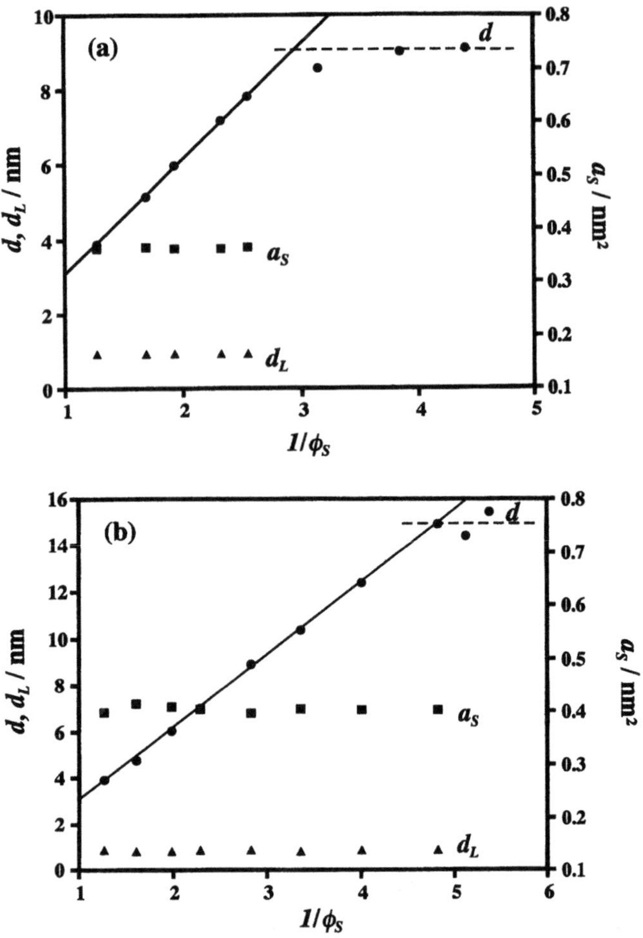

Fig. 3 SAXS data for the L_α phase in water–$C_{12}EO_3$ system (a) and in water–$C_{12}EO_4$ system (b), respectively

length, as described before. In contrast with a_s, d_L in the $C_{12}EO_3$ system is longer than that in the $C_{12}EO_4$ system. Since the hydrophobic chain length, d_L, in both systems is considerably shorter than the dodecyl chain in its extended form (1.67 nm) [11], surfactant molecules in the bilayers are in a disordered state.

The L_α phase in the $C_{12}EO_4$ system can swell with a large amount of water compared with the $C_{12}EO_3$ system as is shown in Fig. 3. It is known that the L_α phase can swell with a much more amount of water in the $C_{12}EO_5$ system than the shorter EO-chain polyoxyethylene-type nonionic surfactant systems, although the measured temperature is different [6]. When a_s is large and the hydrocarbon chain of the surfactant in the bilayer, d_L, is short, the arrangement of surfactant molecules in the bilayer is considered to be in a more disordered state. Hence, the bilayer in the $C_{12}EO_4$ system is more flexible and the solubilization of water in the L_α phase is larger compared with the $C_{12}EO_3$ system.

Conclusion

Phase diagram of polyoxyethylene dodecyl ethers–water system was constructed as a function of the polyoxyethylene (EO) chain length at 25 °C. The influence of HLB (hydrophile–lipophile balance) and the concentration of surfactant on the types of self-organizing structures is clarified. By decreasing HLB of surfactants, the curvature of surfactant layer decreases, and by increasing the concentration of surfactants, steric hindrance is considered to affect the formation of the self-organizing structure. In addition, the effect of the EO-chain length on the structure of L_α phase was also investigated by means of small-angle X-ray scattering. While the EO-chain increases, a_s increases and d_L decreases, i.e. the bilayer of the surfactant system becomes more flexible and the solubilization of water in the L_α phase becomes larger.

References

1. Clunie JS, Goodman JF, Symons PC (1969) J Chem Soc Trans Faraday Soc 65:287
2. Shinoda K (1970) J Colloid Interface Sci 34:278
3. Saito H (1971) Nihon Kagaku Zasshi 92:223
4. Lang JC, Morgan RD (1980) J Chem Phys 73:5849
5. Mitchell DJ, Tiddy GJT, Waring L, Bostock T, McDonald MP (1983) J Chem Soc Faraday Trans 1 79:975
6. Strey R, Schomacker R, Roux D, Nallet F, Olsson U (1990) J Chem Soc Faraday Trans 86:2253
7. Shigeta K, Suzuki M, Kunieda H (1997) Progr Coll Polym Sci, in press
8. Kunieda H, Shigeta K, Ozawa K, Suzuki M (1997) J Phys Chem, in press
9. Fontell K (1974) In: Gray GW, Winsor PA (eds) Liquid Crystals and Plastic Crystals, Chap 4, Vol 2. Wiley, New York
10. Evans DF, Wennerstrom H (1994) The Colloidal Domain, Chaps 11, 6. VCH, New York
11. Tanford C (1972) J Phys Chem 76:3020
12. Kunieda H, Shinoda K (1982) J Dispers Sci Technol 3:233
13. Griffin WC (1954) J Soc Cos Chem 5:1249
14. Jahns E, Finkelmann H (1987) Coll Polym Sci 265:304
15. Kunieda H, Ozawa K, Huang KL, to be published
16. Kunieda H, Rajagopalan V, Kimura E, Solans C (1994) Langmuir 10:2570

Progr Colloid Polym Sci (1998) 110:175–180
© Steinkopff Verlag 1998

F. Mantegazza
V. Degiorgio
M.E. Giardini
A.L. Price
D.C. Steytler
B.H. Robinson

Electric birefringence study of rod-shaped water-in-oil microemulsions

F. Mantegazza*
Istituto di Scienze Farmacologiche
Universitá di Milano
20133 Milano
Italy

Prof. V. Degiorgio* (✉)
Dipartimento di Elettronica
Universitá di Pavia
via Ferrata 1
27100 Pavia
Italy

* INFM
Istituto Nazionale per la
Fisica della Materia
Italy

A.L. Price · D.C. Steytler
B.H. Robinson
School of Chemical Sciences
University of East Anglia
Norwich NR4 7TJ
United Kingdom

Abstract Transient electric-birefringence (TEB) studies have been carried out on w/o microemulsions stabilized by $Ni(AOT)_2$ the nickel salt of di(ethyl) hexylsulphosuccinate at various volume fractions ϕ and water contents $w = [H_2O]/[Ni(AOT)_2]$. The system forms rod-shaped droplets at low w which convert to more spherical aggregates as w is increased. TEB data have been obtained as a function of micro-emulsion volume fraction, water content, and temperature. Relaxation transients of the electric birefringence signal were found to be non-exponential and have been analyzed by assuming that the rods are rigid and polydisperse. The average rod length L_m of the droplets is determined from the initial fast relaxation of the transients. It is found that L_m grows as the square root of ϕ. At fixed volume fraction, the curve L_m vs. w presents a maximum at a value of w around 30.

Key words Electric birefringence – microemulsions – reverse micelles – $Ni(AOT)_2$

Introduction

It has recently been shown that reverse micelles and water-in-oil (w/o) microemulsions may exhibit a wide range of shapes, in addition to a spherical structure. These include discs [1], rods [2–4], and reverse-vesicular assemblies [5]. A technique which is very useful for the study of aniso-tropic aggregates is transient electric birefringence (TEB). Since the amplitude of the TEB signal is very sensitive to the shape anisotropy of the particles, and the temporal behavior of the signal is controlled by rotational diffusion, TEB is a simple and accurate technique to measure the size and polydispersity of anisotropic micellar aggregates [6]. The TEB technique has previously been applied to a num-ber of amphiphile solutions including rod-shaped ionic micelles [7, 8], non-ionic micelles near the cloud point [9], reverse micelles [10], droplet clustering in w/o microemul-sion media [11] and deformable spherical droplets [12].

It is known that the w/o microemulsions stabilized by the surfactant sodium di(ethylhexyl) sulphosuccinate (AOT) form spherical droplets which have been extensive-ly characterized by a wide range of experimental tech-niques [13]. For AOT w/o microemulsions the droplet radius grows in proportion to the water-to-surfactant mo-lar ratio. The system $Ni(AOT)_2$ behaves in a rather differ-ent way. At very low water content, $w = [H_2O]/Ni(AOT)_2]$, small spherical aggregates are present in the oil solution but, as the molar ratio w is increased, rod-shaped microemulsion droplets of short length are first formed. The rod length slightly increases with w up to a maximum at a particular w value after which the length decreases until spherical droplets are again formed close to the solubilization boundary (w_{max}) [3]. In the region of w in which rods are formed the radius increases in proportion to w regardless of rod length.

We have undertaken a series of TEB measurements on w/o Ni(AOT)$_2$ microemulsions, over a range of volume fraction ϕ, water content w, and temperature T. Our primary concern is the determination of the length of the rod-shaped aggregates formed. In this paper we discuss both the steady-state birefringence Δn_{s} and the birefringence decay after the applied electric field is switched off. The rod-lengths obtained from TEB are compared with those derived from small angle neutron scattering (SANS) measurements made on the same system [14].

Experimental

Materials

The Ni(AOT)$_2$ surfactant was prepared from AOT (Sigma) using a liquid–liquid ion exchange process, as reported previously [15]. A saturated solution of nickel nitrate was shaken with AOT in absolute ethanol ($1\,\mathrm{mol\,dm^{-3}}$), and extracted with diethyl ether. Total nickel substitution of sodium was confirmed by UV-VIS spectroscopy. In all the measurements we have used analytical grade cyclohexane (BDH-AnalaR) as solvent (oil) and analytical grade water (BDH-AnalaR). Analysis of the surfactant showed the presence of trace quantities of Ni(NO$_3$)$_2$ which is difficult to remove in the washing stage. This has the effect of reducing the rod length of w/o microemulsions formed. Since the main aim of this work was to test TEB as a technique for determining microemulsion droplet dimensions, care was taken to use the same batch of surfactant for all TEB (and SANS) measurements.

TEB measurements

The TEB experiments were performed by applying a rectangular voltage pulse to the electrodes of a Kerr cell containing the microemulsion and observing the time dependence and the steady-state value of the birefringence signal $\Delta n(t)$ arising from induced anisotropy of the sample. Details of the apparatus employed have been presented previously [16]. The rate of rise and decay of this signal is controlled by the rotational-diffusion time τ_{r} of the droplets. The pulse duration was chosen to be longer than τ_{r}, so that the induced birefringence can reach the stationary value Δn_{s}. Our investigation was limited to the Kerr regime where the stationary induced birefringence Δn_{s} is proportional to the square of the applied electric field E.

The Kerr constant is defined as: $B = \Delta n_{\mathrm{s}}/(\lambda E^2)$, where λ is the wavelength of the laser beam used as birefringence probe.

The results described in this work were obtained by applying pulses with amplitude in the range 10–100 V and duration 10–100 ms to a cell presenting a gap between the electrodes of 1 mm and a geometrical pathlength of 6 cm. It should be noted that, unlike TEB measurements on aqueous solutions, Joule heating is negligible for w/o microemulsions since they present a very low electrical conductivity. The optical source was a He–Ne laser emitting at $\lambda = 0.633\,\mu\mathrm{m}$. The power of the laser beam was kept at a very low level (about 0.2 mW) to prevent heating of the sample due to absorption of laser light. Signal averaging over many pulses was performed by a Data Precision DATA 6100 transient digitiser.

The measurements on the Ni(AOT)$_2$ w/o microemulsions were made at 25 °C over a range of droplet volume fractions, ϕ, from 0.005 to 0.07 and w values ranging from 10 to 40. To estimate the effect of temperature, measurements were also made at 15 °C for a w value of 18.

Theoretical

In principle, it is easily possible to obtain the length L, of monodisperse, rigid rod particles (or aggregates) using TEB in the limit of extreme dilution, where there are no inter-particle interactions. In such a case the normalized time-dependent birefringence decay is an exponential function $\exp(-t/\tau_{\mathrm{r}})$ with a time constant $\tau_{\mathrm{r}} = (6D_{\mathrm{r}})^{-1}$, D_{r} being the rotational diffusion coefficient of the individual rod. D_{r} is related to the rod length L and to the rod radius r, by a well-known expression [6]. It is useful to note that, apart from logarithmic corrections, $D_{\mathrm{r}} \propto L^{-3}$.

Theories of self-aggregation of surfactants [17] predict that rod-shaped aggregates should exhibit a polydispersity in length described by an exponential distribution $P(L)$:

$$P(L) = \frac{1}{L_{\mathrm{m}}} \exp\left(-\frac{L}{L_{\mathrm{m}}}\right) \tag{1}$$

with an average rod length L_{m} proportional to the square root of the volume fraction of rods

$$L_{\mathrm{m}} \propto \phi^{1/2} \exp\left(\frac{W}{2k_{\mathrm{B}}T}\right), \tag{2}$$

where W is the energy required to create a pair of end caps in the middle of a rod of infinite length [18]. As a consequence of polydispersity, the specific Kerr constant of the system can be written as

$$\frac{B}{\phi} = \int_{2r}^{\infty} K(L)P(L)\,\mathrm{d}L, \tag{3}$$

where $K(L)$ is the specific Kerr constant of a system of monodisperse rods having length L. Note that the lower limit of integration is the particle diameter $2r$ and not 0. The normalized relaxation function after the electric field is switched off is given by

$$\frac{\Delta n(t)}{\Delta n_s} = \frac{\int_{2r}^{\infty} K(L) P(L) \exp\left[-\frac{t}{\tau_r(L)}\right] dL}{\int_{2r}^{\infty} K(L) P(L) dL}. \tag{4}$$

We call τ_1 the time constant derived from the initial slope of the logarithm of the relaxation function:

$$\tau_1^{-1} = -\left[\frac{d(\ln\Delta n(t))}{dt}\right]_{t=0} = \frac{\int_{2r}^{\infty} K(L) P(L)\frac{1}{\tau_r(L)} dL}{\int_{2r}^{\infty} K(L)P(L) dL}. \tag{5}$$

The time constant τ_1 can be easily derived from the TEB experiment, and can be related to τ_{rm}, the rotational time of the rod having length L_m, once $K(L)$ is known.

Bellini et al. [7] have examined the effect of polydispersity on the TEB relaxation. By assuming stiff rods with an exponential probability distribution for the rod length as given by Eq. (1), $\Delta n(t)$ is found to behave asymptotically for long times as a stretched exponential:

$$\Delta n(t) \propto \Delta n_0 \exp\left[-\left(\frac{t}{\tau_{SE}}\right)^{\alpha_{SE}}\right], \tag{6}$$

where the stretching exponent takes the value $\alpha_{SE} = \frac{1}{4}$. The time constant τ_{SE} appearing in the stretched exponential is proportional to τ_{rm}. Therefore, in principle, the average length L_m can be derived from either the initial slope or the long-time tail of the relaxation function.

However, the interpretation of the experimental data may require consideration of scission effects and of interparticle interactions. In fact, a self-assembled rod may undergo a reversible breaking anywhere along its length. Cates et al. [18] assume a random scission scheme which assigns a uniform probability for scission at any point along the length of the rod. The effect of scission is that of shortening the rotational relaxation times.

Interactions give rise to hindered rotation: as the concentration of rods is increased above a specific "overlap" concentration (entanglement), $\phi^* \propto 6r^2/L^2$, the excluded volume interaction between rods has the effect of slowing down the rotational Brownian motion so that the effective rotational diffusion coefficient D_r' becomes smaller than D_r. Cates et al. [18] proposed a very interesting extension of the treatment of Ref. [7] by including both entanglement and reversible scission. Scission is characterized by a breaking time τ_b which is taken as inversely proportional to the average rod length. The treatment by Cates et al.

gives explicit solutions only in the limiting cases of very short or very long breaking times relative to the rotational time. Calling τ_{bm} the breaking time of the rod of length L_m, the relaxation function of the dilute polydisperse solution is given by Eq. (6) only in the limit $\tau_{bm} \gg \tau_{rm}$. In the opposite limit, that is, in the situation in which the rods are disrupted during the time scale of the rotational relaxation, an exponential decay is predicted in the dilute regime with τ_{bm} and τ_{rm} contributing to a different extent to the observed relaxation time.

Results and discussion

When the electric field is switched on, the measured birefringence grows monotonically and finally attains the steady-state value Δn_s. After the electric field is switched off, the induced birefringence decays with a fast initial relaxation followed by a slower non-exponential tail. A qualitatively similar behavior was previously observed for rod-shaped ionic micelles in water by Bellini et al. [7] and by Hoffmann et al. [6]. Bellini et al. have focused their attention on the slow tail which was found to be well described by a stretched-exponential function. In our case, we also find that the slower part of the relaxation can be fitted accurately by a stretched-exponential function. In the work by Hoffmann et al. the transients were fitted by the superposition of two exponentials with the fast relaxation ascribed to an initial free rotation of the micelles before collisions and the slow relaxation attributed to entanglement effects. However, both the data of Ref. [7] and the present data show that the non-exponential tail appears also at volume fractions which are certainly below the entanglement volume fraction ϕ^*. This confirms the interpretation of Ref. [7] which attributes the non-exponential behavior to polydispersity, and not to interactions.

We present in the Figs. 1–3 the quantities which can be derived from the measured TEB transients, namely, the specific Kerr constant B/ϕ obtained from the steady-state value Δn_s, the decay time τ_1 of the initial part of the relaxation, and the exponent α derived from the stretched-exponential fit of the slow relaxation.

From Fig. 1 it can be seen that B/ϕ is always negative and takes rather large values. Its absolute value is an increasing function of w and a decreasing function of T. The data at fixed w and T can be approximately described by a power law relation: $-B/\phi \propto \phi^\gamma$ with $\gamma = 1.2$.

We recall that, for a system of non-interacting monodisperse particles having fixed size and shape, the specific Kerr constant represents a single particle property and should be independent of ϕ. In the following section we

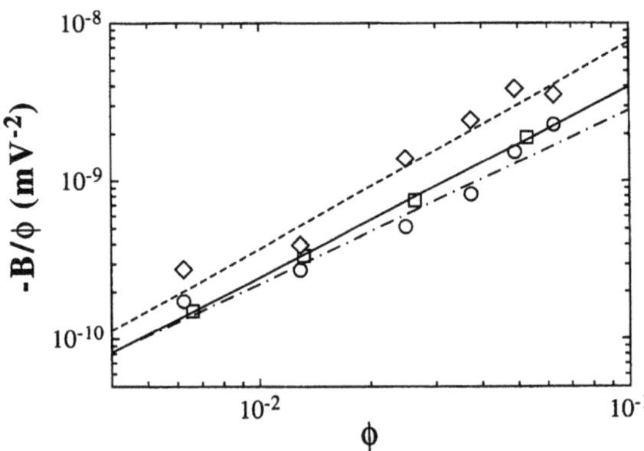

Fig. 1 Log–log plot of the measured specific Kerr constant (B/ϕ) as a function of ϕ. The lines correspond to a fit with a power law $B/\phi \propto \phi^{\gamma}$. The dashed, full, and dashed-dotted line are the results of a least-squares linear fit of the data at $T = 15\,°\text{C}$, $w = 18$ (diamonds), $T = 25\,°\text{C}$, $w = 22$ (squares), $T = 25\,°\text{C}$, $w = 18$ (circles). The best-fit values of γ obtained are 1.3, 1.2 and 1.1, respectively

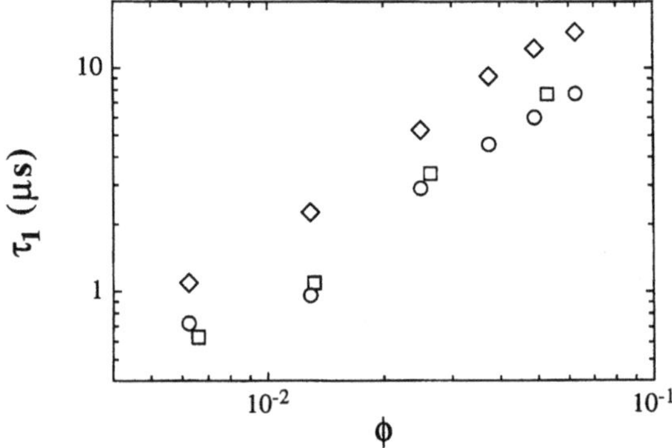

Fig. 2 Log–log plot of the time constant τ_1 measured as a function of ϕ for the data shown in Fig. 1

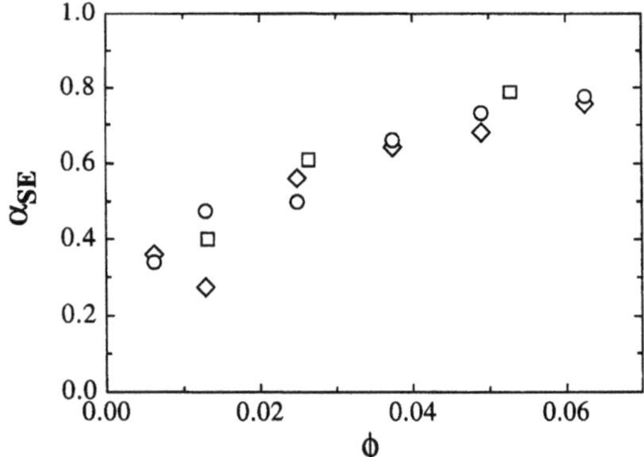

Fig. 3 Stretching exponent α_{SE} measured as a function of the volume fraction ϕ for the data shown in Fig. 1

show that the increase of the quantity $|B|/\phi$ with ϕ can be attributed to the increase in length of the microemulsion droplets and not to a collective effect caused by interparticle interactions (entanglement).

As shown in Fig. 2, τ_1 is a rapidly increasing function of the volume fraction ϕ. Qualitatively, this represents a clear indication of the growth of the average rod length with ϕ. However, in order to derive the average rod length from the relaxation time it is necessary to take into account the size polydispersity and establish whether the scission dynamics and entanglement effects also contribute to the observed TEB relaxation.

Figure 3 shows that the stretching exponent α_{SE} takes a value consistent with the prediction $\alpha_{\text{SE}} = 0.25$ at low ϕ. With increasing ϕ the value of the exponent α_{SE} grows in a systematic fashion from 0.25 to a value around 0.8 indicating an approach to a simple exponential behavior. According to the data presented in Fig. 3, the increase of the stretching exponent α_{SE} seems to be independent of w or T. The presence of the stretched-exponential asymptotic time-decay reflects the size polydispersity of the droplets. In particular, the observation that at low volume fractions the exponent α_{SE} takes the value 0.25 represents a confirmation of the assumption that the probability distribution of rod lengths is an exponential, as discussed in Ref. [7]. Within the formalism of Cates et al., the observed behaviour of α_{SE} vs. ϕ could suggest a crossover from the regime in which the breaking time of the droplets is longer than the rotational time to the fast breaking exponential relaxation. Such a behavior could be plausible on the grounds that L_{m} grows as $\phi^{1/2}$, τ_{bm} scales as the inverse of L_{m} and τ_{rm} (approximately) as L_{m}^{3}. The ratio $\tau_{\text{bm}}/\tau_{\text{rm}}$ is therefore expected to scale as ϕ^{-2} suggesting that it is indeed possible to approach the fast breaking regime on increasing ϕ. However, as discussed below, it is unlikely that this interpretation applies for our case, because the measured length of the rod-like aggregates is rather small in the investigated range of volume fractions.

The shape of the observed relaxation does not contain evidence of entanglement effects: entanglement would give a stretched-exponential behavior with values of α_{SE} decreasing as ϕ increases, whereas we find that α_{SE} grows with ϕ. We have performed an approximate calculation of the entanglement volume fraction for the case $w = 18$, $T = 25\,°\text{C}$, finding $\phi^* = 0.03$. It is known from studies of rotational relaxation of rigid polymer chains [19] that entanglement effects on rotational times become significant only when ϕ/ϕ^* is larger than 10. Since the largest investigated volume fraction is 0.07, it is reasonable to expect that entanglement is not influencing the rotational motion in the case of our experiments.

In order to derive L_{m} from the measured τ_1 we have performed an approximate calculation of the specific Kerr constant $K(L)$. All the details of the calculation will be reported elsewhere [20]. We only mention here that the calculation reproduces correctly the sign and the order of magnitude of the Kerr constant measured in our microemulsions. It should be noted that the magnitude of the Kerr constant is very sensitive to the detailed structure of the droplet: a droplet modelled as an isotropic water particle or as an anisotropic particle of oriented AOT molecules would present values for $K(L)$ which are several orders of magnitude lower than the experimental ones.

In the range of L explored, we find that the dependence of $K(L)$ on L follows approximately a power law with exponent 2.2, that is, $K(L) \propto L^{2.2}$. Therefore, by using Eq. (3), we find that $B/\phi \propto L_m^{2.2}$.

Considering that L_m is proportional to $\phi^{1/2}$, we have a power-law dependence of B/ϕ on ϕ with an exponent $\gamma = 1.1$. This is in good agreement with the value $\gamma = 1.2$ which we derive from the experimental data of Fig. 1. It is very interesting to note that Koper et al. [10] derive, by using a scaling approach, $\gamma = 1.3$, and measure for elongated lecithin inverted micelles values of γ in the range 1.1–1.3.

By using the expressions for $K(L)$ and $\tau_r(L)$, we have calculated numerically the integrals appearing in Eq. (5). From τ_1 and the values of the rod radius r measured by SANS [14] we have derived the average rod length L_m as a function of ϕ. The results are reported in Fig. 4. It is clear that the data follow the law: $L_m \propto \phi^{0.5}$, consistently with the theoretical model and also with the experimentally observed stretched-exponential behavior. Some SANS results are also plotted in Fig. 4: the SANS and TEB lengths are in very good agreement. From the data we can derive the scission energy W which appears in Eq. (2). We find $W = 12\,k_B T$, a value which is smaller than that ($W = 20\,k_B T$) found for sodium dodecyl sulfate micelles.

Fig. 4 Log–log plot of the values of L_m vs. ϕ. L_m is obtained from the experimental values of τ_1 presented in Fig. 2. The lines correspond to a fit with a power law $L_m \propto \phi^\delta$. The best fit values of δ are 0.48, 0.45 and 0.51, respectively. In this plot we show also some data obtained with SANS measurements at different situations: $T = 25\,°C$, $w = 18$ (filled squares), $T = 25\,°C$, $w = 22$ (filled diamonds)

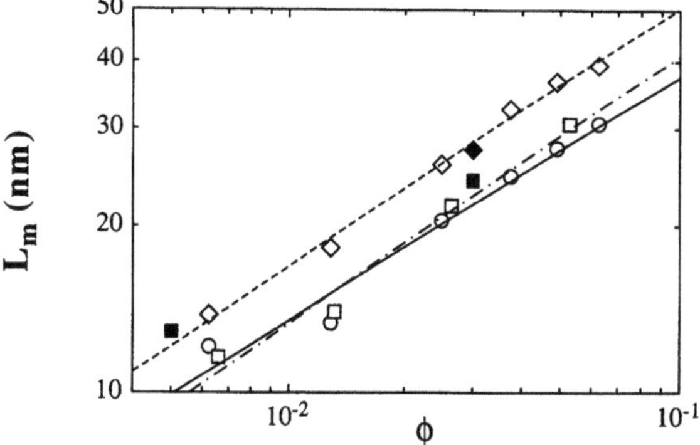

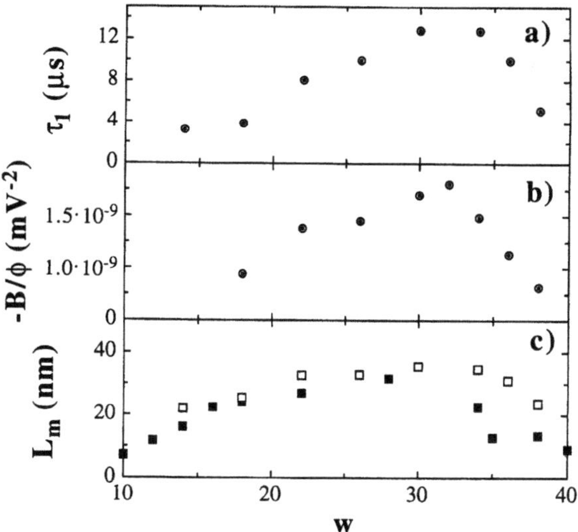

Fig. 5 (a) Time constant τ_1 derived from the initial slope measured as a function of w at $\phi = 0.03$ and $T = 25\,°C$. (b) specific Kerr constant B/ϕ measured as a function of w at $\phi = 0.03$ and $T = 25\,°C$. (c) Values of L_m as a function of w at $\phi = 0.03$ and $T = 25\,°C$, as obtained from TEB data (open squares) and SANS data (filled squares)

The fact that the rod length is in the range of few nanometers casts some doubt on the interpretation of the α_{SE} vs. ϕ behavior discussed at the beginning of this section. If we consider the case of elongated cetyltrimethylammonium bromide (CTAB) micelles in aqueous solutions, where measurements of the breaking time have been performed [21], the rod length at which the breaking time becomes equal to the rotational time is above $0.5\,\mu m$. Although the breaking time of our inverted micelles is not known, it is perhaps possible that the situation may be different form that of CTAB micelles, because of the small value of the scission energy.

We have also investigated the dependence of the average rod length on the water-to-surfactant ratio w. We present in Fig. 5 the behavior of τ_1, $-B/\psi$, and L_m as a function of w for $Ni(AOT)_2$ at $T = 25\,°C$ and $\phi = 0.03$. The three plots are qualitatively very similar. We find that L_m presents a maximum for a value of w between 30 and 35, in agreement with viscosity measurements [14]. In the case of L_m we present in Fig. 5 also the SANS data which appear to be fully consistent with the TEB data.

Conclusions

We have applied TEB for the measurement of the average length of rod-shaped w/o microemulsion droplets of relatively short length. In the volume fraction range $\phi = 0.005$–0.07 the TEB relaxation shows a non-exponential decay which follows asymptotically a stretched-exponential behavior. The value of the stretching exponent α_{SE} at low volume fraction is consistent with the assumption that the length probability distribution is exponential.

A model describing the Kerr response of the microemulsion droplets is developed. By using the model we derive the specific Kerr constant as a function of the volume fraction, finding a good agreement with the experimentally observed behavior and also with literature results referring to a different w/o microemulsion. We also use the model for the interpretation of the dynamic data, deriving, from the intial slope of relaxation, the mean rod length L_m.

It is found that L_m grows approximately as the square root of ϕ. Values for L_m obtained from TEB exhibit essentially the same dependence on ϕ, w, and temperature as those obtained from SANS measurements.

Acknowledgements We would like to thank G. Fragneto, T. Jenta, I. MacDonald, and T. Towey for the help in measurements and sample preparations, and T. Bellini and M.E. Cates for useful discussions. BHR, DCS and ALP acknowledge financial support for travel and consumables from BBSRC to undertake TEB experiments at Pavia (Italy) and EPSRC for SANS measurements at ISIS (UK). A postdoctoral fellowship from BBSRC is gratefully acknowledged by ALP.

References

1. Steytler DC, Sargeant DL, Robinson BH, Eastoe J, Heenan RK (1994) Langmuir 10:2213
2. Schurtenberger P, Scartazzini R, Magid LJ, Leser ME, Luisi PL (1990) J Phys Chem 94:3695
3. Eastoe J, Towey TF, Robinson BH, Williams J, Heenan RK (1993) J Phys Chem 97:1459
4. Steytler DC, Sargeant DL, Welsh GE, Robinson BH (1996) Langmuir 12:5312
5. Nakamura K, Kunieda H, Strey R (1996) Langmuir 12:3045
6. Schorr W, Hoffmann H (1985) In: Degiorgio V, Corti M (eds) Physics of Amphiphiles: Micelles, Vesicles and Microemulsions. North-Holland, Amsterdam, p 160
   Schelly ZA (1997) In: Kumar P, Mittal KL (eds) Microemulsions: Fundamental and Applied Aspects. Marcel Dekker, New York
7. Bellini T, Mantegazza F, Piazza R, Degiorgio V (1989) Europhys Lett 10:499
8. Hoffmann H, Krämer U, Thurn H (1990) J Phys Chem 92:2027; Oda R, Lequeux F, Mendes E (1996) J Phys II (France) 6:1429
9. Degiorgio V, Piazza R (1985) Phys Rev Lett 55:288; Degiorgio V, Piazza R (1987) Progr Colloid Polym Sci 73:76
10. Koper GJM, Cavaco C, Schurtenberger P (1995) In: Brey JJ (ed) 25 Years of Non-Equilibrium Thermodynamics. Springer, Berlin
11. Eicke H-F, Hilfiker R, Thomas H (1985) Chem Phys Lett 120:272; Guering P, Cazabat AM, Paillette M (1986) Europhys Lett 2:953; Tekle E, Schelly ZA (1994) J Phys Chem 98:7657
12. van der Linden E, Geiger S, Bedeaux D (1989) Physica A 156:130
13. Tapas KD (1995) Adv Colloid Interface Sci 59:95
14. Price AL, Robinson BH, Steytler D, MacDonald I, unpublished
15. Partridge JA, Jensen RC (1969) J Inorganic Nucl Chem 31:2587
16. Piazza R, Degiorgio V, Bellini T (1986) Opt Commun 58:400; Piazza R, Degiorgio V, Bellini T (1986) J Opt Soc Am B 3:1642
17. Israelachvili JN (1985) In: Degiorgio V, Corti M (eds) Physics of Amphiphiles: Micelles, Vesicles and Microemulsions. North-Holland, Amsterdam, p 24
18. Cates ME, Marques CM, Bouchaud J-P (1991) J Chem Phys 94:8529
19. Teraoka I, Ookubo N, Hayakawa R (1985) Phys Rev Lett 55:2712
20. Mantegazza F, Degiorgio V, Giardini ME, Price AL, Steytler D, Robinson BH (1998) Langmuir 14:1
21. Candau SJ, Merikhi F, Waton G, Lemaréchal P (1990) J Phys France 51:977; Faetibold E, Waton G (1995) Langmuir 11:1972

Progr Colloid Polym Sci (1998) 110:181–187
© Steinkopff Verlag 1998

G. Palazzo
A. Mallardi

Interaction of photosynthetic reaction centers with hydrophobic quinones in reverse micelles

G. Palazzo (✉)
Dipartimento di Chimica
Università di Bari
via Orabona 4
I-70126 Bari
Italy

A. Mallardi
CNR
Centro Studi Chimico-Fisici Sull'
Interazione Luce-Materia
via Orabona 4
I-70126 Bari
Italy

Abstract An integral membrane protein, the photosynthetic reaction center (RC) from *Rb. sphaeroides*, has been solubilized in *n*-hexane phospholipid reverse micelles. The interaction between this protein and its liposoluble exchangeable cofactor (Ubiquinone-10) has been investigated by means of flash absorption spectroscopy.

The kinetic behavior of the RC has been studied as a function of quinone concentration and temperature, demonstrating that quinone molecules are in fast exchange between the Q_B site of the protein and the bulk organic phase.

In this system a fraction of RC lacking Q_B site functionality is also present and the meaning of this fraction is discussed.

Key words Protein – cofactor interactions – membrane model – charge recombination kinetics *Rhodobacter sphaeroides*

Introduction

The reaction center protein (RC) of purple bacteria provides a useful system for studying the protein–cofactor interactions. When light energy is absorbed by this trans-membrane protein, a series of electron-transfer reactions involving quinones is initiated, resulting in a charge separation across the membrane.

Studies of the electron-transfer reactions have been focused on the RC isolated from *Rb. sphaeroides* R-26 [1]. Electron transfer occurs between several redox sites, which are held by the surrounding protein in a well-defined orientation and separation. The reaction sequence is triggered by the adsorption of a photon by a bacteriochlorophyll dimer (the primary donor, *P*). The first stable charge separation occurs between P and Q_A, the first quinone electron acceptor, which is located in a hydrophobic pocket of the protein. Stabilization of the primary charge separated state $P^+Q_A^-$ is achieved by replacement of the electron on P^+ by a reduced cytochrome c_2 and by

a forward electron transfer. Q_A, which is a one-electron carrier, reduces the second quinone acceptor, Q_B, which, on the contrary, works as a two electron acceptor and binds to a relatively polar protein domain. Q_B is loosely bound to the protein and can exchange with a pool of quinone molecules localized on the inside of the native membrane [1].

Understanding the factors governing the binding of the quinone to the redox catalytic site of the RC is crucial for a complete comprehension of the mechanisms regulating the efficient conversion of the light in photosynthesis and, more in general, the interaction between hydrophobic proteins and their cofactors.

Usually the RC is dispersed in aqueous phase by means of detergent direct micelles. In this system the dynamics of detergent interaction with the solubilized protein can deeply influence both the electron transfer in the acceptor quinone complex and the quinone binding equilibrium constant. In the presence of a topologically disconnected lipophilic domain and of a continuous aqueous bulk, assumptions on the quinone activity coefficients must be made [2], and a distinction between the contribution of solvent water and direct protein–quinone interaction to the binding of these hydrophobic ligands is required [3, 4].

In this work the RC of *Rb. sphaeroides* has been solubilized in *n*-hexane phospholipid reverse micelles (RM), in order to have a system characterized by a continuous lipophilic bulk where the strong interfering contributions of quinone aqueous desolvation effects are absent [3, 4]. Moreover, the presence of a continuous organic phase, in which hydrophobic compounds, like quinones, are soluble allows a detailed study of the interaction of liposoluble molecules with membrane proteins, avoiding possible artefacts present in aqueous solutions (direct micelles of detergent and liposomes).

So, RM offer a useful tool to gain insight about the physico-chemical features of the binding processes of the ubiquinone-10 (UQ_{10}) to the Q_B site of the RC.

Materials and methods

RCs were isolated and purified from *Rhodobacter sphaeroides* R-26 and solubilized in reverse micelles as already described [5]. The RCs were first incorporated into liposomes and the reverse micelle solution was formed by extracting into *n*-hexane the phospholipid–protein complex. In order to preserve the RC photoactivity it is necessary to use proteoliposomes made of a mixture of three phospholipids: phosphatidylcholine, phosphatidylethanolamine and phosphatidylserine in the ratio, respectively, of 1 : 1 : 2. Usually, the RC-containing reverse micelles obtained from different preparations were pooled in order to minimize heterogeneity due to the extraction procedure. Samples of 1 ml each were taken from the pool and used for the kinetic measurements at a given temperature varying the quinone concentration.

Flash-induced redox changes of the primary donor of the RC solubilized in reverse micelles were monitored at 600 nm with a single-beam spectrophotometer of local design [5].

The sequential analysis into multiple exponential decays and the simultaneous global analysis described in the text were performed by using the "STEFIT" program (STELAR s.n.c.).

Results and discussion

The reaction center of *Rb. sphaeroides* has been solubilized in RM, where any exogenous electron donor to P^+ is absent and only a one-electron activity is possible.

In this system the RC photochemistry has been investigated by means of flash spectroscopy. In single turn-over flash experiments, the light-induced electron transfer from P to the acceptor quinone complex (Q_A and Q_B) is obtained, followed by the dark reduction of P^+, which reflects the charge recombination from reduced acceptor quinones [6]. The primary quinone, Q_A, is tightly bound, but the secondary one, Q_B, is readily removed. When Q_B is absent, or the electron transfer from Q_A^- to Q_B is blocked, the light-induced charge separation is limited to the $P^+Q_A^-$ state and the following charge recombination occurs with a lifetime of about 100 ms. When Q_B is present, electron transfer from Q_A^- to Q_B is rapid (compared to the recombination of $P^+Q_A^-$) and, as the charge recombination from Q_B^- occurs by repopulation of Q_A^- [6], the decay of P^+ is slowed down. In general, the binding site of Q_B is not saturated, the residence time for the quinone is not more than a millisecond and the binding equilibrium is established from the immediate quinone pool [7, 8]. So, the kinetics of P^+ reduction can be influenced by the quinone availability [6].

On this basis, the dependence of the kinetics of charge recombination on the quinone concentration ([Q]) has been investigated at different temperatures, in a range between 37 °C and 6 °C. For each temperature the absorption changes at 600 nm (reflecting changes of P^+) followed as a function of the [Q].

The deconvolution procedure, performed over the entire set of experimental traces obtained, shows that under the condition tested, the P^+ decay is always accounted for by two exponential phases, characterized by a different behavior with respect to quinone concentration and temperature. Figure 1 summarizes the results of the deconvolution procedure. In Fig. 1A the lifetime dependence of the slow component (τ_S) on the [Q] is reported. Over the whole temperature range investigated, τ_S increases at increasing quinone concentration. Moreover, as already described [9], at fixed [Q] a decrease in temperature determines an increase in τ_S.

The lifetime of the fast component (τ_F) is, on the contrary, essentially unaffected from both the quinone concentration and temperature, as evidenced in Fig. 1B. The τ_F values vary randomly along the isothermal experiment and do not reveal any trend when both the parameters are changed. The relative amplitude of the fast phase (P_F) obtained by deconvolution procedure and not exceeding 30% of the total remains constant within the experimental error, for each binding isotherm, and seems to be correlated to the extraction procedure (samples from the same pool of *n*-hexane extracts show about the same P_F, data not shown).

As is discussed in detail in literature, the observable kinetics of the $P^+Q_B^-$ recombination depend, in principle, on the time scale of both the pre-flash and the post-flash binding of the quinone to the Q_B binding site in accord-

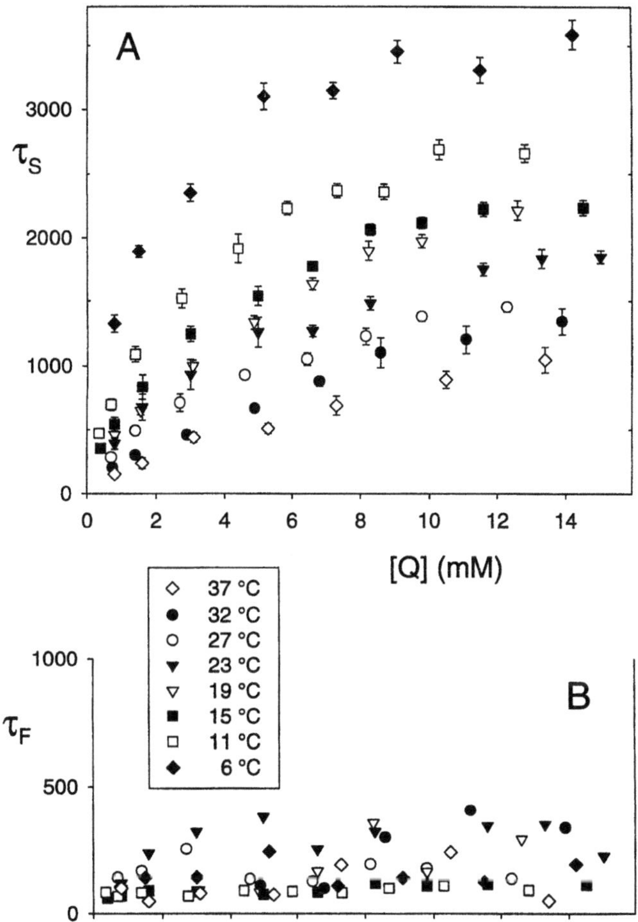

Fig. 1 [Q] dependence of P^+ relaxation in reverse micelles at different temperatures: results coming from a two exponential deconvolution of each decay. Lifetimes of the slow (A) and fast (B) components

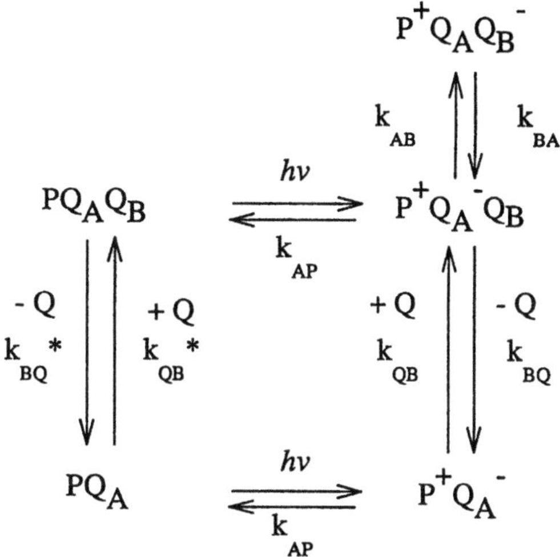

Fig. 2 Kinetic relationships among the photoinduced charge separation, relaxation and the quinone exchange at Q_B site of the RC. The rate constants of the electron transfer between Q_A and Q_B and the quinone binding and release at the Q_B site are reported accordingly to Ref. [6]. From these scheme, it is possible to define the equilibrium constant for one-electron transfer from Q_A to Q_B ($L_{AB} = k_{AB}/k_{BA}$) and the equilibrium binding constants in the "light" ($K_{bind} = k_{QB}/k_{BQ}$) and in the "dark" ($K_{bind}^* = k_{QB}^*/k_{BQ}^*$) state

ance with the scheme of Fig. 2. Under conditions of high quinone concentration, when a pseudo-first-order approximation holds for the binding at Q_B site, the kinetics of P^+ decay, following a flash, can be derived analytically from the scheme of Fig. 2 (for comprehensive review see Ref. [6]).

If the Q_B binding equilibrium is slow compared to the charge recombination rate from $P^+Q_A^-$, the pre-flash (dark) distribution of quinones among the RCs will give rise to the biphasic kinetics of P^+ dark recovery

$$P^+(t) = F e^{-k_{AP}t} + (1 - F)e^{-\lambda t} , \qquad (1)$$

where k_{AP} is the rate constant of $P^+Q_A^-$ recombination ($k_{AP} \approx 10\,\mathrm{s}^{-1}$ in *Rb. sphaeroides* RCs [1]), F is the fraction of RCs without Q_B at the time of the flash ($F = 1/(1 + k_{QB}^*[Q]/k_{BQ}^*)$) and λ is the rate constant of the recombination from Q_B^-, reflecting the electron transfer between Q_A^- and Q_B ($\lambda = k_{AP}/(1 + L_{AB})$; where $L_{AB} = k_{AB}/k_{BA}$ is the equilibrium constant for the electron trans-

fer from Q_A^- to Q_B). The slow exchange behavior is usually observed in RCs reconstituted with long-tail quinones in aqueous detergent systems [2, 6]. This behavior is, however, incompatible with the results obtained by us in *n*-hexane. In fact Eq. (1) predicts biphasic kinetics with rate constants independent of the quinone concentration and with the fraction of the fast phase which decreases with increasing [Q], in contrast with our data, where τ_S increases strongly with the amount of quinone (Fig. 1A), whereas P_F is essentially unchanged.

When the rates of exchange of quinones are of the same order of magnitude of k_{AP}, in order to describe the P^+ dark relaxation, the general solution of scheme of Fig. 2 has to be considered; this yields [6]:

$$P^+(t) = F e^{-\lambda_1(t)} + S e^{-\lambda_2(t)} . \qquad (2)$$

Here F and S are the amplitude of the fast and slow components, respectively, which depend on k_{AP}, k_{BQ}, k_{QB} and [Q]; the values of λ_1 and λ_2 are defined by

$$\lambda_1 + \lambda_2 = \frac{k_{AP}}{1 + L_{AB}} + \frac{k_{BQ}}{1 + L_{AB}} + k_{QB}[Q] + k_{AP} ,$$

$$\lambda_1\lambda_2 = k_{AP}\frac{(k_{AP} + k_{BQ})}{1 + L_{AB}} + \frac{k_{AP}}{1 + L_{AB}}k_{QB}[Q] . \qquad (3)$$

Thus, under these conditions both the sums and the products of the rate constants coming from a two-exponential fit of the experimental P^+ decay should increase with the quinone concentration. In the case of RC in n-hexane exactly the opposite happens, τ_F being almost constant and τ_S increasing when [Q] is increased. The fast and slow phases of our deconvolution cannot, therefore, be identified with phases 1 and 2 of Eq. (2).

Finally, with fast quinone exchange, one can assume quasi-equilibrium between states $P^+Q_A^-$ and $P^+(Q_AQ_B)^-$ during dark relaxation. The kinetics of P^+ recovery can be described by a single exponential decay with an effective rate constant k_p determined by

$$k_p = \frac{k_{AP}}{1 + L_{AB}^{app}}, \qquad (4)$$

where L_{AB}^{app} is the apparent equilibrium constant for the electron transfer between Q_A and Q_B and depends on the quinone concentration and on the equilibrium constant for the binding of quinones (K_{bind}) according to

$$L_{AB}^{app} = \frac{L_{AB} K_{bind} [Q]}{1 + K_{bind} [Q]} . \qquad (5)$$

$K_{bind} = K_{QB}/K_{BQ}$ (see Fig. 2), defines the binding to $P^+Q_A^-$ state. L_{AB}^{app} is lower than the intrinsic equilibrium constant L_{AB} of the electron transfer between Q_A and Q_B, but approaches it at high quinone concentrations. As a consequence, the observed lifetime ($1/k_p$) rises with quinone concentration up to a plateau. This description is quite promising, in order to rationalize the results obtained in reverse micelles, since the τ_s of Fig. 1 is qualitatively in agreement with the behavior predicted by Eqs. (4) and (5).

Otherwise, the fast exchange case predicts a single exponential decay for the reduction of P^+ for the RCs. If, however, in a fraction of RCs the Q_B site functionality has been lost, the charge-separation should be limited to the $P^+Q_A^-$ state. Consequently, the overall experimental P^+ relaxation will consist of the contribution of RCs lacking quinone bound to the Q_B site (with a rate constant k_{AP}) and of a contribution due to the fully competent RCs (with a rate constant k_p which is quinone dependent according to Eqs. (4) and (5)). Furthermore, in this hypothesis, the ratio of the two phases should be independent of both quinone concentration and temperature, as long as these parameters do not affect the Q_B site functionality.

Our data show that the τ_F and P_F values obtained from a two-exponential deconvolution are not constant, although randomly distributed. Since, in our deconvolution, the ratio between the two rates for a two-exponential fit is small (between 2 and 20), it is difficult to obtain safe values for the two rates and for their respective contribution [10]. Nevertheless, the trend of τ_s vs [Q], similar for all the isotherms, the independence of τ_F of the quinone concentration and the fact that experiments performed on samples coming from the same n-hexane extraction reveal

variation on P_F less than 10%, induce one to attribute the fast component to a $P^+Q_A^-$ recombination taking place in a minor fraction of RCs in which the Q_B site functionality has been lost during the extraction in organic solvent.

In the experiments performed at low temperature (where the ratio τ_S/τ_F is relatively high) the fluctuations in the values of both P_F and τ_F are very limited. This evidence further supports our hypothesis and suggests that the variation in P_F and τ_F observed at high temperatures are essentially due to the intrinsic difficulty in resolving two exponential phases with close lifetimes. In order to test this hypothesis we have performed a global analysis [10] of each subset of data taken at a given temperature (in the same sample), fitting the P^+ decay to

$$P^+(t) = A_0 \cdot (P_F e^{-t/\tau_F} + (1 - P_F) e^{-t/\tau_S}) , \qquad (6)$$

where the same parameters P_F (the fraction of fast phase) and τ_F (the lifetime of the fast phase) have been assumed for all the traces at a given temperature, while τ_S, the life-time of the slow phase, is a function of [Q]. The parameter A_0 is the extent of the bleaching at $t = 0$ and acts as a free normalization constant in each signal.

Figure 3 shows the results obtained from this global analysis (for simplicity only the fits relative to two temperatures are reported). It is clear that the agreement between experimental traces and fits is still good under the restrictive hypothesis of Eq. (6).

In Fig. 4A the values of τ_s obtained by global analysis are reported as a function of temperature and [Q]. For all the temperatures the value of the lifetime of the slow component was found to increase at increasing [Q] up to a plateau.

In a previous work, the dependence of the lifetimes of the slow phase on the quinone concentration has been related, by us, to the values of the electron transfer (L_{AB}) and binding (K_{bind}) constants at each temperature, allowing the determination of the enthalpy and entropy changes related to the processes of electron transfer and quinone binding [5]. In the present work the meaning of the fast phase of the P^+ decay is focused and discussed.

The τ_F values obtained by global analysis and shown in Fig. 4B are temperature independent, as expected for the $P^+Q_A^-$ recombination. In order to test the validity of our deconvolution procedure, we have recorded the kinetics of P^+ decay in the presence of $1.5\,mM$ terbutryne (TBT, an inhibitor of the $Q_A^-Q_B \rightarrow Q_AQ_B^-$ electron transfer) at different temperatures and at $[Q] = 13\,mM$. Under these conditions the photoinduced electron transfer, and the following charge recombination, is limited to the $P^+Q_A^-$ state. In Fig. 4B are reported the lifetime values of the recombination from Q_A^-, obtained by fitting the experimental traces to one exponential plus a constant term (the value of the constant is around 2% of the maximum

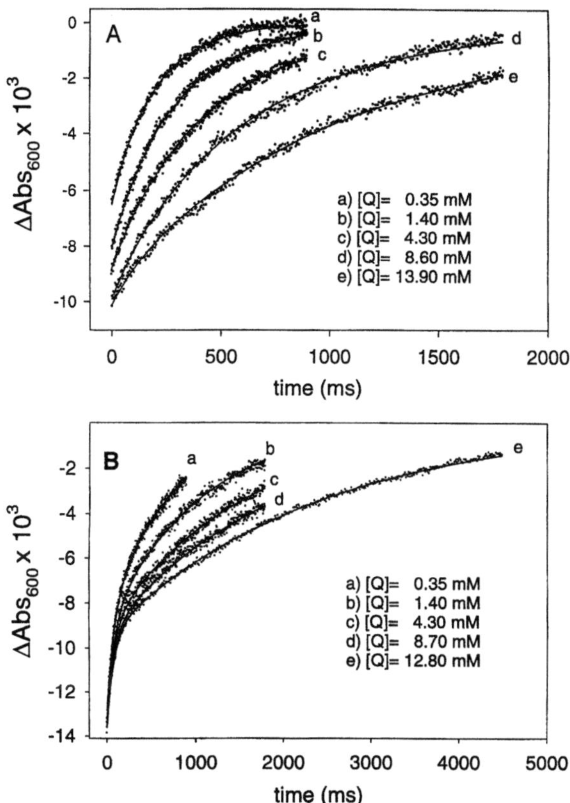

Fig. 3 Comparison between the experimental charge recombination kinetics and the global analysis (see text). The points represent the experimental values of absorbance change measured at different times after a light flash and for different ubiquinone-10 concentrations. The lines represent the best global fit according to Eq. (6). (A) $T = 32\,°C$, $[RC] = 1.7\,\mu M$, the continuous lines correspond to $P_F = 0.10$ and $\tau_F = 145 \pm 52\,ms$. (B) $T = 11\,°C$, $[RC] = 1.7\,\mu M$, the lines correspond to $P_F = 0.31$ and $\tau_F = 80 \pm 11\,ms$

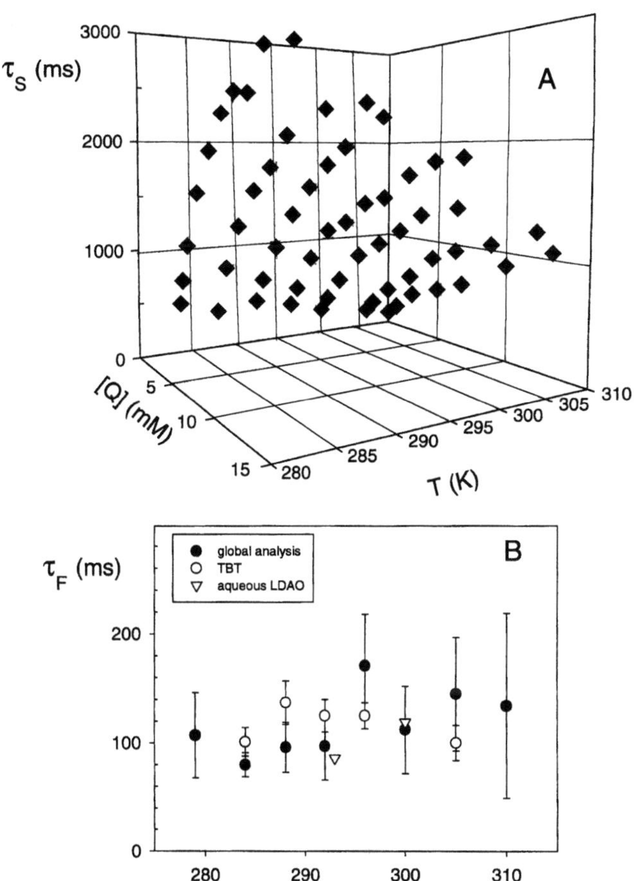

Fig. 4 (A) temperature and quinone concentration dependence of the lifetimes of charge recombination from $P^+Q_B^-$ (τ_S) obtained by global analysis. (B) temperature dependence of the lifetimes of charge recombination from $P^+Q_A^-$ (τ_F) measured in different systems

bleaching). It is evident that the values of τ_F obtained from the global analysis are in agreement with the lifetimes of the $P^+Q_A^-$ recombination obtained by RM in the presence of TBT. Moreover, all these values are close to those found in the same range of temperature in LDAO aqueous systems [11], which are also reported in Fig. 4B.

The values of the fraction of fast phase (P_F) obtained from global analysis of samples are shown in Fig. 5. In the above-reported analysis, samples from two different pools of *n*-hexane extract have been used and measurements performed over a four days period. The variability in the value of P_F appears to be mainly related to the aging of the samples. In fact, within one day from sample preparation, the contribution of the fast phase is about 10% of the total. This value increases up to 30% in samples 4 days old.

As a whole these results confirm that in reverse micellar solutions quinone molecules are in fast exchange between the Q_B sites of RCs and the organic phase, with the

exception of a fraction of RCs lacking the Q_B site functionality.

Recently, in a similar system (lecithin–serine organogels) a decrease of the fast phase amplitude with the increase of the molar ratio water/phospholipid (W_0) has been described, while, the same parameter is unaffected by any variation in the overall water amount, the W_0 remaining constant [12]. The process is fully reversible, i.e. by removing or adding water to the solution it is possible to modulate the fraction of RC undergoing charge recombination from Q_A^-. This suggests that the presence of RC lacking the Q_B site functionality is not an artefact due to the extraction procedure but is related to the intimate properties of the host system.

It is interesting to note that, in the case of organogels, at maximum W_0 value, before the phase separation, the fraction of Q_B sites not accessible to the quinone binding is about 10%, the same value was observed in reverse

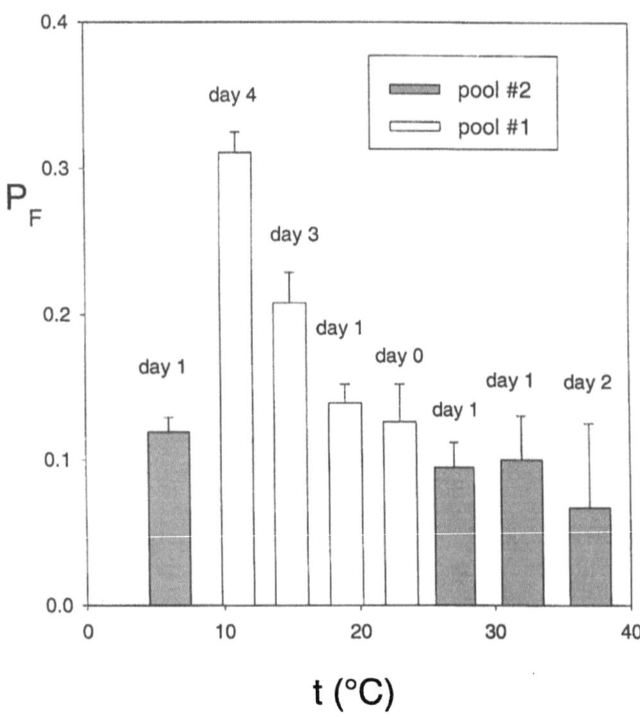

Fig. 5 Influence of temperature, aging and heterogeneity of the preparations on the relative amount of fast phase (P_F). The experiments were performed on two different pools of hexane extracts (see materials and methods). For each temperature the age of the sample is indicated (time between the extraction and the measure). P_F was determined by global fitting of the charge recombination kinetics recorded at a given temperature and at several (at least 7) [Q] according to Eq. (6)

agreement with the X-ray structure [13] of the RC from *Rhodobacter sphaeroides* which reveals several water molecules buried in the core of the protein, some of them being well positioned to play a role in the binding process of the secondary quinone molecule, Q_B. The presence of highly organized micellar water in equilibrium with the water molecules inside the reaction center can induce a partial dehydration of the protein with a subsequent modification of the quinone–protein interaction. In line with this interpretation it should be mentioned that in aqueous systems both detergent depletion [14] and an increase of the osmotic pressure [15] are able to decrease the binding affinity for ubiquinone. In the present study no temperature dependence of P_F has been found in experiments performed at $6\,°C \leq T \leq 37\,°C$, while aging of samples at $4\,°C$ for periods of times longer compared to the time course of an isothermal titration (4–6 h) strongly increases the fraction of the fast phase. Such a feature is not easily understood, but can be, tentatively, related to the degradation of phosphatidylserine which is essential for the integrity of the protein in reverse micelles and is known to be fastly subject to oxidation.

micelles. Both the amount of water solubilized during the extraction in hexane of proteoliposomes and the phase separation boundary in organogels reflect the maximum water chemical potential compatible with a three-components isotropic single phase. In both cases we found the same fraction of RCs lacking the Q_B site functionality. In addition, the degree of organization of the water molecules in RM is, usually, believed to be a function of W_0 (and not of the overall water amount). As a whole, these informations indicate that the activity of the micellar water may play a role in the RC binding affinity for the ubiquinone molecules and/or affect the protein ability to solvate the negatively charged semiquinone. This conclusion is in

Conclusions

The *n*-hexane phospholipid reverse micelles have been used as host system for studying the kinetic behavior of an integral membrane protein. In this system, where the strong hydrophobic effects are absent and where the influence of the micellar dynamics is negligible, it has been possible to perform an accurate analysis of the interaction between the reaction center and its liposoluble native cofactor, ubiquinone-10. The information obtained should reflect the way of the binding and the Q-exchange acting in native membranes. So, reverse micelles have opened the way for the study of integral proteins in a membrane mimetic environment.

Acknowledgements The authors are indebted to Professor Giovanni Venturoli for illuminating discussion. The flash spectroscopy experiments reported in this paper were performed at "Laboratorio di Biochimica e Biofisica – Università di Bologna". This work was supported by Consorzio Interuniversitario per lo Sviluppo dei Sistemi a Grande Interfase (CSGI – Firenze).

References

1. Feher G, Allen JP, Okamura MY, Rees DC (1989) Nature 33:111–116
2. McComb JC, Stein RR, Wraight CA (1990) Biochim Biophys Acta 1015:156–171
3. Warncke K, Dutton PL (1993) Proc Natl Acad Sci, USA 90:2920–2924
4. Warncke K, Gunner MR, Braun BS, Gu L, Yu CA, Bruce M, Dutton PL (1994) Biochem 33:7830–7841
5. Mallardi A, Palazzo G, Venturoli G (1997) J Phys Chem 101:7850–7857
6. Shinkarev VP, Wraight CA (1993) In: Deisenhofer J, Norris JR (eds) The Photosynthetic Reaction Center. Academic Press, New York, Vol I, pp 193–255

7. Wraight CA (1982) In: Trumpower BL (ed) Function of Quinones in Energy Conserving Systems. Academic Press, New York, pp 181–197
8. Crofts AR, Wraight CA (1983) Biochim Biophys Acta 726:149–185
9. Mallardi A, Angelico R, Della Monica M, Giustini M, Palazzo G, Venturoli G (1995) In: Mathis P (ed) Photosynthesis: From Light to Biosphere. Kluwer Academic Publishers, Dordrecht, Vol I, pp 843–846
10. Beechem JM (1992) Methods Enzymol 210:37–54
11. Ortega JM, Mathis P, Williams JC, Allen JP (1996) Biochem 35:3354–3361
12. Palazzo G, Giustini M, Mallardi A, Colafemmina G, Della Monica M, Ceglie A (1996) Progr Colloid Polymer Sci 102:19–25
13. Ermler U, Fritzsch G, Buchanan W, Michel H (1994) Structure 2:925–936
14. Hemelriijk PW, Gast P, van Gorkom HJ, Hoff AJ In: Mathis P (ed) Photosynthesis: From Light to Biosphere. Kluwer Academic Publishers, Dordrecht, Vol I, pp 643–646
15. Larson GW, Wraight CA (1995) Photosynth Res Suppl 1:65

Progr Colloid Polym Sci (1998) 110:188–192
© Steinkopff Verlag 1998

M.L. Curri
G. Palazzo
G. Colafemmina
M. Della Monica
A. Ceglie

Synthesis of cadmium sulfide nanoparticle in four-components microemulsions: effect of the water and alcohol content

M.L. Curri
CS-CFILM
CNR
via Orabona 4
I-70120 Bari
Italy

G. Palazzo (✉) · G. Colafemmina
M. Della Monica
Dipartimento di Chimica
Università di Bari
via Orabona 4
I-70126 Bari
Italy

A. Ceglie
DISTAAM
Molise University
via Tiberio 21/A
I-86100 Campobasso
Italy

Abstract Microdispersed and colloidal semiconductors are increasingly utilized in photochemical and catalytic reactions. Band gap excitation produces electrons in the conduction band and holes in the valence band of semiconductor particles. The photophysical and photochemical properties are drastically dependent on the size of the aggregate which varies with the preparation mode. In the water-in-oil CTAB/pentanol/hexane/water microemulsion the droplet radius has been found to be dependent from the alcohol content. The synthesis of CdS in this microemulsion system can be achieved simply by mixing two microemulsions containing Cd^{2+} and S^{2-}, respectively. We found a size quantization effect of CdS nanoparticles synthesized at various values of [water]/[CTAB] and [pentanol]/[CTAB]. The reverse micelle dimensions were determined by means of pulsed gradient spin-echo NMR measurements for all the cases examined (Cd^{2+}, S^{2-}, and CdS containing microemulsion).

Key words Q-particles – semiconductors – reverse micelles – self-diffusion – CTAB

Introduction

When the characteristic dimension of semiconductor nanocrystals is comparable or smaller than their bulk exciton diameter due to the quantum confinements of electrons, they exhibit size-dependent optoelectronic properties [1, 2], among which the shift in the absorption spectra towards shorter wavelength with the decreasing particle size is the most well-known example. The use of semiconductor nanoparticles has been proposed in photocatalytic applications [3] and as components in optoelectronic devices [1]. In all the cases nanoparticles of controlled dimensions and low degree of polydispersity are needed. Synthetic routes that exploit polymers, glasses, zeolites, reverse micelles, capping molecules, vesicles or coordinating solvents control particle growth were extensively used in recent years. Several studies existing in the literature show that surfactants either in water-in-oil (w/o) [4] either in oil-in-water [5] microemulsions have been used to control particle size. These synthesis methods, however, are only qualitatively understood and rely on trial-and-error to optimizing reaction conditions. In particular, in w/o microemulsions, the role played by different key factors, is not clear, namely: the water droplets size, the degree of occupancy of reverse micelles, the local concentration of the reagent ions (degree of hydration). This is because, in ternary w/o microemulsions (the most well-known example is the system AOT/water/hydrocarbon), these parameters are strongly correlated and it is not possible to make, for example, solutions in which only the droplets dimension are changed, being the water pool concentration, the degree of occupancy and the overall concentration unchanged. As a further complication, usually, stabilizing agents should be added to the system, and more in general one should be aware of the possible perturbation of the reference system, due to the presence of reactants, stabilizers and products. Among the numerous w/o microemulsions known, the system CTAB/n-pentanol/n-hexane/water offers numerous advantages: It has been demonstrated [6] that at certain compositions (CTAB concentration around 0.1 M, mole ratio pentanol/CTAB in the range 8–20, and water content up to 80 moles of water per mole of CTAB) it consists of almost spherical reverse micelles. Furthermore, the effective polar-head area has been proved [7] as being a, growing, function of the alcohol amount. It is thus possible, in this microemulsion system, to modulate the droplets dimension by changing the water and/or the alcohol content. For these reasons we have performed the synthesis of cadmium sulfide (CdS) in the CTAB/n-pentanol/n-hexane/water system in order to obtain insight on the key factors regulating the size of nanoparticles in microemulsions. Obviously, to switch to a quaternary system (to say nothing of the amount of the cadmium and sulfide ions) increases the variables and requires a larger number of data, and the present report, inevitably, should be considered a preliminary one.

Materials and methods

Cetyltrimethylammonium bromide (CTAB) from Fluka, was purified as previously described [6]. n-hexane (C_6) (spectroscopic grade), and n-pentanol (C_5OH), both from Fluka, have been used without any further purification. Cadmium nitrate and sodium sulfide were from Aldrich. Double-distilled water in an all-quartz apparatus ($\chi < 1.30 \times 10^{-4}\,S \cdot m^{-1}$) was used.

Three parameters are needed in order to define the composition of a four-component system in a single-phase region. We have chosen as parameters the two mole ratios water/CTAB (W_0), and n-pentanol/CTAB (P_0) and the molar concentration of CTAB ([CTAB]) at $W_0 = 5$. Systems at greater W_0 were prepared by water (salt solution) dilution of a stock solution at a given P_0 and $W_0 = 5$. Microemulsions at [CTAB] = 0.1 M and $W_0 = 5$ were prepared by weighing in a volumetric flask the appropriate amounts of surfactant, alcohol, water, and oil. The water (salt solution) dilutions were performed by means of Hamilton microsyringes in a thermostatted room ($25 \pm 3\,°C$). Usually, colloidal CdS particles were prepared by mixing rapidly equal volumes of two micellar solutions with the same W_0 value, one obtained by dilution with a sodium sulfide aqueous solution ([Na_2S] = 2×10^{-2} M) of a stock solution of reverse micelles at $W_0 = 5$, the other prepared in the same way using a cadmium nitrate solution ([$Cd(NO_3)_2$] = 2×10^{-2} M). The final (overall) concentration of both the ions was around 10^{-4} M. In few

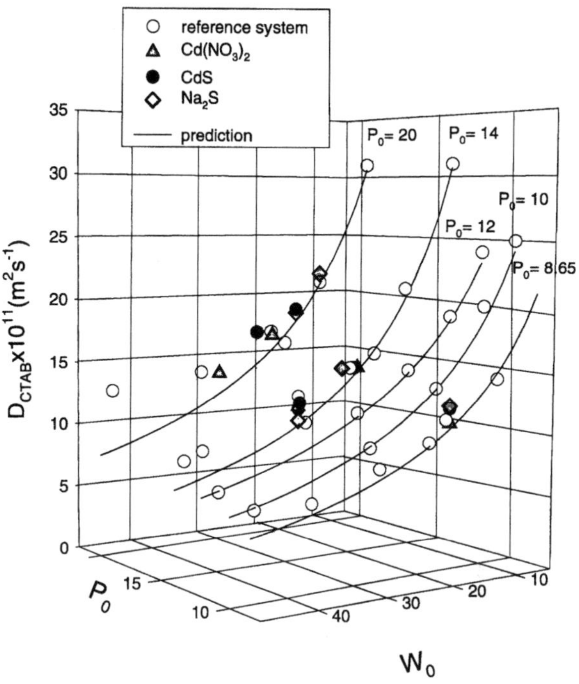

Fig. 1 CTAB self-diffusion coefficients in different microemulsions as a function of the alcohol and water content. The samples were prepared as water or solution dilution of stock solution at [CTAB] = 0.1 M, $W_0 = 5$ and different P_0 values. White circles = CTAB/$C_5OH/C_6/H_2O$; grey triangles = CTAB/C_5OH/C_6/ (cadmium nitrate solution); grey diamonds = CTAB/C_5OH/C_6/(sodium sulfide solution); black circles = CTAB/C_5OH/C_6/(CdS + sodium nitrate solution); solid lines represent the diffusion values calculated using the surfactant geometrical parameters of ref. [7]

experiments reverse micellar solutions of salts at different W_0s were mixed in order to modulate the ratio [Cd^{2+}]/[S^{2-}], the P_0 being unchanged.

Self-diffusion coefficients measurements have been carried out by the Fourier transform NMR Pulsed Field Gradient Spin Echo (PFGSE-NMR) method [8] using a BS-587A NMR (TESLA) spectrometer operating at 80 MHz for the proton, equipped with a pulsed field gradient unit (Autodif 504, STELAR S.n.c.), see ref. [6] for the details. The accuracy of the CTAB and water self-diffusion coefficients was always within 5%. The exponential NMR Echo decay and the conductivity curves were fitted according to three different algorithms (Simplex, Powell and Quadric) by using the "STEFIT" (STELAR S.n.c.) software. In any case the differences among the results coming from different algorithms were within the calculated error.

UV-visible absorption spectra were recordered using a UVIKON 942 KONTRON UV-visible spectrophotometer. The correlation between the position of the absorbance maximum and semiconductor particles radius was made using the data of ref. [9].

Results and discussion

As already demonstrated the L_2 region of the CTAB/ $C_5OH/C_6/H_2O$ microemulsion is constituted of water droplets stabilized by a surfactant/cosurfactant interfacial film [6, 7]. The surfactant, CTAB, is soluble only to a small amount in C_5OH/C_6; a higher solubility is achieved only when aggregates with a minimum amount of water $(W_0) \approx 5$ are formed. The comparison between dynamic light scattering and PGSE-NMR data indicate that CTAB molecules are confined within the aggregates (presumably into the interfacial region) and that they can thus be used as a good probe for the diffusion of the whole aggregate [6]. Otherwise, the cosurfactant molecules are partitioned between the dispersed and the continuous phase [7]. A similar situation occurs for water, although to a less extent [7]. The introduction of sodium sulfide solution $(2 \times 10^{-2} M)$ reduces the maximum water uptake at high P_0 values. Consequently, the maximum W_0 allowed is shifted from about 80 to ≈ 40 and ≈ 30 for $P_0 = 14$ and 20, respectively, presumably due to the strong interactions between the positively charged CTAB polar-heads and the divalent anions. On the contrary, there are no difficulties in preparing transparent microemulsions with cadmium ions. In Fig. 1 are reported the CTAB self-diffusion coeffi-cients (D_{CTAB}) measured at different values of W_0 and P_0 in the reference microemulsion system (made with pure water). In the same figure are reported the D_{CTAB} values measured in microemulsions prepared with $2 \times 10^{-2} M$ salt ($Cd(NO_3)_2$ or Na_2S) solution together with those relative to micellar solutions containing colloidal CdS. Although the presence of salts tends to reduce the water spin-spin relaxation time, it is evident from Fig. 1 that cadmium and sulfide ions as well as CdS nanoparticles do not influence the surfactant diffusion. The same behaviour is observed for water and alcohol diffusion (data not shown), indicating that there are not remarkable structural variations in the salt based microemulsions compared to the reference system. It is, thus, correct to use, also in presence of ions or semiconductor nanocrystals the in-formation previously found on the $CTAB/C_5OH/C_6/H_2O$ system [6, 7]. Briefly, the alcohol partition between dispersed and continuous phase reduce the influence of the C_5OH amount on the viscosity of the continuous medium. On the other hand, the alcohol concentration has a dra-matic effect on the effective polar-head area (S_{CTAB}) of the CTAB molecules. S_{CTAB} was found increasing (almost lin-early) with the P_0 for $W_0 \leq 30$. The D_{CTAB} evaluated by means of the Stokes–Einstein relation (taking into account the obstruction factor for hard-spheres and the thickness

Fig. 2 Absorption spectra of microemulsions at $W_0 = 15$, containing colloidal CdS at different times from synthesis ([CdS] $= 10^{-4}$ M). - - - $P_0 = 8.65$; $\cdots\cdots$ $P_0 = 14$; — $P_0 = 20$

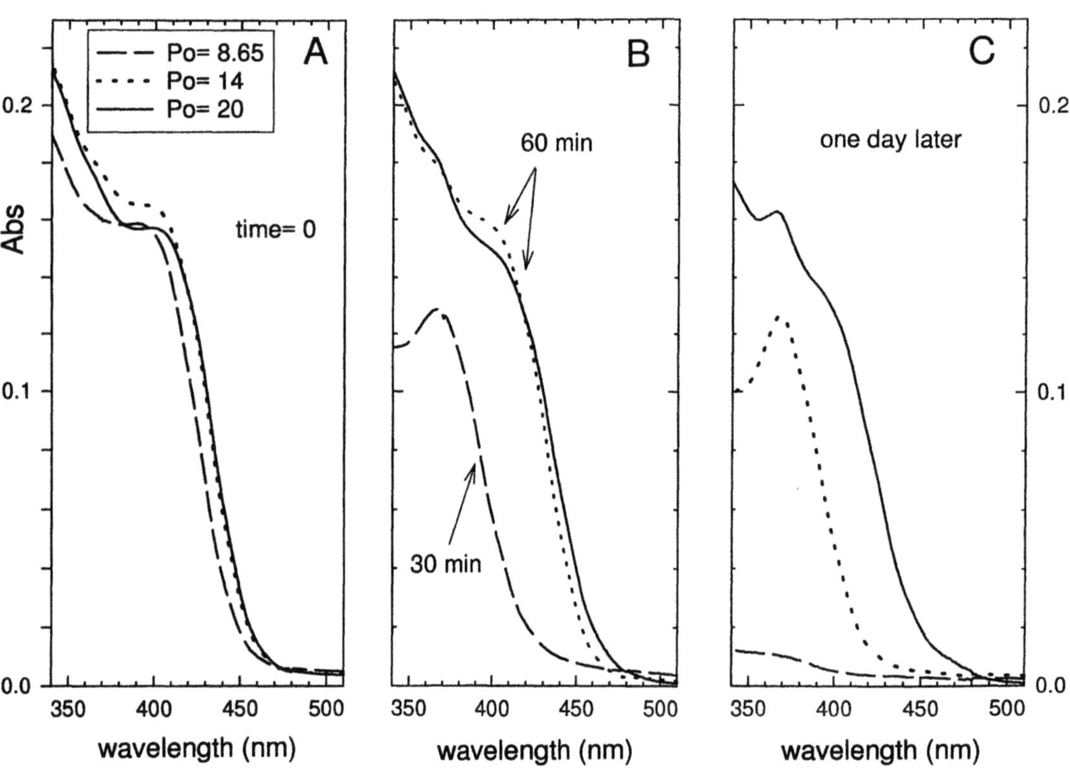

of the interfacial film [7]) using the S_{CTAB} values of ref. [7] are shown in Fig. 1 as continuous lines. An accurate comparison between the predicted and experimental values reveal, however, that the observed D_{CTAB} is greater than expected at $P_0 = 20$ for $W_0 > 30$. The reasons of this behaviour could be explained with a change in the shape of the aggregates and/or in an increase in their size polydispersity. For this reason we have not performed CdS at $P_0 = 20$ and $W_0 > 25$.

In Fig. 2A the comparison among the absorption spectra registered, immediately after the nanoparticles were formed, of colloidal CdS prepared in microemulsion at $W_0 = 15$ at three different P_0 values is shown. All the solutions show an onset of absorption around 470 nm and an enhanced shoulder around 400 nm. As the P_0 is increased the tail of the shoulder is shifted through longer wavelengths and in general the spectra become more structured. Surprisingly such features are nearly independent from the water content. After a time lag (which increase with P_0) a blue shift is observed in all the cases (Fig. 2B and C). Furthermore, the spectra reveal a clearly resolved maximum and, in the case of $P_0 = 14$ and 20 are well structured. Probably, the most striking influence of the alcohol content is on the stability of the nanoparticles. A comparison between Fig. 2A, B and C reveals a dramatic decrease in the optical density of the solution at $P_0 = 8.65$ due to the sedimentation of nanocrystal which form a yellow precipitate at the bottom of the cuvette. On the contrary, reverse micellar solution at high pentanol amount exhibit a relatively constant optical density also at high W_0. This is more clearly visible in Fig. 3 where the time course of the absorption at 470 nm is reported for microemulsions at $P_0 = 8.65$, 14 and 20 at different W_0. The physical explanation of such behaviour is not clear at present. Increasing the alcohol content the hydration of the surfactant polarheads and of the reagent ions remain unchanged. The greater structural change induced by a raise in the C_5OH content is a reduction in the micellar radius (Fig. 1). If the ions are distributed among the micelles following a Poisson distribution one obtain the plots of Fig. 3B, C and D from which we can deduce that only at $P_0 = 8.65$ there is a relevant ($>10\%$) fraction of micelles occupied by more than two ions of a given charge. Any way, an influence of the micellar degree of occupation should be reflected in different kinetic paths (different nucleation radii, different spectra after the mixing). Our results do not show relevant variation in the absorption spectra just after the mixing of the cadmium and sulfur ion-based micellar solutions. The main difference seems to be, instead, the time evolution of the mean dimension. All the solutions evolve from wavelengths around 400 nm (corresponding to a particle radius [9], r_P, of about 18–20 Å) towards a maximum of absorption around 370 nm ($r_P \approx 10 - 12$ Å). This indicates that particles with r_P around 10 Å are kinetically more

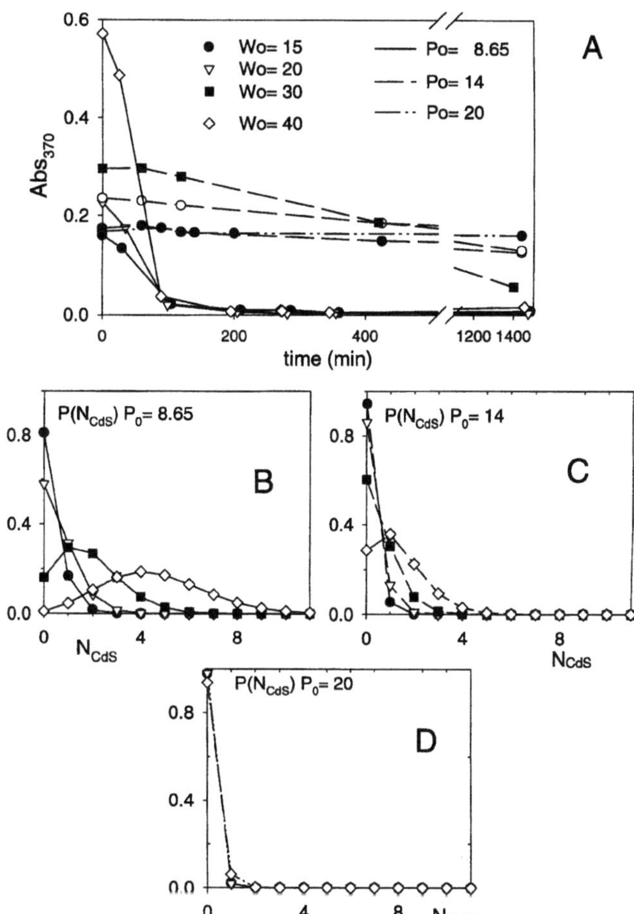

Fig. 3 (A) Time course of the absorption at 370 nm for CdS synthetised in microemulsions at different W_0 and P_0. ● $W_0 = 15$; ▽ $W_0 = 20$; ■ $W_0 = 30$; ◇ $W_0 = 40$; — $P_0 = 8.65$; - - - $P_0 = 14$; — ·· — $P_0 = 20$. (B) Poisson distribution of 10^{-4} M CdS in microemulsions at $P_0 = 8.65$ and different W_0, symbols as in panel A. (C) Poisson distribution of 10^{-4} M CdS in microemulsions at $P_0 = 14$ and different W_0, symbols as in panel A. (D) Poisson distribution of 10^{-4} M CdS in microemulsions at $P_0 = 20$ and different W_0, symbols as in panel A

stable in solution than larger ones and the peak at 370 nm become visible when particles with higher r_P values are precipitated. The systems at low P_0 reveals an extended sedimentation phenomena (almost complete on a time scale of one day) while the reverse micelles at $P_0 = 14$ and 20 show less than 20% of precipitation over the same time window (Figs. 2 and 3A). Furthermore, also the time required for the blue shift in the maximum of absorption is enhanced at high pentanol content as compared to a solution at $P_0 = 8.65$ (Fig. 2). All these features as a whole seem to suggest a direct role of the alcoholic functionality in stabilizing the nanoparticle possibly by a chemical interaction between the semiconductor surface and the hydroxylic moiety of the n-pentanol molecules. A look in the blue region of the spectra reveals, in samples at low

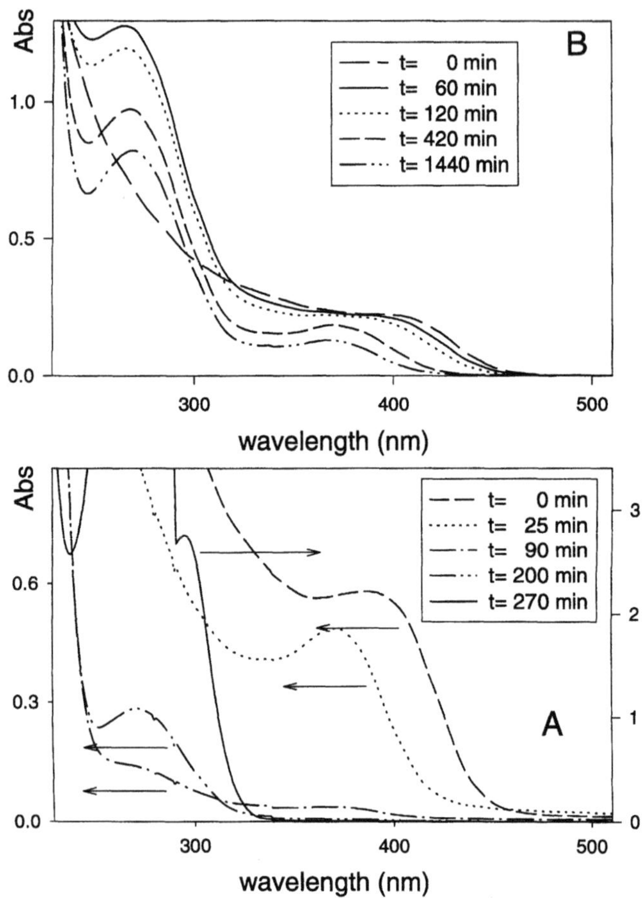

Fig. 4 (A) Absorption spectra of microemulsions at $W_0 = 40$ and $P_0 = 8.65$, containing colloidal CdS at different times from synthesis ([CdS] $= 6 \times 10^{-4}$ M). (B) Absorption spectra of microemulsions at $W_0 = 20$ and $P_0 = 14$, containing colloidal CdS at different times from synthesis ([CdS] $= 3 \times 10^{-4}$ M)

C_5OH content, the interesting phenomenon of precipitation-re-dissolution illustrated in Fig. 4A. When almost all the CdS is precipitated (low absorbance up to 240 nm) a band around 260 nm, indicative of small nanoparticle ($r_P \approx 7$ Å) appears and starts to grow. In samples at high water content ($W_0 = 40$) such a growth is rapid and results in a well defined band (see the minimum at 240 nm)

with a small peak at 293 nm. The extinction coefficient of CdS nanocrystals is known to increase with the decrement of r_P [9] and at our concentration when almost all the particles radii lie around 7 Å the absorption at 260 nm is well above 3. At $P_0 = 14$ and 20 there is no indication of such a process and moreover in general the blue region of the spectrum shows a continuous increase in the absorption without relevant structure (only a small shoulder around 265 nm is visible). A remarkable exception was found at $W_0 = 20$ and $P_0 = 14$ where after few minutes of preparation a well-resolved maximum at 265 nm appears (Fig. 4A). The band around 400 nm blue shifts and decreases in intensity with time as reported above. On the other hand, the position of the maximum of absorption in the blue remains unchanged, notwithstanding the decrease in intensity. This permits to exclude the growth of smaller particles at the expense of larger crystallites.

Conclusion

A first look on the spectral properties of colloidal CdS synthesized in the quaternary microemulsion CTAB/n-pentanol/n-hexane/water has revealed several interesting features. The tendency of the particles to sediment increases when the amount of alcohol decreases. The precipitation seems to be kinetically size selective and in general a relatively narrow absorption maximum centered at 370 nm is found after an appropriate time (which is function of P_0). On the contrary, the W_0 seems to have no effect on the size of the nanocrystal, suggesting that the aqueous droplets dimension plays a minor role on the dimension of the semiconductor nanoparticles. A further, extensive investigation (at present in progress in our laboratory) is, however, required to characterize the influence of the hosting microemulsion structure on the size and photoreactivity of the CdS nanocrystals.

Acknowledgements This work was supported by Consorzio Interuniversitario per lo Sviluppo dei Sistemi a Grande Interfase (CSGI-Firenze).

References

1. Alivisatos AP (1996) J Phys Chem 100:13226–13239 and references therein
2. Brus L (1996) Curr Opinions Colloids Interface Sci 1:197–201 and reference therein
3. Fendler JH (1987) Chem Rev 87:877–899 and reference therein
4. Eastoe J, Warne B (1996) Curr Opinions Colloids Interface Sci 1:800–805 and references therein
5. Petit C, Jain TK, Billoudet F, Pileni MP (1994) Langmuir 10:4446–4450
6. Giustini M, Palazzo G, Colafemina G, Della Monica M, Giomini M, Ceglie A (1996) J Phys Chem 100:3190–3198
7. Colafemmina G, Palazzo G, Balestrieri E, Giomini M, Giustini M, Ceglie A (1997) Prog Colloid Polym Sci 105:281–289
8. Stilbs P (1987) Prog NMR Spectr 19:1–45
9. Vossmeyer T, Katsikas L, Giersig M, Popovic IG, Diesner K, Chemseddine A, Eychmuller A, Weller H (1994) J Phys Chem 98:7665–7673

Progr Colloid Polym Sci (1998) 110:193–198
© Steinkopff Verlag 1998

Photoisomerizable cationic surfactants as microviscosity probes

T. Kozlecki
K.A. Wilk
L. Syper

T. Kozlecki · Prof. K.A. Wilk (✉)
Institute of Organic and Polymer
Technology
Wroclaw University of Technology
Wybrzeze Wyspianskiego 27
50-370 Wroclaw
Poland
E-mail: kaw@itots.ch.pwr.wroc.pl

L. Syper
Institute of Organic Chemistry
Biochemistry and Biotechnology
Wroclaw University of Technology
Wybrzeze Wyspiangkiego 27
50-370 Wroclaw
Poland

Abstract New cationic stilbene amphiphiles, 4-alkyl-4'-trimethyl-ammonium-stilbene bromides (abbreviated as $C_mStbTAB$) and 4-alkyl-4'-(ω-trimethyl ammonium-alkyl) stilbene bromides (abbreviated as C_mStbC_mTAB) were prepared and characterized spectroscopically. The photobehavior (fluorescence and *trans–cis* isomerization) of these molecules in homogeneous and micellar systems permits them to be useful micro-viscosity probes.

Key words Stilbene amphiphiles – cationic surfactants – microenvironment probes

Introduction

A number of anionic amphiphilic molecules containing the *trans*-stilbene chromophore at various distances along a fatty acid chain has been reported in the literature [1, 2]. 4'-Alkyl-4-stilbene carboxylic acids, and ω-(4'-alkyl-stilben-4-yl) alkanecarboxylic acids, called "stilbene fatty acids" [2], were developed as photophysical (fluorescence spectra) and photochemical (*trans–cis* isomerization) structural probes for the rigidity of the inner region of organized media (microviscosity [3]), such as micelles [4], microemulsions [1, 2], vesicles [4], monolayers and multilayers [5], thin films [2, 5], and inclusion complexes [6].

In the present contribution, we have designed and synthesized another group of new photophysical structural probes (abbreviated as $C_mStbTAB$ or C_mStbC_nTAB),

of a general structure given below:

$$C_mH_{2m+1} - \text{(benzene ring)} - CH=CH - \text{(benzene ring)} - (CH_2)_n NMe_3^+ \; Br^-$$

m	n
0	8
2	0
2	4
4	0
4	4
8	0

We have tested then fluorescence and *trans–cis* isomerization quantum yield of the studied stilbene surfactants towards viscosity of homogeneous solution (water, 2-methyl-2-butanol, ethylene glycol) and microviscosity of aggregates formed by standard hexadecyltrimethylammonium chloride (CTAC), hexadecyltrimethylammonium bromide (CTAB), and catalytic [7, 8] or hydrophobic [9, 10] analog of the latter (i.e. chemodegradable [2-tridecyl-(1,3-dioxolan-4-yl)] methyltrimethylammonium bromide (C_{13}DTAB)):

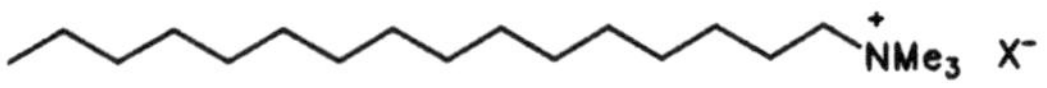

$$X = Cl\ (CTAC);\ Br\ (CTAB)$$

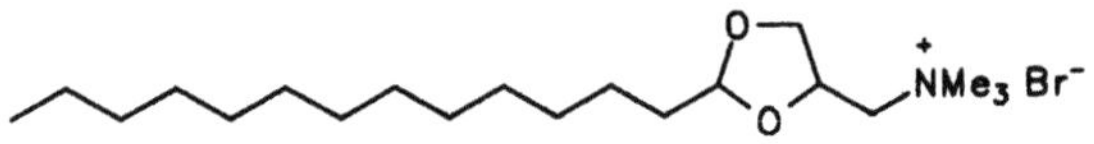

$$C_{13}DTAB$$

Cationic micelles of the selected surfactants affect rates of many organic reactions [7, 11, 12]. Characteristics of the aggregate microenvironments could be very useful for the kinetic quantitative description of micellar catalysis [13, 14].

Experimental

Materials

Water was obtained from a Millipore MilliQ purification system and deoxygenated with argon. Hexadecyltrimethylammonium chloride (CTAC) and bromide (CTAC) were purchased from Merck (p.a. grade) and were recrystallized three times from acetone-diethyl ether. [2-Tridecyl-(1,3-dioxolan-4-yl)]methyl trimethylammonium bromide (C_{13}DTAB) was prepared and purified as previously indicated [15]. 2-Methyl-2-butanol and ethylene glycol were purchased from Fluka (p.a. grade) and purified according to the literature procedures [16].

Stilbene probes

4-Alkyl-4′-trimethylammoniumstilbene bromides, C_mStbTAB were achieved from 4-alkylbromomethylbenzenes [17] via the Arbusov reaction, followed by the Horner–Emmons reaction with 4-dimethylaminobenzaldehyde, and quarternization with methyl bromide. Crude bromides were recrystallized several times from acetone-diethyl ether and dried at 56 °C/1 Torr for 6 h. 4-Alkyl-4′-(ω-trimethylammonium alkyl) stilbene bromides, C_mStbC$_n$TAB, were

synthesized from tetrahydropyranyl ethers of bromo-alcohols [18, 19], via formation of the Grignard reagent, coupling with 1-bromo-4-(1,3-dioxolan-2-yl) benzene in the presence of catalytic amounts of $Ni(PPh_3)_2Cl_2$ [20], followed by acidic work-up, resulting in the formation of aldehyde, the Horner–Emmons reaction with diethyl (4-alkylphenyl) methanephosphonate, replacement of the hydroxyl function with bromine using phosphorus tribromide, and quaternization with trimethylamine. Purification was accomplished analogously to the method described for C_mStbTAB. The physicochemical characteristics of C_mStbTAB and C_mStbC$_n$TAB is indicated in Table 1.

Measurements and procedures

All stilbene probes solutions were 0.008 mM. Typically, the appropriate volume of 1 mM solution of the probe in acetone (Merck, p.a. grade) was evaporated under an argon stream and then used to prepare the working solutions. In a flask coated with the probe, the appropriate amount of CTAC, CTAB or C_{13}DTAB was dissolved in water. Concentration of surfactant was 2.00 mM.

Fluorescence spectra were obtained by means of a Perkin–Elmer LS50B luminescence spectrometer. Solutions were prepared as described above. A reference solution of quinine sulfate in 1 N sufuric acid (Merck, Uvasol) with a known quantum yield of fluorescence (Φ_f) = 0.61 [21]) was also prepared with the same optical density at 313 nm. Fluorescence spectra were measured with excitation at 313 nm. The entire emission band was integrated between 380 and 540 nm to obtain the relative intensities of the solutions. The fluorescence quantum yields were calculated by multiplying the fluorescence quantum yield of the reference by the ratio of the integrated intensities of the sample and the reference.

Trans–cis isomerization quantum yields were measured after irradiating 50 ml samples prepared as described above with a low pressure 16 W mercury lamp (Photochemical Reactors) using the 313 nm line. Isomerization were carried out to 10–20% conversion. Azobenzene actinometry was used [22], and the amounts of isomeric stilbenes formed were determined by ^{1}H NMR according to the previously described procedure [23]. Quantum yields for the isomerization were determined by the absorption spectroscopy by using standard analytical techniques and equations [24].

The critical micelle concentrations of stilbene surfactants C_mStbC$_n$TAB were determined by surface tension measurements on a Krüss K12 microprocessor tensiometer equipped with a du Nuoy ring. The measurements were performed at 25 °C. The surface tension was

Table 1 Physicochemical characteristics of C_mStbTAB and C_mStbC$_n$TAB surfactants

Compound (Formula)	^{1}H NMR ppm	^{13}C NMR ppm	Analysis	UV (water) λ max (ε_{max})	cmc (mM)
C_2StbTAB ($C_{19}H_{24}BrN$)	7.95, 7.70 (dd, J = 8.6 Hz, 4H); 7.35, 7.11 (dd, J = 8.7 Hz, 4H); 7.31, 6.99 (dd, J = 16.3 Hz, 2H); 3.60 (s, 9H); 2.62 (q, 2H); 1.18 (t, 3H)	143.5, 140.6, 136.6, 134.8, 131.0, 128.8, 128.3, 127.0, 126.7, 120.7, 60.4, 29.7, 15.8	Calcd.: C, 65.90; H, 6.98; N, 4.04. Found: C, 66.05; H, 7.11; N, 4.2	315 (21 400)	9.72[a]
C_4StbTAB ($C_{21}H_{28}BrN$)	7.93, 7.71 (dd, J = 8.5 Hz, 4H); 7.33, 7.14 (dd, J = 8.3 Hz, 4H); 7.27, 7.03 (dd, J = 15.6 Hz, 2H); 3.58 (s, 9H); 2.56 (t, 2H); 1.81–0.85 (m, 7H)	143.9, 141.0, 136.1, 135.0, 131.3, 128.5, 128.0, 127.2, 126.4, 119.7, 59.7, 36.0, 34.0, 22.9, 14.9.	Calcd.: C, 67.38; H, 7.54; N, 3.74. Found: C, 67.48; H, 7.84; N, 3.55	314 (20 900)	3.27[a]
C_8StbTAB ($C_{25}H_{36}BrN$)	7.95, 7.70 (dd, J = 8.5 Hz, 4H); 7.30, 7.10 (dd, J = 8.8 Hz, 4H); 7.27, 7.03 (dd, J = 16.0 Hz, 2H); 3.55 (s, 9H); 1.88–0.83 (m, 15H)	143.9, 141.0, 136.1, 135.0, 131.3, 128.8, 128.0, 127.2, 126.4, 119.7, 61.0, 36.0, 34.0, 31.6, 27.5, 25.8, 23.7, 22.9, 14.9	Calcd.: C, 69.76; H, 8.43; N, 3.25. Found: C, 69.94; H, 8.66; N, 3.39	315 (21 000)	0.37
StbC$_8$TAB ($C_{25}H_{36}BrN$)	7.43–6.96 (m, 11H); 3.39 (t, 2H); 3.18, (s, 9H); 2.58 (t, 2H); 1.73–1.02 (m, 12H)	136.7, 136.5, 136.1, 135.1, 128.4, 128.0, 127.7, 127.5, 126.7, 125.0, 60.3, 37.2, 36.3, 34.1, 32.1, 32.0, 29.0, 28.7	Calcd.: C, 69.76; H, 8.43; N, 3.25. Found: C, 69.94; H, 8.66; N, 3.39	307 (19 500)	0.012
C_2StbC$_4$TAB ($C_{23}H_{32}BrN$)	7.33–7.01 (m, 10H); 3.45 (t, 2H); 3.09, (s, 9H); 2.58 (q, 2H); 1.80–1.18 (m, 7H)	136.9, 136.2, 136.0, 134.7, 128.1, 127.8, 127.5, 127.4, 126.9, 125.2, 61.3, 34.8, 29.4, 29.2, 26.3, 15.8	Calcd.: C, 68.55; H, 8.02; N, 3.48. Found: C, 68.74; H, 8.19; N, 3.48	307 (21 000)	0.033
C_4StbC$_4$TAB ($C_{25}H_{36}BrN$)	7.30–7.00 (m, 10H); 3.40 (t, 2H); 3.13, (s, 9H); 2.58 (t, 2H); 1.80–0.90 (m, 11)	137.2, 136.0, 135.7, 134.4, 128.0, 127.9, 127.6, 127.2, 126.5, 125.4, 61.7, 34.8, 30.5, 29.4, 29.2, 26.3, 25.5, 15.8	Calcd.: C, 69.76; H, 8.43; N, 3.25. Found: C, 70.02; H, 8.70; N, 3.43	308 (20 800)	0.014

[a] From ref. [25].

plotted against the surfactant concentration and the cmc was determined from the break point of the plots.

NMR spectra were recorded with a Bruker AMX-300 spectrometer. (^{1}H at 300.13 MHz with the residual CHCl$_3$ signal as reference, ^{13}C at 75.5 MHz with the CDCl$_3$ signal as reference). UV-vis spectra were obtained by means of a Hewlett-Packard HP8752A spectrophotometer. Elemental analyzes were carried out by use of a Perkin–Elmer 2400 CHN analyzer.

Results and discussion

For the studied cationic stilbene amphiphiles, C_mStbTAB or C_mStbC$_n$TAB, we have elaborated convenient methods for both the syntheses, and the purification. Satisfactory NMR and UV spectra, and elemental analyzes were obtained for all the stilbene derivatives (Table 1). The reported compounds exhibit good solubility in water, 2-methyl-2-butanol, and ethylene glycol under the concen-

tration studied. Their solutions at room temperature are pale yellow, transparent, and non-viscoelastic. The critical micelle (aggregate) concentration (cmc) values, determined in water by means of surface tension measurements, are also tabulated in Table 1 [25]. The C_mStbTAB and C_mStbC$_n$TAB amphiphiles easily incorporate into a micelle or a microemulsion droplet of CTAC, CTAB or C_{13}DTAB, due to their structural similarity. Being cationic in nature they are quite compatible with the selected cationic surfactant molecules. Additionally, molecular dimensions of the stilbene chromophore, elongated and rod-like, do not differ substantially from those of an alkyl chain in an extended all-*trans* configuration [4].

Because the photobehavior of *trans*-stilbene and related molecules (evidenced by fluorescence and *trans–cis* isomerization) is particularly sensitive to environmental effects, especially solvent viscosity [26], in the next part of our work we employed the newly synthesized C_mStbTAB and C_mStbC$_n$TAB to probe the microviscosity of several cationic micellar aggregates, which were commonly applied so far as reaction media in micellar catalysis [7, 11]. The experiments were performed in relation to moderately viscous homogeneous solvents. The type of micellar aggregates and the solvents used are indicated in Table 2 (fluorescence quantum efficiencies, Φ_f) and Table 3 (*trans-cis* isomerization quantum yields, Φ_{tc}).

Spectra of solutions of all C_mStbTAB's and C_mStb C_nTAB's contain, within the spectral range covered by our experiments (280–500 nm), distinct band at ca. 310 nm (Table 2), characteristic for the stilbene moiety [27, 28]. Incorporation of C_mStbTAB or C_mStbC$_n$TAB into the CTAC, CTAB or C_{13}DTAB aggregates gives a clear solution having an absorption spectrum red shifted by several nanometers (Fig. 1). Fluorescence of all of the stilbenes studied in the applied media shows characteristic monomer emission with no excimer formation at the concentration level used (Fig. 1).

According to our findings incorporating the stilbene chromophore into the hydrophobic chain of the cationics anticipates – as expected [1, 2] – with the degree of organization for the host medium. The photoreactivity of C_mStbTAB and C_mStbTAB in aqueous micelles and oil in water microemulsions formed by CTAC, CTAB, and C_{13}DTAB in relation to moderately viscous standard media exhibits increased values for Φ_f, and decreased Φ_{tc}. However, the intra-chain compounds, such as C_2Stb C_4TAB and C_4StbC$_4$TAB have considerably higher quantum yields for fluorescence than both StbC$_8$TAB, and *trans*-stilbene (Stb) itself ($\Phi_f = 0.05$ in methylcyclohexane [29]). The single-chain derivatives C_2StbTAB, C_4StbTAB, and C_8StbTAB reveal the Φ_f values 6–7 times lower than those of C_mStbC$_n$TAB.

As can be noted from Table 3 the isomerization yields generally follow inversely the changes previously noted in fluorescence; accordingly isomerization yields for all of the studied probes are greater in homogeneous media, than in micellar aggregates. The StbC$_8$TAB probe has the Φ_{tc} value nearly the same as for Stb ($\Phi_{tc} = 0.50$ in benzene

Table 2 Fluorescence[a] quantum yields for C_mStbTAB and C_mStbc$_n$TAB in water, selected alcohols and aggregated system at 25 °C

m	n								
									$10^2\Phi_f$
		Water[b]	2-Methyl-2-butanol[c]	Ethylene glycol[d]	CTAC micelles[e]	CTAB micelles[e]	C_{13}DTAB micelles[e]	CTAB microemulsions[f]	C_{13}DTAB microemulsions[g]
2	0	3.7	4.5	5.5	7.3	6.6	6.8	6.4	6.3
4	0	3.3	3.9	4.8	7.0	6.3	6.4	6.0	6.0
8	0	2.8	3.6	4.5	6.8	6.1	6.3	5.9	5.8
0	8	9.2	10.5	12	20	18	17	16	16
2	4	23	27	34	47	44	44	42	42
4	4	24	28	35	46	43	44	43	42

[a] The probe concentration 0.008 mM; excitation $\lambda_{ex} = 366$ nm; emission $\lambda_{em} = 380$–540 nm.
[b] Viscosity 0.89 cP [13].
[c] Viscosity 3.2 cP [13].
[d] Viscosity 0.65 cP [13].
[e] Total surfactant concentration 2.0 mM in water.
[f] 18.0 wt% CTAB, 33.9 wt% 1-butanol, 25.5 wt% *n*-octane, 20.0 wt% water.
[g] 17.3 wt% C_{13}DTAB, 34.6 wt% 1-butanol, 26.0 wt% *n*-octane, 19.5 wt% water.

Table 3 Quantum yields for trans–cis photoisomerization[a,b] of C_mStbTAB and C_mStbTAB in water, selected alcohols and aggregated at 25 °C

m	n								
									$10^2\Phi_f$
		Water	2-Methyl-2-butanol	Ethylene glycol	CTAC micelles	CTAB micelles	C_{13}DTAB micelles	CTAB microemulsions	C_{13}DTAB microemulsions
2	0	5.0	4.3	3.8	2.9	3.2	3.1	3.3	3.4
4	0	5.3	4.5	4.0	3.0	3.5	3.4	3.6	3.7
8	0	6.0	5.0	4.3	3.3	3.7	3.5	3.8	3.9
0	8	49	46	42	35	38	39	40	41
2	4	37	33	30	22	25	25	26	27
4	4	37	32	30	23	26	26	27	27

[a] Total probe concentration 0.008 mM; irradiation at $\lambda = 366$ nm.
[b] Concentrations as in Table 2.

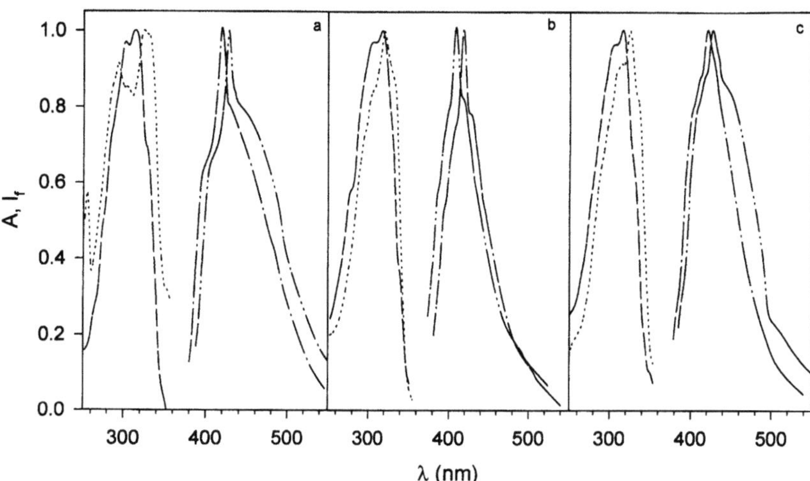

Fig. 1 Normalized absorption and fluorescence spectra (λ_{ex} = 313 nm) for (a) C_8StbTAB, (b) $StbC_8TAB$, (c) C_4StbC_4TAB in aqueous (––,–·–) and CTAB micelles. (···,–··–) at 25 °C

[29]), slightly higher than for the 4,4'-disubstituted (intrachain) stilbene ammonium bromides C_mStbC$_n$TAB, and ca. 10 times higher than for the 4-monoalkylsubstituted derivatives C_mStbTAB. Evidently, the presence of a strong electron-withdrawing group (NMe$_3^+$) at the 4'-position of C_mStbTAB increases radiationless decay of the excited state, and in consequence an efficiency for luminescence and *trans–cis* isomerization decreases.

Given the relationship among the quantum yields of fluorescence and isomerization, and known values of the medium viscosity we can estimate of a fairly precise magnitude of microviscosity. Although, the stilbene chromophore of C_mStbTAB and C_mStbC$_n$TAB occupies a site in the micelle which is simultaneously quite rigid and reasonable nonpolar, at the present stage of studies we can say nothing about the exact location of the Stb probe in the given aggregated structure. The micellar aggregates of the selected cationic surfactants are also shown by the applied probes as moderately viscous fluid media with values for Φ_f and Φ_{tc} comparable to that of ethylene glycol ($\eta = 16.5$ cP at 25 °C [16]). The microviscosity of the studied organized systems for all of C_mStbTAB and C_mStbC$_n$TAB increase in the range η (C_{13}DTAB microemulsions) $\sim \eta$ (CTAB microemulsions) $< \eta$ (CTAB micelles) $\sim \eta$(C_{13} DTAB micelles) $< \eta$(CTAC micelles).

The reported 4-alkyl-4'trimethylammoniumstilbene bromides and 4-alkyl-4'-(ω-trimethylammoniumalkyl) stilbene bromides comprise an interesting class of photoisomerizable cationic surfactants. Their photochemistry (consisting mainly of fluorescence and *trans–cis* isomerization) make them useful indicators of microviscosity for other than micelles and microemulsions cationic organized assemblies.

Acknowledgement Support of this work by the State Committee of the Scientific Research (Grant No. 2P 30307207) is gratefully acknowledged.

References

1. Whitten DG, Russell JC, Schmehl RH (1982) Tetrahedron 39:2455
2. Whitten DG (1993) Acc Chem Res 26:502
3. Kalyanasundaram K (1991) In: Ramamurthy V (ed) Photochemistry is Organized and Constrained Media. VCH, New York, pp 39–77
4. Suddaby BR, Brown PE, Russell JC, Whitten DG (1985) J Am Chem Soc 107:5609
5. Mooney WF, Brown PE, Rusell JC, Costa SB, Pedersen LG, Whitten DG (1984) J Am Chem Soc 106:5659
6. Suddaby BR, Dominey RN, Hui Y, Whitten DG (1985) Can J Chem 63:1315
7. Wilk KA, Bieniecki A, Matuszewska B (1994) J Phys Org Chem 7:646
8. Bieniecki A, Gapiński J, Wilk KA (1997) J Phys Chem 101:871
9. Sokołowski A, Bieniecki A, Wilk KA, Burczyk B (1995) Colloids Surfaces A 98:73
10. Bieniecki A, Gapiński J, Wilk KA (1997) Progr Colloid Polym Sci 105:in press
11. Wilk KA (1992) J Phys Chem 96:901
12. Bunton CA (1991) In: Grätzel M, Kalyanasundaram KA (eds) Kinetics and Catalysis in Microheterogeneous Systems. Marcel Dekker, New York, pp 13–47
13. Bunton CA, Savelli G (1986) Adv Phys Org Chem 22:213
14. Romsted LS (1977) In: Mittal KL, Lindman B (eds) Micellization, Solubilization and Microemulsions, Vol 2. p 509
15. Wilk KA, Bieniecki A, Burczyk B, Sokolowski A (1994) J Am Oil Chem Soc 71:81
16. Riddick JA, Bunger WB (1970) Organic Solvents Physical Properties and Methods of Preparation. Wiley Interscience, New York
17. Mitchell RH, Iyer VS (1989) Synlett:55
18. Tran-Chi NH, Falk H (1995) Monatsh Chem 126:565
19. Miyashita N, Yoshikoshi A, Grieco PA (1977) J Org Chem 42:3772
20. Felkin H, Swierczewski G (1975) Tetrahedron 31:2735

21. Guilbault GG (ed) (1973) Practical Fluorescence. Marcel Dekker, New York, pp 12–15
22. Gauglitz G, Hubig S (1984) Z Phys Chem (Wiesbaden) 139:237
23. Koźlecki T, Wilk KA (1996) J Phys Org Chem 9:645
24. Rau H, Greiner G, Gauglitz G, Meier H (1990) J Phys Chem 94:6523
25. Sokołowski A, Wilk KA, Masiowska A (1996) Proc 4th Worlds Surfactant Congress, Barcelona, pp 274–282
26. Saltiel J, D'Agostino JT (1972) J Am Chem Soc 94:6445
27. Waldeck DH (1991) Chem Rev 91:415
28. Meier H (1992) Angew Chem Int Ed Engl 31:1399
29. Brown PE, Whitten DG (1985) J Phys Chem 89:1217

Progr Colloid Polym Sci (1998) 110:199–203
© Steinkopff Verlag 1998

Synthesis and surface properties of *N*-alkylaldonamides

L. Syper
K.A. Wilk
A. Sokołowski
B. Burczyk

L. Syper
Institute of Organic Chemistry
Biochemistry and Biotechnology
Wrocław University of Technology
Wybrzeze Wyspianskiego 27
50-370 Wrocław
Poland

K.A. Wilk · Prof. A. Sokołowski (✉)
B. Burczyk
Institute of Organic and Polymer
Technology
Wrocław University of Technology
Wybrzeze Wyspianskiego 27
50-370 Wrocław
Poland
E-mail: sokolowski@itots.ch.pwr.wroc.pl

Abstract New types of ecologically benign surfactants with sugar head groups have been described, namely: *N*-dodecyl-*N*-methyllactobionamide (C_{12}-MLA), *N*-dodecyl-*N*-methylglucoheptonamide (C_{12}-MGHA), *N*-dodecyl-*N*-methylgluconamide (C_{12}-MGA), *N*-dodecyl-gluconamide (C_{12}-GA) and *N*-octyl-glucoheptonamide (C_8-GHA). Their surface properties, such as static surface tension (γ), critical micelle concentration (CMC), surface area demand per molecule (A) and foaming and wetting properties have been analyzed in relation to lactobionamides (C_n-LA) and well-known oligooxyethylene alkyl ethers. The presence of both the amide methyl group, and the additional 1.4-glucose segment, increases significantly the surfactant solubility and decreases its hydrophobicity. The performing properties (foaming and wetting behavior) of *N*-alkyl-*N*-methylaldonamide-type surfactant prove that they can make suitable candidates as ecologically favorable surfactants.

Key words Nonionic saccharide surfactants – lactobionamides – glucoheptonamides – gluconamides – surface tension – critical micelle concentration – foaming properties – wetting properties

Introduction

Saccharide-based surfactants in which a relatively long fatty alkyl chain is hydrophilized by carbohydrates instead of ethylene oxide have recently awaked attention from both the ecological and dermatological points of view [1, 2]. The linkage between the alkyl chain and the carbohydrate grouping can be established by ester [3, 4], ether [5], amide [6] or amine bridges [7]. Shinoda et al. have demonstrated the amphiphilic nature of single-chain *n*-alkyl- and alkyl-aryl-glucosides via interfacial tension measurements [8]. Other research groups have focused their interest on D-glucitol derivatives [7]. More recent papers describe the solid and liquid crystal structures and gel formation of sparingly soluble single-chain glucamides and methyl glucamides [9–12].

In the present contribution we report on synthesis of a new series of water-soluble sugar surfactants of the alkylaldonamide-type (structures given below) as well as on their adsorption, micellization, and some performance properties (see Fig. 1).

The solubility of the studied structures has been improved either by the methyl group incorporation into the amide linkage or by additional 1.4-glucose unit. Moreover, the length of the sacharide hydrophilic group has been modified by the additional H–C–OH moiety.

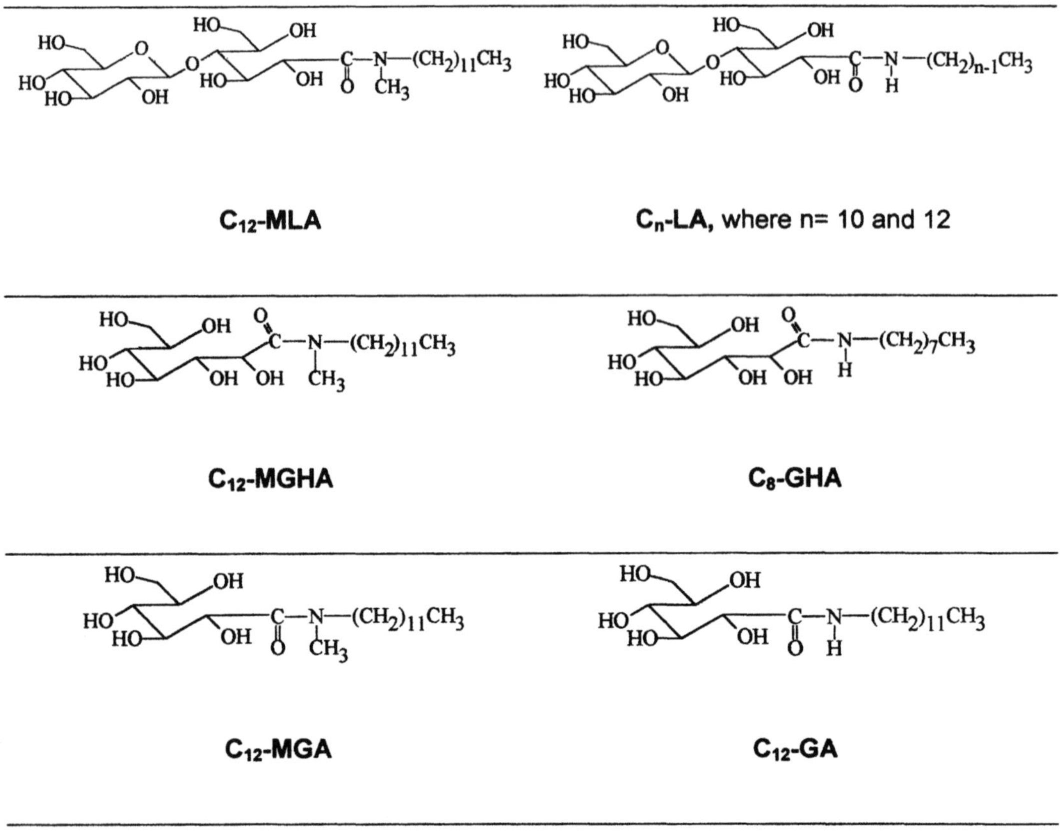

C$_{12}$-MLA

C$_n$-LA, where n= 10 and 12

C$_{12}$-MGHA

C$_8$-GHA

C$_{12}$-MGA

C$_{12}$-GA

Fig. 1 Structures of *N*-alkylaldonamides

Experimental

Materials

Reagents and solvents were of commercial grade and were not additionally purified before use. Water used for all experiments was distilled twice and purified using a Millipore Milli-Q purification system.

Synthesis of *N*-dodecyl-*N*-methyllactobionamide (C$_{12}$-MLA)

The mixture of lactobionic acid (20 mmol), dodecyl-methylamine (20 mmol) and methanol (100 cm^3) was vigorously stirred for 19 h at room temperature. The solid was filtered off, washed several times with small quantities of methanol, dried in vacuo, purified by repeated recrystallization from methanol–water mixtures of various composition and dired in vacuo over solid KOH to give C$_{12}$-MLA.

Yield 70.5%. ^{1}H NMR δ 0.83 (t, 3H, $\underline{CH_3}$, –(CH$_2$)$_{11}$, $J = 6$ Hz), 1.18–1.32 (m, 18H, CH$_3(\underline{CH_2})_9$–) , 1.4–1.65 (m, 2H, –$\underline{CH_2}$CH$_2$N(CH$_3$)–), 2.96 (m, 2H, –$\underline{CH_2}$N(CH$_3$)–), 2.85 and 3.02 (s, $\Sigma = 3$H, –N(CH$_3$)–), 2.6–4.8 (sugar part). Anal. Calcd. for C$_{25}$H$_{49}$O$_{11}$N: C, 55.64; H, 9.15; N, 2.59; Found: C, 55.87; H, 9.18; N, 2.64.

Synthesis of *N*-dodecyl-*N*-methylglucoheptonamide (C$_{12}$-MGHA)

The above described method for C$_{12}$-MLA has been applied. Thus, 40 mmol of α-D-glucoheptonic γ-lactone and 40 mmol of dodecylmethylamine afford C$_{12}$-MGHA. Yield 54.0%. ^{1}H NMR δ 0.85 (t, 3H, $\underline{CH_3}$, –(CH$_2$)$_{11}$, $J = 6$ Hz), 1.18–1.30 (m, 18H, CH$_3(\underline{CH_2})_9$–), 1.35–1.38 (m, 2H, –$\underline{CH_2}$CH$_2$N(CH$_3$)–), 3.25 (m, 2H, –$\underline{CH_2}$N(CH$_3$)–), 2.82 and 2.99 (s, $\Sigma = 3$H, –N(CH$_3$)–), 3.3–4.8 (sugar part). Anal. Calcd. for C$_{20}$H$_{41}$O$_7$N: C, 58.94; H, 10.14; N, 3.44; Found: C, 59.02; H, 10.19; N, 3.42.

Synthesis of *N*-dodecyl-*N*-methylgluconamide
(C_{12}-MGA)

The method described for C_{12}-MLA has been applied.
40 mmol of D(+)-gluconic acid δ-lactone and 40 mmol of
dodecylmethylamine afford C_{12}-MGA. Yield 91.5%.
^{1}H NMR δ 0.86 (t, 3H, $\underline{CH}_3$, –$(CH_2)_{11}$, $J = 6$ Hz),
1.13–1.32 (m, 18H, $CH_3(\underline{CH}_2)_9$–), 1.364–1.50 (m, 2H,
–$\underline{CH}_2CH_2N(CH_3)$–), 3.28 (m, 2H, –$\underline{CH}_2N(CH_3)$–), 2.80
and 2.98 (s, $\Sigma = 3$H, –N(CH_3)–), 3.3–5.0 (sugar part). Anal.
Calcd. for $C_{19}H_{39}O_6N$: C, 60.45; H, 10.41; N, 3.71; Found:
C, 60.53; H, 10.49; N, 3.69.

N-alkyllactobionamides (C_n-LA), *N*-octylglucohepto-
namide (C_8–GHA) and *N*-dodecylgluconamide (C_{12}-GA)
were synthesized by reaction of alkylamine with lactobionic acid, α-D-glucoheptonic γ-lactone and D(+)-
gluconic acid δ-lactone, respectively. The procedures used
for carrying out the reactions have been described elsewhere [13, 14].

Methods

The ^{1}H NMR measurements were performed at 298 K
with a Bruker DRX-300 NMR spectrometer. ^{1}H chemical
shifts are referenced in DMSO-d_6 to residual protons of
DMSO-d_6 (2.50 ppm). Elemental analyses were performed
on a Perkin–Elmer CHN Analyser. Surface tensions of
aqueous solutions were measured continuously during
a few hours using a Krüss K12 automatic tensiometer
equipped with a du Nuoy Pt–Ir ring (resolution
± 0.01 mN/m) at 298 ± 0.1 K. Sets of measurements were
taken at intervals until no significant change occurred in
the tension. The pull method was employed with this
tensiometer, and the ring did not become detached from
the surface. The γ-concentration data were the averages of
two independent runs. The equilibrium surface tension
data were reproducible within 0.2 mN/m. One-minute
values of contact angles were measured on drops by means
of a TM system (Technicome S.A.). Drops on a parafin
wax surface were studied for concentrations below and
above the CMC. Each reading was repeated for at least 10
times. Wetting properties were determined as the concentration (g dm^{-3}) of surfactant solution necessary for immersion of a cotton-fabric ring (3.5 cm in diameter) in
100 s, according to the Polish Standard PN-74/C-04800.
The solutions were stored for 24 h before determinations.
Foaming properties were measured by the beating method
with a perforated disk of 5.2 cm diameter with 24 symmetrical holes of 0.2 cm inside diameter oscillated vertically in
a glass cylinder. The frequency of beating was 1 s^{-1} and
the foam generation time was 60 s. The foaming ability
was defined as the volume of foam formed over 100 ml of
surfactant solution. The measurements were carried out
for 0.1 wt% solutions in distilled water at 25 °C.

Results and discussion

Surface properties of *N*-alkyl-*N*-methylaldonamides

The dodecylgluconamide (C_{12}-GA) and octylglucohep-
tonamide (C_8-GHA) are insoluble in water even at
0.1 wt%. They are dissolved only on heating to 60–80 °C,
which indicates strong hydrogen bonding via the amide
group. Introducing a methyl group into the amide linkage
(compounds C_{12}-MGA and C_{12}-MGHA) increases the
surfactant's solubility. Additionally, the solubility of C_{12}-
MGHA is lower than that of C_{12}-MGA; it turned out that
the introduced H–C–OH moiety decreases hydrophilic
character of surfactant head group.

The air–water surface tension γ vs log [aldonamide
surfactant] exhibits classic behavior for surfactants;
a steady, linear decrease in γ, and a clean break with the
absence of a minimum at the CMC was observed. The
post-cmc tension for all surfactants under study is from
32.2 to 35.1 mN m^{-1}. The use of the Gibbs equation in
order to calculate the mean area demand per molecule at
the interface (A) from surfacce tension measurements is not
without errors [15, 16]. To extract A it is usual to fit a
polynomial (quadratic) to the pre-CMC data and calculate
the derivative $d\gamma/d\ln c$ at the cmc. This method was tested
for the data shown in Fig. 2 and owing to the large range of

Fig. 2 Surface tension of aqueous *N*-alkylaldonamides solution at
25 °C

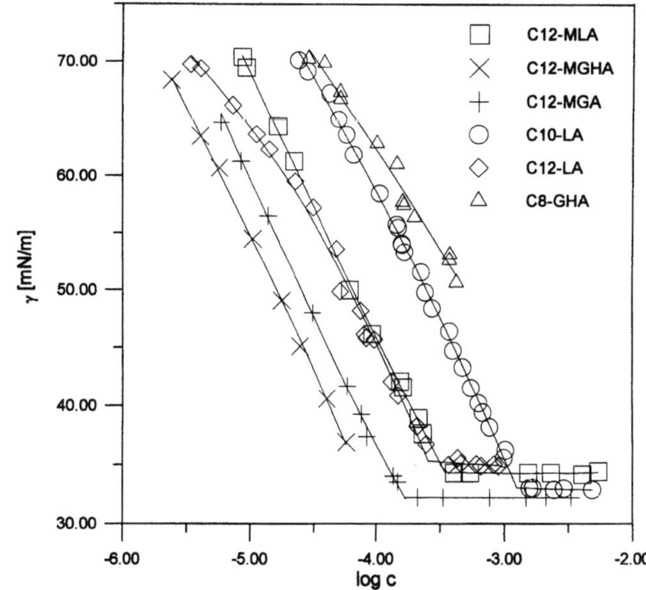

linear behavior, no major differences were found as compared with a linear least-squares fit.

As in the case of alkyl monoglycosides [17], the CMC values of the studied aldonamide-type surfactants having $C_{10}H_{21}$ and $C_{12}H_{25}$ alkyl chain length is significantly higher than those of $C_{10}E_4$ (CMC 0.68 mM [18]) and $C_{12}E_5$ (CMC 0.04 mM [18]), respectively. The lactobionamides, C_{10}-LA and C_{12}-LA, showed the CMC values different than those recently reported [19] although the A values are comparable.

Table 1 includes adsorption parameters for the aldonamide surfactants. The structure and number of hydroxyl groups of the hydrophilic part of sugar molecules has a small effect upon the surface excess concentration, Γ, of surfactant. The introduction of an additional H–C–OH group into the head grouping of C_{12}-MGA causes an hydrophobic effect comparable to the effect of 0.2 methylene group ($-CH_2-$) in the main alkyl chain in C_{12}-MGA homologous series of surfactants. The differences between γ_{cmc} for studied surfactants are negligible. It seems probably that the structure of the film formed is predominantly modified by the electrostatic dipole–dipole interactions of the hydrophilic part of aldonamides with water.

The hydrophilic moiety of saccharide surfactants has a smaller cross-sectional area than the oligo(oxyethylene) unit in $C_{12}E_m$ surfactants, whose number of oxyethylene units, m, is above 3. For the nonionic $C_{12}E_3$, Rosen et al. [20] and Thomas et al. [21] have shown that at the CMC the values of A are equal to $42 \times 10^{-20}\,m^2$ and $(36 \pm 2) \times 10^{-20}\,m^2$, respectively. Polynomial fit of the surface data for C_{12}-MLA, C_{12}-LA and C_{12}-MGA gave a value of ca. $40 \times 10^{-20}\,m^2$, in agreement with some earlier data published for *N*-alkylmaltonamides [22] and dodecyllactobionamide [19]. This problem can be understood on the ground of the results presented in earlier papers. The hydrophilicity of oligooxyethylene alkyl ethers arises from the formation of two hydrogen bonds per oxyethylene group with water molecules. On the other hand, a hydroxyl group of the hydrophilic part of sugar surfactant can form three hydrogen bonds with water molecules [23]. This situation can make the sugar fragment to remain extended into the bulk phase in contrary to oligooxyethylenated chain.

Relative positions of aldonamide and C_nE_m molecules in monolayer (and also in mixed monolayer) cannot be elucidated definitely by surface tension measurements but require other techniques, i.e., neutron reflection technique with selective isotopic labeling of the alkyl chain length of both the long chain sugar surfactant and oligooxyethylene alkyl ether.

Performance properties

Among the performance properties, we have studied the foaming and wetting behavior of *N*-dodecyl-*N*-methylaldonamides. The foam volumes obtained just after the end of beating and still maintained after 1 and 10 min are presented in Table 2. Comparing to nonionic fatty alcohol oxyethylates, both *N*-dodecyl-*N*-methyllactobionamide (C_{12}-MLA), and *N*-dodecyl-*N*-methylgluconamide (C_{12}-MGA) show high foaming power and a high foam-stabilizing effect which in the case of C_{12}-MGA in particular, approaches that of anionic surfactants (SDS). *N*-alkyl-*N*-methylaldonamide foams are more round, wet, and creamier than those of SDS.

The wetting ability is expressed as the concentration of solution necessary to wet a standard cotton disc in 100 s. Table 2 shows that the wetting behavior of studied surfactants decreases along with increasing size of the surfactant head group. Furthermore, the immersion test is a function of the contact angle of the solution on paraffin wax. Compound C_{12}-MGA is less effective wetting agent than the standard oligooxyethylenated dodecanol $C_{12}E_5$.

Table 1 Surface properties of aqueous *N*-alkylaldonamides at 25 °C

Surfactant	$10^6\Gamma_{CMC}$ [mol m^{-2}]	$10^{20}A$ [m^2]	10^3CMC [M]	γ_{CMC} [mN m^{-1}]	ΔG°_{CMC} [kJ mol^{-1}]
C_{12}-MLA	4.22	40.6	0.338	34.3	− 19.8
C_{12}-MGHA	3.8	44	0.06[b]	36.9[b]	—
C_{12}-MGA	4.09	39.3	0.148	32.2	− 21.9
C_{10}-LA	4.28	40	1.32	33.0	− 16.4
	3.78[a]	44[a]	3.40[a]	33.4[a]	
C_{12}-LA	4.39	39	0.251	35.1	− 20.6
	3.78[a]	44[a]	0.18[a]	36.1[a]	
C_8-GHA	2.95	58	0.4[b]	50.7[b]	

[a] From ref. [19].
[b] Solubility point.

Table 2 Foaming and wetting properties of N-alkyl N-methylaldonamides, oligo-oxyethylenated alcohols and sodium dodecyl sulfate at 25 °C

Surfactant	Foam volume, V [cm^3]			Wetting ability [g dm^{-3}]	Contact angle, Θ	
	0 s	After 1 min	After 10 min		0.1 CMC	1.1 CMC
C_{12}-MLA	950	830	780	0.60	88.7°	78.8°
C_{12}-MGHA	—[a]	—	—	—	—	—
C_{12}-MGA	1200	1150	1080	0.26	84.9°	78.8°
$C_{12}E_5$	40	0	0	0.12	72.3°	56.0°
$C_{10}E_4$	0	0	0	>3	80.6°	66.4°
Triton X-100	100	20	0	0.7	79.5° (at 0.2 CMC)[b]	
					54.0° (at 0.8 CMC)[b]	
SDS	1800	1730	1580	0.6	—	—

[a] Solubility in water 0.02 g dm^3.
[b] From ref. [24].

Conclusions

(1) From surface tensiometry it was shown that the substitution of the glucoside head group by lactobionamide one has no significant influence on surface area demand per molecule, A, but the CMC is shifted to higher value (from 1.48×10^{-4} to 3.38×10^{-4} M. A similar trend, i.e., a week increase of the CMC with increasing head group size is also well-known for oxyethylenated alcohols.

(2) The introduction of an additional CHOH group into the head grouping of C_{12}-MGA causes an hydrophobic effect comparable to the effect of 0.2 methylene group ($-CH_2-$) in the main alkyl chain. Additionally, this group has a small effect on the surface excess concentration of surfactant in the saturated monolayer.

(3) The foaming-stabilizing effect and rate of wetting of cotton suggest that the studied nonionic sugar surfactants share some common properties that are different from those with an oligooxyethylene grouping. The soluble aldonamide derivatives possess high foaming power and show a high foam-stabilizing tendency.

Acknowledgements Support of this work by the Polish State Committee for Scientific Research, Grant No. 3-TO9B 057 13 is gratefully acknowledged.

References

1. Weigandt H (ed) (1985) Glycolipids. Elsevier, Amsterdam
2. Frindi M, Michels B, Zana R (1992) J Phys Chem 96:8137
3. Otey FH, Mehltretter CL (1958) J Am Oil Chem Soc 35:455
4. Raaijmakers H, Zwanenburg B, Chittenden GJF (1993) J Carbohydr Chem 12:1117
5. Miethchen R, Holz J, Prade H (1992) Tetrahedron 48:3061
6. Zhang T, Marchant RE (1994) Macromolecules 27:7302
7. van Doren HA, Terpstra KR (1995) J Mater Chem 5:2153
8. Shinoda K, Kamanaka T, Kinoshita K (1959) J Phys Chem 63:648
9. Pfannemüller B, Welte W (1985) Chem Phys Lipids 39:313
10. Hall C, Tiddy GJT, Pfannemüller B (1991) Liq Cryst 9:527
11. Morley WG, Tiddy GJT (1993) J Chem Soc Faraday Trans 89:2823
12. Fuhrop J-H, Helfrich W (1993) Chem Rev 93:1565
13. Fuhrop J-H, Schnieder P, Rosenburg J, Boekema E (1987) J Am Chem Soc 109:3387
14. Fuhrop J-H, Schnieder P, Boekema E, Helfrich W (1988) J Am Chem Soc 110:2861
15. Sokołowski A (1991) J Colloid Interface Sci 147:496
16. Simister EA, Thomas RK, Penfold J, Aveyard R, Binks BP, Cooper P, Fletcher PDI, Lu JR, Sokołowski A (1992) J Phys Chem 96:1383
17. Nickel D, Nitsch C, Kürzendorfer P, von Rybinski W (1992) Prog Coll Polym Sci 89:249
18. Becher P (1967) In Schick MJ (Ed) Nonionic Surfactants. Marcel Dekker, New York, p 478
19. Arai T, Takasugi K, Esumi K (1996) Colloids Surfaces A: Physicochem Eng Aspects 119:81
20. Rosen MJ, Cohen AW, Dahanayake M, Hua XY (1982) J Phys Chem 86:541
21. Lu JR, Lee EM, Thomas RK, Penfold J, Flitsch SL (1993) Langmuir 9:1352
22. Zhang T, Marchant RE (1996) J Colloid Interface Sci 177:419
23. Ookawauchi M, Hagio M, Ikawa Y, Sugihara G, Murata Y, Tanaka M (1987) Bull Chem Soc Japan 60:2718
24. Fowkes FM (1953) J Phys Chem 57:98

Progr Colloid Polym Sci (1998) 110:204–207
© Steinkopff Verlag 1998

Formation of the lamellar phase of alkyl-benzenesulphonates from surfactant/water/electrolyte solutions

Đ. Težak
N. Jalšenjak
N. Ljubešić

Đ. Težak (✉) · N. Jalšenjak
Department of Chemistry
Faculty of Science
University of Zagreb
Marulicev trg 19
P.O. Box 163
10001 Zagreb
Croatia
E-mail: djurdjica.tezak@zg.tel.hr

N. Ljubešić
Rudjer Bošković Institute
Bijenička c.54
P.O. Box 1016
10001 Zagreb
Croatia

Abstract In the phase diagram of sodium 1'-(4)-alkylbenzenesulfonate (4SDBS)/water/$Mg(NO_3)_2$ in the diluted surfactant regime, the appearance of three regions exhibiting various micro-structures of isotropic phases is shown. The visualization of the dynamics of the lamellar phase formation from the saturated viscous isotropic surfactant/electrolyte solution was done by the light microscopy operated with differential interference contrast (DIC) using video camera. The first stages of the liquid crystalline nucleation presented the spontaneous transition from the saturated isotropic solution to the bilayer structures. The formation of bilayer cylinders, flat lamellar, and bent lamellar phases have been found. In the biphasic *isotropic + lamellar* region, a growth of the lamellar microtubuli into cylinders from the viscous isotropic phase, and afterwards the distortion of the cylinders into cones, has been observed as a spontaneous nucleation process, if the isotropic solution taken over the lamellar phase has been kept between microscopic object and cover glass at room temperature.

Critical phenomena of the partial miscibility in the ternary surfactant/water/electrolyte isotropic solution in this paper have been compared to those explained in binary surfactants/water micellar solutions in the literature.

Key words Critical Phenomena – dynamics – microheterogeneity – surfactant

Introduction

Critical phenomena of macroscopically homogeneous phases show their critical behavior concerning the formation of microheterogeneities in the micellar [1] and isotropic phases. The defects in lyotropic liquid crystal textures between the glass walls have been also described caused by changed orientation across the wall [2]. The first stages of a liquid crystalline phase nucleation as represented in contact preparations by polarized light microscopy [3], show very similar textures to the microstructures represented by transmission electron microscope (TEM) [2]. The fractal nature of the aggregation process for the molecular organization in the alkylbenzenesulphonate/electrolyte/water system was investigated using an indirect Fourier transform study of the light scattering data [4].

In this paper the results can be considered as the preliminary report of the kinetically presented behavior of mesophases in the ternary system of sodium 1'-(4)-dodecylbenzenesulfonate (4SDBS)/water/$Mg(NO_3)_2$. The kinetics of the first stages' nucleation was followed by using light microscopy operated with differential interference contrast optics (DIC LM). In comparison to the binary system without $Mg(NO_3)_2$ [5], these systems showed a shift of the biphasic region boundary, i.e. of the *isotropic + lamellar phase* region, towards lower concentration of surfactant.

In this paper, in the biphasic region, a spontaneous growth of the lamellar microtubuli from the viscous isotropic phase into cylinders, and afterwards, their distortion into cones, has been observed caused by homogeneous nucleation from the highly viscous isotropic phase, if the samples have been kept between microscopic object and cover glass at room temperature. It can be considered that the viscous phase is macroscopically isotropic, but it shows the heterogeneity at the micro-level.

Experimental

Materials

The analytically pure sodium 1'-(4)-alkylbenzenesulphonate (4SDBS) was obtained by courtesy of the Henkel Co.; it was used without further purification. Magnesium-nitrate was used from "Merck", Darmstadt; the water solution of magnesium-nitrate was standardized using standard procedure. Doubly distilled water was used for all of the experiments.

Methods

The samples for phase diagram were prepared in 5 mm NMR-tubes, which were then flame sealed; they were mixed by continuous shaking in a bench-type electric shaker, vortexed, and then kept standing in a thermostat at the desired temperature (between 20 and to 70 °C) for several days. The isotropic and opalescent phases were visually characterized with crossed polars. The determination of liquid crystalline phases was done by microscopic textures using a Zeiss polarization light microscope with a Mettler FP 82 hot stage and FP 80 central processor.

The isotropic part of the biphasic *isotropic + lamellar* samples were observed at room temperature using an Axiovert 35 Zeiss polarized light microscope operated with differential interference contrast optics (DIC LM). The viscous isotropic solution was pipetted onto a glass slide and covered with a cover-slip, i.e., it was sandwiched between two glasses to form a thin film. Several minutes after the sample was insert into the light beam in the microscope, the nucleation of the liquid crystalline phase started, and it was recorded by a video-camera.

Results and discussion

The phase diagram of (4SDBS)/water/Mg(NO$_3$)$_2$ in the diluted surfactant regime, including the constant electrolyte concentration of $[Mg(NO_3)_2] = 6 \times 10^{-2}$ mol dm^{-3},

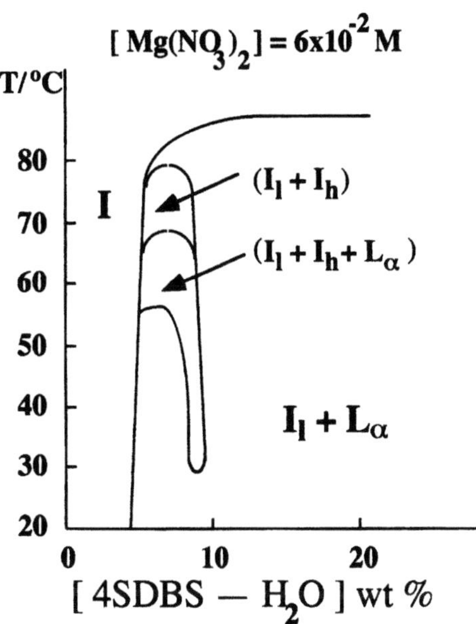

Fig. 1 Phase diagram of the system (4SDBS)/water/Mg(NO$_3$)$_2$ in the diluted surfactant regime

is presented in Fig. 1. The water-rich corner is characterized by an isotropic phase, that borders on the *isotropic + lamellar*, $(I_l + L_\alpha)$, phase region. The appearance of three regions exhibiting the various micro structures of isotropic phases is temperature dependent. By increasing the temperature, the textures of two isotropic phases can be observed in the polarizing microscope, presumably a low-viscous (I_l), and a high-viscous phase (I_h).

To be sure that the micro-phases are not nematic, the ^{2}H NMR pattern was investigated, but only the isotropic maximum was recorded. It can be assumed that one of the isotropic phases, I_h, is supersaturated. Then, it would be reasonable, besides of the concentration and temperature extent, to consider the critical limit for the formation of a lamellar phase could be also the neighborhood of a solid interface.

The isotropic phase over the lamellar phase was sandwiched between two glasses into a DIC LM. The sample shows a fast nucleation of the microtubuli when observed between object and cover glass. The close neighborhood of two parallel glass surfaces, that could represent the solid/liquid boundaries, to the thin film of the saturated micellar solution in between them, could be considered as a nucleation seed. Before the nucleation of microtubuli starts, the parts of the micellar solution generate as a different phase, also isotropic but presumably supersaturated. The nucleation of the new heterogeneous phase begins from these high viscous (I_h) supersaturated parts within the isotropic solution. These parts are represented

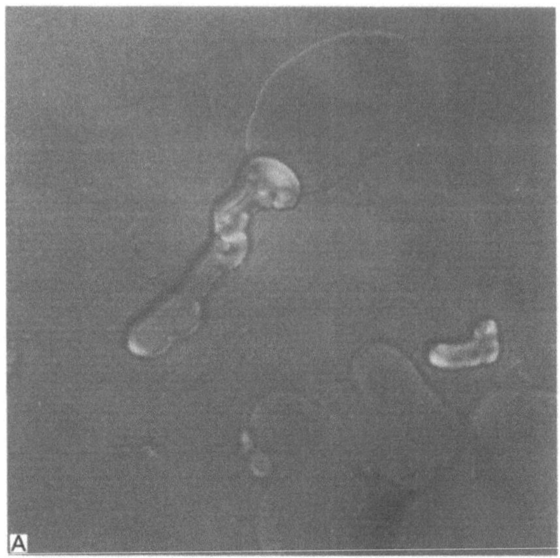

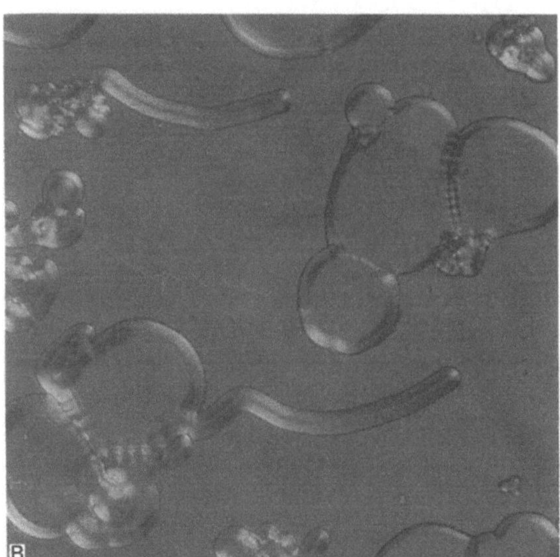

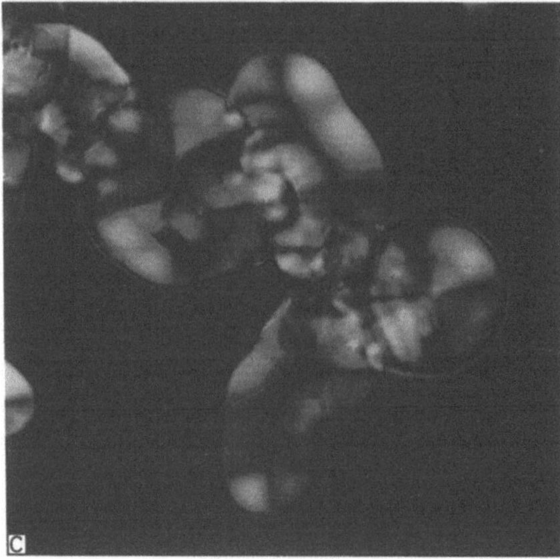

in Fig. 2A showing different structures within the isotropic phase. The first appearance of microtubuli from these supersaturated regions can be seen several minutes after the sample preparation. Figure 2B represents already well-formed cylinders. Two hours after preparation of samples, well-developed lamellar cylinders with focal cones formed at the bending parts of the cylinders, are represented in Fig. 2C. The microtubuli show the formation of focal cones at the bending positions, due to the folding of the lamellar bilayers.

These microheterogeneous isotropic phases can be considered as supersaturated parts within the solution, i.e., it can be considered macroscopically homogeneous, but microscopically heterogeneous. This does not mean that the system is unequilibrated, since such an effect appeared after repeating the experiment several times in the flame sealed NMR-tubes.

The dynamics of the formation of lamellar phases followed from the aqueous solutions of simple-, or double-tailed alkylbenzenesulphonates (ABS) showed that the nucleation of lamellar phases undergoes the following process: the association of surfactant molecules into micelles, transition either of flat lamellar to bent lamellar bilayers forming cylinders, or bending of flat lamellar into focal conic structure.

Conclusion

The presented results concerning the growth of lamellar phases from the homogeneous supersaturated solution have to be considered as preliminary, and it seems to be worthwhile to investigate the dynamics of ABS liquid crystal formation in the terms of micro heterogeneous structures within a macroscopically imaged systems. Although this is the first report of the influence of micro heterogeneity of the micellar solution on the formation of a new phase, it is likely that such phenomenon can occur in the phases with other surfactants. The further investigations have to be done.

Fig. 2 DIC LM photomicrographs of the sample of 4SDBS (20%) in water containing $6 \times 10^{-2}\,\mathrm{mol\,dm^{-3}}$ [Mg(NO$_3$)$_2$] at room temperature: (A) several minutes after the sample preparation. Crossed polars, λ-plate, magnification $120\times$; (B) after 13.5 min. Crossed polars, λ-plate, magnification $120\times$; (C) 2 h after the sample preparation. Crossed polars, λ-plate, magnification $300\times$

Progr Colloid Polym Sci (1998) 110:204–207
© Steinkopff Verlag 1998

References

1. Ockelford J, Timini BA, Narayan KS, Tiddy GJT (1993) J Phys Chem 97: 6767–6769
2. Huang Y, Shen JR (1995) Liquid Crystals 19:313–318
3. Težak Đ, Martinis M, Punčec S, Fischer-Palković, Popović S (1996) Prog Colloid Polym Sci 100:136–138
4. Težak Đ, Punčec S, Martinis M (1997) Liquid Crystals 23:17–25
5. Težak Đ, Hertel G, Hoffmann H (1991) Liquid Crystals 10:15–27

Progr Colloid Polym Sci (1998) 110:208–213
© Steinkopff Verlag 1998

F. Bordi
C. Cametti
A. Di Biasio
G. Onori

Effect of different solvents on the low-frequency electrical conductivity of water-in-oil microemulsions

F. Bordi*
Sezione di Fisica Medica
Dipartimento di Medicina Interna
Università di Tor Vergata
Rome, Italy

C. Cametti* (✉)
Dipartimento di Fisica
Università di Roma "La Sapienza"
Rome, Italy
E-mail: cesare.cametti@roma1.infn.it

A. Di Biasio*
Dipartimento di Matematica e Fisica
Università degli Studi di Camerino
Italy

G. Onori**
Dipartimento di Fisica
Università degli Studi di Perugia
Italy

*Istituto Nazionale di Fisica della Materia
(INFM) Unitá di Roma I
**Istituto Nazionale di Fisica della Materia
(INFM) Unitá di Perugia

Abstract The low-frequency electrical conductivity of water-in-oil micro-emulsions, built up with different solvents (carbon tetrachloride, n-pentane, n-heptane and n-decane), has been measured in the temperature interval from 5 to 60 °C, for different values of the water-to-surfactant ratio W, from 2 to 20. The results have been analyzed on the basis of the charge fluctuation theories recently proposed by Eicke, Hall and Halle and the effects of short-range attractive interactions due to surfactant tail interpenetration between two adjacent water droplets have been evaluated.

Key words Microemulsions – electrical conductivity – charge fluctuation

Introduction

Microemulsions are homogeneous, transparent, isotropic and thermo-dynamically stable solutions composed of water and oil phases separated by a monomolecular layer of amphiphilic molecules [1]. These systems exhibit a large variety of structural rearrangements, consisting in an inverted micellar phase (W/O microemulsions), a bicontinuous phase due to water-in-oil channels, a lamellar phase, up to an inverted bicontinuous phase due to oil-in-water channels and including clustering, percolation and critical phenomena [2].

Within certain ranges of composition and temperature, a water-in-oil microemulsion can be effectively considered as a heterogeneous two-component system, made up of conducting spherical droplets of water, coated by a mono-layer of surfactant molecules and uniformly dispersed in a continuous, non-conducting, oil medium.

The electrical conductivity of these systems has been extensively studied both below the percolation threshold [3], where the charged water droplets contribute to conduction by means of Brownian movement, and close and above percolation [4], where charges jump between different clusters or transient merging of connected droplets occurs.

Below percolation and at low-to-moderate volume fractions, where the microemulsion is viewed as an ensemble of charged water droplets, the transport mechanism responsible of the observed electrical conductivity (of the order of 10^{-4}–$10^{-6}\,\Omega^{-1}\,\text{m}^{-1}$) is the migration of charged droplets under the influence of an external electric field [5] and reflects characteristics inherent to the dynamics of the single droplet.

There is strong support [6, 7] that the mechanism of charge fluctuation, which produces charged droplets starting from a complete electroneutral system, is due to transient fusion of two adjacent droplets to form a short-lived "dimer droplet", where ions (counterions) can randomly redistribute, giving rise to separate charged droplets, when two new isolated droplets are formed.

Within this context, different treatments of charged fluctuation, including the effects due to droplet–droplet interactions, have been proposed [8–11], which account for the observed electrical conductivity, both in large and small droplet regime.

Although these theories provide a quite reasonable description of the electrical conductivity observed in microemulsion systems, the effect of an attractive short-range droplet–droplet interaction due to the overlapping of surfactant tails is not generally taken into account. As pointed out by Luisi et al. [12], these short-range attractive forces, which increase as the temperature limit of the single-phase region is approached, depend on the size of the droplet and on the chemical nature of the oil phase (essentially, the hydrocarbon chain length of alkanes) and cause a greater penetration of the two droplets, resulting in a longer contact time and thus favoring the charge exchange.

In this note, we report on low-frequency electrical conductivity measurements of water-in-oil microemulsion systems, formed by using four different organic solvents (n-pentane, n-heptane, n-decane and carbon tetrachloride) at different water-to-surfactant molar ratio W, in the temperature range from 5 to 60 °C (up to the stability limit of the single-phase system).

The experimental data have been analyzed in the light of the currently stated charge fluctuation theories and the parameters characterizing the short-range interactions due to tail interpenetration have been evaluated for the different solvents investigated.

The results presented here give further support to the identification of the short-range interdroplet interaction as due to the penetration of the surfactant tails of adjacent droplets.

A summary of charge fluctuation models: theoretical background

Below the percolation threshold, where the system is composed of isolated water droplets uniformly dispersed in the oil phase, the electrical conductivity in the low-frequency regime is due to charge fluctuation, which leads to deviation from electroneutrality between two transiently approaching droplets. Although the net charge rarely exceeds one or two elementary charges, charge fluctuation, according to the equilibrium

$$A + A \Leftrightarrow A^+ + A^- \tag{1}$$

allows the conductivity of the microemulsion (of the order of 10^{-4}–$10^{-6}\,\Omega^{-1}\,\text{m}^{-1}$) to be several folds higher than that of the oil phase (of the order of 10^{-10}–$10^{-13}\,\Omega^{-1}\,\text{m}^{-1}$).

Considering only the Born energy contribution, Eicke and coworkers [8] predicted that the reduced conductivity σ/Φ (normalized to the fractional volume Φ of the dispersed water droplets) follows the equation

$$\frac{\sigma}{\Phi} = \frac{\varepsilon_0 \varepsilon K_{\mathrm B} T}{2\pi\eta R_{\mathrm d}^3}, \tag{2}$$

where ε and η are the permittivity and the viscosity of the oil phase, respectively, $K_{\mathrm B}T$ the thermal energy and $R_{\mathrm d}$ the hydrodynamic radius of the surfactant-coated water droplet and ε_0 the dielectric constant of free space.

A refinement of this model has been proposed by Kallay et al. [11], by distinguishing between the Born radius $R_{\mathrm c}$ that corresponds to the water core, where ions are confined, and the Stokes radius $R_{\mathrm d}$ that defines the size of the moving droplet. As pointed out by Kallay et al. [11], these radii differ of the thickness δ of the surfactant tail, causing a significant better agreement with experimental results, especially for small droplet radius. As a consequence, Eq. (2) is replaced by

$$\frac{\sigma}{\Phi} = \frac{\varepsilon_0 \varepsilon K_{\mathrm B} T}{2\pi\eta} \frac{R_{\mathrm d} - \delta}{R_{\mathrm d}^4}, \tag{3}$$

that predicts the conductivity maximum as a function of droplet size, as observed experimentally.

Following the thermodynamic fluctuation theory of Eicke et al. [8], Hall [9] proposed a more general theory which includes Eq. (2) as a limiting case. Within this model, the reduced electrical conductivity σ/Φ is given by

$$\frac{\sigma}{\Phi} = \frac{e^2}{8\pi^2\eta R_{\mathrm d}^4} \frac{\sum_{z=-\infty}^{z=+\infty} z^2 \exp(-z^2 e^2/8\pi\varepsilon_0 \varepsilon K_{\mathrm B} T R_{\mathrm d})}{\sum_{z=-\infty}^{z=+\infty} \exp(-z^2 e^2/8\pi\varepsilon_0 \varepsilon K_{\mathrm B} T R_{\mathrm d})}, \tag{4}$$

where ze is the electric charge on each droplet. As a result, the conductivity passes through a maximum and then decreases as the inverse droplet volume. This effect may explain the observed decrease of the microemulsion conductivity for small droplets.

We have already shown [3, 4] that in the case of water/AOT/decane microemulsions, Eq. (4) predicts values in very good agreement with the measured ones in the region of low Φ, below the percolation threshold. In particular, we found that, on average, each droplet bears one or two electrical charge so that only small deviations from an overall neutrality are expected in order to justify the observed behavior.

Finally, following a more rigorous approach, Halle et al. [10, 13] proposed a more general theory of charge fluctuations in the presence of interactions, where the droplet charge distribution was obtained by minimizing the free energy density of the system, resulting in a distribution $P(z)$ of the net droplet charge ze given by

$$P(z) = \frac{\exp(-(\mu_\mathrm{I}(z) + \mu_\mathrm{S}(z))/K_\mathrm{B}T)}{\sum_z \exp(-(\mu_\mathrm{I}(z) + \mu_\mathrm{S}(z))/K_\mathrm{B}T)}, \tag{5}$$

where μ_I and μ_S are two different contributions to the chemical potential. While $\mu_\mathrm{S}/K_\mathrm{B}T$ is exactly the term that appears in the Hall expression, [Eq. (4)], μ_I is calculated by Halle et al. [13] by describing the microemulsion as a "conventional" electrolyte and using the restricted primitive model [RPM] of an electrolyte to obtain the thermodynamics of the system. Within this framework, μ_I is given by

$$\mu_\mathrm{I} = -z^2 B \frac{l_\mathrm{B}}{2R_\mathrm{d}} K_\mathrm{B}T, \tag{6}$$

where l_B is the Bjerrum length

$$l_\mathrm{B} = \frac{e^2}{4\pi\varepsilon_0\varepsilon K_\mathrm{B}T} \tag{7}$$

and B a characteristic parameter of RPM theory that depends on the ratio $R_\mathrm{d}/l_\mathrm{D}$, with l_D the Debye length.

Substituting μ_I and μ_S into Eq. (5) yields

$$P(z) = \frac{\exp(-z^2\alpha)}{\sum_z \exp(-z^2\alpha)}, \tag{8}$$

where α is the dimensionless coupling parameter given by

$$\alpha = \frac{l_\mathrm{B}}{2}\left(\frac{1}{R_\mathrm{c}} - \frac{B}{R_\mathrm{d}}\right). \tag{9}$$

However, at small droplet distances, droplets interact with a short-range attractive potential that can be modeled in various ways. The physical mechanism responsible for the attractive potential [14] can be due to the amphiphile chain interpenetration of two adjacent droplets, the penetration length being limited by the double-chain structure of the surfactant. As pointed out by Huang et al. [14], the attractive interaction arises when the surfactant molecules on adjacent droplets interpenetrate each other realeasing solvent molecules and thus resulting in a entropic gain. A particularly simple model that repro-

duces in a rather good quantitative way the experimentally measured percolation line [15] has been proposed by Baxter [16]. In this model, the interaction potential is given by an hard-sphere repulsion and an attractive square-well potential. More recently, Rasaiah et al. [17–20] studied the thermodynamic properties of a system governed by a potential built up by adding to the Baxter potential a long-range Coulombic potential (sticky electrolyte model [SEM]). In this case, the chemical potential μ_I can be written as

$$\mu_\mathrm{I} = -z^2 B' \frac{l_\mathrm{B}}{2R_\mathrm{d}} K_\mathrm{B}T - \frac{\langle N\rangle\zeta}{2} K_\mathrm{B}T, \tag{10}$$

where $\langle N\rangle$ is the average number of droplets "sticked" to each other, $\zeta K_\mathrm{B}T$ the depth of the attractive well and B' is a quantity depending, in a rather complicated way, on the Debye length l_D, the droplet radius R_d and $\langle N\rangle$.

Since SEM model reduces to the RPM model in absence of the short-range attractive potential, in this limit Eq. (10) reduces to Eq. (6) and B' equals B of Eq. (6).

In Eq. (10), the second term represents the contribution due to the short-range attractive potential, but, since ζ is independent of the charge ze, this term does not contribute to the distribution function $P(z)$ [Eq. (5)], while the first one contains both the Coulombic interaction contributions and those deriving from the short-range interaction, through B'. On the other hand, these microemulsion systems show a lower critical solution temperature and this means that short-range interactions should increase the temperature [14]. This fact can be used to separate in the quantity B', in a phenomenological way, the contributions due to the short-range attractive potential.

By expanding the quantity B' to the first order in T, the short-range interactions are included in the coefficient of the linear term, only, so that the coupling parameter α of Eq. (9) can be rewritten as

$$\alpha = \frac{l_\mathrm{B}}{2}\left(\frac{1}{R_\mathrm{c}} - \frac{a + bT}{R_\mathrm{d}}\right) \equiv \frac{l_\mathrm{B}}{2}\left(\frac{1}{R_\mathrm{c}} - \frac{a}{R_\mathrm{d}}\right) - \Psi. \tag{11}$$

If the charge fluctuation is limited to a single elementary charge ($z = \pm 1$), the final expression of the reduced conductivity σ/Φ is given by

$$\frac{\sigma}{\Phi} = \frac{e^2}{8\pi^2\eta R_\mathrm{d}^4}$$

$$\times \frac{2\exp(-(e^2/8\pi\varepsilon\varepsilon_0 K_\mathrm{B}T)(1/R_\mathrm{c} - a/R_\mathrm{d}) + \Psi)}{1 + 2\exp(-(e^2/8\pi\varepsilon\varepsilon_0 K_\mathrm{B}T)(1/R_\mathrm{c} - a/R_\mathrm{d}) + \Psi)}. \tag{12}$$

From a direct comparison with experiments, the two parameters a and Ψ can be properly evaluated. In particular, by varying the chemical characteristics of the oil phase of the microemulsion, the short-range interdroplet interaction parameter Ψ can be studied in different experimental conditions.

Experimental

The microemulsions were prepared by mixing appropriate amounts of water and oil in presence of sodium bis(2-ethylhexyl) sulfosuccinate [AOT]. This surfactant is an anionic molecule consisting of a double-tailed alkyl chain and a polar head-group composed of a sulfonate ion, two ether groups and a sodium ion as a counterion. AOT was purchased from Sigma Chemical Co. and used without further purification.

Microemulsion were prepared by using four different solvents, i.e. carbon tetrachloride, n-pentane, n-heptane and n-decane in order to investigate the effect of different disperse-phase solvents on the water droplet interactions.

Distilled and deionized water (electrical conductivity was about $1–5 \times 10^{-6}\,\Omega^{-1}\,cm^{-1}$ at room temperature) was used throughout the experiments. For each solvent investigated, the appropriate amount of water was added to the solution with a microsyringe to obtain a value of the molar ratio of water-to-surfactant, $W = [H_2O]/[AOT]$, ranging from 2 to about 15 (in the case of decane, W has been varied up to 25).

All the microemulsions were prepared at a constant water + surfactant volume fraction Φ ($\Phi = 0.10$), calculated by assuming an ideal mixing behavior.

The electrical conductivity of the microemulsions was measured at a frequency of 10 kHz (low-frequency limit) by means of a Hewlett–Packard mod. 4192A Impedance Analyzer, in connection with a temperature-controlled parallel-plane capacitance cell. The cell constants were determined by calibration using standard solutions of known conductivity and permittivity. The temperature was varied from 5 °C to 60 °C within 0.1 °C.

In these systems, the radius R_c of the water core can be systematically varied by varying the molar ratio W of water to surfactant, according to the relationship [12, 21]

$$R_c = \left(\frac{3v}{a_0}\right)W + \frac{3v_H}{a_0},\tag{13}$$

where v is the volume of an individual water molecule, v_H the average volume occupied by a head group and a_0 is the area occupied by the surfactant head group.

The validity of the above equation, assuming that the surfactant is totally located at the oil–water interface, has been verified, up to $W = 50$, by means of different experimental techniques, such as small-angle scattering measure-

ments [22, 23], X-ray diffraction [24] and fluorescence quenching kinetic studies [25].

The Stokes radius R_d assumes the value

$$R_d = R_c + \delta\tag{14}$$

with $\delta = 1.05$ nm in the case of AOT surfactant (length of an AOT molecule without SO_3^- group) [26, 27]. Although at low values of W (low water content), water molecules are immobilized in the neighborhood of the surfactant head groups and the validity of Eq. (9) can be questionable, as the amount of water is increased, over the hydration of the head groups, water begin to form a pool in the cores of the droplets and their size become independent of the water concentration.

Results and discussion

Figure 1 shows a typical behavior of the conductivity σ as a function of temperature, from 5 °C to 60 °C, for the microemulsions investigated at a fixed water-to-surfactant molar ratio, $W = 8$. As W increases, in the high-temperature range, the conductivity deviates from an approximately linear behavior, indicating the approaching of the percolation threshold. In the case of n-pentane, the measurements were confined at a narrow temperature range, owing to the lower boiling point of this disperse-phase solvent.

In order to investigate the effect of different solvents on the conductivity properties of the four microemulsion systems investigated, the reduced conductivities σ/Φ as

Fig. 1 The electrical conductivity σ as a function of temperature T for different microemulsion systems: ($\triangledown$): $H_2O/AOT/n$-decane; ($\square$): $H_2O/AOT/n$-heptane; ($\triangle$): $H_2O/AOT/n$-pentane; ($\circ$): $H_2O/AOT/$ carbon tetrachloride. The fractional volume of coated water droplet is $\Phi = 0.1$ and the water-to-surfactant molar ratio is $W = 8$

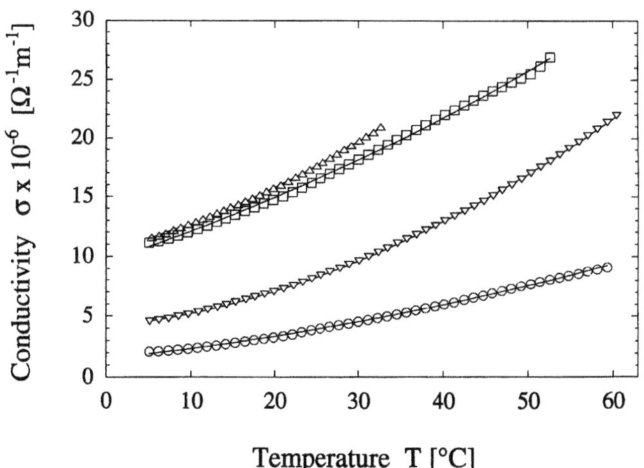

a function of water-to-surfactant molar ratio W, at some selected temperatures, are shown in Figs. 2 to 5. The predictions on the basis of Eq. (12), shown as the solid lines in Figs. 2 to 5, are calculated using the experimental temperature dependence of ε and η and values of R_c and R_d as a function of W given by Eqs. (13) and (14).

The attractive interaction contribution due to the mutual interpenetration of surfactant tails is taken into account by the parameter Ψ (in Eqs. (10) and (11)), whose value has been derived by means of a non-linear least-squares fitting minimization of Eq. (12) to the measured conductivity values. As expected, this parameter, that describes the attractive interaction, depends on the solvent characteristics and, in particular, on the length of the solvent chains, whereas the parameter a, that takes into account the Coulombic interactions, is largely independent of the solvent.

The behavior of the interaction parameter Ψ as a function of W for the four different solvents investigated is shown in Fig. 6. As can be seen, the different solvents behave differently and induce different short-range interactions between water droplets. It must be noted that, whereas the alkanes used show a maximum in the interaction, whose strength and position depend on the molar ratio W, in the case of carbon tetrachloride, there is a small, approximately linear, increase of Ψ as the droplet sizes are increased. The values of Ψ suggest that the well depth in the short-range potential increases approximately with the length of the molecule solvent chains, owing to the increase in interpenetration volumes, and increases with temperature. These findings are in agreement with the identification of the interdroplet interaction as due to the penetration of the surfactant tails over short distance. As pointed out by Huang et al. [14], the

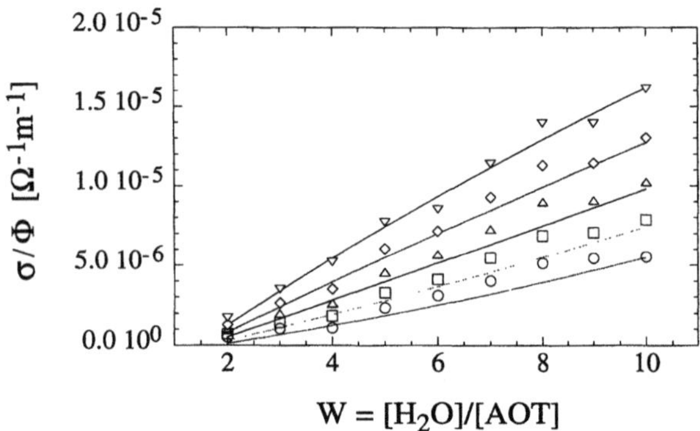

Fig. 2 The reduced conductivity σ/Φ as a function of the water-to-surfactant molar ratio W for H_2O/AOT/carbon tetrachloride microemulsion at different temperatures: (○): $T = 10\,°C$; (□): $T = 20\,°C$; (△): $T = 30\,°C$; (◇): $T = 40\,°C$; (▽): $T = 50\,°C$. The full lines represent the calculated values according to Eq. (12)

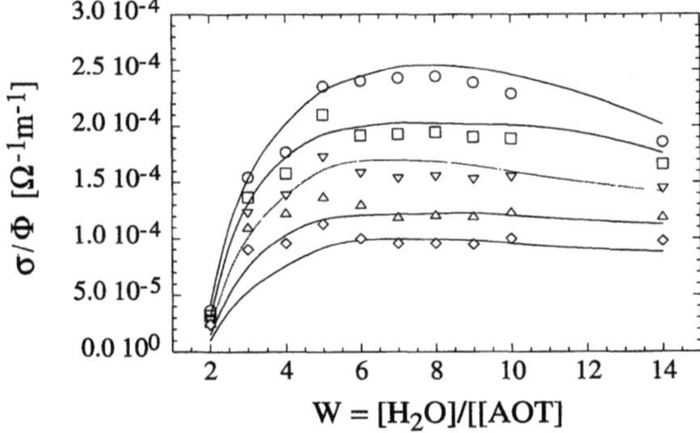

Fig. 4 The reduced conductivity σ/Φ as a function of the water-to-surfactant molar ratio W for H_2O/AOT/n-heptane microemulsion at different temperatures: (◇): $T = 10\,°C$; (△): $T = 20\,°C$; (▽): $T = 30\,°C$; (□): $T = 40\,°C$; (○): $T = 50\,°C$. The full lines represent the calculated values according to Eq. (12)

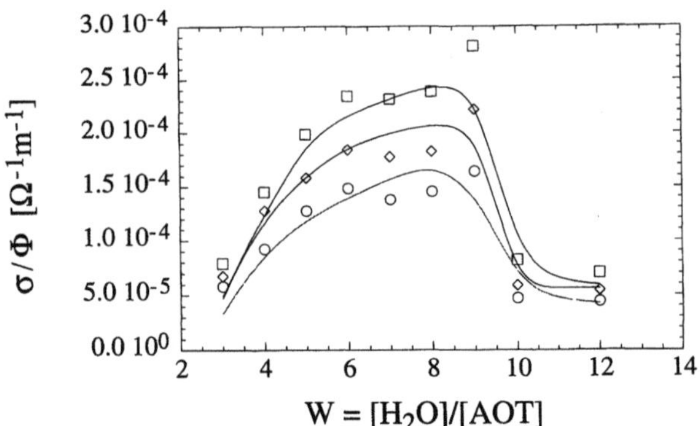

Fig. 3 The reduced conductivity σ/Φ as a function of the water-to-surfactant molar ratio W for H_2O/AOT/n-pentane microemulsion at different temperatures: (○): $T = 10\,°C$; (◇): $T = 20\,°C$; (□): $T = 30\,°C$. The full lines represent the calculated values according to Eq. (12)

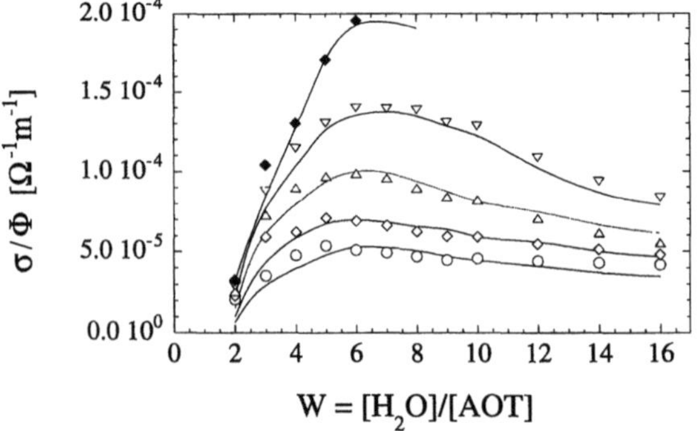

Fig. 5 The reduced conductivity σ/Φ as a function of the water-to-surfactant molar ratio W for H_2O/AOT/n-decane microemulsion at different temperatures: (○): $T = 10\,°C$; (◇): $T = 20\,°C$; (△): $T = 30\,°C$; (▽): $T = 40\,°C$; (◆): $T = 50\,°C$. The full lines represent the calculated values according to Eq. (12)

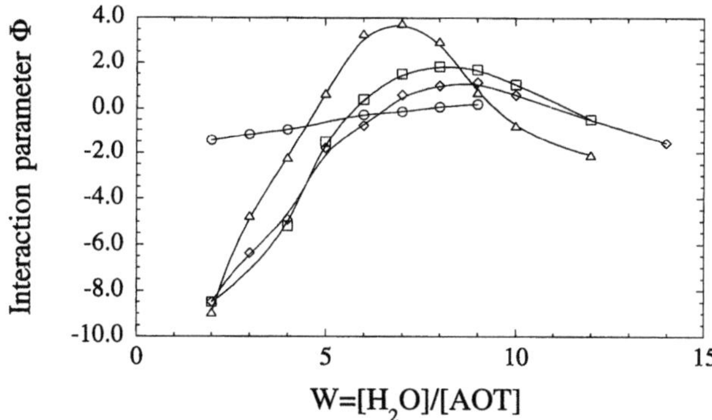

Fig. 6 The short-range interaction parameter Ψ as a function of the water-to-surfactant molar ratio W for the different microemulsion systems investigated: ($\triangle$) $H_2O/AOT/n$-decane; ($\diamond$) $H_2O/AOT/n$-heptane; ($\square$) $H_2O/AOT/n$-pentane; ($\circ$) H_2O/AOT/carbon tetrachloride

surfactant molecules are appropriately packed to interact with those of an adjacent droplet, in contrast to the solvent molecules that are neither optimally oriented nor packed to interact with the surfactant tails. As the temperature is increased, surfactant molecules tend to interact with each other, rather that with the solvent molecules and this is favored as much as the length of solvent chain is small.

The presence of a maximum in the behavior of Ψ as a function of W can be due to the fact that the interactions depend on the droplet distance and, in our measurements, we have followed a pathway where the fractional volumes Φ is maintained constant and hence the distance between water droplets varies with the values of W, the volume of a droplet increases as W^3 and the number of droplets decreases as W^{-2}. Moreover, in the region of low W, in correspondence with small droplets, the interaction parameter increases rather sharply, starting from negative values. This behavior, almost partially, can be due to droplet size polydispersity or to the fact that water at low W is highly immobilized and the ion exchange is favored only when the water mobility approaches that of the bulk water with increasing W [28].

References

1. Chen SH, Huang JS, Tartaglia P (eds) (1991) Structure and dynamics of strongly interacting colloids and supramolecular aggregates in solution. Nato ASI Series, Vol 369; Degiorgio V, Corti M (eds) (1985) Physics of amphiphiles: Micelles, vesicles and microemulsions. Proc Int School of Physics, E. Fermi Course XC, North-Holland, Amsterdam
2. Jada A, Lang J, Zana R (1989) J Phys Chem 93:10
3. Di Biasio A, Cametti C, Codastefano P, Tartaglia P, Rouch J, Chen SH (1993) Phys Rev E 47:4258; Cametti C, Codastefano P, Tartaglia P, Chen SH, Rouch J (1992) Phys Rev A 45:5358
4. Bordi F, Cametti C, Codastefano P, Sciortino F, Tartaglia P, Rouch J (1996) Progr Colloid Polym Sci 100:170
5. Meier W, Eicke HF (1996) Curr Opinion Colloid Interface Sci 1:279
6. Eicke HF, Shepherd JCW, Steinemann A (1976) J Colloid Interface Sci 56:68

7. Fletcher PDI, Howe AM, Robinson BH (1987) J Chem Soc Faraday Trans I 83:985
8. Eicke HF, Borkovec M, Das-Gupta B (1989) J Phys Chem 93:314
9. Hall DG (1990) J Phys Chem 94:429
10. Halle B (1990) Progr Colloid Polym Sci 82:211
11. Kallay N, Chittofrati A (1990) J Phys Chem 94:4755
12. Luisi PL, Giomini M, Pileni M, Robinson BH (1988) Biochim Biophys Acta 947:209
13. Halle B, Björling M (1995) J Phys Chem 103:1655
14. Huang JS, Safran SA, Kim MW, Grest GS (1984) Phys Rev Lett 53:592
15. Chiew YC, Glandt ED (1983) J Phys A 16:2599
16. Baxter RJ (1968) J Chem Phys 49:2770
17. Lee SH, Rasaiah JC, Cummings PT (1985) J Chem Phys 83:317
18. Rasaiah JC, Lee SH (1985) J Chem Phys 83:5870

19. Rasaiah JC, Lee SH (1985) J Chem Phys 83:6396
20. Lee SH, Rasaiah JC (1987) J Chem Phys 86:983
21. Kotlarchyk M, Chen SH, Huang JS, Kim MW (1984) Phys Rev A29:2054
22. Kotlarchyk M, Chen SH, Huang JS (1982) J Phys Chem 86:3273
23. Robinson C, Toprakcioglu C, Dore JC, Chieux P (1984) J Chem Soc Faraday Trans I 80:13
24. Pileni MP, Zemb T, Petit C (1985) Chem Phys Lett 118:414
25. Bridge NJ, Fletcher PDI (1983) J Chem Soc Faraday Trans I 79:2161
26. Kotlarchyk M, Chen SH (1985) J Phys Chem 89:4382
27. Dijk MAV, Joosten JGH, Levine YK, Bedeaux D (1989) J Phys Chem 93:2506
28. Wong M, Thomas JK, Nowak T (1977) J Am Chem Soc 99:4730

Progr Colloid Polym Sci (1998) 110:214–219
© Steinkopff Verlag 1998

V. Buckin
E. Kudryashow
S. Morrissey
T. Kapustina
K. Dawson

Do surfactants form micelles on the surface of DNA?

V. Buckin (✉)· E. Kudryashow
S. Morrissey · K. Dawson
Centre for Colloid Science and Biomaterials
Department of Chemistry
University College Dublin
Belfield, Dublin 4
Ireland

T. Kapustina
Institute of Theoretical and Experimental
Biophysics of Russian Academy of Sciences
Pushchino, Moscow region, 142292
Russia

Abstract The complex formation between the cationic surfactant, dodecyltrimethylammonium bromide ($C_{12}TAB$), and short fragments of Calf Thymus DNA, 220 base pairs, has been studied by using a combination of high-precision ultrasonic velocity and density measurements. The compressibility and the volume effects of the binding of $C_{12}TAB$ with DNA are 139 $cm^3 mol^{-1} Gpa^{-1}$ and 9.0 $cm^3 mol^{-1}$, respectively. In addition to that we obtained the compressibility effect in the micelle formation for three surfactant molecules $C_{12}TAB$, $C_{14}TAB$ and $C_{16}TAB$. These effects are 127.5, 148.9 and 169.2 $cm^3 mol^{-1} Gpa^{-1}$ for $C_{12}TAB$, $C_{14}TAB$ and $C_{16}TAB$, respectively. The high value of the compressibility effect in the binding of the $C_{12}TAB$ surfactant to DNA cannot be explained by hydration changes only and indicates the formation of a structure with a highly compressible core. This conclusion, and the close values of the compressibility effect in the formation of the $C_{12}TAB$ micelles in solution, and in the binding of $C_{12}TAB$ to DNA, demonstrates the formation of micelle-like aggregates of $C_{12}TAB$ molecules on the DNA surface.

Key words DNA – dodecyltrimethylammonium bromides – DNA–surfactant complex – ultrasonic velocity – compressibility – hydration

Introduction

The complexes of cationic surfactants with DNA play an important role in the construction of liposomal genetic delivery systems [1–3]. In previous studies it was demonstrated that the cationic surfactants bind to the negatively charged DNA molecules in dilute solutions cooperatively [4–7]. This phenomenon is similar to the binding of surfactants to synthetic polymers where the process of association of the surfactants with opposite charged polymers is considered to be driven by electrostatic interactions between charged groups of the surfactant and polyelectrolyte, and by hydrophobic interactions between the surfactants bound with the polymer. This results in a formation of micelles bounded with synthetic polymers [8,9]. However, the structure of the complexes between DNA and surfactants is not clear. The fundamental difference between the DNA molecule and most of the other synthetic polymers is its stiffness. This molecule is a rigid cylinder with a diameter of about 2 nm and with persistence length of more than 30 nm. Therefore, it cannot be wrapped easily around the surfactant micelles as in the case with the synthetic polymers [9].

In our present work high-resolution ultrasonic velocity measurements were used to evaluate the structure of the complex of DNA with the cationic surfactant, dodecyltrimethylammonium bromide ($C_{12}TAB$). The concentration increment of ultrasonic velocity in solutions is determined by the compressibility of water in the hydration shell of the solute molecules and, in the case of molecular aggregates, by their intrinsic compressibility. The ultrasonic velocity measurements have been successfully applied to characterize the physical state of water in the hydration shells of biopolymers and their components [10–15]. This technique was also used in studies of the physical state of liposomes [15] and in the incorporation of protein molecules into bilayer lipid membranes [14]. In the present work we evaluated, from the ultrasonic velocity measurements, the compressibility effect on the binding of $C_{12}TAB$ to the short fragments of DNA molecules. we found that this effect is the same as the compressibility change in the micelle formation of the same surfactant. This fact as well as the value of the compressibility effect in the surfactant–DNA binding can be explained only by the formation of micelle-like aggregates of surfactant molecules on the DNA surface, which have a highly compressible internal core.

Materials

The surfactants dodecyltrimethylammonium bromide ($C_{12}TAB$), tetradecyltrimethylammonium bromide ($C_{14}TAB$) and cetyltrimethylammonium bromide ($C_{16}TAB$) were obtained from Lancaster Synthesis Ltd (UK). The samples of the surfactants were recrystallized at least twice from an ethanol-ethyl acetate mixture. They were dried in a vaccum oven at 333 K over 48 h before use. All solutions were prepared by weight using degassed water obtained from a Millipore Super-Q-System. Calf thymus DNA from Sigma was used in the current work. Additional purification of the DNA was made using the standard phenol–ethanol extraction procedure. The samples of short fragments of DNA were obtained as described previously [16]. Briefly, the sonication of the DNA solutions after purification, at a concentration of $2\ mg\ cm^{-3}$, in 50 mM sodium citrate buffer, 2 mM EDTA, and at pH 7.0 has been made using an Ultrasonic Processor VC 50 (Sonics & Materials Inc.) at low intensity. The sonicated solution of DNA was precipitated into fractions with acetone. Finally the DNA solution was dialysed against 5 mM HEPES buffer, 1 mM EDTA and 10 mM NaBr, and was then used for the experiments reported here. The average length of the fragments determined by static light scattering and by agarose gel electrophoresis was 220 base pairs. The concentration of DNA was determined by a spectroscopic method, using the molar extinction coefficient $\varepsilon_{260} = 6600\ l\ mol^{-1} cm^{-1}$. All measurements were performed at 298.15 K.

Ultrasonic measurements

Ultrasonic velocity and absorption measurements were made at 7.5 MHz using the resonator technique as described earlier [11, 13]. Ultrasonic resonator cells with a minimum sample volume of $0.8\ cm^3$ were thermostated at 298.15 K with a temperature stability of 0.01 K. The reproducibility of the relative measurements of ultrasound velocity was $\pm 10^{-4}\%$.

The main experimental observable in ultrasonic measurements is the concentration increment of ultrasonic velocity, A, which is determined by the relation:

$$A = (u - u_0)/(u_0\ c\rho_0)$$

Where u and u_0 are the ultrasonic velocities in the solution and pure solvent, respectively, c is the concentration of the solute in $mol\ g^{-1}$, and ρ_0 is the density of water at 298.15 K in $g\ cm^{-3}$.

For the ultrasound velocity and absorption measurements the aqueous solutions of pure surfactants of different concentration were prepared by weight. The ultrasonic titrations of the DNA solutions by surfactants were performed as described earlier [11, 12]. Firstly, the measuring and reference cells were filled with the same volumes (about 0.85 ml), of DNA solution and buffer respectively. Then the concentrated solution of the surfactant was added into both cells, 10–25 μl stepwise by a 25 μl Hamilton syringe with a Chaney adapter, through small holes in the lid of the cells. The initial DNA concentration was ranged from 2 to 3 mM in phosphate.

At the same experimental conditions the UV/VIS titrations of DNA solutions by $C_{12}TAB$ were performed in a 1 cm path length cell at Lamda 40 UV/VIS spectrophotometer in the wavelength range of 200 –800 nm so as to evaluate possible aggregation of the DNA molecules.

Density measurements

The solution densities were measured using a differential vibrating tube densiometer (DMA-602, Anton Paar, Austria). The precision of the density measurements, including the reproducibility of refilling the cell, was $\pm 1.5 \times 10^{-4}\%$. The apparent molar volumes, Φ_v, were calculated from the following well-known equation:

$$\Phi_v = M/\rho - (\rho - \rho_0)/(\rho_0\rho c)$$

were ρ is the density of the solution and M is the molecular weight of the solute.

The solutions of DNA and surfactant, used for the density measurements, were prepared by mixing corresponding volumes of DNA, surfactant solutions and buffer (which was used to keep the concentration of the polymer constant about 2 mM) by weight.

The values of the apparent molar adiabatic compressibilities of the surfactant and the DNA solutions, Φ_{kS}, and their changes due to micelle formation or binding of the surfactant with DNA, were determined from corresponding experimental values of the concentration increment of ultrasonic velocity, A, and the values of apparent molar volume $(\Phi_v)_S$, as follows [17]:

$$(\Phi_{kS})_S = 2\beta_{so}(\Phi_v - A - M/2\rho_0) \tag{1}$$

Where β_{so} is the coefficient of adiabatic compressibility of the solvent.

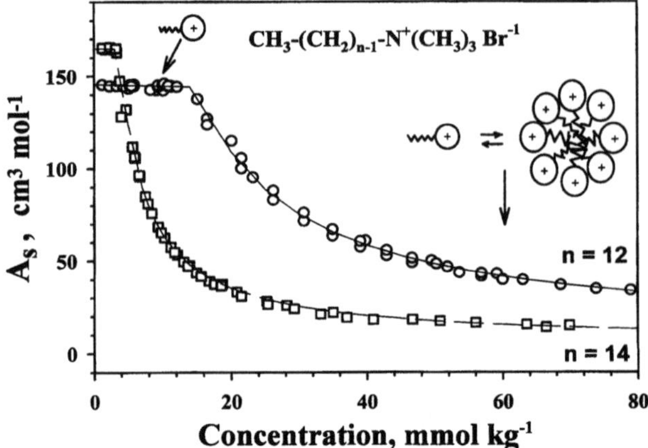

Fig. 1 The dependencies of concentration increment of ultrasonic velocity, A_S, on the concentration of the surfactants in aqueous solutions at 298.15 K: (o) dodecyl-, (□) tetradecyltrimethylammonium bromides. The lines are theoretical curves obtained according to Eq. (2)

Results

Micelle formation

The dependencies of the concentration increment of ultrasonic velocity, A_S, of the surfactants C$_{12}$TAB and C$_{14}$TAB, plotted as a function of the surfactant concentration, are shown in Fig. 1. At low surfactant concentration the value of A_S is constant which corresponds to the monomer form of the surfactant. At concentrations higher than the CMC, the value of the concentration increment of ultrasonic velocity decreases as a result of the micelle formation. In all systems no excess ultrasonic absorption, in comparison with the buffer, was observed.

The experimental results were analyzed using an expression derived in the framework of a pseudo-phase model for micelle formation [18, 19]. In dilute solutions the concentration increment of ultrasonic velocity, A, is an additive parameter because it is a linear function of the additive characteristics Φ_{kS}, Φ_v, and M (Eq. (1)). The estimated deviation from linearity at our concentrations is less than experimental error in the measurements of the A value. Therefore, it can be given as a sum of contributions from the surfactants in monomer and micellar states:

$$A = A_f, \quad c < \text{CMC}$$
$$\tag{2}$$
$$A = A_f + \Delta A_{\text{mic}}(1 - \text{CMC}/c), \quad c > \text{CMC}$$

where the value of A_f is the concentration increment of ultrasonic velocity of the surfactant in the monomer state at infinite dilution and ΔA_{mic} is the change of this value in micelle formation. Equation (2) is true in the absence of

a relaxation contribution to the value of ultrasonic velocity due to a pressure-induced redistribution of the surfactant molecules between the free and micellar states. The absence of the excess absorption of ultrasound in our measurements, as well as literature data for the relaxation times of this process for surfactants, with a chain length of 12–16 carbon atoms (18), indicates the negligibility of relaxation contribution in our systems.

The values of A_f, ΔA_{mic} and CMC have been determined by fitting the concentration dependencies of A using Eq. (2). The results of the fitting are summarized in Table 1. Figure 1 shows the consistence of the fitted curves with the experimental data. The obtained CMC values are in a good agreement with the literature data. Using the values of ΔA_{mic} and the literature data on the apparent molar volumes of free surfactants, $(\Phi_v)_S$ and volume effects on micelle formation, $\Delta(\Phi_v)_{\text{mic}}$, the values of apparent molar adiabatic compressibility of free surfactants $(\Phi_{kS})_S$, and the compressibility effects of micelle formation, $\Delta(\Phi_{kS})_{\text{mic}}$, have been obtained.

Binding of the surfactant with DNA

The dependence of the concentration increment of ultrasonic velocity of DNA, A_{DNA}, on the total concentration of the surfactant in solution is given in Fig. 2. At low concentration of the surfactant in solution, the value of A_{DNA} is constant. In the concentration range 0.5–2 mM the concentration increment of ultrasonic velocity decreases indicating the binding of the surfactant to DNA. At a concentration of more than 2 mM, corresponding to a 1:1 ligand per phosphate ratio, a plateaux is observed as a result of

Table 1 Critical micellar concentrations, CMC, concentration increment of ultrasonic velocity, A_S, and the changes of concentration increment of ultrasonic velocity, ΔA_S, in the micelle formation and binding with DNA of alkyltrimethylammonium bromides at 298.15 K

Surfactants	CMC[*] [mmol kg^{-1}]	A_S[*] [cm^3 mol^{-1}]	ΔA_S[*] [cm^3 mol^{-1}]
Micelle formation			
C$_{12}$TAB	14.8 (0.2) 15.3[a]; 15.8[b]	145.6 (0.7)	−134.4 (1.0)
C$_{14}$TAB	3.7 (0.1) 3.6[a]; 4[b]	164.6 (1.0)	−157.5 (1.0)
C$_{16}$TAB	0.9 (0.1) 0.9[a]; 1.2[b]	183.5 (1.5)	−178.6 (2.0)
Binding with DNA			
C$_{12}$TAB		145.6 (0.7)	−146 (3.0)

[*] The values in the parentheses are standard deviations.
[a] [20].
[b] [21].

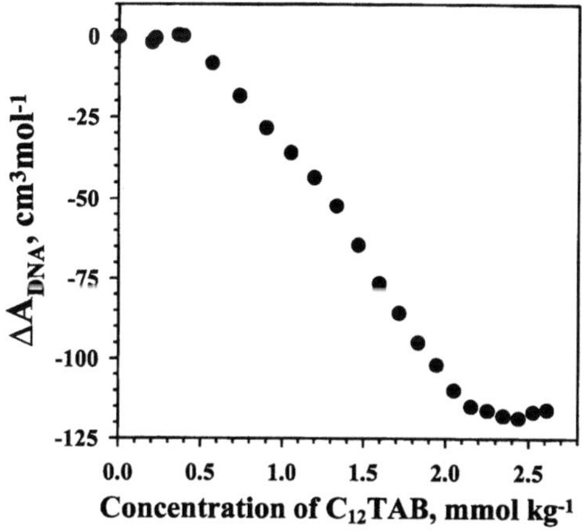

Fig. 2 The dependencies of concentration increment of ultrasonic velocity of DNA, A_{DNA} on the concentration of the dodecyltrimethylammonium bromides in solution in 10 mM NaBr, 5 mM Heppes, 1 mM EDTA, pH 7.5 at 298.15 K

the saturation of all the binding sites on the DNA surface. The shape of the curve and the concentration range where the binding occurs is in good agreement with the binding isotherms obtained earlier by a surfactant-selective electrode technique [5, 6, 16]. The total change of the concentration increment of ultrasonic velocity, ΔA_{DNA}, on binding is -115 cm^3 mol^{-1} (per mole of phosphate). The maximum binding degree obtained earlier by the surfactant-selective electrode technique is 0.8 molecules of ligand per phosphate [5, 6, 16]. This results in -146 cm^3 mol^{-1} of the change of the concentration increment of ultrasonic velocity per mole of surfactant bound, ΔA_S.

The dependence of apparent molar volume of DNA, $(\Phi_v)_{DNA}$, on the total concentration of the surfactant in solution is similar to the dependence of the value of A_{DNA}. The value of $\Delta(\Phi_v)_{DNA}$ is 7 cm^3 mol^{-1}. The corresponding value of volume effect per mole of surfactant bound, $\Delta(\Phi_v)_S$, is 9 cm^3 mol^{-1}. The changes of apparent molar adiabatic compressibility of the surfactant due to the complex formation with DNA, $\Delta(\Phi_{kS})_S$, have been calculated using Eq. (1), is 139×10^{-4} cm^3 mol^{-1} GPa^{-1}.

We did not observe any increase in the ultrasonic absorption or in the light absorption in the wavelength range 300–800 nm (outside the absorption band of nucleotides) in the binding of C$_{12}$TAB molecule to DNA. This indicates the absence of any aggregation of DNA molecules in our measurements. This observation is in a good agreement with the previous results of light scattering measurements for the short fragments of DNA (about 200 base pairs) [16].

Discussion

Compressibility effects in micelle formation

In dilute solutions the apparent molar volume, Φ_v, and the apparent molar adiabatic compressibility, Φ_{kS}, are given by the following two expressions [13]:

$$\Phi_v = V_m + \Delta V_h$$

$$\Phi_{ks} = K_m + \Delta K_h \tag{3}$$

where V_m is the intrinsic volume of the solute molecule, which is inaccessible to the solvent, and K_m is the intrinsic molar compressibility of this solute volume; ΔV_h and ΔK_h represent, respectively, the difference between volume and compressibility of the hydration shell of a solute and the volume and compressibility of the bulk water. The intrinsic compressibility of low molecular weight compounds, including our surfactants, K_m, is determined by the compressibility of covalent bonds and van der Waals radii, which is small and can be neglected within the limits of experimental error [13]. For high molecular weight molecules, such as globular proteins [10] and also for molecular aggregates, such as micelles or liposomes [14, 15], the contribution of the value of the internal compressibility can be significant. Apparent molar adiabatic compressibility of our surfactant in a free state is negative. It results from the fact that at room temperatures water in the coordination shell of the low molecular weight molecules is normally less compressible than the bulk water [10].

According to Eq. (3), the compressibility effect upon micelle formation is a sum of two values: compressibility

Table 2 Apparent molar adiabatic compressibilities of alkyltrimethylammonium bromides in monomer form, $(\Phi_{kS})_S$; compressibility, $\Delta(\Phi_{kS})_S$, and volume, $\Delta(\Phi_v)_S$, effects in micelle formation and binding of alkyltrimethylammonium bromides with DNA at 298.15 K

Surfactants	$(\Phi_{kS})_S$ $[\mathrm{cm}^3\,\mathrm{mol}^{-1}\,\mathrm{GPa}^{-1}]$	$\Delta(\Phi_{kS})_S$ $[\mathrm{cm}^3\,\mathrm{mol}^{-1}\,\mathrm{GPa}^{-1}]$	$\Delta(\Phi_v)_S$ $[\mathrm{cm}^3\,\mathrm{mol}^{-1}]$
Micelle formation			
C_{12}TAB	−9.9 (0.8)	127.5 (1.0)	6.6[*]
C_{14}TAB	−11.8 (0.7)	148.9 (1.0)	8[*]
C_{16}TAB	−13.9 (2.5)	169.2 (3.0)	9.4[*]
Binding with DNA			
C_{12}TAB		139 (5)	9 (2)

The values in the parentheses are standard deviations.
[*] Values are given from ref. [19].

change resulting from the dehydration of the surfactants which is due to their transfer from aqueous environment to the micellar core, $\Delta(\Delta K_h)$, and the compressibility of the micellar core, K_m. The first contribution can be evaluated from the apparent molar adiabatic compressibility of the surfactants in the free state. In this case $\Phi_{kS} = \Delta K_h$, and the maximal possible hydration contribution corresponding to the total dehydration of the surfactant molecule is equal to $-\Phi_{kS}$ (of the free surfactant). This gives $10\,\mathrm{cm}^{-3}\,\mathrm{mol}^{-1}\mathrm{GPa}^{-1}$ for C_{12}TAB, $12\,\mathrm{cm}^{-3}\,\mathrm{mol}^{-1}\mathrm{GPa}^{-1}$ for C_{14}TAB, $14\,\mathrm{cm}^{-3}\,\mathrm{mol}^{-1}\mathrm{GPa}^{-1}$ for C_{16}TAB. This is less than 10% of the compressibility effect of micelle formation of the same molecules. In reality, the hydration contribution to the compressibility effects of micelle formation is less than the estimated maximal values, as the most hydrated part of the surfactant (the charged head group) is exposed to water in a micelle and hence has some level of hydration. Finally, we can conclude that the obtained compressibility increase on micelle formation is mainly determined by the intrinsic compressibility of the micellar core. Therefore the coefficient of compressibility of the internal core of micelle, β, can be estimated from: $\beta \approx \Delta(\Phi_{kS})_S/V_{core}$, where V_{core} is the volume of the hydrophobic core of micelles. The V_{core} value can be obtained from the literature data on the contribution of –CH_2– group to the apparent molar volume of the surfactant in micelle, $(\Phi_v)_{CH_2}$, which is $15.7\,\mathrm{cm}^{-3}\,\mathrm{mol}^{-1}$ [19]: $V_{core} \approx N(\Phi_v)_{CH_2}$ where N is the number of carbon atoms in the alkyl chain of the surfactant. This gives the β values of $67.2 \times 10^{-2}\,\mathrm{GPa}^{-1}$ for C_{12}TAB, $66.95 \times 10^{-2}\,\mathrm{GPa}^{-1}$ for C_{14}TAB and $66.5 \times 10^{-2}\,\mathrm{GPa}^{-1}$ for C_{16}TAB. These values are close to the previous estimation of the compressibility coefficients of the hydrophobic part of the lipid bilayer membrane [15] which is about 10–20% lower than compressibility coefficient of pure linear hydrocarbons with the same number of –CH_2– groups [22]. The obtained β value does not depend on the length of surfactant in contrast to the coefficient of compressibility of linear hydrocarbons, which decreases with length of the molecule [22].

Compressibility effect in the binding of the surfactant with DNA

As can be seen from Table 2 the compressibility effect in the binding of C_{12}TAB molecules to DNA is nearly the same as the compressibility effect of the micelle formation. From the previous discussion it follows that the hydration contribution to the effect of binding is small. It is supported by the fact that even in the case of binding of highly hydrated Mg^{2+} ions to DNA the compressibility change is 10 times less than we obtained for the C_{12}TAB molecule [12]. The intrinsic compressibility of the DNA double helix is small [11] and its change cannot contribute to the observed effects. Therefore the large compressibility effect in the binding of the C_{12}TAB molecule to the DNA can be explained only by the formation of a structure with a highly compressible internal core. We can suggest then, that the C_{12}TAB molecules form structures on the DNA surface with the intrinsic compressibility close to the compressibility of C_{12}TAB micelles in solution. From the sterical considerations it is clear that the C_{12}TAB molecules cannot cover the whole surface of the DNA molecules at a maximum binding ratio 0.8 surfactant per phosphate. The C_{12}TAB molecules have large charged head and narrow hydrophobic tail. Thus it suggests, that the surfactant binding with DNA results in the formation of micelle-like aggregates on the DNA surface.

Acknowledgements We acknowledge support from the Forbait Basic Research Grant SC/97/529 and Alexander von Humboldt Fellowship (V.B). V.B. and T.K. and are grateful to Professor L. De Maeyer from the Max-Planck Institute for Biophysical Chemistry, for his support, including that of his Department and for his helpful discussions, to Dr. A. Gorelov from our Department for his helpful discussions and his help in experimental part of the work and to Dr. Th. Funck from the Max-Planck Institute for Biophysical Chemistry for helpful discussions and technical support.

References

1. Felgner A, Ringold GM (1989) Nature 337:387–388
2. Behr JP (1994) Bioconjugate Chem 5:382–389
3. Pinnaduwage P, Schmitt L, Huang L (1989) Biochim Biophys Acta 985:33–37
4. Chatterjee R, Chattoraj DK (1979) Biopolymers 18:147–166
5. Hayakawa K, Santerre JP, Kwak JCT (1983) Biophys Chem 17:175–181
6. Shirama K, Tashima K, Takisawa N (1987) Bull Chem Soc Jpn 60:43–47
7. Mel'nikov SM, Sergeyev VG, Yoshikawa K (1995) J Am Chem Soc 117:9951–9956
8. Goddard ED (1986) Coll Surf 19:301–329
9. Hayakawa K, Kwak J (1991) In: Rubingh DN, Holland PM (eds) Cationic Surfactants: Physical Chemistry. Marcel Dekker, New York, pp 189–248
10. Buckin VA, Sarvazyan AP, Kharakoz DP (1989) In: Deryagin BV, Churaev NV, Ovcharenko FD (eds) Water in Disperse System. Khimiya, Moscow, pp 45–63
11. Buckin VA, Kankiya BI, Sarvazyan AP, Uiedaira H (1989) Nucleic Acids Res 17:4189–4203
12. Buckin VA, Kankiya BI, Rentzeperis D, Marky LA (1994) J Am Chem Soc 116:9423–9429
13. Buckin VA, Kankiya BI, Kazaryan RL (1989) Biophys Chem 34:211–223
14. Hianik T, Buckin VA, Piknova B (1994) Gen Physiol Biophys 13:493–501
15. Buckin VA, Sarvazyan AP, Pasechnik VI (1979) Biofizika (Russian) 24:61–66
16. Gorelov AV, Kudryashov ED, Dawson KA (1997) J Phys Chem (B) submitted
17. Owen BB, Simons HL (1957) J Phys Chem 61:479–482
18. Zana R (1991) In: Rubingh DN, Holland PM (eds) Cationic Surfactants: Physical Chemistry. Marcel Dekker, New York, pp 41–85
19. De Lisi R, Milloto S, Verrall RE (1990) J Solution Chem 19:665–691
20. Lianos P, Zana R (1981) J Colloid Interface Sci 84:100–107
21. Lee YS, Woo KW (1995) J Colloid Interface Sci 169:34–38
22. Sachdeva VK, Nanda VS (1981) J Chem Phys 75:4745–4746

Progr Colloid Polym Sci (1998) 110:220–224
© Steinkopff Verlag 1998

E. Mendes
U. Wiesner
V. Schädler
P. Lindner

Zwitterionic and monofunctional block copolymers in a selective solvent: Model macromolecular surfactants

E. Mendes (✉)
Laboratoire de Dynamique
des Fluides Complexes
4, rue Blaise Pascal
F-67070 Strasbourg
France

U. Wiesner · V. Schädler
Max-Planck-Institut für Polymerforschung
Postfach 3148
D-55021 Mainz
Germany

P. Lindner
Institut Laue-Langevin
Ave. des Martyrs, BP 156
F-38042 Grenoble Cedex 9
France

Abstract Solutions of block copolymers containing ionic groups at the chain end in a selective solvent condition are investigated using small-angle neutron scattering (SANS). The self-assembling as a consequence of the competition between electrostatic and hydrophobic interactions is studied in those model macromolecular surfactants. The influence of the position of the ionic group on the block-copolymer chain on the structure and size of the aggregates is considered. Dilute solutions containing spherical micelles are investigated. The molecules considered are polystyrene–polyisoprene (PS–b–PI) block-copolymers in di-methyl acetamide which is a very bad solvent for the PI block. At low concentrations, block-copolymer micelles with a PI core are formed. At ambient temperature, the PI core is liquid and this ensures thermodynamic equilibrium. Three charged systems are studied: (1) sample S – polystyrene–polyisoprene (PS–PI) block copolymers containing a sulfonate ionic group at the PI end (non-soluble block); (2) sample Q – another monofunctional PS–b–PI containing, this time, a quaternized amine at the PS end (swollen polymer shell) and (3) sample Z – zwitterionic (PS–b–PI) with ionic groups of opposite charges at each chain end. It is found that the aggregation number of the charged micelles is smaller than that of the neutral micelles. Micelles formed by zwitterionic copolymers (sample Z) present an aggregation number which is similar to that observed for micelles formed by monofunctional block copolymers containing ionic groups at the core (sample S). The zwitterionic (Z) and the mono-functionalized (S) samples exhibits a soft correlation maximum at dilute concentrations, while sample Q, behaves much more like a colloidal particle with a pronounced structure factor.

Key words Block copolymers – surfactants – micelles – SANS – self-assembling

Introduction

For concentrations higher than the critical micelle concentration [1], low-molecular weight surfactant molecules self-assemble into different geometries which depend on the surfactant concentration and molecular shape. Block-copolymer molecules in a selective solvent also self-assemble [2] into different phases which strongly depend on chain length and polymer concentration [3, 4]. Such assemblies, specially spherical micelles, have been widely studied in the last decades by means of very different experimental techniques, such as static and dynamic light

Progr Colloid Polym Sci (1998) 110:220–224
© Steinkopff Verlag 1998

scattering, small-angle X-ray and neutron scattering, sedimentation velocity, size exclusion chromatography, etc. In most of the cases, the samples considered presented a glassy core, the final state of the system being a kind of "kinetically frozen thermodynamic equilibrium" which depends on sample preparation [5]. Low-molecular surfactant molecules, however, have short hydrophobic tails, and thermodynamic equilibrium can be easily achieved.

Despite of the experimental difficulties due to thermodynamic equilibrium achievement in block copolymers, self-assembling into spherical micelles in both systems, that is, block copolymers and low-molecular weight surfactants, present many similarities [6]. A full analogy, however, cannot be complete if the electrostatic interaction is not considered in the self-assembling of block-copolymers. Most of commercial low-molecular surfactants have an ionic head group which ensures the solubility of the molecule, the hydrophobic tail being responsible for the self-assembling behavior. The competition between these two tendencies, monitored by the molecular shape and electrostatic interaction, governs the geometry, and consequently, the macroscopic properties of the observed phases in surfactant dispersions.

In order to investigate the self-assembling as a consequence of the competition between electrostatic and hydrophobic interactions in block copolymers, and also to push further the analogy between block-copolymers and ionic surfactant dispersions, we decided to synthesize a model system, containing, as in the case of surfactant molecules, one single ionic group per chain. The system chosen is polystyrene–polyisoprene (PS–b–PI) block-copolymers in a polar selective solvent for PS, di-methyl acetamide (DMAc). The ionic group is, in a first time, placed either at the PI chain end (micelle core) or at the PS chain end (outer shell). The high dielectric constant of the solvent ($e \sim 39$) ensures a high dissociation ratio of the ionic groups. In order to broaden the analogy described above between the two systems, another interesting molecular geometry is also considered: a zwitterionic block-copolymer sample. In this case, the electric charge at the end of the outer shell chain behaves as a bond counter-ion that cannot leave the micelle towards the solution. The details of sample geometries and molecular masses are described in the paragraph below. The aim of this paper is, therefore, to investigate how the position of the ionic group on the block copolymer chain influences the self-assembling and the structure of dilute solutions in the case of the monofuntionalized samples and in the case of "zwitterionic" micelles, to investigate the influence of the bond counter-ion. The experimental technique employed here is small-angle neutron scattering (SANS) where the solvent is labeled with respect to the polymer.

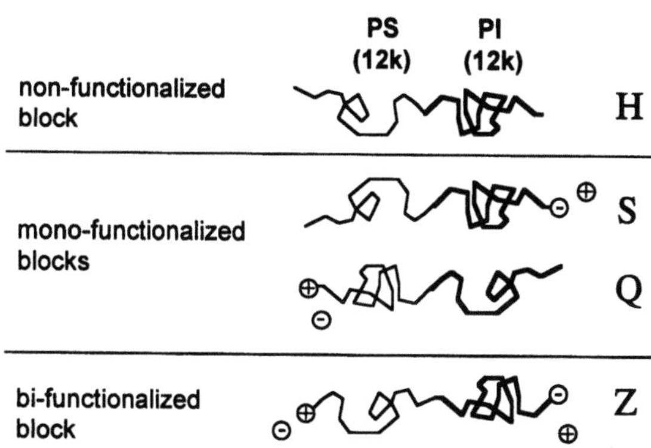

Fig. 1 Schematic representation of studied samples. The nomenclature used is also indicated

Experimental

samples

Symmetrical PS–b–PI diblock copolymers were obtained by sequential anionic polymerization using [2-N,N-dimethylaminomethylphenyl]lithium as initiator and 1,3-propanesultone as terminating reagent. The mass of each block was $M_w \sim 12\,000$. Details of the synthesis as well as the characterization are described in Ref. [7]. The functionality of the sulfonate and ammonium chain end of the telechelic species was shown to be higher than 95%. Solutions of PS–b–PI diblock copolymers in deuterated dimethylacetamide (DMAc) were prepared using DMAc as purchased from Aldrich. All solutions were allowed to stand for > 48 h before measurement, in order to ensure complete equilibrium. In Fig. 1, the studied samples are represented schematically, together with the nomenclature employed. Three charged systems are studied: (1) sample S – polystyrene–polyisoprene (PS–PI) block copolymers containing a sulfonate ionic group at the PI end (nonsoluble block); (2) sample Q – another monofunctional PS–PI block copolymer containing, this time, a quaternized amine at the PS end (swollen polymer shell) and (3) sample Z – zwitterionic (PS–PI) block-copolymers with ionic groups of opposite charges at each chain end. The results for the systems containing electric charges are systematically compared to those of neutral di-blocks (sample H) of same molecular masses and block sizes. The anionic polymerization was carried out in "one batch" ensuring the same molecular masses for all the four species.

Neutron scattering

Small-angle neutron scattering experiments have been performed at spectrometer D11 at Institut Laue-Langevin

E. Mendes et al.
Model macromolecular surfactants

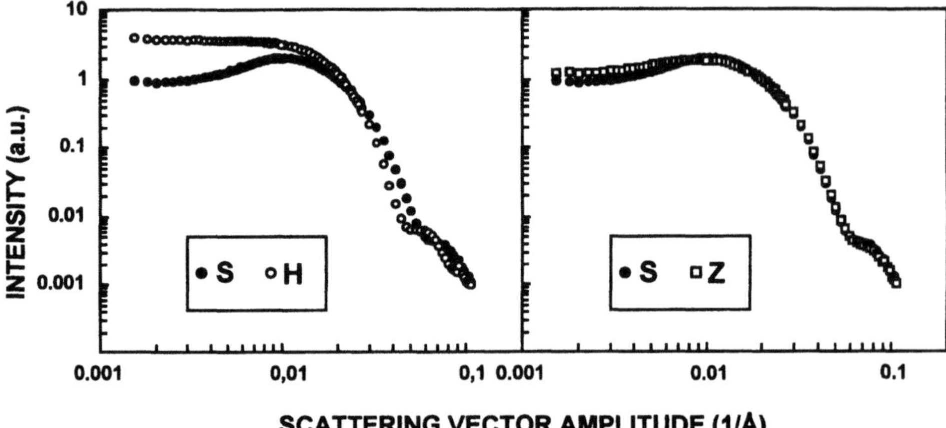

Fig. 2 (a) Scattering intensity for sample S at a concentration of 1%. The scattering from neutral micelles (H) is also plotted for comparison. (b) Comparison between the scattering from sample S and Z at the same concentration of 1%

(Grenoble-France) and the spectrometer configurations used were: $\lambda = 10$ Å and sample detector distances of 40, 10 and 3 m. Standard procedures of data treatment have been applied to raw data.

Results

In Fig. 2, the absolute scattered intensity at small angles as a function of the scattering vector is plotted in a log–log scale for solutions of H, S and Z block copolymers at a concentration of 1%. The scattering function of the H sample is a typical signature of a spherical micelle [8]. No maximum in the structure factor of the H solution is observed at this concentration, revealing clearly the dilute regime. At very small angles, an Ornstein–Zernicke function describes data for sample H, with a scattered intensity given by $I(q) \sim 1/(1 + R_G^2 q^2)$, where R_G is the gyration radius of the micelle. At high q values, the scattering function exhibits a strong oscillation due to the scattering of the PI micelle core and PS corona. At the same concentration (1%), sample S (Fig. 2a) which contains an ionic group at the PI chain end, presents a scattering spectrum with a soft maximum at a finite q-value, indicating correlation between neighboring micelles. This long-range correlation is a strong indication that dissociation of the sulfonate group at the PI end is taking place. At the same time, an oscillation at the high q-values is still observed, indicating that micellization is also taking place for sample S. Note, however, that at high q-values, the scattering from S sample is shifted towards higher q's with respect to the spectrum of the neutral micelles. This indicates that the micelle core of S micelles is smaller than that of neutral micelles H [9]. Hence, aggregation number diminishes when an ionic group is added to the PI end. Fitting of the scattering functions, under certain simplifying assumptions [10] gives a micelle core radius equal to $R_H \sim 83$ Å

and $R_S \sim 65$ Å, for sample H and S, respectively. The values of R_H and R_S above are in very good agreement with recent light scattering experiments in hydrogenated solvent [10]. This decrease of micellar core radius due to the presence of sulfonate group at the core chain end together with the observation of the soft maximum in the structure factor of S solutions is a strong indication that ionization is taking place [11]. The fact that ionization occurs suggests, at least partially, that the PI chain ends are well solvated by DMAc. Since the micelle core is liquid, one can expect that PI chain ends on sample S are preferentially placed at the core surface. This would force the PI chain inside the core to fold back to the surface, leading to the formation of a loop. The micellization of ionically end-capped block copolymers taking into account the electrostatic energy due to charges around the micelle core as well as the "loop" energy inside the core has been considered in Ref. [9].

The sulfonated (S) and zwitterionic (Z) scattering functions are qualitatively similar. They are plotted together in Fig. 2b. Micellization, therefore, also takes place for the Z polymer. It can be seen from Fig. 2b that the scattering at very small angles is stronger for the Z sample. This is partially related to the fact that, for the same polymer concentration, sample Z has more counter-ions, of both signs, and screening of electrostatic interactions are stronger than in sample S. Therefore, the interaction between neighboring micelles is weaker in sample Z than in sample S, increasing $I(q \rightarrow 0)$. Almost no difference between the two curves of Fig. 2b can be observed at higher vectors, where the oscillation on the scattering curve occurs. Detailed fitting, however, gives a micellar core radius for the Z sample a few percent bigger than that of the S sample. This result is confirmed with much more precision by recent static light scattering experiments on the same samples [10]. Although the conformation of the outer shell chains has not yet been investigated, one can speculate on

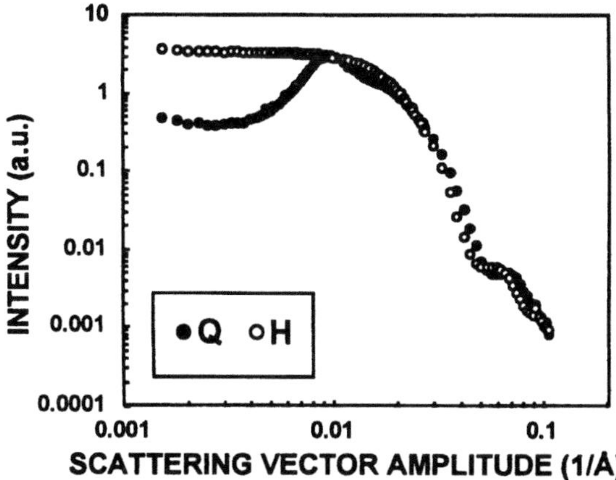

Fig. 3 Scattering intensity from sample Q at a concentration of 1% together with the scattering from neutral micelles (H)

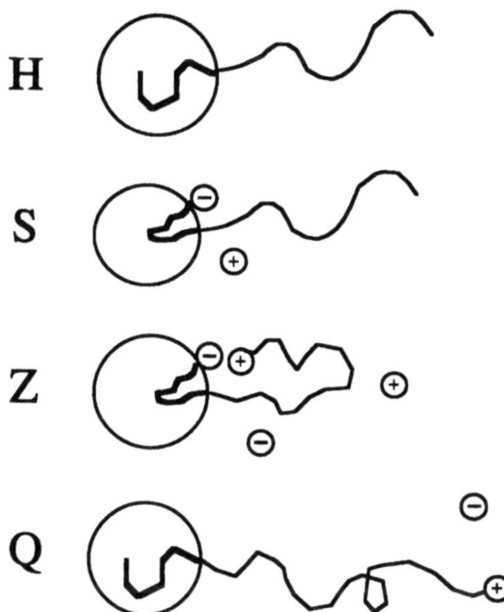

Fig. 4 Schematic proposed representation of the micellar structure for differently charged block copolymer micelles. A neutral micelle is also represented. Micelles dimensions are drawn in qualitative agreement with the discussion in the text

the possibility that some of the outer shell chains are folding back towards the micelle core attracted by the inner charges of opposite signs, screening the electrostatic interactions, and consequently, increasing the aggregation number.

In Fig. 3, the scattering from sample Q, charged at the end of the outer shell chain, is plotted together with the scattering from the neutral polymer. It can be seen from the figure that the structure factor of the charged copolymers exhibits strong correlations between neighboring micelles. A very strong scattering maximum at a finite q value is observed together with a strong decrease at very small angles. Contrary to the other two charged micelles, when charges are placed at the outermost end of the shell, a structure factor very similar to that of colloidal particles is observed. This characteristic spectrum has its origin probably on the charge distribution around the micelle. The electric charges, repelling each other, are preferentially placed on the outermost region surrounding the polymer shell. The neighboring micelles interact strongly, crudely speaking, like hollow charged spheres. The counter-ions being not enough to screen the electrostatic interactions. Note also that at higher q vectors, the scattering from the charged polymer resembles that of the neutral one. At this q range, the difference between the two spectra comes again from the shifting of the charged micelle spectrum towards higher q's. As in the case of the other charged micelles, this suggests a smaller aggregation number for the charged micelles with respect to the neutral ones. In the present case, however, higher-order oscillations in the structure factor can be pronounced, making it more difficult to obtain a precise value of the micelle core radius. In any case, one can imagine that the electric charges at the outer shell, repelling each other, can pull the PI chains from the core, decreasing the aggregation num-

ber. The strength of this possible mechanism remains to be calculated.

Finally, to summarize the above discussion, we display in Fig. 4, a schematic representation of the micelles structures according the position of the ionic group in the original block copolymer. The micelles core and corona sizes are drawn qualitatively according to the above discussion.

Conclusions

Using small-angle neutron scattering we have investigated the aggregation behavior of model macromolecular surfactant systems. Dilute solutions of symmetrical polystyrene–block–polyisoprene (PS–b–PI) micelles dispersed in di-methyl acetamide (DMAc) were considered. The low glass temperature of the PI block ensures for thermodynamic equilibrium and the ionic character of the copolymers is provided by the addition of a single group at one of the free chain ends of the block or at both ends. When charges were present at both ends, they had opposite signs. For PI blocks of 12 k, aggregation into spherical micelles was observed for all charged polymers, the aggregation number being smaller than that found for neutral micelles. Self-screening was suggested to occur for the zwitterionic micelle, the charge at the outer shell behaving like a bound counter-ion. It was also shown that the interaction between neighboring micelles depends strongly on the position of the electric charges on the micelle, the structure factor of the solutions varying from a soft to a very pronounced maximum if the charges are preferentially placed at the micelle core or at the micelle outermost region.

References

1. Mukerjee P, Mysels KJ (1971) Critical Micelle Concentration of Aqueous Surfactant Systems. NSRDS-NBS 36, Governement Printing Office, Washington, DC
2. Galot Y, Franta E, Remp P, Benoît H (1964) J Pol Sci 4:473
3. Halperin A, Tirrel M, Lodge T (1992) Adv Polym Sci 31:100
4. Marques CM, Joanny JF, Leibler L (1988) Macromolecules 21:1051
5. Zhang L, Eisenberg A (1995) Science 268:1728
6. Furukawa J (1993) Colloid Polym Sci 271:852
7. Schädler V, Spickermann J, Räder HJ, Wiesner U (1996) Macromolecules 29:4865
8. Pedersen JS, Gerstemberg MC (1996) Macromolecules 29:1363
9. Mendes E, Schädler V, Marques L, Lindner P, Wiesner U (1998) Europhys Lett 40:521
10. Schädler V, Wiesner U, Mendes E. Macromolecules, submitted
11. The degree of ionization of charged samples was not available at the time this paper was written

Progr Colloid Polym Sci (1998) 110:225–229
© Steinkopff Verlag 1996

T. Iwanaga
Y. Shiogai
H. Kunieda

Phase behavior of polyoxyethylene modified silicone with water

T. Iwanaga (✉) · Y. Shiogai
Shiga Central Research Laboratory
Noevir Co., LTD
Okada-cho 112-1
Youkaichi-shi 527
Japan

H. Kunieda
Graduate School of Engineering
Yokohama National University
Tokiwadai 79-5
Hodogaya-ku
Yokohama 240
Japan

Abstract The phase behavior of polyoxyethylene modified silicone (PEOS) in water and water–oil system was investigated. In a water system, hydrophilic PEOS, whose HLB value is 13, forms only a hexagonal liquid crystal phase, while lamellar and reverse hexagonal phases are produced in a lipophilic PEOS (HLB value = 6) system. In the latter system, these liquid crystals are formed at a rather high surfactant concentration compared to an ordinary polyoxyethylene nonionic surfactant–water system. The detailed structures of these liquid crystals were determined by small-angle X-ray scattering (SAXS).

The ternary phase diagrams of water/lipophilic PEOS/silicone oil or hexadecane were also constructed at constant temperature (25 °C). In the case of silicone oil, the reverse micellar solution or W/O microemulsion (Om) solubilizes a large amount of water whereas the surfactant is practically insoluble in hexadecane.

Key words Polyoxyethylene modified silicone – phase behavior (PEOS)

Introduction

Polyoxyethylene modified silicones have been widely used as emulsifying agents and detergents in cosmetics, toiletries, etc. [1]. Silicone surfactants can reduce the surface tension of water upto 20–25 mN/m because the silicone (polydimethylsiloxane) chain contains a number of methyl group [2]. Since the solubility parameter of the silicone chain is approximately estimated to be 6 $(cal^{1/2}/cm^{3/2})$ while that of normal hydrocarbons is 7–8 [3], the compatibility of the silicone chain and hydrocarbon is rather low. The molecular structures of polyoxyethylene-modified silicones are typically classified into a branch-type, a triblock-type and a comb-type and/or a pendant type. Although the ternary phase behavior of the triblock-type modified silicone has been already studied [4, 5], there is hardly any basic studies on the phase behavior of the comb- (graft-) type silicone surfactants and the structural analysis of liquid crystals formed in water or water–oil.

In this context, we investigated the phase behavior of the comb-type polyoxyethylene-modified silicones in water and in the presence of silicon oil or hydrocarbon oil. Furthermore, the structural analysis and the phase transition of the observed liquid crystals were also studied by means of small-angle X-ray scattering.

Materials

Two graft-type polyoxyethylene-modified silicones, PEOS-13 (MW = 2600, HLB number = 13) and PEOS-6 (MW = 5400, HLB number = 6), and silicone oil, octa-methyl-cyclotetrasiloxane (OMCTS), were obtained from Dow Corning Toray Silicone Co. Ltd. Their chemical structures are shown in Fig. 1. Extra-pure grade hexadecane was obtained from Tokyo Kasei Kogyo Co. Doubly distilled water was used.

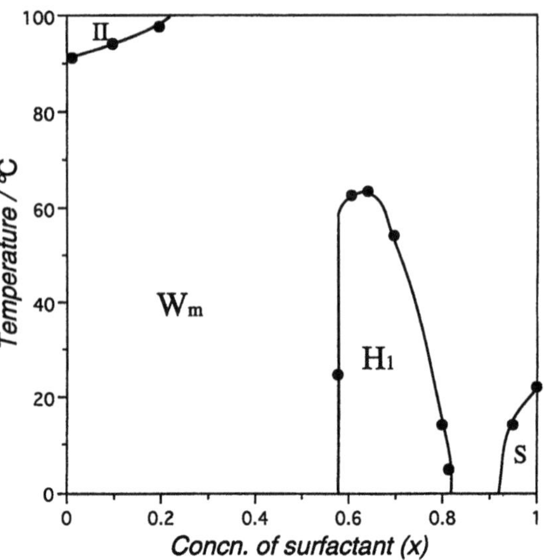

PEOS-13 : m=3~10, n=1~5, a=10~17, R=H
PEOS-6 : m=20~70, n=1~5, a=7~15, R=H

Fig. 1 The molecular structures of PEOS-13 and PEOS-6

Procedures

Determination of phase diagram

Series of water, oil and surfactant were sealed in ampoules. These ampoules, which were kept in a thermostated water bath, were well shaken and were left at constant temperature from several hours to few weeks depending on the phase separation. Phase equilibria were determined by visual observation. If necessary, a polaring microscope and small-angle X-ray scattering (SAXS) were used to determine the phase boundaries and types of liquid crystals.

Small-angle X-ray scattering (SAXS)

Interlayer spacings of liquid crystals were measured using small-angle X-ray scattering (SAXS), performed on a small-angle scattering goniometer with an 18 kW Rigaku Denki rotating anode goniometer (RINT-2500) at about 25 °C. The samples of liquid crystals were lapped by the plastic films for the measurement (Mylar seal method).

Results and Discussion

Phase behavior of a water/PEOS-13 system

Phase diagram of a water/PEOS-13 system was constructed as a function of temperature and is shown in Fig. 2. PEOS-13 is very hydrophilic and the cloud temperature of the aqueous solution (91 °C) is higher than that of octaoxyethylene dodecyl ether aqueous solution (77 °C) [6]. On the other hand, a discontinuous cubic phase, which is usually observed in a hydrophilic ordinary polyoxyethylene-type nonionic surfactant system, does not form, and only hexagonal liquid crystalline (H_1) phase appears in the concentration range of 60–80% surfactant. The maximum temperature of the H_1 phase is considerably low (63 °C) although PEOS-13 has a bulky and long hydrophobic

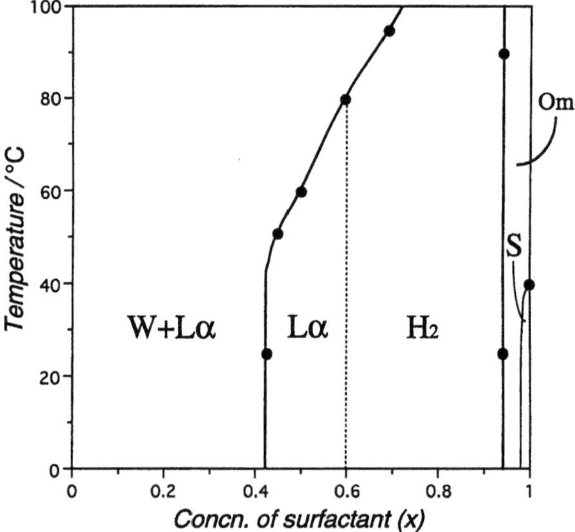

Fig. 2 Phase diagram of water/PEOS-13 system as a function of temperature. Wm, H_1, and S indicate aqueous micellar solution, normal hexagonal liquid crystalline phase and solid-present region, respectively. II means a two-phase region consisting of two liquid phases

Fig. 3 Phase diagram of water/PEOS-6 system as a function of temperature. Om indicates reverse micellar solution or surfactant liquid. H_2 and $L\alpha$ mean reverse hexagonal and lamellar liquid crystalline phases, respectively

chain. This tendency of a small liquid-crystal, region may be attributed to the weak cohesive energy of the silicone chain.

In the case of lipophilic PEOS-6 system, the formation of vesicles (the dispersion of the lamellar ($L\alpha$) phase in water) was observed in a dilute region ($L\alpha + W$) as is shown in Fig. 3. With increasing surfactant concentration, the Lα phase is changed to the reverse hexagonal (H$_2$) phase. This Lα-H$_2$ phase transition is very rare in an ordinary hydrocarbon surfactant–water system except double-chain ionic surfactant systems [7]. Judging from Fig. 3, the H$_2$ phase coexists with an excess water phase (W) at high temperature although it was not verified because emulsions are stable.

SAXS analysis

We measured the interlayer spacings (d) of liquid crystals in Figs. 2 and 3 by means of SAXS and the results are shown in Figs. 4 and 5. The radius of lipophilic cylinder in

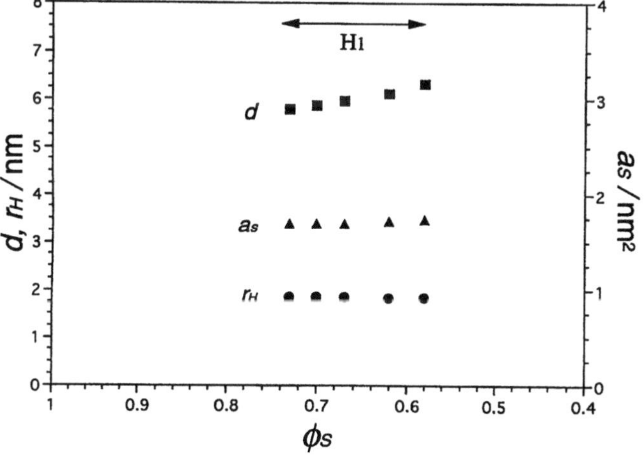

Fig. 4 Change in interlayer spacing (d), effective cross-sectional area (a_S), and radius of cylinder (r_H) in H$_1$ phase as a function of the volume fraction of surfactant in water/PEOS-13 system

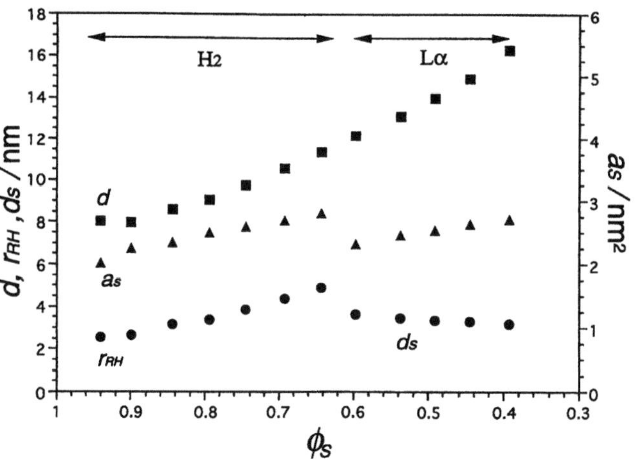

Fig. 5 Change in interlayer spacing (d), effective cross-sectional area (a_S), and radius of cylinder (r_{RH}) in H$_2$ phase or half-thickness of bilayer (d_S) in Lα phase as a function of the volume fraction of surfactant in water/PEOS-6 system

the H$_1$ phase (r_H), the half thickness of the surfactant bilayer in the Lα phase (d_S), the radius of hydrophilic cylinder (r_{RH}) in the H$_2$ phase, and the effective cross-sectional area (a_S) of one surfactant molecule are approximately calculated by the following equations and the results are also shown in Figs. 4 and 5: For normal hexagonal liquid crystal (H$_1$), we obtain,

$$r_H = \left\{ \frac{2}{\sqrt{3}\pi} (1 - A\phi_S - \phi_W) \right\}^{1/2} d, \tag{1}$$

$$a_S = \frac{2v_S}{r_H} \left(\frac{1 - A\phi_S - \phi_W}{\phi_S} \right). \tag{2}$$

For lamellar liquid crystal (Lα) of a surfactant

$$d_S = \frac{d}{2} \phi_S, \tag{3}$$

$$a_S = \frac{v_S}{d_S}. \tag{4}$$

For reverse hexagonal liquid crystal (H$_2$), we obtain, radius of cylinder

$$r_{RH} = \left\{ \frac{2}{\sqrt{3}\pi} (A\phi_S + \phi_W) \right\}^{1/2} d. \tag{5}$$

Effective cross-sectional area

$$a_S = \frac{2v_S}{r_{RH}} \left(A + \frac{\phi_W}{\phi_S} \right), \tag{6}$$

where v_S is the volume of one surfactant molecule, ϕ_S and ϕ_W are the volume fractions of surfactant and water, respectively. A is the volume ratio of hydrophilic EO chain to the total surfactant. The A is related to the Griffin HLB number [8]. Although the present silicone surfactants are the mixtures of different hydrophobic and hydrophilic chains, the average A is calculated to 0.6 for PEOS-13 and 0.24 for PEOS by using the average molecular weights, the densities, and the EO contents [8].

In the PEOS-13 system, r_H and a_S are almost constant in the H$_1$ phase whereas the interlayer spacing gradually decreases with increasing surfactant concentration. Hence, it is considered that only the distance between the cylindrical aggregates is separated by dilution. On the other hand, all the values increase in the H$_2$ phase in PEOS-6 system on increasing the water content. In the H$_2$ phase, even at very high surfactant concentration (0.5 mol of water is attached to 1 mol of EO unit), the reverse hexagonal phase is formed. In a polyoxyethylene oleyl ether–water system, 2 mol of water is necessary to form a H$_2$ phase [8]. As described before, the solubility parameter of the silicone chain is considerably low and there is a segregation tendency between the silicone and EO chains. This may cause the formation of the H_2 phase at very high surfactant concentration.

On increasing the water content, the hydration of the EO chain is enhanced and the effective cross-sectional area is increased. However, the cross-sectional area is suddenly decreased at the H_2-$L\alpha$ transition point although the inter-layer spacing is continuously increased with increasing water content. This phenomenon has not been reported in polyoxyethylene nonionic surfactant systems. Since PEOS-6 has a very large lipophilic moiety, the rearrangement of the silicone chain may cause the reduction in a_S in the $L\alpha$ phase.

Ternary system

Figure 6 shows the ternary phase diagram of water/PEOS-6/OMCTS system at 25 °C. The PEOS-6 is soluble in OMCTS and the reverse micellar solution or W/O microemulsions (Om) can solublize water. In a dilute region, the Om phase coexists with an excess water phase and the two-phase region (Om + W) is formed. In a water-rich region, a lamellar liquid crystal coexists with an excess water phase, and the formation of vesicles were observed. Although it is considered that there is a three-phase region between the $W + L\alpha$ and $W + Om$ regions, the phase boundary was not determined because stable W/O macro-emulsions are formed.

On the other hand, the ternary phase diagram of water/PEOS-6/hexadecane system is shown in Fig. 7. Different from Fig. 6, the solubility of PEOS-6 in hexadecane

is extremely small and a surfactant phase containing hexadecane (Om) coexist with excess oil phase near the PEOS-6-hexadecane axis. The H_2 and $L\alpha$ regions are also small and these phases also coexist with excess oil phase (O) as is shown in Fig. 7. As a result, in a dilute region, a wide three-phase region of $W + L\alpha + O$ is present. As desribed before, since the solubility parameter of silicone chain is considered to be much lower than that of hydrocarbon, the compatibility of both is low. This causes a big difference in phase behavior of the ternary systems.

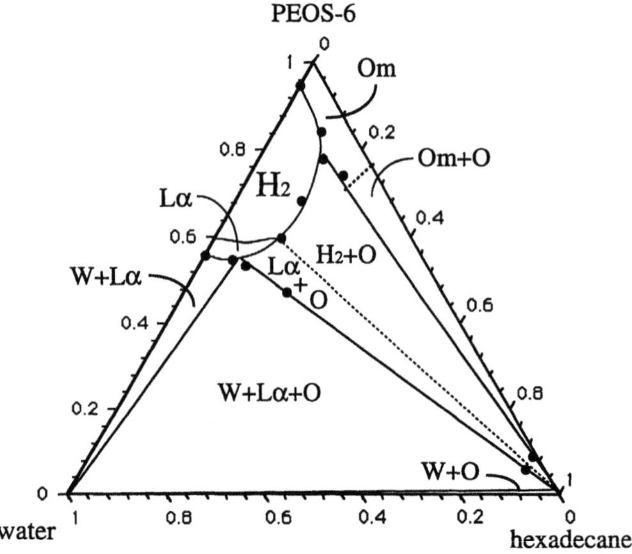

Fig. 7 Ternary phase diagram of water/PEOS-6/hexadecane system at 25 °C

Conclusion

Phase behavior of hydrophilic and lipophilic comb-(graft-) type polyoxyethylene-modified silicone in water and in water–oil systems was investigated. Although the present silicone surfactants have large lipophilic moieties, liquid-crystal regions (H_1 in the hydrophilic PEOS-13 system, and H_2 and $L\alpha$ in the lipophilic PEOS-6 system) are rather small. Since they are a comb-type polymer surfactant, the effective cross-sectional areas in liquid crystals are considerably larger than that of ordinary hydrocarbon surfactant.

The lipophilic PEOS-6 is soluble in silicone oil but insoluble in hexadecane. The difference reflects on the phase behavior in ternary water/PEOS-6/oil systems.

Acknowledgment The authors wish to thank Dow Corning Toray Silicone Co. Ltd for technical information.

Fig. 6 Ternary phase diagram of water/PEOS-6/OMCTS system at 25 °C

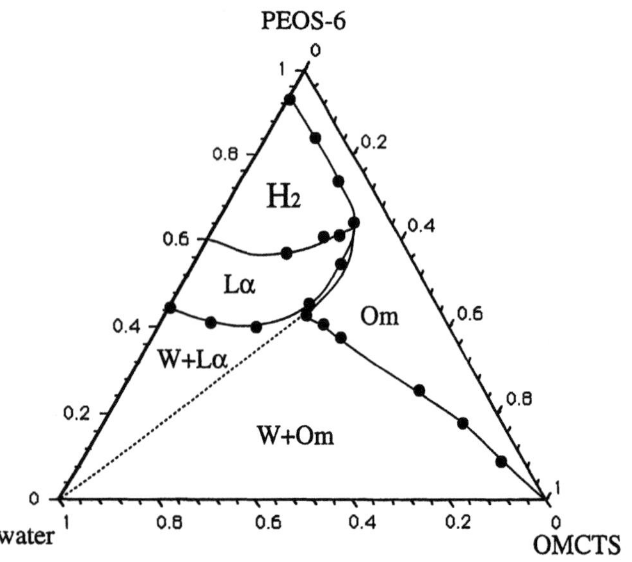

Progr Colloid Polym Sci (1998) 110:225–229
© Steinkopff Verlag 1998

Reference

1. Riess G, Hurtrez G, Bahadur P (1985) Encyclopedia of Polymer Science and Engineering. 2nd ed, Vol 2, p 324. Wiley, New York
2. Rosem MJ (1989) Surfactants and Interfacial Phenomena. p27. Wiley, New York
3. Hildebrand JH, Prausnitz JM, Scott RL (1970) Regular and Related Solutions, p213. Van Nostrand Reinhold, New York
4. Yang J, Wegner G, Koningsveld R (1992) Colloid Polym Sci 270:1080
5. Haesslin HW, Eicke HF (1984) Makromol Chem 185:2625
6. Mitchell DJ, Tiddy GJT, Waring L, Bostock T, McDonald MP (1983) J Chem Soc Faraday Trans I 79:975
7. Rogers J, Winsor PA (1967) Nature 216:477
8. Kunieda H, Shigeta K, Ozawa K, Suzuki M (1997) J Phys Chem (in press)

Progr Colloid Polym Sci (1998) 110:230–234
© Steinkopff Verlag 1996

B.K. Jha
M. Svensson
K. Holmberg

A titration calorimetry study of a technical grade APG

B.K. Jha · M. Svensson
K. Holmberg (✉)
Institute for Surface Chemistry
P.O. Box 5607
SE 11486 Stockholm
Sweden

Abstract This paper reports titration microcalorimetric measurements on micellization of a technical grade alkyl polyglucoside (APG) surfactant. The dilution enthalpy was recorded at three different temperatures and the curves obtained were compared to those of pure β-octyl glucoside. For comparative purposes titration microcalorimetry was also conducted with the conventional nonionic surfactant octa (ethylene glycol) monododecyl ether and with the anionic surfactant sodium dodecylsulfate, SDS. The results indicate that, contrary to the alcohol ethoxylate, both glucoside surfactants undergo micellization without much loss of water of hydration. Compared to the pure octyl glucoside, the technical grade APG exhibited a lower cmc and a less endothermic enthalpy of micellization at room temperature.

Key words Microcalorimetry – enthalpy – micellization – surfactants – alkyl polyglucosides

Introduction

Surfactants have the tendency to assemble at interfaces and to form multi-molecular aggregates, micelles, in solution. Aggregation of surfactants in aqueous solution has been investigated by means of a variety of experimental techniques. In particular, the critical micelle concentration (cmc) has been determined by several methods [1, 2]. One of these is titration microcalorimetry which is an attractive method since (a) it measures bulk aggregation not a phenomenon at the surface as is the case with surface tension measurements; (b) no probe, which may affect the aggregation process, is required [3–14]. This technique has been used as a method to determine the cmc and the micellization enthalpy, and also to quantitatively test the various theories and models for the micellization [13, 15].

During the 1990s, alkyl polyglucosides (APGs) have become available as a new class of nonionic surfactants. These surfactants are attractive because of excellent biodegradability and the fact that they are made from renewable raw materials [16]. Surfactants within this class contain m number of carbon atoms in the hydrophobic alkyl chain and n number of anhydroglucose units in the hydrophilic head group and may be denoted $C_m G_n$, with a commercial blend typically containing nonintegral values of both m and n. In recent years the physico-chemical behavior of pure and technical grade APGs have been investigated. A characteristic difference between APGs and the traditional nonionic surfactant class, alcohol ethoxylates, is the head group interaction with water as a function of temperature [17–19]. The phase behavior of aqueous solutions of APGs shows only a weak temperature dependence [20]. It is known that the extent of hydration and, thus, the contribution to polarity of the sugar moiety depends upon the stereochemistry of the molecule. For instance, α- and β-linked oligosaccharide surfactants show large difference with regard to both headgroup–headgroup and headgroup–water interactions [21, 22]. The importance of the hydrogen-bonding properties of different sugars has also been demonstrated by the effects of carbohydrates on membrane stability at low water activities [23]. Since commercially produced APGs are anomeric mixtures with regard to the alkyl substituent and the linkage between the anhydroglucose rings may be either α or β, their physico-chemical properties are likely to differ from those of anomerically pure monosaccharide compounds.

Several studies have dealt with the micellization and phase behavior of well-defined alkyl glucosides with a monosaccharide head group [24–26]. In a recent study titration calorimetry was employed to measure the cmc and the heat of micellization (ΔH_{mic}) of octyl glucoside and other surfactants at different temperatures [4]. The entropy was found to be the dominant factor responsible for micellization at room temperature, whereas at elevated temperatures the contribution from enthalpy dominated. To our knowledge, no such data are available for micellization of a technical grade APG in water despite their commercial importance. As mentioned above, a commercial APG is a mixture of homologues and anomers and this makes its solution chemistry complex. In the present paper we report the results of calorimetric experiments to determine the dilution enthalpy of a technical grade APG with a C_8–C_{10} alkyl chain and a slightly oligomerized sugar head group at three different temperatures, 20, 25 and 40 °C. The values obtained are compared with those of an anomerically pure β-octyl glucoside. For comparison, titration microcalorimetry was also made with a standard nonionic surfactant, octa(ethylene glycol) monododecyl ether ($C_{12}E_8$), and with the anionic surfactant, sodium dodecylsulfate (SDS).

Materials and methods

Chemicals

Octa(ethylene glycol) monododecyl ether ($C_{12}E_8$) and sodium dodecylsulfate (SDS) were purchased from Nikko Chemicals, Japan and BDH, Germany, respectively, and were used without purification. SIMULSOL SL8, a technical grade alkyl polyglucoside was a gift from Seppic, France. It has an alkyl chain length of C_8–C_{10} and a degree of polymerization (average number of condensed glucose rings) of 1.3. Surfactant solutions were always made immediately before each measurement. Millpore filtered deionized water was used for all experiments.

Isothermal titration microcalorimetry (ITC)

Titration microcalorimetry was made with a 2277 Thermal Activity Monitor System from Thermometric, Sweden with a power sensitivity of 1 μW [27]. The temperature was kept constant in a waterbath of 25 l with thermal fluctuations within 0.05°. The experiments were performed in a 25 ml stainless-steel vessel with a solid stainless-steel vessel used as reference. The surfactant solutions were added by means of a thin stainless-steel capillary, fastened on a 1 ml Hamilton gas-tight syringe. The capillary tube passes through a tube with 1 mm inner diameter in a slit at the inside of the stirrer holder and ends in the reaction solution. A microprocessor-controlled motor-driven high-quality syringe was used for the injections, and samples were injected at a rate of 1 ml s^{-1}. The stirring speed was 50 rpm. The microcalorimetric experiments consisted of measuring the enthalpy changes occurring upon consecutive additions of a concentrated aqueous surfactant solution to the calorimetric vessel, which initially contained pure water. The molar enthalpy of micelle formation in aqueous solution was calculated from the enthalpy recorded in the dilution measurements.

Results and discussion

The chemical structures of the surfactants studied by titration microcalorimetry are shown in Fig. 1. To study the enthalpy of micelle formation of the surfactants in water, concentrated surfactant solutions were added in increments to pure water. The results are presented in Figs. 2–5, where the total heat of dilution (ΔH_{obs}) for each injection is plotted against the total surfactant concentration at different temperatures. The plots for the technical grade APG, as well as for $C_{12}E_8$ and SDS, were determined in this work. The curves shown for β-octyl glucoside (Fig. 3), which span a broader temperature range than for the other surfactants, are taken from the literature. In all figures the plots can be subdivided into two concentration ranges in which the reaction enthalpies are relatively constant. For

Fig. 1 Structures of the surfactants used. I: Technical grade APG (SL8), II: Octa(ethylene glycol) monododecyl ether ($C_{12}E_8$), and III: Sodium dodecylsulfate (SDS)

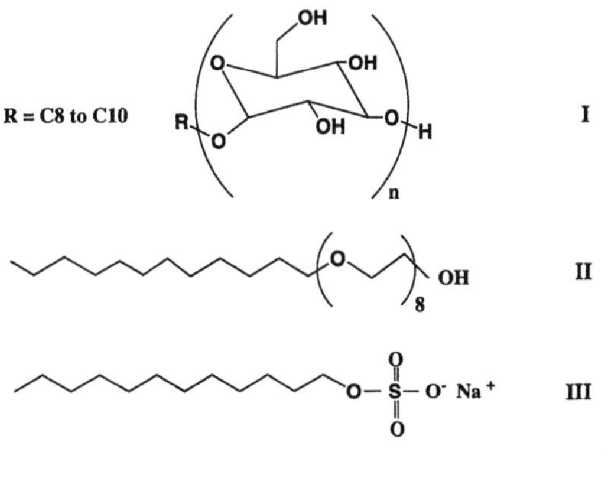

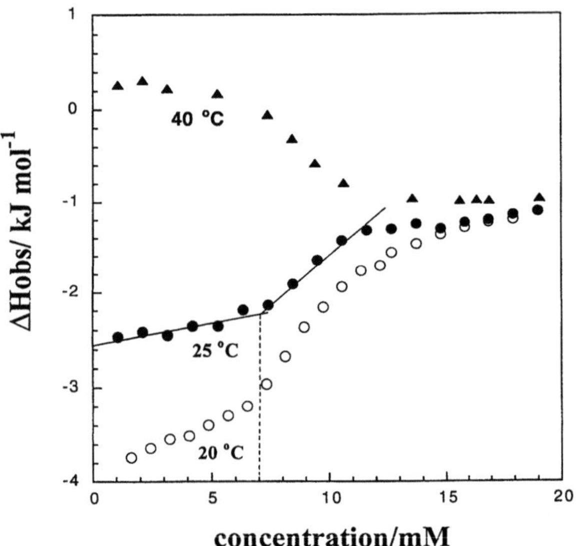

Fig. 2 Differential enthalpies of solution (ΔH_{obs}) of SL8, a technical grade APG, vs concentration at different temperatures

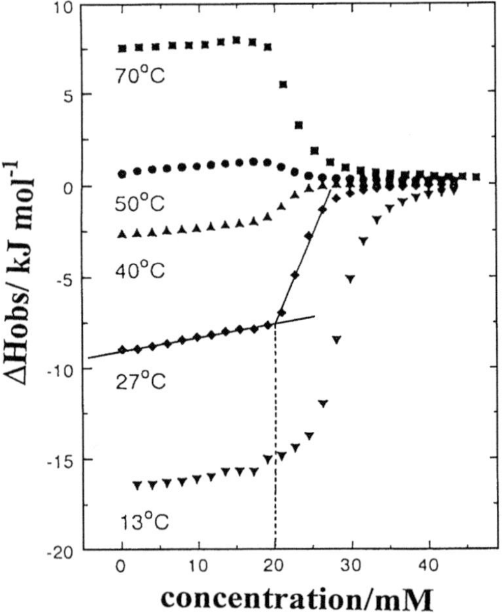

Fig. 3 Differential enthalpies of solution (ΔH_{obs}) of β-octyl glucoside vs concentration at different temperatures (adapted from Ref. [4])

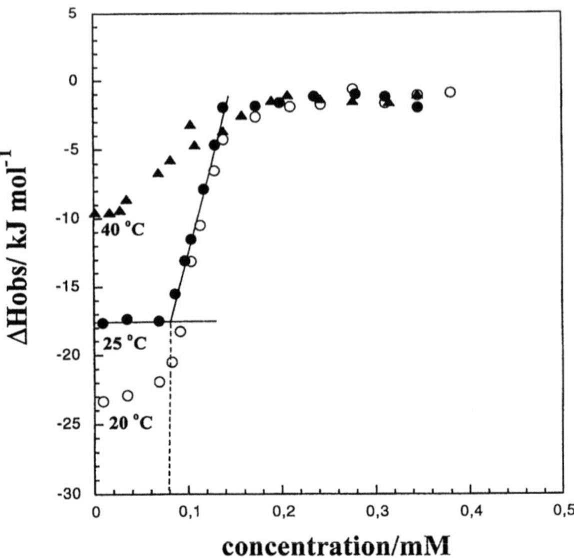

Fig. 4 Differential enthalpies of solution (ΔH_{obs}) of octa(ethylene glycol) monododecyl ether ($C_{12}E_8$) vs concentration at different temperatures.

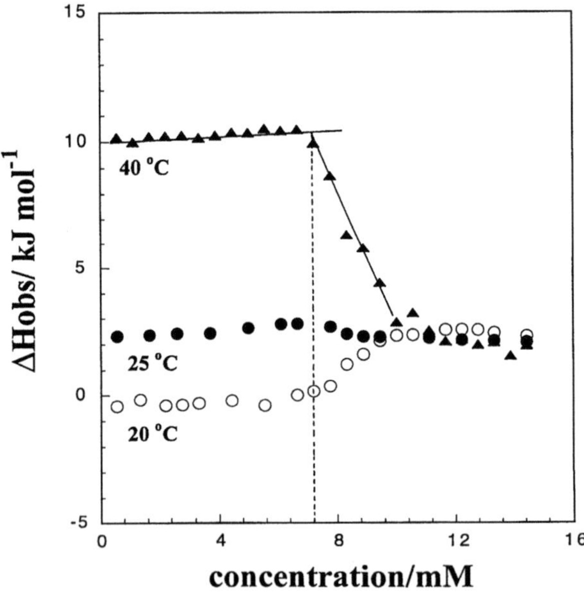

Fig. 5 Differential enthalpies of solution (ΔH_{obs}) of sodium dodecylsulfate (SDS) vs concentration at different temperatures

the first injections, the final concentrations in the sample cell are below the cmc. Hence, all added micelles will break up into monomers. The enthalpic effects observed are due to dilution of micelles, the demicellization process ($-\Delta H_{mic}$), and dilution of resultant monomers. In the intermediate region, obtained after addition of more micellar solution, a fraction of the micelles will turn into monomers while the rest will remain as aggregates. The concentration range over which demicellization occurs is a measure of the cooperativity of the micellization process. Thus, the slope of the ΔH_{obs} vs surfactant concentration curve is an indication of the aggregation cooperativity, i.e. of the aggregation number, N_s. In principle, the steeper the curve, the larger the N_s. A vertical slope indicates an infinite N_s, i.e. a phase separation. Further addition of concentrated surfactant solution will move the final concentration beyond the cmc region. In this concentration range the micelles are not dissolved and the heat measured is caused only by the dilution of micelles. The difference in dilution enthalpy between the high- and the low-concentration regimes then equals the sum of the enthalpy of demicellization ($\Delta H_{\text{demic}} = -\Delta H_{\text{mic}}$) and the enthalpy of monomer dilution. The latter term is negligible in the region close to the onset of the slope; hence ΔH_{mic} can be determined from the difference in ΔH_{obs} measured on the two sides of, but adjacent to, the cmc region [13, 28, 29].

The titration curve can be used to determine the cmc for the surfactant. This is normally done by using the crossing point between extrapolated pretransitions and linear ascent lines (see figures) [30] but the inflexion point of the curve has also been used as the point of cmc [31]. The former practice corresponds to the onset of micellization whereas the latter records the concentration where approximately equal amount of surfactant molecules exist as monomers and as micelles. For a highly cooperative micellization, which gives rise to a steep slope, the difference between the two values of cmc is small.

Figure 2 shows dilution enthalpy plots of SL8, the technical grade APG. It can be seen that ΔH_{mic} is endothermic at 20 °C and 25 °C and exothermic at 40 °C. The results indicate that micelle formation of this surfactant is entropy driven at room temperature and that the enthalpy starts to contribute at between 30 °C and 35 °C. The enthalpic term becomes increasingly important (and the entropic term diminishes in importance) with further rise in temperature. Comparing the curves with the dilution enthalpy plots of the pure octyl glucoside (Fig. 3), one can see that the general trend is the same but that the transition from endothermic to exothermic occurs at a considerably lower temperature for SL8. For octyl glucoside micellizaton is entirely entropy-driven up to 45–50 °C; only above this temperature the enthalpic terms become helpful. The smaller value of ΔH_{mic} at ambient temperature for SL8 than for octyl glucoside is not easy to predict from the difference in molecular structure. SL8 has a somewhat longer alkyl chain, a C_8–C_{10} mixture versus C_8 for the pure glucoside surfactant, and a slightly larger polar head group. It is known from microcalorimetric studies on alcohol ethoxylates that ΔH_{mic} decreases (becomes less positive) as the alkyl chain increases while keeping the size of the polar group constant and increases with the size of the polar group at constant alkyl chain length [10].

From Figs. 2 and 3 values of cmc were determined from the onset of the steep slopes of the 25 °C and 27 °C plots, respectively. The values obtained, 7 and 20 mM for SL8 and octyl glucoside, respectively, indicate that the former surfactant has a more negative free energy of

micellization ($\Delta G^\circ_{\text{mic}}$). This is to be expected in view of the fact that ΔH_{mic}, which is positive for both surfactants at ambient temperature, is smaller for SL8 and that the entropy term is not likely to differ much for the two very similar sugar surfactants. By comparing Figs. 2 and 3 it can also be seen that the concentration range over which micellization starts to occur (the steep part of the curves) is about the same, around 5 mM, for the technical surfactant and for the pure octyl glucoside. The fact that the former surfactant consists of a mixture of surface active species with different cmc values seems not to influence the cooperativity of the micellization process.

Figure 4 shows results from microcalorimetry with the pure alcohol ethoxylate $C_{12}E_8$. The enthalpy of micellization is endothermic at all temperatures used and the large value of ΔH_{mic} at room temperature is indicative of a strong entropic driving force for micellization. This is in accordance with the observation of others [7, 10, 32]. The strong endothermic heat of micellization is probably caused by the enthalpically unfavorable dehydration of the polyoxyethylene chains which takes place during the close-packing of surfactant molecules into aggregates.

Comparing Fig. 4 with Figs. 2 and 3 it can be seen that at room temperature the ethoxylated surfactant, $C_{12}E_8$, has a considerably larger value of ΔH_{mic} than both glucoside surfactants. Since the large ΔH_{mic} value of the alcohol ethoxylate is believed to be due to dehydration of polyoxyethylene chains during micellization, the results seem to indicate that micelle formation of alkyl glucosides does not involve much dehydration of the polar group. This is the expected behavior because, unlike ethoxylates, sugar based surfactants have little conformational freedom in the polar head.

Dilution enthalpy plots of SDS are shown in Fig. 5. As can be seen ΔH_{mic} at room temperature is relatively small and the transition from an endothermic to an exothermic enthalpy of micellization occurs around 25 °C. The cmc as determined from the plots is around 7 mM, a value in good agreement with values obtained by other methods. One may note that the cmc of SDS is in the same range as the cmc values of the two sugar-based surfactants with much shorter hydrocarbon tails and two orders of magnitude higher than the cmc of $C_{12}E_8$ which has the same alkyl chain length. It is well known that entropic effects make micelle formation much less favorable for ionic than for nonionic surfactants [33]. It is then interesting to observe that the cmc of SL8 is lower than that of pure octyl glucoside. It has been claimed that technical grade APGs may have a partial anionic charge, i.e. have a fraction of the surfactant molecules carrying a negative charge [34]. The bleaching process which is part of the after-treatment in the surfactant production is the probable cause of that charge. If the APG of this work had carried a partial charge, the cmc would have been expected to be much higher; thus, there are no indications of oxidized species in SL8.

Acknowledgement The authors are grateful for the valuable discussions with Drs. Gerd Olofsson and Bengt Kronberg. BKJ thanks the Swedish Institute for a research grant.

References

1. Rosen MJ (1989) Surfactants and Interfacial Phenomena. Wiley, New York
2. Moroi Y (1992) Micelles: Theoretical and Applied Aspects. Plenum, New York
3. Zajac J, Chorro C, Lindheimer M, Partyka S (1997) J Phys Chem 13: 1486–1495
4. Paula S, Sus W, Tuchten H, Blume A (1995) J Phys Chem 99:11742–11751
5. Aratono M, Ohta A, Ikeda N, Matsubara A, Motomura K, Takiu T (1997) J Phys Chem B 101:3535–3539
6. Zajac J, Chorro M, Chorro C (1995) Prog Colloid Polym Sci 98:199–302
7. Weckström K, Rosenholm JB (1997) J Chem Soc Faraday Trans 93:569–578
8. Olofsson G (1985) J Phys Chem 89:1473–1477
9. Birdi KS (1983) Colloid Polym Sci 26:45–48
10. Andersson B, Olofsson G (1988) J Chem Soc Faraday Trans 84(1):4087–4095
11. Lisi RD, Inglese A, Milioto A, Pellerito A (1997) Langmuir 13:192–202
12. Gu G, Yan H, Chen W, Wang W (1996) J Colloid Interface Sci 178:614–619
13. Johnson I, Olofsson G, Jönsson B (1987) J Chem Soc Faraday Trans 83(1): 3331–3344
14. Nusselder JJH, Engberts JBFN (1992) J Colloid Interface Sci 48:353–361
15. Desnoyers JE, Perron G (1996) Langmuir 12:4044–4045
16. Nilsson F (1996) INFORM 7:490–497
17. Kameyama K, Takagi T (1990) J Colloid Interface Sci 137:1–10
18. Marl DB (1991) Tenside Surf Det 28:419–427
19. Förster T, Guckenbiehl, Hansen H, von Rybinski W (1996) Prog Colloid Polym Sci 101:105–112
20. Mitchell DJ, Tiddy GJT, Waring L, Bostock T, Bcdonald MP (1983) J Chem Soc Faraday Trans 79:975–1000
21. Hato M, Minamikawa H (1996) Langmuir 12:1658–1665
22. Dupuy C, Auvray X, Petipas C, Rico-Lattes I, Lattes A (1997) Langmuir 13:3965–3967
23. Crowe LM, Mouradian R, Crowe JH, Jackson SA, Womersely C (1984) Biochim Biophys Acta 769:141–150
24. Nilsson F, Söderman O (1996) Langmuir 12:902–908
25. Drummond CJ, Warr GG, Griesser F, Ninham BW, Evans DF (1985) J Phys Chem 89:2103–2107
26. von Rybinski W (1996) Current Opinion Colloid Interface Sci 1:587–597
27. Suurkuusk J, Wadsoe I (1982) Chem Scr 20:155–163
28. Bach J, Blandamer MJ, Burgess J, Cullis PM, Soldi LG, Bijma K, Engberts JBFN, Kooreman PA, Kacperska A, Rao KC, Subha MCS (1995) J Chem Soc Faraday Trans 91:1229–1235
29. van Os NM, Daane GJ, Haandrikman G (1991) J Colloid Interface Sci 141: 199–217
30. Olofsson G, Wang G (to be published)
31. Desnoyers JE, Caron G, DeLisi R, Roberts D, Roux A, Perron G (1983) J Phys Chem 87:1397–1406
32. Corkill JM, Goodman JF, Harrold SP (1964) Trans Faraday Soc 60:202–207
33. Jönsson B, Lindman B, Holmberg K, Kronberg B (1998) Surfactants and Polymers in Aqueous Solution. Wiley, New York, Ch. 2
34. Balzer D (1996) Tenside Surf Det 33:102–110

Progr Colloid Polym Sci (1998) 110:235–239
© Steinkopff Verlag 1998

J. Esquena
C. Solans

Study of low energy emulsification methods for the preparation of narrow size distribution W/O emulsions

J. Esquena (✉) C. Solans
Departament de Tecnologia de Tensioactius
CID/CSIC
Jordi Girona 18-26
E-08034 Barcelona
Spain

Abstract The main aim of this work was to prepare W/O emulsions by low-energy emulsification methods to be used as reaction media for the preparation of silica particles. In this reaction work the system $H_2O/C_{12}E_5$/isooctanol/isooctane was chosen as a model system. The emulsification methods assayed were: (1) addition of water to a surfactant solution in oil and (2) addition of oil to an aqueous surfactant solution. Emulsions prepared according to method (1) were highly polydisperse, both droplet size and polydispersity were found to be dependent on mechanical stirring. In contrast, monodisperse emulsions could be obtained with method (2). The influence of parameters such as stirring speed and the order of addition of the components (in method (2)), were evaluated. The results have been interpreted according to the phase transitions induced during emulsification, that is the changes in the natural curvature of the surfactant during the process.

Key words Emulsions – monodispersity – nonionic surfactants – HLB-temperature – spontaneous emulsification

Introduction

The preparation of narrow size distribution emulsions has attracted a great deal of attention in recent years because of the applications of such emulsions as a reaction medium for the preparation of monodisperse particles, with technological applications in the fields of ceramics, magnetic particles, semiconductors, supporters for catalysts, chromatographic absorbents, etc.

The Hydrophile–lipophile Balance (HLB), concept first described by Griffin [1], correlates the chemical structure of surfactant molecules with their surface activity and provides a simple way to choose the surfactant for the preparation of emulsion, but does not take into consideration the effects of the surfactant on the properties of a given system. Temperature can play an important role in determining the surface activity where hydration is the principal mechanism of solubilization. Therefore, the activity of the surfactants is sensitive to temperature. The temperature at which the emulsion was found to be inverted from O/W to W/O (or *vice versa*) was defined as the Phase Inversion Temperature (PIT) or HLB temperature [2]. At this temperature, the hydrophile and lipophile properties of the surfactant in a given system are balanced.

Emulsions are non-equilibrium systems. Therefore, their properties are dependent on the method of preparation and on the order of addition of the different components.

The preparation methods are usually classified in two main groups:

– *Mechanical methods.* The energy required to emulsify is introduced using mechanical devices, for example, by stirring. These methods usually have several inconveniences. Firstly the consumption of energy is high, and generally produces a big polydispersity [3].

– *Low-energy emulsification methods.* The energy input, not introduced by mechanical devices can be very low. The energy to emulsify comes from the chemical activity of the components. These methods allow to obtain narrow size distributions with lower energy input.

Some of the low-energy methods consist in producing a phase transition by changing some parameters like temperature, salinity, etc. Shinoda [4] describes a method of preparation of monodisperse emulsions by changing the temperature from HLB temperature, where the interfacial tensions are very low. Another method of preparation of emulsions was first described by Marszall [5], studied by Lin [6] and modified by Sagitani [7,8]. Homogeneous O/W emulsions are obtained by phase inversion by addition of water. The surfactant diffuses from the oil to the aqueous solution, where it is more soluble.

Narrow size distribution emulsions can be obtained without energy input by diffusion of one of the components into the others [9–11]. The emulsification is produced when one of the components comes into contact with the others. Therefore, these processes are also called "diffusion emulsification", "spontaneous emulsification", or "self-emulsification".

In more recent works it has been shown that interfacial turbulence is a necessary but not sufficient condition for spontaneous emulsification [12]. Models for such kind of emulsification have been presented [13], explaining the droplet surface growth as due to the diffusion flux of surfactant to the interface. The dynamics has been studied by particle-size analysis and low-frequency dielectric-spectroscopy techniques. It has been shown that self-emulsification may be associated with liquid crystal formation [14].

Moreover, new self-emulsification methods have been described, using a gel-like phase [15], and spontaneous emulsification methods have been used for the preparation of polymer particles [16, 17].

In the present research work, the effect of the order of addition of the different components has been studied. The polydispersity of the obtained emulsions has been interpreted by studying the phase behavior and the phase transitions during emulsification.

Experimental

Materials

Pentaethyleneglycol dodecyl ehter ($C_{12}E_5$) was supplied by Nikko Chemicals and was used without further purifi-

cation. Isooctane (>99%) and isooctanol (>99%) were from Merck and Fluka, respectively. Water was deionized and Millipore filtered by a Milli-Q system.

Methods

Phase diagrams

The phase boundaries were determined weighing samples in glass ampoules which were sealed and after stirring were kept in thermostated baths.

Optical videomicroscopy

The samples were observed with a Reichert Polyvar 2 optical microscope supplied by Leica, equipped with video and interference contrast prism.

Results and discussion

Formation of W/O emulsions was studied at 25 °C in the $H_2O/C_{12}E_5$/isooctanol/isooctane system. At constant temperature, the HLB temperature of the system can be tuned by changing isooctanol concentration. Figure 1 shows that for isooctanol concentration higher than 1 wt%, the HLB temperature is lower than 25 °C. In this

Fig. 1 HLB temperature as a function of isooctanol concentration. $H_2O/C_{12}E_5$ /isooctanol/ isooctane ($C_{12}E_5 = 5$ wt%, H_2O : isooctane = 1 : 1)

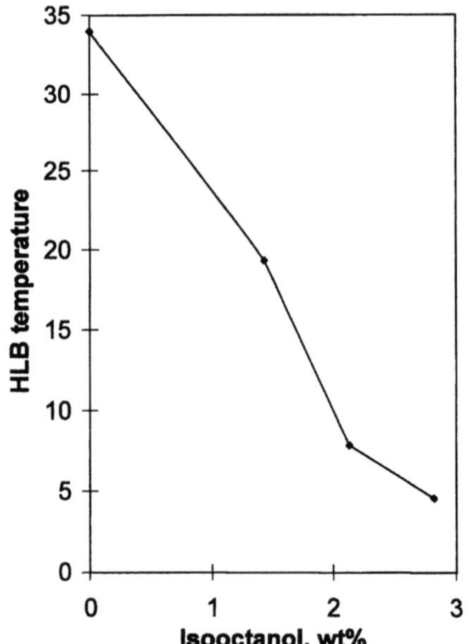

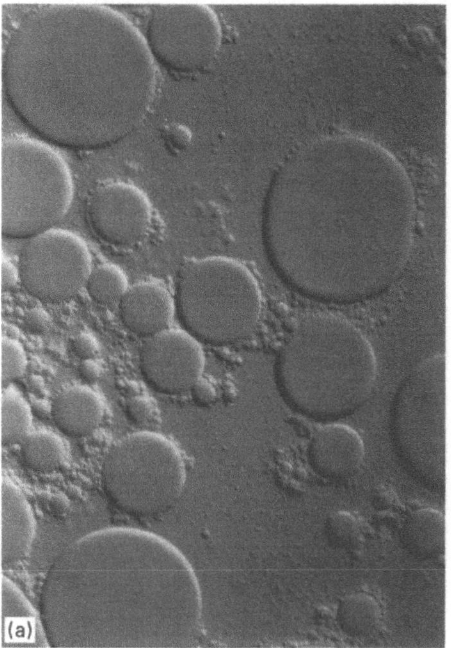

Fig. 2 Emulsions obtained by the method of addition of water to a surfactant solution in oil. $H_2O:C_{12}E_5:$ isooctane: isooctanol $= 4:1:16:1$ (weight ratios), at 25 °C. Magnification X405.(A) 700 rpm, (B) 1600 rpm

work, W/O emulsions were prepared in systems containing 2.5 and 4.5 wt% isooctanol. The following emulsification methods have been carried out.

Addition of water to a surfactant solution

Figure 2 shows W/O emulsions obtained by this emulsification method. The emulsions are polydisperse and droplet size depends on stirring speed. Droplets with diameters between approximately 1 and 10 μm are obtained at 1600 rpm (Fig. 2B) and diameters up to 100 μm at 700 rpm (Fig. 2A).

Considering the phase behavior during the emulsification process (arrow in Fig. 3) it is observed that when water is added to the surfactant/oil solution, firstly a W/O microemulsion (I) is obtained, and when the solubilization limit is reached a water in oil emulsion (II) is formed. Therefore, during the emulsification no significant change in the natural curvature of surfactant is produced.

Addition of oil to aqueous surfactant solutions

In a quaternary water/surfactant/alcohol/hydrocarbon, the alcohol can be considered as a cosurfactant or as a cosolvent. Figure 4 shows emulsions with a $H_2O:C_{12}E_5$ weight ratio equal to 4 obtained by adding: (A) Isooctanol + isooctane to the aqueous surfactant solution and (B) isooctane to the surfactant aqueous solution containing isooctanol. At this water:surfactant ratio the emulsion

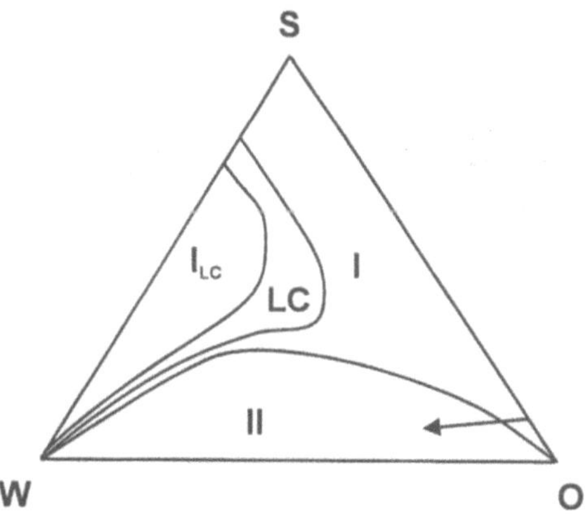

Fig. 3 Phase transitions carried out during the emulsification

obtained by method A has a droplet size about 1 μm, with a narrow size distribution, while that obtained by method B is highly polydisperse.

These emulsification paths are indicated by the arrows in Fig. 5. When the mixture of isooctane and isooctanol (arrow A) is added to the rest of the components, a phase transition from lamellar liquid crystal to W/O emulsion through a narrow area of microemulsion is produced during emulsification. In contrast, no such phase transition takes place when adding isooctane to a mixture of water $C_{12}E_5$ and isooctanol (arrow B).

By decreasing the $H_2O:C_{12}E_5$ weight ratio of the emulsions from 4 to 2, the type of phase transitions are identical independent of the order of addition of the components,

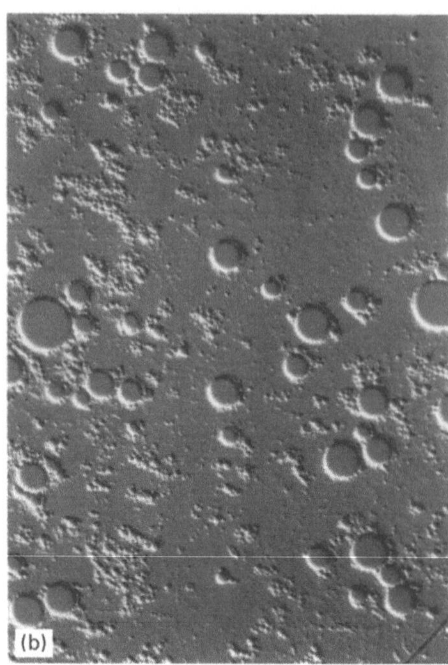

Fig. 4 Emulsions obtained at 25 °C: (A) Adding isooctane + isooctanol to $H_2O + C_{12}E_5$ (B) Adding isooctane to $H_2O + C_{12}E_5$ + isooctanol. Emulsion composition: 68 wt% isooctane, 2.5 wt% isooctanol, $H_2O : C_{12}E_5 = 4$, Magnification: X405

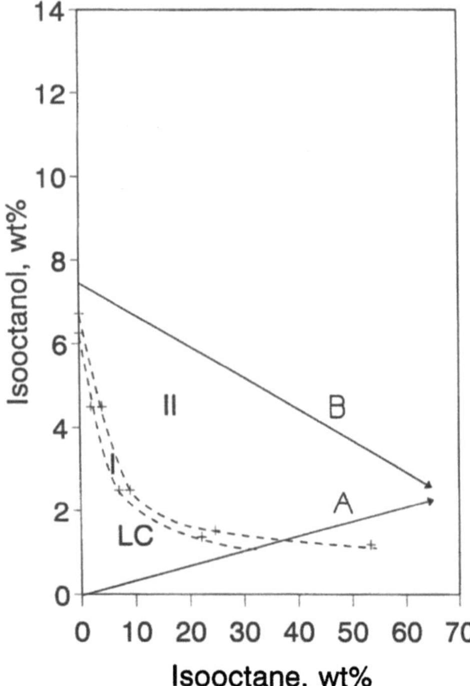
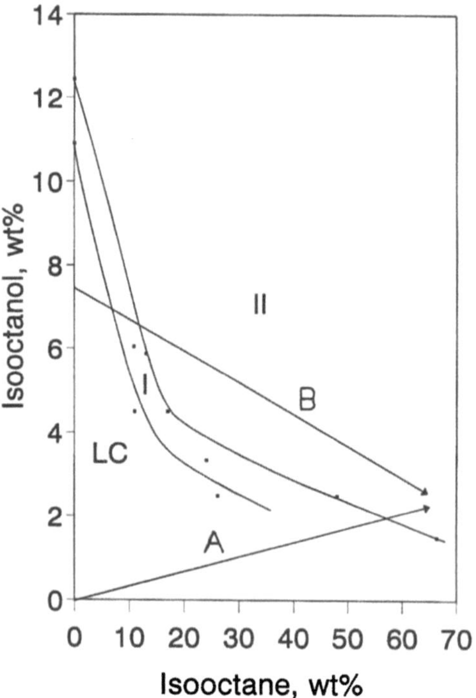

Fig. 5 Phase diagram as a function of isooctane and isooctanol concentrations at $H_2O/C_{12}E_5 = 4$ and 25 °C., indicating the emulsification paths. (A) Adding isooctane + isooctanol to $H_2O + C_{12}E_5$ (B) Adding isooctane to $H_2O + C_{12}E_5$ + isooctanol

Fig. 6 Phase diagram as a function of isooctane and isooctanol concentrations at $H_2O : C_{12}E_5 = 2$ and 25 °C., indicating the emulsification paths. (A) Adding isooctane + isooctanol to $H_2O + C_{12}E_5$ (B) Adding isooctane to $H_2O + C_{12}E_5$ + isooctanol

as shown in Fig. 6. Therefore, during emulsification the natural curvature of the surfactant changes significantly in both methods. Accordingly, the obtained emulsions should show narrow size distributions. Figure 7 displays such emulsions, with an average droplet size about 1 μm.

Fig. 7 Emulsions obtained at 25 °C: (A) Adding isooctane + isooctanol to H_2O + $C_{12}E_5$ (B) Adding isooctane to H_2O + $C_{12}E_5$ + isooctanol. Emulsion composition: 68 wt% isooctane, 2.5 wt% isooctanol, $H_2O:C_{12}E_5 = 2$, Magnification: X405

Conclusions

– Emulsions prepared by addition of water to the surfactant solution in oil (method A) were highly polydisperses and both droplet size and polydispersity were found to be dependent on mechanical stirring.
– The polydispersity of emulsions prepared by addition of isooctane to the aqueous solution of surfactant and isooctanol was found to be dependent on the $H_2O:C_{12}E_5$ weight ratio. Emulsions with narrow size distributions were always obtained when a mixture of isooctane and isooctanol were added to the rest of the components.
– Phase behavior studies showed that narrow size distributions are obtained when a phase transition involving a significant change in the natural curvature of surfactant is produced during emulsification.

Acknowledgements The authors gratefully acknowledge financial support by CICYT (Grant QUI960454) and Generalitat de Catalunya (Grant 1997SGR-00105).

References

1. Griffin WC (1949) J Soc Cosmet Chem 1:311
2. Shinoda K (1967) J Colloid Interface Sci 24:4
3. Walstra P (1983) In eds Becher P Marcel Dekker Encyclopedia of Emulsion Technology Vol 1 New York, P. 217
4. Shinoda K, Saito H (1969) J Colloid and Interface Sci 30:258
5. Marszall L (1975) Cosmet Perf 90 2:37
6. Lin TJ, Kurihara H, Ohta H (1975) J Soc Cosmetic Chem 26:121
7. Sagitani H, Hirai Y, Nabeta K, Nagai M (1986) J Japan Oil Chemists' Soc, 35:102
8. Sagitani H (1988) J Dispersion & Technology, 9, 2:115
9. Benton WJ, Miller CA, Fort T (1982) J Dispersion Sci and Tech 3, 1:1
10. Miller CA (1988) Colloids & Surfaces, 29(1):89
11. El Aasser MS, Lack CD, Vanderhoff JW, Fowkes FM (1988) Colloids & Surfaces 29(1):103
12. Rudin J, Wasan DT (1993) Chemical Engineering Sci, 48, 12:2225
13. Granek R, Ball RC, Cates ME (1993) J de Physique II 3(6):829
14. Craig DQM, Barker SA, Banning D, Booth SW (1995) Intern J Pharm 114(1):103
15. Fuji N, Hamano M, Yuasa K (1995) Biosci Biotech & Biochem 59(4):667
16. Niwa T, Takeuchi H, Hino T, Kunou N, Kawashima Y (1994) J Pharm Sci 83(5):727
17. Leroux JC, Alleman E, Doelker E, Gurny R (1995) Euro J Pharm Biopharm 41(1):14

Progr Colloid Polym Sci (1998) 110:240–244
© Steinkopff Verlag 1998

L. Ramos
P. Fabre

Elasticity of a swollen hexagonal phase

L. Ramos (✉)
Laboratoire de Physique de la Matiére
Condensée
Collège de France
CNRS, URA 792
11 Place Marcelin Berthelot
F-75231 Paris Cedex 05
France

P. Fabre
UMR CNRS-Elf Atochem
95 rue Danton
F-Levallois-Perret Cedex
France

Abstract The systems under study are doped hexagonal phases, realized by incorporating solid magnetic particles of nanometric size inside the non-polar cylinders of a swollen lyotropic hexagonal phase. The elasticity of these swollen hybrid systems is explored via the determination of a penetration length and the evaluation of the two elastic constants for the deformation of the triangular lattice of cylinders. We show that the solid particles induce a hardening of the hybrid phase and then discuss the elastic constants of swollen hexagonal phases in comparison with conventional non swollen hexagonal phases and swollen lamellar phases.

Key words hexagonal phase– elasticity – swollen lyotropic phases

Introduction

Hexagonal phases are lyotropic systems of particular interest for both fundamental and applied reasons. On the one hand, they present a highly ordered liquid crystalline structure which is still largely unexplored, due in particular to the difficulty to handle them. On the other hand, they are already used in such applications as mesoporous materials, which are expected to develop even more in the near future. Recently, the elaboration of new hexagonal phases, swollen and doped with magnetic particles [1], has opened the way to a better understanding of these systems. Indeed, two of the major obstacles to their study: the narrow extension of the hexagonal region in the phase diagram, and the difficulty to orient them, are overcome by the hybrid system. By offering the possibility to have monocrystalline samples via a magnetic field [2], and to vary the periodicity of the lattice via the radius of the oil-swollen cylinders [3], this phase is thus not only the first example of a unidimensional liquid magnetic object, but allows also to gain precious information on hexagonal systems. For instance, dynamic light scattering experiments [4] have already evidenced unexpected properties of these phases, and in particular a very different behavior from equivalent lamellar systems.

In this paper, we focus on the elastic properties of swollen hexagonal phases, and discuss the evolution of their elastic constants with their composition and particles content. After describing the phase, we explain the experimental determination of a penetration length, and present results on the elastic constants. They are discussed in comparison with non-swollen hexagonal phases on the one hand, and swollen lamellar phases on the other.

The systems are prepared in the hexagonal phase region of the phase diagram of a quaternary mixture of sodium dodecyl sulfate (sds), pentanol, cyclohexane and brine (with a sodium chloride concentration ranging between 0.1 and 0.5 M) or pure water. They are constituted of non-polar cylinders arranged in an hexagonal array in an aqueous medium. By varying concomitantly the salt concentration of the aqueous solvent and the oil content, it has been shown [3] that one can tune the radius of the oil cylinders over one decade, from 15 to 168 Å, while the intercylinder spacing is kept small and constant (25 Å). On the other hand, the doping objects are solid ferrimagnetic (γ-Fe_2O_3) particles, which are stabilized against van der Waals attraction by surfaction. The size distribution is

Progr Colloid Polym Sci (1998) 110:240–244
© Steinkopff Verlag 1998

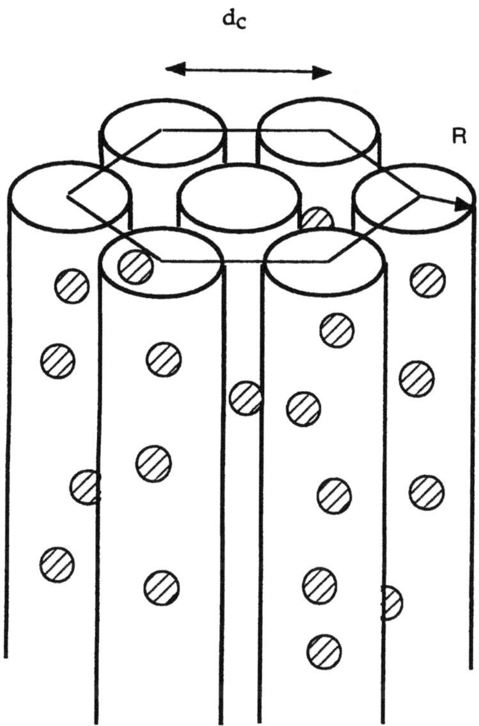

Fig. 1 Schematic representation of the swollen lyotropic hexagonal phase doped with magnetic particles suspended in oil

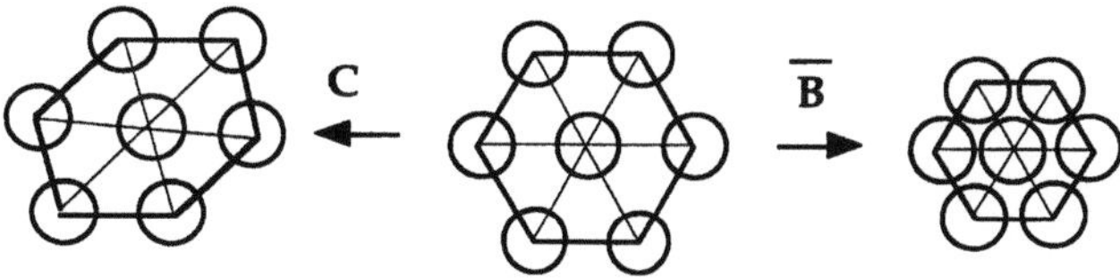

Fig. 2 Schematic representation of unstrained and deformed states of a hexagonal phase; the compression stress is controlled by $\bar{B}$ and the shear stress by C

determined by small-angle X-ray scattering (SAXS) and is characterized by a mean radius $a_0 = 35$ Å with a standard deviation $\sigma = 0.35$. The doped hexagonal phases are realized by incorporating the particles inside the oil cylinders of the swollen hexagonal phase, up to a volume fraction (relative to cyclohexane) of 2%. We use samples prepared at a composition (weight percent) of 24.1% brine 0.4 M, 61.5% cyclohexane and 4.7% pentanol for which the radius of the cylinders is equal to 150 Å and the lattice parameter is equal to 325 Å. The doped system has been completely characterized elsewhere [1]; a sketch is given in Fig. 1.

Elasticity of a hexagonal phase

On general grounds the expression for the density of distorsion energy of a hexagonal phase [5, 6] expanded to lowest order is expressed as

$$F_{\text{elast}} = \tfrac{1}{2}\,\bar{B}(\partial_x u_x + \partial_y u_y)^2 + \tfrac{1}{2}\,C[(\partial_x u_x - \partial_y u_y)^2$$
$$+ (\partial_y u_x + \partial_y u_x)^2] + \tfrac{1}{2}\,K_3[(\partial_z^2 u_x)^2 + (\partial_z^2 u_y)^2]. \qquad (1)$$

In this expression, the z-axis is parallel to the columns and u_x and u_y are the column displacements along x and y-axis. Three constants characterize the elasticity of the system,

namely $\bar{B}$, the elastic constant corresponding to a homogeneous compression of the triangular lattice, C, the constant for a shear deformation of the triangular lattice (with constant average surface) (Fig. 2) and K_3, the column bending constant.

The constant K_3 can be evaluated for non-doped swollen phases via a simple microscopic model [3, 7], which takes into account the curvature energy of a cylindrical surfactant film of radius R [8]:$K_3 \approx (2\pi/\sqrt{3})(\kappa R/d_c^2)$, where d_c is the lattice parameter, and κ, is the bending rigidity of the surfactant monolayer. In the present case, a reasonable value for κ is kT, since the interfacial film is majoritarily composed of alcohol [9]. With $R = 150$ Å, one gets $K_3 \approx 2 \times 10^{-13}$ N. Besides, previous experiments on the doped phases have allowed, through the measurement of the "Helfrich-Hurault" critical magnetic field [2, 6], to determine the following combination of the elastic constants: $K_3(\bar{B} + C)$. In order to go further and evaluate separately the two constants for the deformation of the triangular lattice $\bar{B}$ and C, we have thus measured another characteristic elastic parameter of these phases: a penetration length. These measurements are explained below.

Determination of a penetration length

Theoretical background

The quantity $\lambda = \sqrt{K_3/C}$ has the dimension of a length and is called a penetration length. Physically, a linearly polarized deformation of wave vector q parallel to the cylinders decays on a length $1/\lambda q^2$ in a direction perpendicular to the cylinders displacements. A high value of λ is then the indication of a "soft" hexagonal phase. It is possible through the analysis of the textures of a hexagonal sample to have access to this penetration length. Indeed, thermomechanical instabilities corresponding to small undulations of the columns around their mean axis often occur in oriented samples, leading to very regular striation patterns observed by polarized light microscopy [10, 11]. The periodicity Λ of these undulations is simply related to the penetration length λ along $\Lambda = \sqrt{4\pi\lambda h}$, where h is the

Fig. 3. Striation pattern of an oriented sample of a doped ($\Phi = 2\%$) lyotropic hexagonal phase. The striations are perpendicular to the average direction of the lyotropic cylinders and correspond to undulations of the cylinders. bar $= 100\,\mu$m

thickness of the sample [12]. A typical striation pattern for our samples is shown in Fig. 3.

Experimental results

The samples are held in glass capillaries of rectangular section 0.1×1 mm^2. The determination of the penetration length requires well-oriented samples. Hexagonal phases are extremely difficult to orient and with no particles, we used those of the samples which have spontaneously oriented. On the contrary, owing to the magnetic nature of the solid particles, doped hexagonal samples are rather easily oriented with the following process: the hexagonal phase is first cooled down to an isotropic state (at about 8 °C) and then slowly heated up to room temperature under a homogeneous and constant magnetic field, along which the cylinders align.

Measurements of λ were carried out for swollen hexagonal phases, with a volume fraction of particles Φ ranging between 0 and 2%, and are reported in Fig. 4a. The first thing to notice is that in these systems the values of λ are of the same order of magnitude as the periodicity of the hexagonal array. However, there is a decrease of λ, from about 330 to 180 Å, with increasing Φ, which clearly demonstrates an influence of the particles: the presence of these solid objects results in a hardening of the system. In order to compare to non-doped systems, previous results [3] are given in Fig. 4b, which concern the variation of λ versus the radius of the cylinders: they show an increase of the penetration length when the system swells. Two antagonist effects are thus responsible for the value of λ in hexagonal phases doped with particles. Indeed, on the one hand, a swollen hexagonal phase is much softer than a non-swollen one, but on the other hand, the presence of

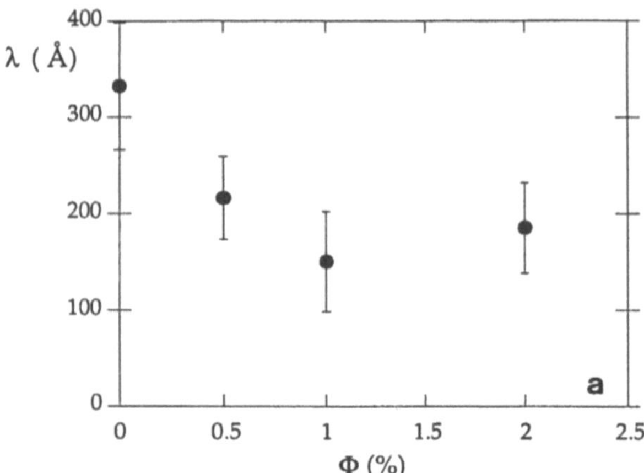

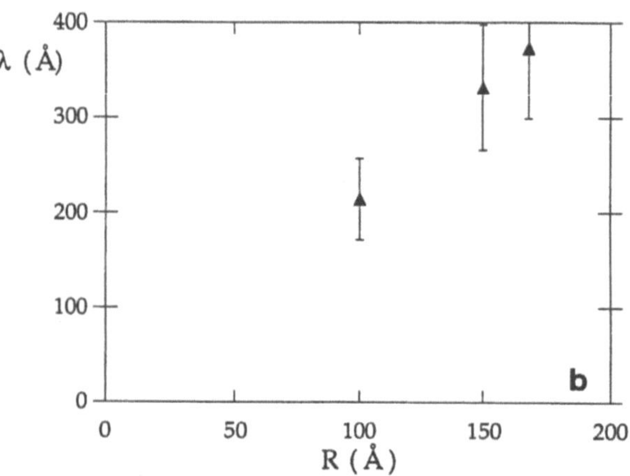

Fig. 4. Variations of the penetration length λ as obtained by optical measurements; (a) versus Φ the volume fraction in particles (for $R = 150$ Å); (b) versus R the radius of the cylinders (for non-doped systems, $\Phi = 0$)

particles increases its hardness. On these two accounts, the behavior of this hybrid system appears very similar to the behavior of doped lamellar phases [13].

The interpretation of the results is delicate for there are no microscopic models available that would account for the behavior of the two elastic constants with the particle concentration. However, to the first order, both a shear deformation which occurs at constant cylinders surface, and a bending deformation which keeps the cylinders volume constant, do not change the space available for the particles. At this order of approximation, the elastic constants K_3 and C are therefore not expected to depend on the colloid concentration. The rather weak decrease of λ with Φ observed experimentally could result from a coupling between the solid particles and the lyotropic matrix, which may be related to the fact that the particles are depleted from the walls of the cylinders, as previously demonstrated [1].

Progr Colloid Polym Sci (1998) 110:240–244
© Steinkopff Verlag 1998

Determination of the elastic constants—Discussion

Because the effect of the particles have been found to be weak, we will in a first order of approximation neglect it and assume that the three elastic constants do not vary with Φ. The evaluations of the two elastic constants $\bar{B}$ and C, through the combination of the measurement of λ determined here and of previous magnetic results mentioned earlier, is straightforward. One obtains $C = (500 \pm 160)\,$Pa and $\bar{B} = (3.0 \pm 1.6) \times 10^4\,$Pa.

The results obtained for the first time for swollen hexagonal phases call for several remarks. First of all, the two elastic constants, $\bar{B}$ and C, are very different from the one another. The much larger value of $\bar{B}$ indicates that a compression of the triangular lattice costs much more elastic energy than a shear deformation of this lattice. Moreover, the comparison of these results with results for a direct non-swollen hexagonal phase [14] is interesting. Indeed, a bending constant K_3 of $4 \times 10^{-12}\,$N has been experimentally determined, which is 20 times larger than the value we have calculated ($K_3 \approx 2 \times 10^{-13}\,$N). Concerning the measured uniaxial compression modulus ($\bar{B} + C \approx 1.2 \times 10^7\,$Pa), it is more than two orders of magnitude larger: the swelling of a hexagonal phase thus implies a very large "softening" of the system. Unfortunately, it is not possible to go further into the comparison and the evolution of the elastic constants with swelling since $\bar{B}$ and C are not separately known for non-swollen phases.

From a different point of view, it appears interesting to discuss the evolution of the elastic properties versus the topology of the phase, at a given composition. This is made possible by the existence of swollen hexagonal and lamellar phases of equivalent swelling ratios in the same chemical system and the knowledge of the elastic bending and compression constants. First of all, the values for the bending constants,. $K \approx 1$–$3 \times 10^{-13}\,$N for a lamellar phase with a periodicity around 300 Å [15, 16] and $K_3 \approx 2 \times 10^{-13}\,$N for the hexagonal phase, are of the same order. This is consistent with the microscopic models where the bending constant depends essentially on the amount of interfacial film and not on the topology of the surfactant film. On the contrary, what is noteworthy is that the compression moduli of swollen lamellar and hexagonal phases are dramatically different. Indeed, for a lamellar phase swollen with oil still with a periodicity around 300 Å, the compression modulus is of the order of 100 Pa [17], i.e. more than two orders of magnitude lower than the one we found for the swollen hexagonal phase.

The geometrical difference between the two phases is thus extremely crucial for the two elastic constants. A reason for such a behavior can be put forward with the

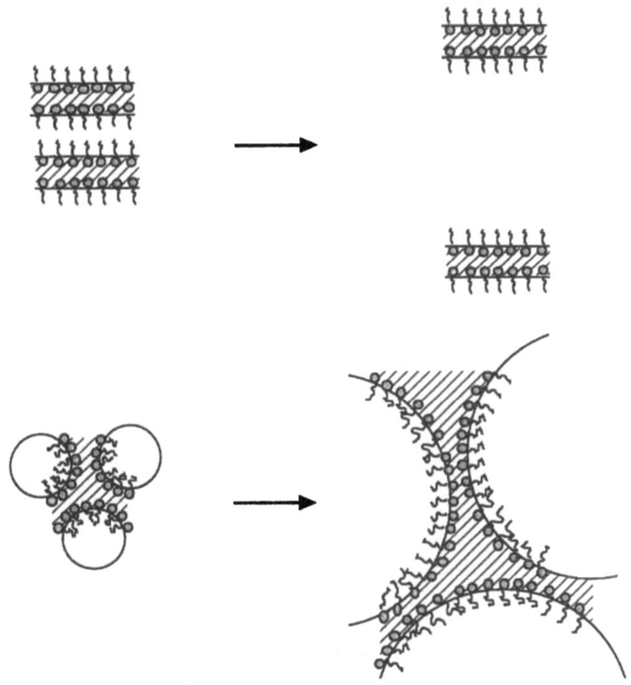

Fig. 5 Schematic representation of the swelling of a lamellar phase (top) and of a hexagonal phase (bottom). In both cases, it consists of diluting an inverse membrane of surfactant and water (dashed area), curved and connected in the hexagonal case.

following hand-waving arguments, which point out an important distinction between the compression mechanisms for lamellar and hexagonal phases. Indeed, to compress a swollen lamellar phase consists in bringing closer together two "membranes", constituted of a *bilayer*, which are far apart. In contrast, in the hexagonal phase, this mechanism of compression would represent a very high energetic cost: the interfacial film ("membrane") between two cylinders is not flat but closed, and connected to its neighbors, which implies a variation of the area per surfactant in a process of bringing membranes closer. A less costly mechanism is then to bring closer together two *monolayers* coating neighboring cylinders, which means to change the thickness of the bilayer membrane. This mechanism involves the strong short-range repulsions between two surfactant layers, which are always close to each other regardless of the swelling, in contrast to the lamellar phase where the relevant interactions for $\bar{B}$ are the weak long-range Helfrich repulsions [18]. For this reason, the value of $\bar{B}$ in a swollen hexagonal phase should be much closer to the compression modulus for packed monolayers than to the one for swollen lamellar phases. This is indeed what one observes since for a non-swollen lamellar phase $\bar{B}$ has been found equal to $10^5\,$Pa [19]: this value is compatible with the mechanism put forward for the hexagonal phase. In order to further develop these qualitative arguments, and make quantitative predictions, one should dispose of specific microscopic models linking the elasticity of the system to the short-range interactions, which are complex to describe.

Conclusion

We have investigated the elastic properties of swollen hexagonal phases. These lyotropic systems consist of oil cylinders at constant and small interspacing, but whose radius can be tuned over one decade. We have shown that the shear modulus is low, while the compression modulus is very large, despite the high swelling ratio of the hexa- gonal system. These results call for a very interesting comparison with swollen lamellar phases. Indeed, a direct analogy can be drawn between the swelling of lyotropic hexagonal phases and lamellar phases: both consist in diluting by oil an inverse membrane of constant thickness, made of surfactant molecules and a small amount of water (Fig. 5). However, the difference in the topology of the membrane (the membrane is connected in the hexagonal case) has strong repercussions on elasticity, as illustrated by a dramatically enhanced compression modulus for hexagonal system compared to lamellar system.

Acknowledgements We thank F. Nallet for a critical reading of the manuscript and M. Veyssié for fruitful comments.

References

1. Ramos L, Fabre P, Ober R (1998) EPJB 1:319–326
2. Ramos L, Fabre P, Fruchter L, submitted
3. Ramos L, Fabre P (1997) Langmuir 3:682–686
4. Ramos L, (1997) Thèse Paris VI
5. Kléman, M, Oswald P (1982) J Phys 43:655–662
6. De Gennes P-G, Prost J (1993) The Physics of Liquid Crystals. 2nd ed. Clarendon Press, Oxford, pp 337–357
7. Selinger JV, Bruinsma RF (1991) Phys Rev A 43:2910–2921
8. Servuss RM, Harbich W, Helfrich W (1976) Biochim Biophys Acta 436:900–903
9. Di Meglio J-M, Dvolaitzky M, Taupin C (1985) Phys Chem 89:871–874
10. Rogers J, Winsor PA (1969) J Colloid Interface Sci 30:500–510
11. Livoland F, Bouligand Y (1986) J Phys 47:1813–1827
12. Oswald P, Géminard J-C, Lejeck L, Sallen L (1996) J Phys II France 6:281–303
13. Quilliet C, Fabre P, Veyssié M (1993) J Phys II France 3:1371–1386
14. Sallen L, Oswald P, Géminard J-C, Malthête J (1995) J Phys II France 5:937–961
15. Halle B, Quist P-O (1994) J Phys II France 4:1823–1842
16. Ponsinet V, Fabre P (1996) J Phys Chem 100:5035–5038
17. Quilliet C (1993) Thèse Paris VI; see also Nallet F, Roux D, Prost J (1989) J Phys France 50:3147–3165 for a system of very close composition (dodecane instead of cyclohexane)
18. Helfrich W (1978) Z Naturforsch 33a:305–315
19. Richetti P, Kékicheff P, Parker JL, Ninham B (1990) Nature 346:252–254

Progr Colloid Polym Sci (1998) 110:245–250
© Steinkopff Verlag 1998

Oscillatory structural interactions in thin emulsion films containing micelles of ionic surfactant

K.G. Marinova
T.D. Gurkov
T.D. Dimitrova
R.G. Alargova
D. Smith

K.G. Marinova · T.D. Gurkov (✉)
T.D. Dimitrova · R.G. Alargova
Laboratory of Thermodynamics and
Physico-Chemical Hydrodynamics
University of Sofia
Faculty of Chemistry
James Bourchier Avenue 1
Sofia 1126
Bulgaria

D. Smith
DOW Deutschland, Inc.
Industriestrasse 1
Postfach 20
D-77836 Rheinmuenster
Germany

Abstract We study thin aqueous films sandwiched between two oil phases. The system is stabilized by anionic surfactant. These films exhibit step-wise thinning (stratification), due to the presence of micellar aggregates. By means of dynamic and static light scattering we determine the hydrodynamic diameter, and the aggregation number of the micelles. Using effective micellar volume fraction, we calculate the contribution of the oscillatory stuctural forces to the energy of interaction between the two film surfaces. Adding also the van der Waals and the electrostatic interactions, we are able to predict the contact angles of films which contain one layer of micelles inside. These angles are measured by interferometry, and the obtained experimental values agree very well with the thcorctical cstimatcs. It is proved that the oscillatory stuctural energy dominates in the total energy of interaction. The addition of electrolyte leads to smaller contact angles. This is explained by the fact that at higher salt content the effective micellar diameter (and volume fraction) decreases due to shrinkage of the electric double layer around each micelle. Therefore, the magnitude of the oscillatory stuctural energy diminishes, thus reducing the overall attraction and the contact angle.

Experiments with batch emulsions confirm that the micelles play a stabilizing role: at higher surfactant concentration the rate of creaming is lower. The repulsive maxima in the oscillatory energy of interaction can serve as potential barriers which impede the process of flocculation and lead to more loosely packed clusters of droplets.

Key words Oscillatory structural forccs – Thin liquid films – micelles in – Thin liquid films – contact angles of – Emulsion stability – effect of surfactant micelles on

Introduction

Step-wise thinning of thin liquid films ("stratification") is known to be due to layer-by-layer destruction of a colloidal crystal of spherical particles inside the film [1]. This phenomenon represents a manifestation of the so-called structural forces. In general, whenever a surface bounds a liquid phase, ordering is induced among the particles neighboring the wall [2]. In the case of a film, the structured regions around the two opposing surfaces overlap, which gives rise to an oscillatory disjoining pressure and interaction energy. The amplitude of the latter decays exponentially with the increase of the film thickness [2].

Recently, Kralchevsky and Denkov [3] proposed convenient explicit equations for calculation of the oscillatory structural contribution to the energy of interaction between the film surfaces, f_{osc}:

$$f_{osc} = \begin{cases} F(h) & \text{for } h \geq d, \\ F(d) - (d - h)P_0 & \text{for } 0 \leq h \leq d, \end{cases} \tag{1}$$

where

$$F(h) \equiv \frac{P_0 d_1 \exp[d^3/(d_1^2 d_2) - h/d_2]}{4\pi^2 + (d_1/d_2)^2}$$

$$\times \left\{ \frac{d_1}{d_2} \cos\left(\frac{2\pi h}{d_1}\right) - 2\pi\sin\left(\frac{2\pi h}{d_1}\right) \right\}. \tag{2}$$

Here h is the film thickness, d denotes the diameter of the particles (hard spheres), and P_0 is their osmotic pressure in the bulk solution. The oscillatory period, d_1, and the decay length, d_2, are found from the following empirical relations [3]:

$$\frac{d_1}{d} = \sqrt{\frac{2}{3}} + 0.23728\Delta\varphi + 0.63300(\Delta\varphi)^2, \tag{3}$$

$$\frac{d_2}{d} = \frac{0.48663}{\Delta\varphi} - 0.42032, \quad \Delta\varphi \equiv \frac{\pi}{3\sqrt{2}} - \varphi. \tag{4}$$

φ is the particle volume fraction in the bulk.

Our aim in this work is to assess quantitatively the contribution of the oscillatory structural forces to the interaction energy in stratifying emulsion films. We check the predictions of Eqs. (1)–(4) by comparison with experiment. For this purpose we measure by light scattering the hydrodynamic diameter, d_h, and the aggregation number of anionic surfactant micelles. As usual, an effective sphere diameter, d, is defined taking into account the existing ionic atmosphere around the micelles: $d = d_h + 2\kappa^{-1}$, where κ^{-1} is the Debye length. Thus, d and φ are determined, and f_{osc} is calculated as a function of h, using Eqs. (1)–(4). Interferometric measurements provide data for the thickness, h, of films containing different number of micellar layers. The total energy of interaction between the film surfaces, f, is represented as a sum of van der Waals, electrostatic, and oscillatory structural components:

$$f = f_{vw} + f_{el} + f_{osc}, \quad f = 2\sigma(\cos\theta - 1). \tag{5}$$

All contributions in f are calculated explicit for films that comprise one layer of micelles. Having found a theoretical value for f, one may predict the contact angle, θ (Eq. (5)). The equilibrium interfacial tension is denoted here by σ.

We compare the contact angle obtained through Eqs. (5) with the value measured directly from the positions of the interference fringes around the circular film. Theory and experiment are found to agree very well. For films that contain one layer of surfactant micelles the leading term in Eq. (5) is f_{osc}.

Experimental part

We used the anionic water-soluble surfactant sodium nonylphenol polyoxyethylene (25) sulfate (SNP-25S), supplied by the Dow Chemical Company. The sample contains 0.153 moles of Na_2SO_4 per one mole of surfactant. The exact ratio was determined by conductivity measurements. Aqueous solutions were prepared with deionized water from a Milli-Q purification system (Millipore). In some cases we added inorganic salt, NaCl (Merck, p.a. grade). The oil phase was styrene (Merck, stabilized against polymerization).

Dynamic and static light scattering measurements were carried out on Autosizer 4700C apparatus (Malvern Instruments), supplied with an argon laser operating at light wavelength 488 nm. All measurements were performed at $25 \pm 0.1\,°C$. By means of dynamic light scattering we determined the hydrodynamic diameter, d_h, of the micelles of SNP-25S. In order to suppress the undesirable electrostatic repulsion we added 0.1M NaCl to all solutions. Under such conditions d_h did not depend on the surfactant concentration (within the range 0.5–2 wt%). We obtained the value $d_h = 6.0 \pm 0.3$ nm. The aggregation number of SNP-25S micelles, v_m, was found from static light scattering data, applying the known Debye plot [4]. We determined $v_m \approx 26$; this number remained constant in the studied interval of surfactant concentrations.

Interferometric measurements with thin aqueous films sandwiched between styrene phases were carried out. Films were formed by sucking out aqueous phase from a biconcave meniscus held in a glass capillary of inner radius 1.60 mm. The cell was mounted on the table of a microscope (Zeiss Axioplan). The films were observed in reflected monochromatic light (with wavelength $\lambda = 546$ nm), through the optically clear cover of the cell. Images were recorded by means of a CCD camera. We connect a VCR and a computer supplied with Targa + grabbing board. Using a special software developed by us, we register the time changes in the intensity of light which is reflected from a small fixed rectangular piece of area in the film (the size and the position of this parcel can be adjusted). The film thickness, h_f, is calculated from the following relation [5]:

$$h_f = \frac{\lambda}{2\pi n_s}(l\pi \pm \arcsin\sqrt{\Delta}), \quad \Delta = \frac{I - I_{min}}{I_{max} - I_{min}}. \tag{6}$$

Here $\lambda = 546$ nm, l is the order of the interference, n_s is the refractive index of the liquid in the film, I is the instantaneous value of the reflected light intensity in the selected area parcel, and I_{min}, I_{max} denote the minimum and the maximum values of I. Thus, we determine h_f during the process of film thinning.

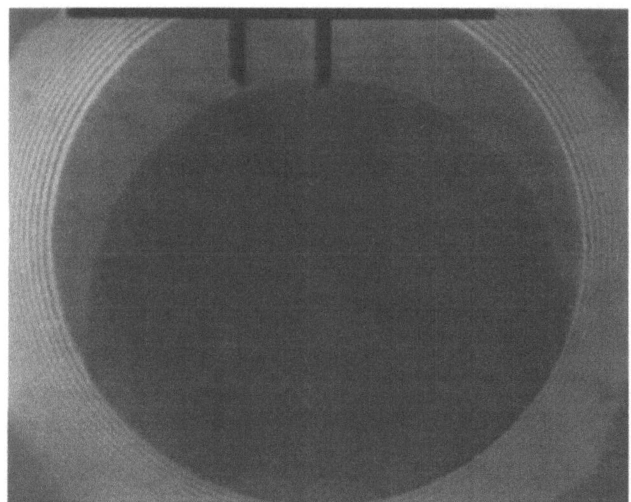

Fig. 1 Interference picture of stratifying aqueous emulsion film, stabilized by 3.35×10^{-2} mol/l SNP-25S in the presence of 0.1 M NaCl. Black and white fringes around the periphery correspond to thicknesses in multiples of $\lambda/(4n_s)$. The darker central region represents a Newton black film, the brighter part comprises one layer of micelles. Note the different contact angle (i.e., the distances between the Newton fringes) around the two parts of the film. The reference distance between the two vertical bars is $31.25\,\mu$m

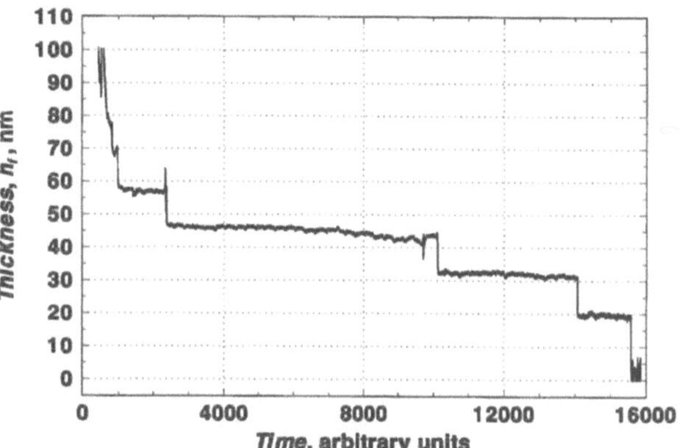

Fig. 2 Step-wise change of thickness vs time, $h_f(t)$, calculated from the experimentally measured reflected light intensity, $I(t)$. The film is made from aqueous solution of 3.35×10^{-2} mol/l SNP-25S (without added NaCl)

The contact angle, θ, is found from the positions of the Newton fringes around the film periphery (see Fig. 1). We have made a computer program that fits the observed fringe locations with the numerical solution of the Laplace equation of capillarity. The contact angle, the exact film radius, and the capillary pressure are determined from the best fit between the experimental data and the calculation. We average data for different radial directions in a given film, as well as for a number of films made separately. The film diameter is kept in the range from 220 to 300 μm. We normally have θ within $\sim \pm 0.1°$, which is sufficient for figuring out the studied effects.

Results and discussion

We performed thin film experiments with two different systems:

(A) The aqueous phase was a solution of 4.76 wt% (0.0335 M) SNP-25S, which contained also 0.00512 M Na_2SO_4 included with the surfactant; the oil was styrene;

(B) The same as (A), with the addition of 0.1 mol/l NaCl to the water phase.

In both cases we observed stratification and the films finally thinned down to Newton black films (NBF). The latter were stable without NaCl (system A), but ruptured within 1–2 min in the presence of NaCl (B). Figure 1 shows

a sample for the interference pattern in a film. The step-wise transition in the thickness is clearly distinguished.

The curve in Fig. 2 represents an interferogram which was obtained by means of Eq. (6), processing experimental data for $I(t)$. The interference maximum and minimum of the reflected light intensity, I_{max} and I_{min}, are measured directly. We take the difference between the thicknesses of the layers, i.e., the height of the step (Fig. 2), which will be denoted hereafter by h_m. The values of h_m are averaged using several curves like that in Fig. 2, obtained by independent runs. The final result is $h_m = 12.7 \pm 1.3$ nm for the system without excess inorganic electrolyte (A), and $h_m = 10.9 \pm 1.3$ nm for the films with added NaCl (B).

We define an effective micellar diameter which is identified with d in Eqs. (1)–(4):

$$d = d_h + 2\kappa^{-1} \quad \text{where} \quad \kappa^2 = \frac{8\pi e^2}{\varepsilon kT}I, \quad I = \frac{1}{2}\sum_i Z_i^2 C_i^o.$$

$$(7)$$

Here $d_h = 6$ nm, ε is the dielectric permittivity of the aqueous solution, e is the elementary charge, and I is the ionic strength. Z_i denotes the valency of the ions i, and C_i^o is their concentration in the bulk solution from which the film is made. The applicability of Eq. (7) has been widely discussed in literature [6, 7]. Our aqueous phase contains micelles that are partially dissociated and have released some counterions (Na^+). Such systems have been studied by other authors [6, 7], and the question of how to find the ionic stregth was clarified. We follow the approach proposed in Ref. [7], and calculate I according to Eq. (3) therein. The degree of micellar dissociation (α_m) is taken into account by us. From conductivity measurements at surfactant concentrations above the CMC we determine

K.G. Marinova et al.
Micelles in emulsion films

$\alpha_m = 0.585$. Then, using Eqs. (7) one estimates $d = 7.72$ nm in the presence of 0.1 M NaCl (system B), and $d = 9.81$ nm without NaCl (system A)-cf. Table 1.

Knowing that the micellar aggregation number is 26, we obtain the concentration of particles in the solution. The corresponding volume fraction of the micelles, φ, is also shown in Table 1. Note the big difference between the values of φ with and without NaCl. Obviously, the effect of the shrinking of the ionic atmosphere is quite essential. As long as the miscelles are considered to behave as hard spheres, it is reasonable to adopt the Carnahan–Starling [8] formula for the osmotic pressure, P_0 (Eqs. (1) and (2)) :

$$P_0 = \rho kT \frac{1 + \varphi + \varphi^2 - \varphi^3}{(1 - \varphi)^3}, \quad \rho = 6\varphi/(\pi d^3). \tag{8}$$

Numerical results for $f_{osc}(h)$, obtained from Eqs. (1) and (2), are plotted in Fig. 3, for the system without added NaCl. The curve has a typical shape, characterized by the existence of two regions: $0 \leq h \leq d$, and $h \geq d$. Switching over from the oscillatory regime ($h \geq d$) to the depletion ($0 \leq h \leq d$) can have important implications in view of the stability of dispersions. In particular, the maxima can serve as potential barriers which prevent flocculation (see below).

Now we can explore quantitatively the case when the film contains one layer of micelles. The thickness, h, which matters for the oscillatory forces, corresponds to the distance between the ends of the two hydrophilic layers of polar heads anchored on the surfaces. Actually, this is h_m; it has been determined in the thin film experiments as the step difference in h_f upon expulsion of the micellar layer. h_f (Eq. (6)) is to be perceived as the distance between the two oil/water surfaces, i.e., the effective aqueous layer thickness. We compare the result $h_m = 12.7 \pm 1.3$ nm with Fig. 3 (for the system A, without NaCl). The experiment reveals that the film thickness lies very close to the *first local minimum* of the oscillatory structural energy. The same is true for the system B (containing NaCl). The calculated values of f_{osc} at the first local minimum are listed in Table 2, and can be identified as the oscillatory structural contributions to the respective interaction energies in our films (systems A and B).

The final goal is to find the contact angle, θ, according to Eqs. (5). The van der Waals interaction energy, f_{vW}, is

Table 1 Characteristic parameters of the studied micellar solutions of SNP-25S

Added NaCl [M]	κ^{-1} [nm]	φ	d [nm]	d_1 [nm]	d_2 [nm]
0	1.91	0.380	9.81	9.66	9.13
0.1	0.86	0.185	7.72	8.83	3.51

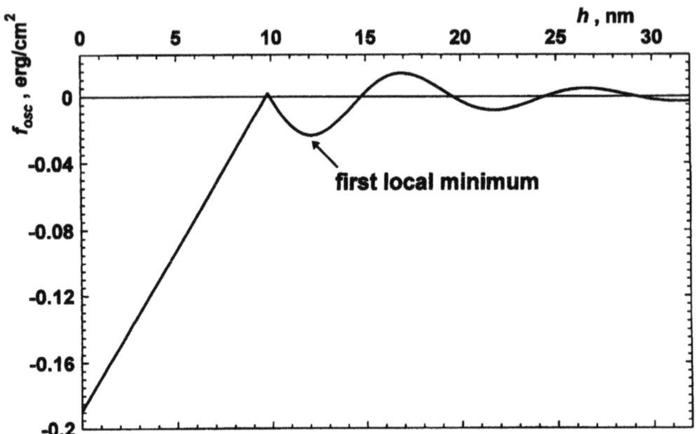

Fig. 3 Calculated interaction energy due to oscillatory structural forces in a film stabilized by 3.35×10^{-2} mol/l SNP-25S (system A). The parameters are taken from Table 1

estimated from the relation [2]

$$f_{vW} = -\frac{A}{12\pi h_f^2}, \tag{9}$$

where A is the compound Hamaker constant, $A = 5 \times 10^{-14}$ erg in oil/water/oil systems [2]. Given the molecular structure of the oxyethylene chains of SNP-25S, we evaluate $h_f \approx h_m + 7$ nm. Then, Eq. (9) provides results for f_{vW}, which are presented in Table 2.

The electrostatic contribution to the interaction energy between the film surfaces, f_{el}, is calculated from the conventional double layer theory [9]. We write

$$f_{el}(h_m) = \int_{h_m}^{\infty} \Pi_{el}(h)\,dh,$$

$$\Pi_{el}(h) = kT \sum_i C_i^0 \left\{ \exp\left[-\frac{Z_i e \Psi_0(h)}{kT} \right] - 1 \right\}, \tag{10}$$

Table 2 Interaction energies and contact angles of emulsion films

Added NaCl [M]	$f_{osc} \times 10^3$ [erg/cm²]	$f_{vW} \times 10^3$ [erg/cm²]	$f_{el} \times 10^3$ [erg/cm²]	$f \times 10^3$ [erg/cm²]	θ calcul. [Deg]	θ measured [Deg]
0	− 23.51	− 0.37	15.47	− 8.41	1.92	1.89 ± 0.08
0.1	− 1.99	− 0.40	0.026	− 2.36	1.02	1.0 ± 0.04

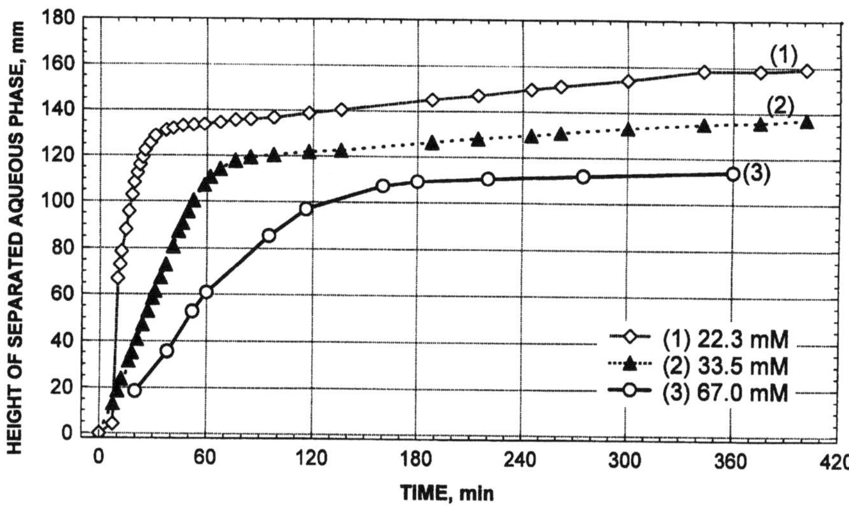

Fig. 4 Role of the oscillatory structural forces for the emulsion stability. The height of the water column separated from a 20 vol% styrene-in-water emulsion is plotted as a function of time. The system is stabilized by SNP-25S, with three different concentrations (cases 1, 2, and 3), all of them above the CMC

with $\Pi_{el}(h)$ being the electrostatic disjoining pressure [9]. The latter is connected with the electric potential at the midplane of the film, Ψ_0. The summation in Eq. (10) is carried out over all ionic species, i. Since our films are thick, in the sense that $h_m \gg \kappa^{-1}$, it is reasonable to adopt the approximation for weak overlapping of the two diffuse ionic layers developed around the charged interfaces: $\Psi_0(h) \approx 2\Psi_1(z = h/2)$. Ψ_1 here refers to the case of a single charged plane (not a film); z denotes distance to that surface. $\Psi_1(z = h/2)$ can be computed by integration of the non-linear Poisson–Boltzmann equation [9]. The final results for f_{el} are listed in Table 2. In the presence of excess NaCl (system B), f_{el} is completely negligible.

The equilibrium interfacial tension, σ, was measured at the boundary between styrene phase and aqueous solution of SNP-25S. Du Nouy platinum ring method was used. From the slope of the plot of σ vs the logarithm of surfactant concentration we found the area per molecule to be 0.76 nm^2 at complete coverage (very close to the CMC).

Let us now sum up the oscillatory, van der Waals, and electrostatic contributions in the interaction energy (Eq. (5)). The total quantity, f (Table 2), together with $\sigma = 7.5$ dyn/cm, measured above the CMC, gives the predicted value of the contact angle, θ (Table 2). Independent measurements of θ were performed with films containing one layer of surfactant micelles (Fig. 1 represents an example). The experimental results agree very well with the theoretical predictions (Table 2). Therefore, the concept of accounting for f_{osc} by means of Eqs. (1)–(4) turns out to provide a realistic description.

It is important to emphasize that the oscillatory structural energy, f_{osc}, is the main component in the total interaction energy, f, cf. Table 2. This fact can be invoked to explain the influence of the inorganic electrolyte upon the value of the contact angle. The conventional theory for the interactions in thin films [9] does not consider the oscillatory structural forces, but is restricted to the van der Waals and electrostatic contributions only. In view of this theory, addition of electrolyte leads to suppression of the electrostatic repulsion between the two film surfaces. Consequently, increase of θ is expected, at rising $|f|$, $f < 0$ (cf. Eq. (5)). However, exactly the opposite trend is observed experimentally (Table 2). It is evident from Table 1 that the effective micellar volume fraction, φ, drops considerably upon addition of NaCl, due to the shrinkage of the ionic atmosphere. This turns out to be the main effect, as it brings about an order of magnitude decrease of the attraction energy, f_{osc} (Table 2).

The relevance of the oscillatory forces to the stability of real dispersions is illustrated in Fig. 4. We present experimental results for the rate of creaming of batch emulsions stabilized by SNP-25S. Three samples have been prepared in such a way that the droplet size distribution is identical, only the micellar content in the aqueous phase is different. Obviously, higher total surfactant concentration promotes the stability and leads to more loosely packed cream (note the initial slopes and the positions of the plateaux in Fig. 4). These effects are presumably brought about by the oscillatory structural forces due to the presence of micelles. Particularly important for the creaming rate are the potential barriers (such as those in Fig. 3, for instance), which impede the flocculation.

Concluding remarks

In this work we investigate stratifying emulsion films stabilized by ionic surfactant. The oscillatory structural energy, f_{osc}, is calculated theoretically. For this purpose the diameter and the aggregation number of the micelles were measured by light scattering methods. Interferometric studies of thin emulsion films made in a capillary provide data for the step-wise transitions in the thickness, and for the contact angles. Detailed interpretation is developed for films containing one layer of micelles inside. We compare the experimental values of the contact angles with those calculated by taking into account the oscillatory structural energy. Good agreement between theory and experiment is observed. It turns out that the film thickness lies close to the first local minimum of f_{osc}, which corresponds to attraction between the two film surfaces. We have proved that f_{osc} dominates in the total energy of interaction. This fact explains the observed decrease of the contact angle upon addition of inorganic electrolyte. The shrinkage of the ionic atmosphere leads to decrease of the effective micellar volume fraction, φ. Consequently, the depth of the first minimum of f_{osc} diminishes by an order of magnitude.

Acknowledgements This work was financially supported by the DOW chemical company, and by the Bulgarian Ministry of Education, Science and Technologies.

References

1. Nikolov AD, Wasan DT, Kralchevsky PA, Ivanov IB (1988) In: Ise N, Sogami I (eds) Ordering and Organisation in Ionic Solutions, p. 302. World Scientific, Singapore.
2. Israelachvili JN (1992) Intermolecular and Surface Forces. Academic Press, London
3. Kralchevsky PA, Denkov ND (1995) Chem Phys Letters 240:385–392
4. Debye P (1949) Ann New York Acad Sci 51:575
5. Traykov TT, Manev ED, Ivanov IB (1977) Int J Multiphase Flow 3:485–494
6. Schmitz KS (1996) Langmuir 12:3828–3843
7. Richetti P, Kekicheff P (1992) Phys Rev Lett 68:1951–1954
8. Carnahan NF, Starling KE (1969) J Chem Phys 51:635
9. Derjaguin BV (1989) Theory of Stability of Colloids and Thin Films. Plenum Press, New York

Progr Colloid Polym Sci (1998) 110:251–254
© Steinkopff Verlag 1998

A.V. Zvelindovsky
G.J.A. Sevink
B.A.C. van Vlimmeren
N.M. Maurits
J.G.E.M. Fraaije

Lamellar phase of diblock copolymer melt under shear: kinetics and coformational analysis

Dr. A.V. Zvelindovsky (✉) · G.J.A. Sevink
B.A.C. van Vlimmeren · N.M. Maurits
J.G.E.M. Fraaije
Dept. Biophysical Chemistry
University of Groningen
Nijenborgh 4
9747 AG Groningen
The Netherlands

Abstract 3D computer simulations of lamellar phase of a diblock copolymer melt under simple steady shear are performed. The polymer system is described by the density functional as a collection of Gaussian chains in a mean field environment. The 3D mesophase morphology is obtained as a solution of a diffusion–convection equation with random noise field. Here, an analysis of polymer chain conformations in lamellae is considered.

Key words Shear – polymer – lamellar – density functional

Because of the wide area of industrial applications the behavior of complex polymer systems in shear flows is intensively studied both experimentally and theoretically [1–3]. The stability of morphology structures and phase transitions in complex polymer liquids in shear have been investigated by many authors [4–7]. During the last years the kinetics of morphologies of complex fluids in applied external shear has also been studied by computer simulation techniques using time dependent Landau–Ginzburg models [8–13]. These models are based on traditional free energy expansion methods [14–16] which contain only the basic physics of phase separation [17]. In contrast to these phenomenological theories we do not truncate the free energy at a certain level, but rather retain the full polymer path integral by a numerical procedure [17–23]. Recently, Kawakatsu and Doi started to use a similar approach [24, 25].

Calculation of polymer path integrals is computationally very intensive and requires efficient numerical techniques. However, it allows us to describe mesoscopic dynamics of a *specific* complex polymer liquid.

We model the polymer melt as a compressible system, consisting of collection of Gaussian chains in a mean field environment. The free energy functional for copolymer melts has the form [19, 20].

$$F[\{\rho\}] = -kT\left(\ln\frac{\Phi^n}{n!} + \sum_I \int_V U_I(\mathbf{r})\rho_I(\mathbf{r})\,d\mathbf{r}\right)$$

$$+ \frac{1}{2}\sum_{I,J}\int_{V^2} \varepsilon_{IJ}(\mathbf{r}-\mathbf{r}')\rho_I(\mathbf{r})\rho_J(\mathbf{r}')\,d\mathbf{r}\,d\mathbf{r}'$$

$$+ \frac{\kappa_H}{2}\int_V\left(\sum_I v_I(\rho_I(\mathbf{r}) - \rho_I^0)\right)^2 d\mathbf{r} \tag{1}$$

where n is the number of polymer molecules, Φ is the intra-molecular partition function for ideal Gaussian chains in an external field U, I is a component index and V is the system volume. The external potentials U_I are conjugate to the densities ρ_I via de Gaussian chain density functional. The average concentration is ρ_I^0 and v_I is the molecular volume. The cohesive interactions have kernels ε_{IJ}. The Helfand compressibility parameter is κ_H [20]. For our system (1) the statistical distribution function of a copolymer chain of N beads, in a certain conformation specified by the coordinates of the beads $\{\mathbf{R}_1, \ldots, \mathbf{R}_N\}$, is

$$\Psi(\mathbf{R}_1, \ldots, \mathbf{R}_N) =$$

$$\exp\left(-\left[\frac{3}{2a^2}\sum_{s=2}^{N}(\mathbf{R}_s - \mathbf{R}_{s-1})^2 + \sum_{s'=1}^{N} U_{s'}(\mathbf{R}_{s'})\right]\right), \tag{2}$$

where the external fields $U_{s'}$ are in units kT and a is the Gaussian bond strength parameter. The ensemble average particle density $\rho_s(\mathbf{r})$ of a certain bead s at position $\mathbf{r}$ in space is

$$\rho_s[U](r) \propto \int_{V^N} \Psi(\mathbf{R}_1, \ldots, \mathbf{R}_N)\,\delta(\mathbf{r} - \mathbf{R}_s)\,d\mathbf{R}_1 \ldots d\mathbf{R}_N. \tag{3}$$

Since the sequence space is discrete, the density functional can be calculated via Green propagators [23]

$$\rho_s(\mathbf{r}) \propto G_s(\mathbf{r}) \times \sigma[G^{\text{inv}}_{s+1}](\mathbf{r}). \tag{4}$$

The set of (once integrated) Green functions $G_s(\mathbf{r})$ and $G^{\text{inv}}_s(\mathbf{r})$ are related by the recurrence equations

$$G_s(\mathbf{r}) = e^{-U_s(\mathbf{r})}\sigma[G_{s-1}](\mathbf{r}),$$

$$G^{\text{inv}}_s(\mathbf{r}) = e^{-U_s(\mathbf{r})}\sigma[G^{\text{inv}}_{s+1}](\mathbf{r}), \tag{5}$$

with $G_0(\mathbf{r}) = G^{\text{inv}}_{N+1}(\mathbf{r}) = 1$. The linkage operator σ is defined as a convolution with a Gaussian kernel:

$$\sigma[f](\mathbf{r}) \equiv \left(\frac{3}{2\pi a^2}\right)^{3/2} \int_V e^{-(3/2a^2)(\mathbf{r}-\mathbf{r}')^2} f(\mathbf{r}')\, d\mathbf{r}'. \tag{6}$$

As for the dynamic model, the time evolution of the density field $\rho_I(\mathbf{r})$ under simple shear flow, $v_x = \dot{\gamma}y,\ v_y = v_z = 0$, can be described by the diffusion–convection equation with a stochastic term [8, 12]

$$\dot{\rho}_I = M_I \nabla \cdot \rho_1 \nabla \frac{\delta F}{\delta \rho_I} - \dot{\gamma}y\nabla_x\rho_I + \eta_I, \tag{7}$$

where M_I is a mobility parameter, $\dot{\gamma}$ is the shear rate (the time derivative of the strain γ), η_I is a thermal noise field which is distributed according to the fluctuation-dissipation theorem [22]. For a $L \times L \times L$ cubic grid we use a sheared periodic boundary condition [26]: $\rho(x, y, z, t) = \rho(x + iL + \gamma jL, y + jL, z + kL, t)$. Here we consider an approximation in the sense that the form of the chain distribution function (2) is not changed in shear.

The density functional technique allows us also to perform conformation analysis of polymer chains. For example, the probability of finding bead s at coordinate $\mathbf{r}$ while the bead s_0 of the same molecule fixed at $\mathbf{r}_0$ is

$$\rho^{(2)}_{ss_0}[U](\mathbf{r}, \mathbf{r}_0) \propto \int_{V^N} \Psi(\mathbf{R}_1, \ldots, \mathbf{R}_N)$$

$$\times \delta(\mathbf{r} - \mathbf{R}_s)\delta(\mathbf{r}_0 - \mathbf{R}_{s_0})\, d\mathbf{R}_1 \ldots d\mathbf{R}_N. \tag{8}$$

In the Green propagator algorithm (5) this leads to replacement of $\exp(-U_{s_0}(\mathbf{r}))$ by $\delta(\mathbf{r} - \mathbf{r}_0)$.

We simulate the behavior of a model polymer which is represented by an A_8B_8 Gaussian chain. The dimensionless parameters that are used in the numerics are chosen as (see [19] for details): the dimensionless time $\tau = \beta^{-1}Mh^{-2}t$, the grid parameter $d = ah^{-1} = 1.1543$, the noise scaling parameter Ω equal to 100, the exchange parameter $\chi_{AB} = \beta\varepsilon_{AB}v^{-1} = 1.0$ ($\chi_{AA} = \chi_{BB} = 0$) and the compressibility parameter $\kappa' = \beta\kappa_H v = 12.0$. The new dimensionless parameter $\tilde{\gamma}$ for the shear is chosen equal to $\tilde{\gamma} = \Delta t\dot{\gamma} = 10^{-3}$. In our simulations $\beta^{-1}Mh^{-2}\Delta t = 1/2$.

We performed a 3D simulation ($L = 64$) starting with the initial structure at $\tau = 500$ (Fig. 1) where no lamellar structure can be observed. During the application of

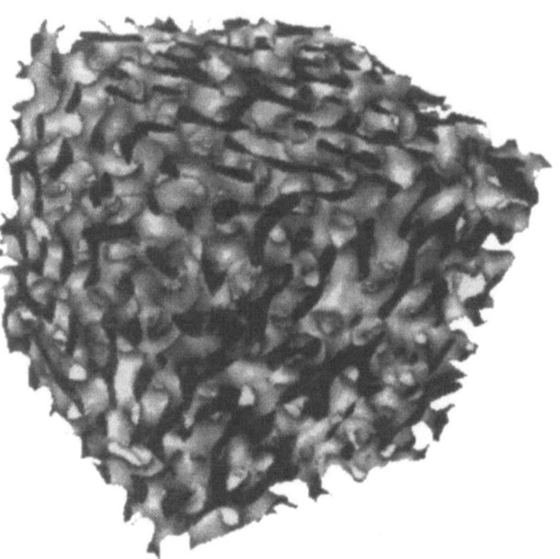

Fig. 1 Isosurface representation of the A_8B_8 melt at $\tau = 500$ with diffusion only. The isolevel is $\theta_A = v\rho_A = 0.3$. This structure is used as an initial structure for shearing

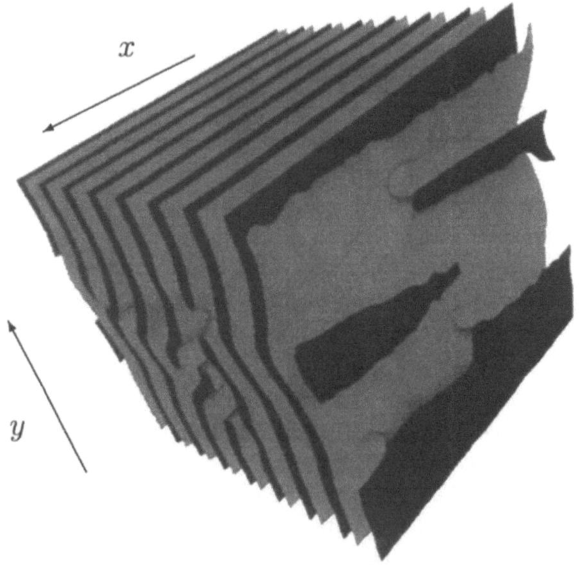

Fig. 2 Isosurface representation of the A_8B_8 melt at $\tau = 7500$ and shear flow $v_x = \dot{\gamma}y,\ v_y = v_z = 0$, using the structure at $\tau = 500$ (Fig. 1) as a starting structure. The isolevel is $\theta_A = v\rho_A = 0.3$. One can clearly observe the global lamellar orientation. The x and y-axis are indicated

steady simple shear between $\tau = 500$ and $\tau = 7500$ a global lamellar orientation appears. The structure essentially differs from one without shear, where only local lamellar clusters with many defects were observed. Thus, application of shear speeds up lamellar formation in a diblock polymer melt. The observed alignment is so called "perpendicular" lamellae (Fig. 2). From experiments and stability analysis this orientation is well known to be the most stable one at high shear rates [3].

Conformational analysis provides the detailed view on the density map of polymer chains in observed morphologies. Two examples are shown in Figs. 3 and 4. In the first case we fixed one bead from a central part of a chain at the surface of perfect lamella. The symmetric shape of the density probability isosurface corresponds to almost undeformed polymer chains. This is not the case in the vicinity of defects, like inter-lameller necks (Fig. 4).

Note that lamellar structure is a stable configuration of a symmetric diblock copolymer melt. After reaching almost perfect lamellar structure we have applied alternative simple steady shear flow with a velocity gradient perpendicular to lamellar plane (Fig. 5). The system has not been destroyed, which indicates that the structure is enough stable. The number of defects did not increase as well.

Our approach can be used for the investigation of shear induced ordering in *specific* polymer systems. The simulation of morphology evolution of Pluronic (lamellar, hexagonal, micellar, bicontinuous phases) under shear is in progress.

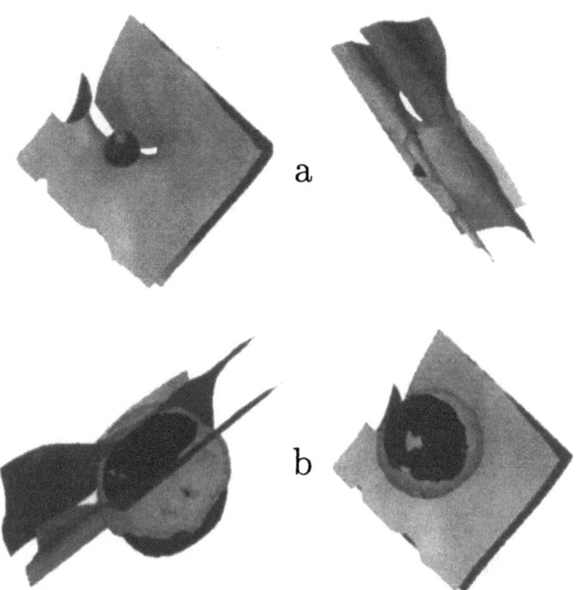

Fig. 4 Conformational analysis of A_8B_8 molecule while bead $A_{n=7}$ fixed at the surface of an interlamellae neck. This position indicated by small square in (a) for two rotations of the same 3D picture. (b) show the isosurface of the density probability (8) of finding A – (light gray) and B – (dark) tails of the chain in space. The density map has clearly non-symmetric shape

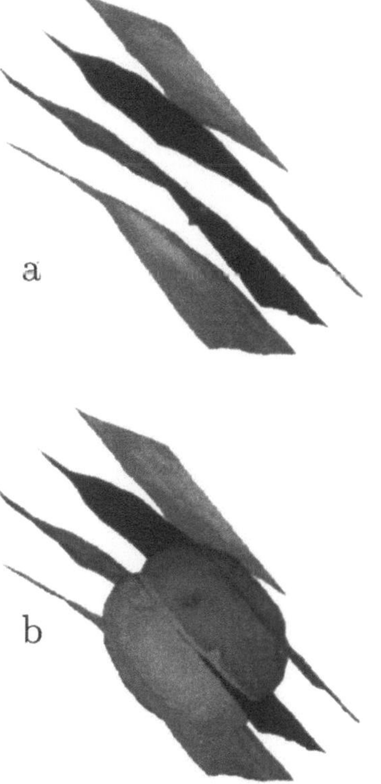

Fig. 3 Conformational analysis of A_8B_8 molecule while bead $A_{n=7}$ fixed at the surface of perfect lamella. This position indicated by small gray square in (a). (b) shows the isosurface of the density probability (8) of finding A – (light gray) and B – (dark grey) tails of the chain in space

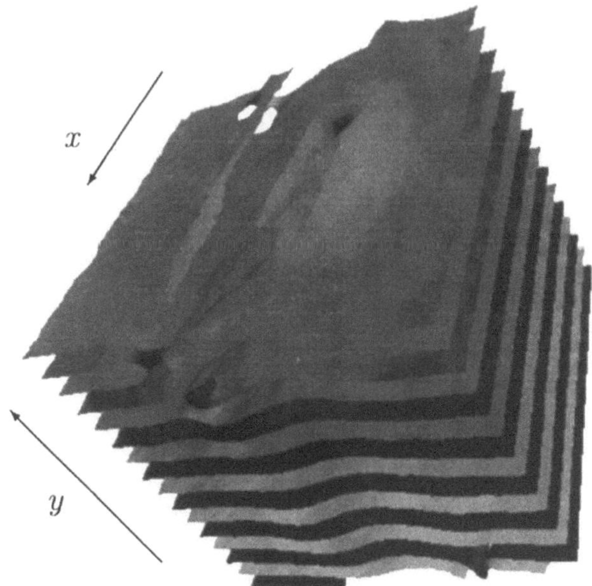

Fig. 5 Isosurface representation of the A_8B_8 melt in alternating shear flow of the same shear rate as in Fig. 2. The structure of the Fig. 2 has been taken as initial one after rotating around x-axis over $\pi/2$. The shear has been applied during time $\tau = 1325$

A.V. Zvelindovsky et al.
Lamellar phase of diblock copolymer melt under shear

References

1. Bird RB, Hassager O, Armstrong RC (1987) in "Dynamics of polimeric liquids". Wiley, New York
2. Schmittmann B, Zia RKP (1994) In: Domb C, Lebowitz J (eds) Phase Transitions and Critical Phenomena. Academic, London
3. Chen ZR, Kornfield JA, Smith SD, Grothaus JT, Satkowski MM (1997) Science 277:1248–1253
4. Fredrickson GH (1994) J Rheol 38: 1045–1067
5. Goulian M, Milner ST (1995) Phys Rev Lett 74:1775–1778
6. Cates ME, Milner ST (1989) Phys Rev Lett 62:1856–1859
7. Balsara NP, Dai HJ (1996) J Chem Phys 105:2942–2945
8. Kodama H, Doi M (1996) Macromolecules 29:2652–2658
9. Qiwei He D, Nauman EB (1997) Chem Engineering Sci 52:481–496
10. Gonnella G, Orlandini E, Yeomans JM (1997) Phys Rev Lett 78:1695–1698
11. Pätzold G, Dawson K (1996) J Chem Phys 104:5932–5941
12. Kodama H, Komura S (1997) J Phys II France 7:7–14
13. Ohta T, Nozaki H, Doi M (1990) Phys Lett A 145:304–308
14. Cahn JW, Hilliard JE (1958) J Chem Phys 28:258–267
15. Oono Y, Puri S (1987) Phys Rev Lett 58:836–839
16. de Gennes PG (1980) J Chem Phys 72:4756–4763
17. Maurits NM, Fraaije JGEM (1997) J Chem Phys 106:6730–6743
18. Fraaije JGEM (1993) J Chem Phys 99:9202–9212
19. Fraaije JGEM, van Vlimmeren BAC, Maurits NM, Postma M, Evers OA, Hoffmann C, Altevogt P, Goldbeck-Wood G (1997) J Chem Phys 106:4260–4269
20. Maurits NM, van Vlimmeren BAC, Fraaije JGEM (1997) Phys Rev E 56:816–825
21. Maurits NM, Fraaije JGEM (1997) J Chem Phys 107:5879–5889
22. van Vlimmeren BAC, Fraaije JGEM (1996) Comp Phys Comm 99:21–28
23. Maurits NM, Altevogt P, Evers OA, Fraaije JGEM (1996) Comp Pol Sci 6:1–8
24. Kawakatsu T (1997) Phys Rev E 56:3240–3250
25. Hasegawa R, Doi M (1997) Macromolecule 30:3086–3089
26. Doi M, Chen D (1989) J Chem Phys 90:5271–5279

Progr Colloid Polym Sci (1998) 110:255–257
© Steinkopff Verlag 1998

G.J.M. Koper
P.A. Cirkel

Do lecithin worm-like micelles behave as equilibrium polymers?

G. J. M. Koper (✉) P. A. Cirkel
Leiden Institute of Chemistry
Leiden University
PO Box 9502
2300 RA Leiden
The Netherlands

Abstract Data collected by means of a number of different experimental techniques on lecithin water-in-oil microemulsions clearly indicate that for water-to-surfactant ratios below 2 and for moderately low concentrations the system is most likely a rod-like micellar solution. For very low concentrations, the micelles are sphere-like. The concentration range in which the sphere-like micelles grow towards rod-like micelles has been alternatively interpreted as indicative of the growth of flexible worm-like micelles. Both the fact that the micelles are here found to be rod-like instead of flexible and the fact that the alternative interpretation yields the anomously large value of 1.2 for the growth exponent, lead to the conclusion that lecithin worm-like micelles do not behave as equilibrium polymers.

Key words worm-like micelles – equilibrium polymers – polymers

Introduction

Lecithin solutions in alkanes, such as isooctane and cyclohexane, with trace amounts of water are microemulsions in which the micelles have a long and cylindrical shape. These systems have recently been subject to intensive research because they are believed to resemble equilibrium polymer solutions [1]. In contrast to conventional polymers, of which the size is fixed, the size distribution of equilibrium polymers is dictated by thermodynamic conditions: in a mean-field picture the size distribution follows from a balance between the enthalpic contribution of forming end caps and the entropic contribution of the various species in the solution [2]. Sofar, most research has concentrated on the scaling regime above the, due to the rather broad size distribution ill-defined, overlap concentration and the agreement with scaling theory [3] has been demonstrated by extensive static and dynamic light scattering experiments [4].

In this contribution, we report on experiments that have been mainly performed in the dilute regime and for low values of the water-to-surfactant ratio and we show that the thus obtained experimental results lead to another picture of these microemulsions in which the micelles appear to be rod-like which hardly, if at all, vary in size with concentration as would be expected on the basis of the equilibrium polymer model. For larger values of the water-to-surfactant ratio the picture is less clear, probably because the micelles have a branched structure [5].

Materials

Samples were prepared with soybean lecithin from Fluka, type Epikuron 200. It was used without further purification so that it consists of a specific mixture of variable chain length, n, and degree of saturation, m (distribution of fatty acids, $C_{n:m}$: $C_{16:0} = 13.3\%$, $C_{18:0} = 3.0\%$, $C_{18:1} = 10.2\%$, $C_{18:3} = 66.9\%$, and $C_{18:3} = 6.6\%$) [6]. The samples are identified by two parameters: w_0, the molar ratio of *added* water to surfactant and C, concentration of surfactants in g/l (useful data to compare these with other quantities in use are: average molar weight of lecithin: 760 g/mol, the density of lecithin: 1.056 kg/l). The water already present per mole of lecithin was determined to be approximately 0.7 mol by IR-spectroscopy and Karl–Fisher titration [7]. This value should be added to w_0 in order to obtain the actual water-to-surfactant ratio in the sample. Isooctane was of analytical grade and obtained from Merck. Water was purified with a Millipore Milli-Q filtering apparatus. Special attention was given to the control of the water content in the samples: A concentration series at one w_0 was performed diluting batches from a large stock solution (0.5 or 0.25 l), to make rather large samples (25, 50 or 100 ml).

Results and discussion

In Fig. 1 the results are shown of static and dynamic light scattering experiments for solutions with water-to-surfactant ratio $w_0 = 1.5$ and lecithin concentrations ranging from 0 to 6 g/l. In the lecithin concentration range from 1 to 6 g/l both the radius of gyration R_g and the hydrodynamic radius R_h do not systematically depend on concentration and their ratio is approximately 1.84. This indicates that, in this regime, the micellar size is independent of lecithin concentration. Similar indications for a concentration-independent micellar size were obtained from experiments using viscometry, dielectric spectroscopy, transient electric birefringence, and dynamic sedimentation [5]. The average values for the hydrodynamic radius and the radius of gyration in this concentration range are, together with the results from these other experiments, displayed in Table 1. Also, in this Table 1 are the theoretical values of these parameters as calculated for a rod-like [8] and for flexible chains. To obtain the parameters for the rod, a value of 30 Å, as obtained from neutron-scattering experiments [9], was used for the cylinder radius and for the contour length 1800 Å, as calculated from the micellar mass obtained by sedimentation experiments. The latter value is in agreement with what can be found from the light scattering experiments. For flexible chains two different contour lengths were taken, 1800 Å as obtained from the micellar mass and 5835 Å as obtained by light scattering, and the Kuhn length, of 300 Å, was taken from neutron-scattering experiments [10]. From the comparison of these parameters in Table 1 it becomes clear, that the rod-like chain model fits the experimental data best. Similar results are obtained for w_0 values 0.5 and 1.

The ratio of the radius of gyration R_g and the hydrodynamic radius R_h is in the concentration regime from 1 to

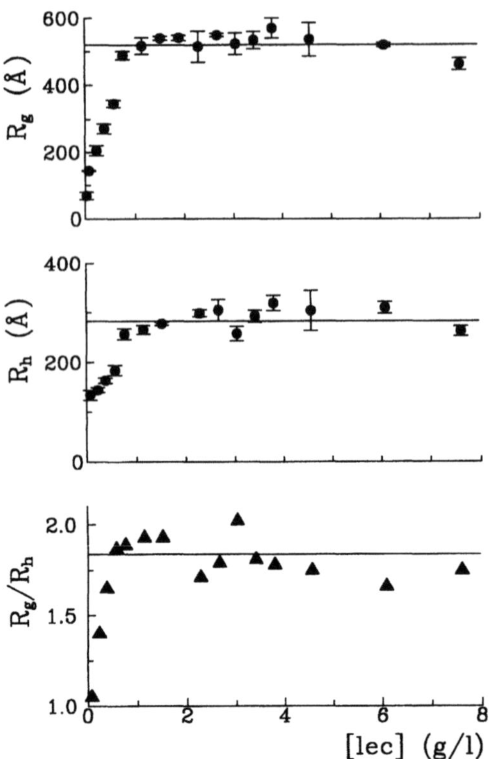

Fig. 1 The radius of gyration, R_g, hydrodynamic radius, R_h, and their ratio as determined by light scattering for the $w_0 = 1.5$ microemulsion

6 g/l equal to 1.84, see Fig. 1, and this is compatible with a rod-like micellar shape. Upon lowering the concentration, this ratio drops to 1 which is close to the value for sphere-like micelles. We therefore conclude that a sphere-to-rod transition takes place in this concentration regime. This conclusion is further supported by the vanishing of the transient electric birefringence signal and the reduction of the radius of gyration to a value of 60 Å which is exactly twice the value for the cylindrical diameter as obtained by neutron scattering [10].

In summary, the data collected by means of a number of different experimental techniques clearly indicates that, in the regime of w_0 values below 2, the system is most likely a rod-like micellar solution for moderately low concentrations. For very low concentrations, the micelles are sphere-like. The concentration range in which the sphere-like micelles grow towards rod-like micelles has been alternatively interpreted as indicative of the growth of flexible worm-like micelles [11]. Both the fact that the micelles are here found to be rod-like instead of flexible and the fact that the alternative interpretation yields the anomously large value of 1.2 for the growth exponent [9] lead to the conclusion that lecithin worm-like micelles do not behave as equilibrium polymers.

Table 1 Experimental data and theoretical predictions for worm-like micellar solutions at $w_0 = 1.5$. Experimentally, the radius of gyration R_g is determined by static light scattering, the diffusion coefficient D by dynamic light scattering, the specific viscosity $[\eta]$ by viscometry, the rotation correlation time by dielectric spectroscopy, τ_{diel}, and by transient electric birefringence, τ_{Kerr}. Theoretical values for rods were calculated using 30 Å for cylinder radius and a contour length L of 1800 Å from the molar mass determined by sedimentation experiments. For flexible worm-like chains also 5835 Å was taken for the contour length as determined by light scattering. In the case of rod-like chains the molar mass determined by sedimentation and by light scattering agreed

	R_g (Å)	D ($10^{-11}\,\mathrm{m^2/s}$)	$[\eta]$ (ml/g)	τ_{diel} ($10^{-4}\,\mathrm{s}$)	τ_{Kerr} ($10^{-4}\,\mathrm{s}$)
Experimental results	520 ± 25	1.63 ± 0.12	54.6 ± 0.6	1.34 ± 0.25	1.8 ± 0.4
Rod-like chain, $L = 1800$ Å	520	1.74	54.0	1.356	2.1*
Flexible worm-like chain $L = 1800$ Å	266	2.64	17.7	0.177	0.3*
Flexible worm-like chain $L = 5834$ Å	520	1.37	48.8	0.911	1.4*

* These data need to be corrected by a factor of order 1 which sensitively depends on the actual size distribution of the micelles

References

1. Schurtenberger P, Scartazzini R, Magid LJ, Leser ME, Luisi PL (1990) J Phys Chem 94:3695–3701
2. For a review see: Cates ME, Candau SJ (1990) J Phys: Condens Matter 2:6869–6892
3. de Gennes P-G (1979) Scaling Concepts in Polymer Physics. Cornell University Press, Ithaca, NY
4. Schurtenberger P, Cavaco C (1994) Langmuir 10:100–108
5. Cirkel PA, Koper GJM (1998), to be published
6. Shinoda K, Araki M, Sadaghiani A, Khan A, Lindman B (1991) J Phys Chem 95:989–993
7. Scartazinni R (1990) PhD thesis nr. 9186, ETH Zürich
8. Doi M, Edwards SF (1986) The Theory of Polymer Dynamics. Clarendon Press, Oxford
9. Jerke G (1997) PhD thesis nr 12168, ETH Zürich
10. Jerke G, Pedersen JS, Egelhaaf SU, Schurtenberger P (1997) Physica B 234:273–275

Progr Colloid Polym Sci (1998) 110:258–262
© Steinkopff Verlag 1998

Curvature energy for droplet dimerization and aggregation in microemulsions

W.F.C. Sager
E.M. Blokhuis

W.F.C. Sager* (✉)
Faculty of Chemical Technology
Inorganic Materials Science
University of Twente
P.O. Box 217
7500 AE Enschede
The Netherlands

E.M. Blokhuis
*Department of Physical and
Macromolecular Chemistry
Leiden Institute of Chemistry
P.O. Box 9502
2300 RA Leiden
The Netherlands.

Abstract We present results of a recent theoretical study on the onset of droplet aggregation in microemulsion systems. Aggregation in microemulsions studied so far, has mainly been analyzed within the concept of liquid state theories. In this work we use the Helfrich free energy expression, which has widely been used to understand microemulsion phase diagrams and to calculate form fluctuations and polydispersity for microemulsion droplets, to compute the free energy for the formation of dimeric droplet aggregates. The curvature energy of the dimers could be analytically determined in the limit that the rigidity constant for bending, k, is small; i.e., when $(k/\sigma)^{1/2} \ll R$, with σ the surface tension and R the radius of the microemulsion droplet. This allowed us to determine the dimerization transition within the one-phase region of, e.g., (ionic) water-in-oil microemulsions. Dimers are favored when either the inverse radius of spontaneous curvature $1/R_0$ decreases, which goes in this case along with, e.g., an increase in temperature, or when the radius of the microemulsion droplets and thus the water/surfactant ratio increases.

Key words Microemulsions – aggregation – Helfrich free energy – sticky hard sphere model – curvature energy

Introduction

Aggregation processes and structural changes in microemulsions have been widely studied over the last two decades by a variety of different methods, for a recent overview see, e.g., Refs. [1,2]. Nevertheless, there is still no uniform physical picture available and the driving force for aggregation remains unknown. Experimentally, special attention has been paid to water-in-oil (w/o) microemulsion droplet phases stabilized by an ionic surfactant, e.g. sodium di-(2-ethylhexyl) sulfosuccinate (AOT) on the oil-rich side of the phase diagram and oil-in-water (o/w) microemulsion droplet phases stabilized by a nonionic surfactant, e.g., longer chain ethyleneoxide monoalkyl-ethers (C_iE_j), on the water-rich side. For both microemulsion systems, aggregation phenomena and changes from spherical to elongated structures occur as temperature is increased [3–7]. At low temperatures, close to the lower phase boundary, below which excess water (for ionic w/o microemulsions) or oil (for nonionic o/w microemulsions) is expelled, the so-called solubilization limit boundary, singly dispersed droplets exist, see the left-hand side of Fig. 1 [8]. The radius of the droplets is given by the water/surfactant ratio for w/o microemulsions and the oil/surfactant ratio for the case o/w microemulsions and the droplets behave like hard spheres [7,9]. Upon raising the temperature an increasing attractive interaction between the droplets has been found using, e.g., static light scattering as well as X-ray or neutron scattering [10,11].

Progr Colloid Polym Sci (1998) 110:258–262
© Steinkopff Verlag 1998

259

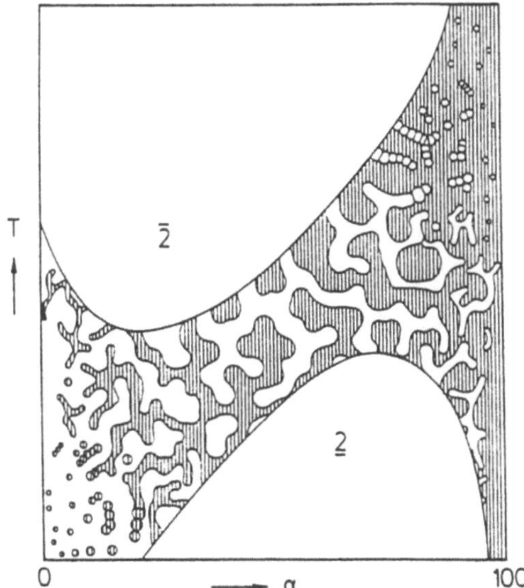

Fig. 1 Schematic phase diagram for a water–oil–nonionic surfactant system as a function of the oil weight fraction α and temperature at a surfactant concentration high enough that a channel by a one-phase microemulsion is formed reaching from the water ($\alpha = 0$) to the oil side ($\alpha = 100$) [17]. At low and high temperatures two-phase regions are present in which an o/w microemulsion is in equilibrium with an excess oil phase ($\underline{2}$) or a w/o microemulsion with an excess water phase ($\bar{2}$), respectively. Within the one-phase region the microstructure is visualised in the way that the white areas represent water domains and the shaded ones oil domains. The surfactant is situated at the water/oil interfaces, depicted by a black line. Droplet phases are stable on the water side (o/w) at low, and on the oil side (w/o) at high, temperatures. At intermediate temperatures and $\alpha \approx 50$ bicontinuous structures form, see also Ref. [28]. In this region the mean curvature of the interfacial surfactant film is approximately zero. Aggregation processes thus occur in the direction of zero mean curvature. A similar diagram could be obtained for a water–oil–ionic surfactant system, except that the temperature dependence is reversed due to the different nature of the surfactant [16]. A one-phase channel visualizing the structures detected by small angle neutron scattering is given in Ref. [6] for the system brine–decane–AOT

The formation of droplet dimers with increasing temperature and droplet volume fraction has been evidenced as the first aggregation step by a variety of different techniques, including dielectric permittivity, electro-optical birefringence and viscosity measurements, see, e.g., Ref. [2]. At temperatures close to the upper phase boundary cylindrical structures have been detected [12–14]. On the other hand, aggregation phenomena occurring with decreasing temperature have been observed [6,15] for ionic o/w microemulsions and nonionic w/o microemulsions, due to the opposite temperature behavior of ionic and nonionic surfactants [16], see right side of Fig. 1. Figure 1 displays a schematic phase diagram for a nonionic microemulsion system at constant surfactant concentration as a function

of temperature and oil weight fraction [17]. Within the one-phase region sketches of the different structures measured are presented.

Temperature induced aggregation of microemulsion droplets has so far, within the concepts of liquid state theories, been dealt with by assuming that the droplets interact via a hard-core plus a short-range attractive pair potential, which strength increases with temperature. In the so-called sticky hard sphere model the interaction potential is modelled by a square well potential with hard-core radius a, attractive well depth $-\varepsilon$, and width Δ, in the limit that ε tends to infinity and Δ to zero. In this limit, Baxter showed [18] that the Ornstein–Zernike equation can be solved analytically using the Percus–Yevick approximation. Chen et al. [7] used Baxter's model to analyse the phase diagram for the water–decane–AOT system obtained at a molar water/surfactant ratio of 40.8 as a function of temperature and droplet volume fraction ϕ, in terms of a liquid–gas transition. This system shows a lower critical point at $T_c = 40\,°\text{C}$ and $f_c = 0.1$. The two phases coexisting above T_c consist of a concentrated and a dilute microemulsion phase of the same droplet size found within the experimental error of small angle neutron scattering (SANS) measurements. They were able to fit the experimentally obtained phase diagram (including the percolation line) when the so-called stickiness parameter $1/\tau$, related to the square well parameters by

$$1/\tau = 12\frac{\Delta}{a}(e^{\beta\varepsilon} - 1), \tag{1}$$

with $\beta = 1/k_3 T$, was chosen appropriately as a function of temperature. The temperature dependence of the stickiness parameter was shown to be consistent with an analysis of the structure factor obtained using SANS (for more details see also Ref. [19]).

Within the "hard-sphere concept" it is, however, not possible to understand the occurrence of increased attractive interactions (and aggregate formation) as the interfacial layer surrounding the droplets is approaching the status of zero mean curvature (see Fig. 1). In this paper we emphasize the correlation between the formation of aggregates and changes in the spontaneous curvature of the surfactant film. This enables us to address droplet aggregation in the same framework as the (temperature) evolution of phases occurring in ternary mixtures of surfactant, water and oil is analyzed; namely by looking at the changes in the properties of the interfacial film separating water and oil domains. Recently, Fletcher and Petsev [20] considered the effect of the full interaction potential (van der Waals), including increased droplet deformability with vanishing spontaneous curvature, on the aggregation of microemulsion droplets. In this paper we use the Helfrich free energy [21] to calculate the curvature energy for the formation of microemulsion droplet dimers and the dimerization transition within the one-phase region of the phase diagram. The Helfrich free energy, originally introduced to describe the nonspherical shapes of membranes and vesicles, is now widely used [22–25] to understand microemulsion phase diagrams and to calculate form fluctuations and polydispersity for microemulsion droplets. The essential feature of this approach is to treat the structures formed in microemulsion systems as flexible objects which are not necessarily spherical. It therefore seems natural to combine both approaches and calculate the change in shape and free energy when two microemulsion droplets form a dimer. This allows to express, for instance, the stickiness parameter in terms of the spontaneous curvature and rigidity constants.

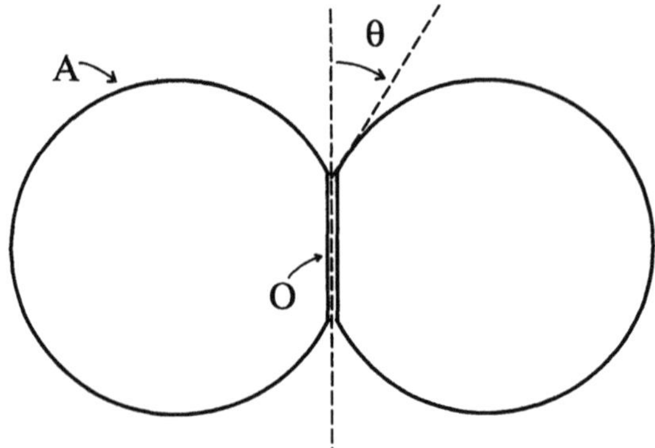

Fig. 2 Two aggregated microemulsion droplets forming a dimer. A is the surface area of the whole droplet while O is the flat surface area formed in between the two droplets. θ is the contact angle

Helfrich free energy for droplet dimerization

In this section, we use the Helfrich free energy [21] to show that a dimerization transition occurs with increasing radius of spontaneous curvature or increasing microemulsion droplet radius. We thus limit ourselves to the calculation of the free energy of dimers, and not larger aggregates, but it should be clear that the curvature energy supplies the underlying mechanism for the observed aggregation phenomena in microemulsion systems.

To calculate the free energy for the aggregation of two microemulsion droplets (see Fig. 2) we add to the usual Helfrich expression for the curvature free energy of the microemulsion droplet surface, a term describing the (contact) interaction energy between the two droplets

$$F = \int dA \left[\sigma - 2\frac{k}{R_0} J + \frac{k}{2} J^2 + \bar{k} K \right] + \int dO\, \sigma_b, \qquad (2)$$

where σ is the surface tension, R_0 is the radius of spontaneous curvature, k the rigidity constant associated with bending, and $\bar{k}$ the rigidity constant associated with Gaussian curvature. The first term features an integral over the dimer surface area, A, of the total curvature, $J = 1/R_1 + 1/R_2$ and Gaussian curvature, $K = 1/(R_1 R_2)$ with R_1 and R_2 the principal radii of curvature at a certain point on the surface A. The second term integrates the bilayer surface tension σ_b over the planar bilayer, with surface area O, formed in between the two droplets (see Fig. 2). We have thus assumed that the interaction energy between the two droplets does not extend beyond the surface area of contact.

The free energy in Eq. (2) has to be minimized with respect to the whole shape of the dimer assuming certain constraints, e.g. constant volume versus constant pressure or constant surface area versus constant surface tension. Unfortunately this minimization cannot be carried out analytically in general, and one has to resort to solving the shape equations, that follow from the minimization, numerically. In view of the large number of parameters (e.g. σ, σ_b, R_0, k) this is however extremely tedious and the need arises for limiting analytical results. We were able to obtain such a limiting result in the limit that the rigidity constant for bending, k, is small; i.e. when $(k/\sigma)^{1/2} \ll R$. This condition is obeyed for values of the parameters typical in microemulsions. We find that the leading order contribution to the free energy due to the presence of rigidity scales as $k^{1/2}$, i.e., the free energy is *not analytic* in k. This is due to the fact that near the edges, where the bilayer surface and monolayer surface meet, the curvature changes abruptly from zero (bilayer) to $\approx 1/R$ (monolayer).

To calculate the dimerization transition we now discuss the conditions under which microemulsions form. The total amount of internal component (the component *inside* the microemulsion droplet) fixes the available volume V_{tot}, while the total amount of available surfactant constraints the total surface area A_{tot}. Let N_m and N_d denote the number of monomers and dimers with radius R_m and R_d, respectively. Then $V_{\text{tot}} = N_m V_m + N_d V_d$ and $A_{\text{tot}} = N_m A_m + N_d A_d + \beta N_d O_d$, where V_m, A_m are the volume and surface area of one droplet, respectively, and V_d, A_d, O_d, are the volume, surface area of the monolayer and surface area of the bilayer in the dimer, respectively. The ratio of the amount of surfactant per unit area in the bilayer compared to the monolayer, realistically expected to be close to 2, appears as the parameter β in the expression for A_{tot}.

We can now investigate the occurrence of a monomer to dimer transition by minimizing the curvature free energy with respect to N_m, R_m, N_d, R_d, and $x \equiv \cos\theta$, keeping V_{tot} and A_{tot} fixed and neglecting any entropic

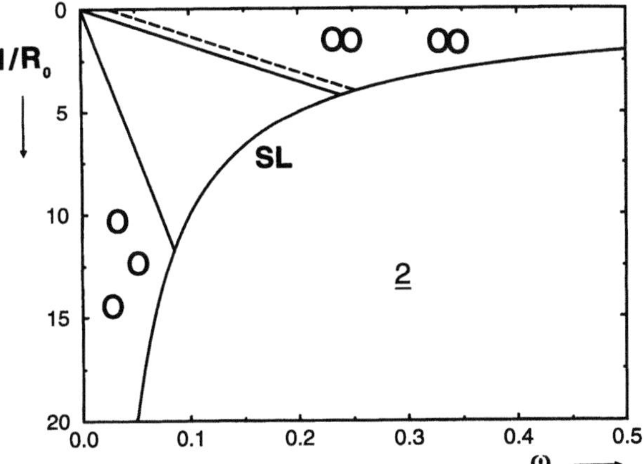

Fig. 3 Phase diagram for the dimerization transition in a microemulsion system as a function of $1/R_0$ and $\omega \equiv 3V/A$. The solid curve SL is the solubilization limit beyond which the internal component is present as an excess phase. We have chosen $\sigma_b/\sigma = 1$, $(k/\sigma)^{1/2} = 0.1$ and two values for β. The solid curves denote the dimerization transition for $\beta = 2.1$ (left) and $\beta = 6$ (right) without rigidity. The dashed curve is the dimerization transition for $\beta = 6$ *with* rigidity

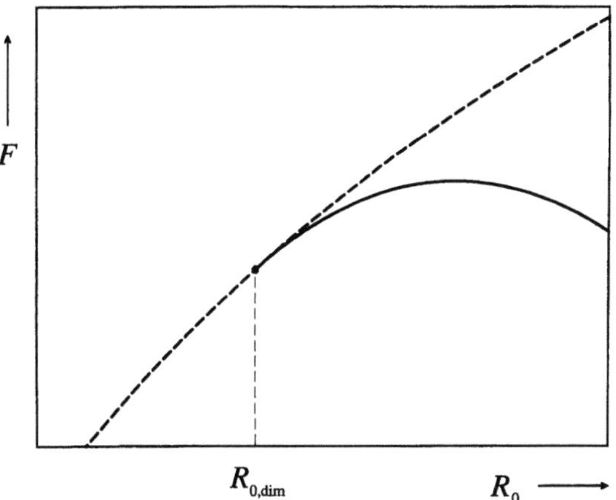

Fig. 4 Sketch of the competing free energies of singles droplets (dashed curve) and dimers (solid curve) as a function of R_0 (or equivalently radius R, or temperature T). To the left of the (continuous) dimerization transition $R_{0,\mathrm{dim}}$, the free energy of the single droplets is lower than that of the dimers. To the right of the dimerization transition, the dimers have the lowest free energy

contributions. Details of the calculation will be presented elsewhere [26] but typical results are shown in Fig. 3. Here the phase diagram is depicted as a function of the inverse radius of spontaneous curvature $1/R_0$ and the internal phase (e.g. water in the case of w/o microemulsions) to surfactant ratio parameter $\omega = 3V_{\mathrm{tot}}/A_{\mathrm{tot}}$. Such a diagram is experimentally accessible as a function of T and ω for ionic w/o and nonionic o/w microemulsions. The solid curve SL, defined by $1/R_0 = 1/\omega$, is the solubilisation limit beyond which ($\omega > \omega_{\mathrm{SL}}$) the microemulsion droplet phase coexists with an excess water (for ionic w/o emulsions) or oil phase (for nonionic o/w emulsions). The two solid curves are the loci of the monomer–dimer transition for $\beta = 2.1$ (left curve) and $\beta = 6$ (right curve) without rigidity. We are, of course, able to use the same approach to calculate the dimerization transition in a system for which the solubilization boundary is the *upper* phase boundary of the one-phase region, e.g. in nonionic w/o or ionic o/w microemulsions.

The transition from monomers to dimers occurs with decreasing $1/R_0$ as observed in experiments, or increasing ω. Since the monomer droplet radius $R_m = \omega$, the latter condition indicates that dimerization occurs with increasing droplet radius. This has indeed been observed by Huang et al. [27] in small angle neutron scattering experiments. The dimerization transition is continuous (second order) when $\beta < \beta_c$ and first order when $\beta > \beta_c$ ($\beta_c \approx 4.30 \ldots$). The continuous transition is illustrated by a sketch of the free energy of single droplets and dimers in Fig. 4 as a function of R_0 (or equivalently radius R, or temperature T). The free energy of the dimers (solid curve)

smoothly joins the free energy of the single droplets (dashed curve) at the dimerization transition $R_{0,\mathrm{dim}}$.

The locus of these continuous transitions is determined by

$$\frac{1}{R_{0,\mathrm{dim}}} = \frac{3}{8}\frac{\omega}{k}\frac{\beta\sigma - \sigma_h}{\beta - 2}. \tag{3}$$

The presence of rigidity shifts the first order transition to lower $1/R_0$ (higher ω) as can be seen by the dashed curve in Fig. 3. To leading order in k, however, the presence of rigidity does not affect the location of the continuous transition.

Conclusions

In this paper we showed that droplet aggregation can lower the free energy if the spontaneous curvature of the interfacial film approaches zero. Our approach allows to treat dimer formation and droplet aggregation under the same framework of the Helfrich curvature energy as the phase evolution in microemulsions. It therefore allows the discussion of aggregation and structural transitions, such as the formation of cylinders and elongated micelles, on equal footing.

Acknowledgements The research of E.M.B. has been made possible by a fellowship of the Royal Netherlands Academy of Arts and Sciences. The work of W.F.C.S. has been supported by the Netherlands Foundation for Chemical Research (SON) in collaboration with the Netherlands Technology Foundation (STW).

References

1. Chevalier Y, Zemb T (1990) Rep Prog Phys 53:279–371
2. Koper GJM, Sager WFC, Smeets J, Bedeaux D (1995) J Phys Chem 99:13291–13300
3. Kahlweit M, Strey R, Haase D, Kunieda H, Schmeling T, Faulhaber B, Borkovec M, Eicke H-F, Busse G, Eggers F, Funck T, Richmann H, Magid L, Söderman O, Stilbs P, Winkler J, Dittrich A, Jahn W (1987) J Colloid Interface Sci 118:436–453
4. Fletcher PDI, Morris JS (1995) Colloids Surfaces A 98:147–154
5. Leaver M, Furo I, Olsson U (1995) Langmuir 11:1524–1529
6. Chen SH, Chang SL, Strey R (1990) J Chem Phys 93:1907–1918
7. Chen SH, Ku CY, Rouch J, Tartiaglia P, Cametti C, Samseth J (1993) J de Physique IV 3:143–161
8. Leaver MS, Olsson U, Wennerström H, Strey R (1994) J Phys II France 4:515–531
9. Menon SVG, Kelkar VK, Manohar C (1991) Phys Rev A 43:1130–1133
10. Cazabat AM (1985) In: Degiorgio V, Corti M (eds) Physics of Amphiphiles: Micelles, Vesicles and Microemulsions, North-Holland, Amsterdam, pp 723–753
11. Hayter B (1985) In: Degiorgio V, Corti M (eds) Physics of Amphiphiles: Micelles, Vesicles and Microemulsions, North Holland: Amsterdam, pp 59–88
12. Glatter O, Strey R, Schubert K-V, Kaler EW (1996) Ber Bunsenges Phys Chem 100:323–335
13. Barnes IS, Hyde ST, Ninham BW, Derian P-J, Drifford M, Zemb TN (1988) J Phys Chem 92:2286–2293
14. Olsson U, Würz U, Strey R (1993) J Phys Chem 97:4535–4539
15. Leaver MS, Olsson U, Wennerström H, Strey R, Würz U (1995) J Chem Soc Faraday Trans 91:4269–4274
16. Kahlweit M, Strey R, Busse G (1990) J Phys Chem 94:3881–3894
17. Schomäcker R (1991) J Phys Chem 95:451–457
18. Baxter RJ (1968) J Chem Phys 49:2770–2774
19. Robertus C, Philipse WH, Joosten JGH, Levine YK (1989) J Chem Phys 90:4482–4490
20. Fletcher PDI, Petsev DN (1997) J Chem Soc Faraday Trans 93:1383–1388
21. Helfrich W (1973) Z Naturforsch. C 28:693–703
22. Andelman D, Cates ME, Roux D, Safran SA (1987) 12:7229–7241
23. Olsson U, Wennerström H (1994) Adv Colloid Interface Sci 49:113–146
24. Huang JS, Millner ST, Farago B, Richter D (1987) Phys Rev Lett 59:2600–2603
25. Safran SA (1994) Statistical Thermodynamics of Surfaces, Interfaces, and Membranes. Addison-Wesley, Reading, MA
26. Blokhuis EM, Sager WFC (1998) to be published.
27. Huang JS, Safran SA, Kim MW, Grest GS, Kotlarchyk M, Quirke N (1984) Phys Rev Lett 53:592–595
28. Strey R (1994) Colloid Polym Sci 272:1005–1019

Progr Colloid Polym Sci (1998) 110:263–268
© Steinkopff Verlag 1998

T.D. Gurkov
K.G. Marinova
A. Zdravkov
C. Oleksiak
B. Campbell

Gradual disintegration of protein lumps contained in thin emulsion films: Role of the surface diffusion

T.D. Gurkov (✉) K.G. Marinova
A. Zdravkov
Laboratory of Thermodynamics and
Physico-Chemical Hydrodynamics
University of Sofia
Faculty of Chemistry
James Bourchier Avenue 1
Sofia 1126
Bulgaria

C. Oleksiak · B. Campbell
Kraft General Foods, Inc.
Technology Center
801 Waukegan Road
Glenview, Illinois 60025
USA

Abstract We study thin aqueous films immersed in oil phase. The system is stabilized by globular protein (BSA), adsorbed on the liquid interfaces. Reversible surface aggregation of protein is observed. Big lumps remain entrapped within the film, where they are squashed by the pressing action of the curved wrapping menisci. The lumps gradually disintegrate and finally disappear. We propose a theoretical explanation of this phenomenon. Molecular migration by the mechanism of surface diffusion is found to be responsible for removing the material from the aggregates. The latter shrink isotropically (at least in a certain range of sizes). The particle radius was measured by interference microscopy, as a function of time. The data were fitted with theoretical computations. The partial differential equation of non-stationary surface diffusion was solved numerically, with boundary condition imposed on the moving circumference of the big protein cluster. Very good agreement between theory and experiment is obtained. From the best fit we find one parameter, whose value gives the maximum adsorption along the rim where the lump is attached to the interface. The results suggest existence of protein multilayers on the liquid surface.

Key words Surface diffusion – Protein aggregation – Thin liquid films – Emulsion-type interfaces

Introduction

Here we present a theoretical explanation of the experimentally observed phenomenon of progressive squashing and disappearance of protein aggregates pressed between the surfaces of thin liquid films. We have previously studied aqueous films, sandwiched between styrene phases, in the presence of bovine serum albumin (BSA) [1]. Reversible surface aggregation of the protein was discovered: Films made in aged systems, more than about 30 min after loading of the two phases in the cell, comprised large protein clusters. No aggregation in the bulk of the solution was detected. Therefore, we could conclude that the slow process of lump formation was promoted by the conformational changes and partial denaturation of the protein molecules on the interface [1].

The aggregation is seen to be reversible. The excess material, entrapped in Newton black films of emulsion type, is gradually squeezing out, and the particles are destroyed. Finally, the whole film becomes uniformly thin and black, as the lumps disappear. This process takes a relatively short period of time – usually about 2–3 min. A question can be raised about the kind of physical mechanism which is responsible for removing the protein from the clusters. Dissolution of molecules into the bulk aqueous phase that fills up the film interior should be ruled out: the Newton black films virtually do not thin, and there is very little liquid (if any) between the two opposing interfacial layers of protein. Moreover, the process of protein redispersion to the bulk is likely to be very slow. The physical state of the molecules in the lump (partially denatured and entangled with their neighbours) is quite different from that in the water solution.

In the present work we shall investigate the possibility that the macromolecules migrate by surface diffusion, and in this way they are removed from the big particles. It is well known that proteins, similarly to lipids and low molecular weight surfactants, are involved in surface diffusion on liquid boundaries [2]. The process was studied by the method of fluorescent recovery after photobleaching (FRAP), applied to thin foam films. Systems stabilized by proteins [2–4] and phospholipids [5] were investigated. The surface diffusion coefficient, D_s, was determined to be about 1×10^{-7} cm^2/s for BSA [2]. We shall use this value in our analysis below. In the frames of a simple model, the lateral distribution of the surface concentration of protein will be calculated as a function of position and time, for a lump whose size gradually diminishes. The time dependence of the particle diameter will be computed and the results will be compared with direct microscopic observations in thin emulsion films containing BSA.

Theoretical model

Let us consider the idealised geometry depicted in Fig. 1. The protein lump is represented as a cylinder with radius a and height H (functions of time). For the sake of simplicity, only one film surface, S, is shown. In reality, the particles captured within a thin liquid film are subjected to increased internal pressure, Δp, due to the squeezing action of the curved wrapping menisci. Explicit calculation of Δp will not be attempted here. Instead, we shall focus on the changes in the adsorption.

When there is no film, the aggregates attached to the interface would be resting in equilibrium with the surrounding surface layer of protein. One can write the condition for equality of the chemical potentials:

$$\mu_{\text{lump}}(T, p) = \mu_{\text{surf}}^0(T, p) + kT \ln \Gamma_0. \tag{1}$$

The expression in the right-hand side of Eq. (1) represents the surface chemical potential in the form used by Gurkov et al. [6], where Γ_0 is the equilibrium adsorption. $\mu_{\text{lump}}(T, p)$ refers to the cluster. When the latter is entrapped in the film, *local* equilibrium is maintained between the material inside the particle and the nearest points on the interface, in close contact with the lump (i.e., along the circumference $r = a$ in Fig. 1). The physical state of the protein in the aggregate is very close to that on the surface.

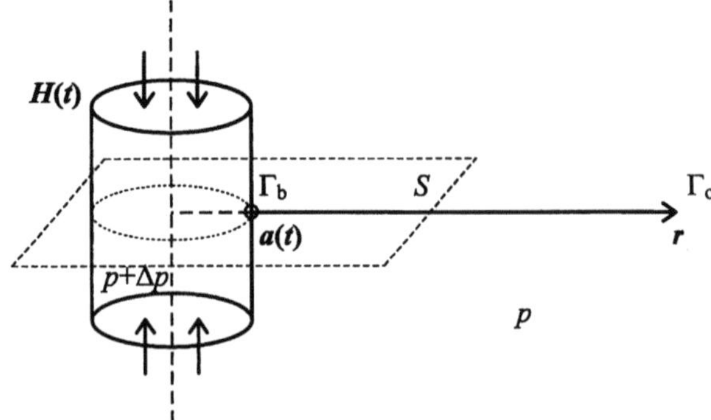

Fig. 1 Simplified geometric model used for solution of the diffusion problem. The cylindrical lump has radius $a(t)$ and height $H(t)$

The counterpart of Eq. (1) for a pressed lump reads

$$\mu_{\text{lump}}(T, p + \Delta p) = \mu_{\text{surf}}^0(T, p) + kT \ln \Gamma_b, \tag{2}$$

at $r = a$. Here we have by definition $\Gamma_b \equiv \Gamma(r = a)$, and this quantity will be supposed to remain constant during the process of squashing of the cluster. Indeed, experimental observations have shown that the particle shrinks isotropically, its shape does not change with time. Hence, Δp will stay more or less the same, at least for not very small sizes.

The mass balance of protein at the interface, for $r > a$ (in cylindrical symmetry, see Fig. 1), is expressed by the following equation [7]:

$$\frac{\partial \Gamma}{\partial t} = D_s \left[\frac{\partial^2 \Gamma}{\partial r^2} + \frac{1}{r} \frac{\partial \Gamma}{\partial r} \right], \tag{3}$$

where Γ is the adsorption (number of molecules per unit area). The respective boundary conditions are

$$\Gamma(r = a(t), t) = \Gamma_b = \text{const.}, \quad \Gamma(r \to \infty, t) = \Gamma_0 = \text{const.}, \tag{4}$$

and the initial condition (at the moment of film formation) is

$$\Gamma(r, t = 0) = \Gamma_0 \quad \text{for } r > a_0, $$
$$\Gamma(a_0, t = 0) = \Gamma_b \quad \text{for } r = a_0, \tag{5}$$

where a_0 denotes the initial value of the lump radius, $a_0 \equiv a(t = 0)$.

Equation (3) implies absence of surface convection (zero macroscopic velocity on the fluid boundary). Experimentally, one does not notice any motion within the interfaces, and moreover, the films practically do not thin, so it is legitimate to restrict our considerations to surface diffusion only.

The main mathematical complication in this problem arises from the fact that the first condition (4) is imposed on a moving circumference, $a(t)$. We introduce new dimensionless radial coordinate

$$y \equiv \frac{a(t)}{r}, \quad 0 \leqslant y \leqslant 1. \tag{6}$$

Then, Eq. (3) acquires the form

$$a^2 \frac{\partial G}{\partial t} + a \frac{da}{dt} y \frac{\partial G}{\partial y} = D_s y^3 \left[y \frac{\partial^2 G}{\partial y^2} + \frac{\partial G}{\partial y} \right], \tag{7}$$

where the dimensionless adsorption is

$$G(y, t) \equiv \frac{\Gamma - \Gamma_0}{\Gamma_b - \Gamma_0}, \quad 0 \leqslant G \leqslant 1. \tag{8}$$

Now, it is convenient to define

$$b(t) \equiv \frac{a(t)}{a_0}, \quad 0 \leqslant b \leqslant 1, \quad \tau \equiv \frac{D_s}{a_0^2} t, \quad 0 \leqslant \tau < \infty. \tag{9}$$

Transforming Eq. (7), we finally arrive at the following differential equation for the two unknown functions, $G(y, \tau)$ and $b(\tau)$:

$$b^2 \frac{\partial G}{\partial \tau} + b \frac{db}{d\tau} y \frac{\partial G}{\partial y} = y^3 \left[y \frac{\partial^2 G}{\partial y^2} + \frac{\partial G}{\partial y} \right], \tag{10}$$

with the corresponding boundary and initial conditions:

$$G(y = 1, \tau) = 1; \quad G(y = 0, \tau) = 0; \tag{11}$$

$$\begin{aligned} G(y, \iota = 0) = 0 \quad \text{for } y < 1 \\ G(1, \tau = 0) = 1 \quad \text{for } y = 1 \end{aligned}, \quad b(\tau = 0) = 1. \tag{12}$$

A second equation, required for solution of this problem, can be derived from the mass balance at the rim of the shrinking lump. There is a surface diffusion flux through the circumference $r = a$: $j_a = - D_s(\partial \Gamma / \partial r)|_{r=a}$. It carries away the material from the particle, whose volume, V, decreases

$$\frac{dV}{dt} = - 2 \times 2\pi a v_1 j_a = 4\pi a v_1 D_s \frac{\partial \Gamma}{\partial r}\bigg|_{r=a}. \tag{13}$$

Here v_1 represents the volume of one protein molecule. The multiplier 2 in Eq. (13) accounts for the presence of two film surfaces attached to the cluster, and two rims through which molecules are leaving.

We can find a connection between dV/dt and da/dt, using the experimentally established fact that the aggregates contract isotropically. Let us introduce a shape factor

$$c \equiv \frac{H(t)}{a(t)} = \text{const}. \tag{14}$$

(Fig. 1). Then, simple geometric considerations yield

$$\frac{dV}{dt} = 3\pi a^2 c \frac{da}{dt}. \tag{15}$$

Combining Eqs. (13) and (15), one obtains

$$\frac{db}{d\tau} = - \frac{A}{b^2} \frac{\partial G}{\partial y}\bigg|_{y = 1}, \quad \text{where } A = \frac{4}{3} \frac{v_1}{a_0 c} (\Gamma_b - \Gamma_0). \tag{16}$$

The system of two equations, (10) and (16), together with the conditions (11), (12), represent the complete mathematical formulation of the problem.

Results and discussion

We performed numerical solution of the set of differential equations (10), (16). Standard procedure was used, based on an implicit scheme [8]. The spatial and time intervals were discretised with increments Δy, $\Delta \tau$. The derivatives with respect to y in Eq. (10) were expressed at the time moment $(\tau + \Delta \tau)$, with central differences of accuracy $(\Delta y)^2$. The resulting linear set of equations was solved by Gauss–Jordan elimination with full pivoting [8]. For the time derivatives we used forward finite differences. In this way we computed the distribution of the adsorption throughout the interface at any moment, $G(y, \tau)$, and the lump radius as a function of time, $b(\tau)$. In dimensionless variables only one parameter, A, is needed for the calculation.

Our goal will be to compare the theoretical results with experimentally measured size of diminishing protein aggregates entrapped in emulsion films. For that purpose we carried out microscopic observations in reflected monochromatic light (wavelength 546 nm). The films were formed in a glass capillary immersed into the oil. The experimental method is described in details elsewhere [1, 9]. The interference picture (see Fig. 2) was recorded by videocamera, and the images were processed with the help of Targa + 16/32P grabbing board.

We investigated aqueous solution of 0.015 wt% lyophilised BSA (fatty acid free, purchased from Sigma), in the presence of 0.15 M NaCl, at pH = 6.4. This value of pH is reached when BSA is dissolved in water, without any addition of acid or base. The oil phase was soybean oil, preliminarily purified by passing it through a column packed with chromatographic adsorbent (Florisil, see Ref. [10] for details).

Figure 2 shows a Newton black film which contains big protein lump. The latter gradually diminishes and finally disappears. We measured the lump diameter, $2a$, accounted at the outer bright ring. Initially, just after formation of the film, the protein cluster contains some liquid. At that stage the boundary between the aggregate and the surrounding thin black film is not sharp, but is diffuse, and the diameter cannot be determined with good precision. For this reason, we do not start measuring $a(t)$ from time

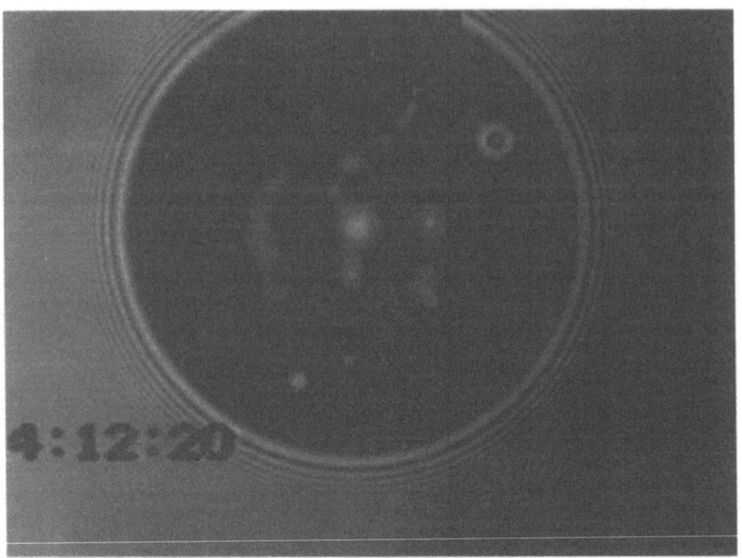
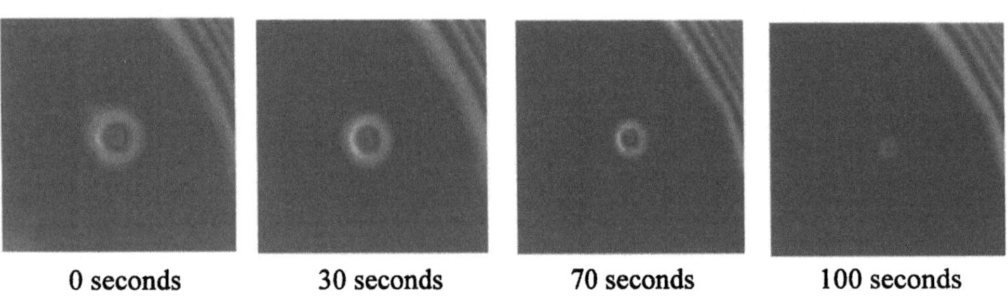

Fig. 2 Interference picture of a Newton black film containing big protein aggregate (above); images of the shrinking lump in the upper part of the film, at four consecutive time moments (below). The observation was carried out in reflected monochromatic light. Black and white zones correspond to thicknesses in multiples of $\lambda/(4n)$ (λ is the wavelength of the light, n is the refractive index of the aqueous phase). The reference distance between the two vertical bars is equal to $50\,\mu m$

zero (i.e., from the moment of film formation). Instead, we begin at a later time, t^*, when the solvent has already drained from the lump. (t^* is accounted from $t = 0$; at the moment $t = 0$ the hypothetical initial adsorption distribution, Eq. (5), is assumed to hold.) If the lump had been compact all the time, then at $t = 0$ there would have been $a = a_0$ (but a_0 and t^* are unknown quantities). The experimental running time can be denoted by $(t - t^*)$; the first measured point is $a = a^*$ at $t = t^*$.

The height of the particle, H, is measured at its center, using the conditions for interference and the intensity of the reflected light. The shape factor, c (Eq. (14)), remains almost constant, which means that the protein cluster contracts isotropically.

Figure 3 presents experimental data for the radius, a, as a function of time, for three different lumps. We shall fit these points with our theory. However, as discussed above, the parameter a_0 is unknown and therefore, conversion to

dimensionless variables is impossible. In order to overcome this difficulty, we developed the following procedure: A portion of the experimental dependence (Fig. 3) is fitted with a straight line. One writes

$$a = a^* - u(t - t^*),\tag{17}$$

where u is found from the slope. Switching over to the variables b and τ in Eq. (17) yields

$$b = \frac{1}{a_0}(a^* + ut^*) - \frac{ua_0}{D_s}\,\tau.\tag{18}$$

Next, arbitrary number is assigned to A (Eq. (16)), and the computation is run, providing the whole curve $b(\tau)$. This theoretical curve also has a linear portion, whose slope and intercept are determined. a_0 and t^* are then calculated according to Eq. (18), for the given A (with $D_s = 1 \times 10^{-7}\,cm^2/s$, Ref. [2]). Using a_0 and t^* we convert the

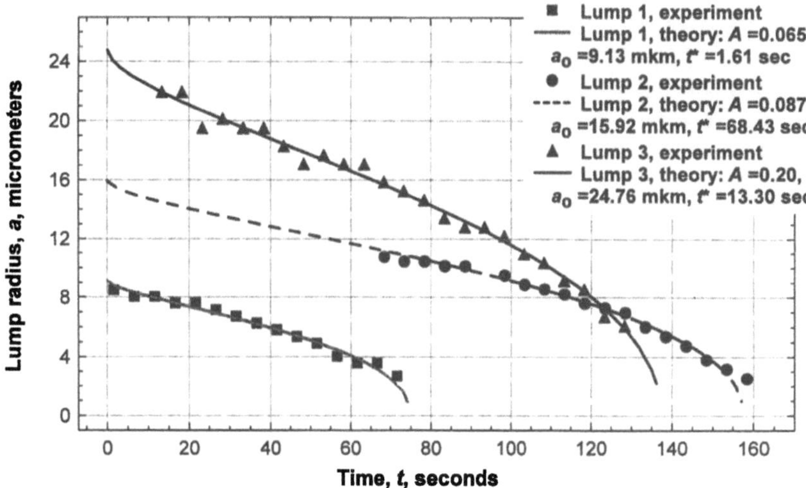

Fig. 3 Experimentally measured (symbols) and theoretically calculated (lines) values of the radii of three different protein aggregates whose size gradually diminishes during the process of squashing

computational results to physical variables and compare with the experimental data. The procedure is repeated iterating for A, until the best fit through the points is finally achieved (Fig. 3). The values of A, a_0 and t^* for the three particular aggregates are indicated on Fig. 3. Very good agreement between theory and experiment is observed. Actually, there is only one free parameter, A, to be varied independently. Although a_0 and t^* are also unknown a priori, they are explicitly calculated for the respective values of A at each step of the fitting procedure.

Thus, we can conclude that our simple model, which accounts for the surface diffusion of protein molecules, describes adequately the process of lump decomposition in thin films. From Eq. (16) it is possible to estimate the excess adsorption, Γ_b. The dimensions of one BSA molecule are $4.16 \times 4.16 \times 14.09$ nm [11], so that the volume is $v_1 = 243.84$ nm^3. Let us take as an example the lump #1 from Fig. 3. We have $c = 0.033 \pm 0.004$ and $A = 0.065$. Then, from Eq. (16) one deduces $\Gamma_b - \Gamma_0 = 6.02 \times 10^{12}$ cm^{-2}. Measurements of protein adsorption on oil/water interfaces were carried out by other authors [12]. As an order of magnitude, Γ_0 is about 4 mg/m^2 at bulk concentrations close to 0.01 wt% [12]. This value of Γ_0 is equivalent to 3.62×10^{12} cm^{-2} (for the molecular weight of BSA we take ~ 66500 g/mol from literature data [13]). Hence, it turns out that $\Gamma_b \approx 2.6\Gamma_0$. In much the same way one obtains $\Gamma_b \approx 5.1\Gamma_0$ for the lump #2, and $\Gamma_b \approx 9.8\Gamma_0$ for #3 (Fig. 3). These results demonstrate that (i) BSA forms multilayers on the liquid boundary (which is a well known

fact [12]); (ii) bigger aggregates are more compressed, and consequently, the adsorption along the periphery, Γ_b, acquires higher values — cf. Fig. 3.

The calculation of $\Gamma(r, t)$ reveals that the disturbance $(\Gamma - \Gamma_0)$ becomes more far-reaching (i.e., extends to larger distances), as time goes by. This is in accord with the expectations.

Conclusions

We propose a theoretical model which describes the process of gradual decomposition of big protein aggregates pressed between the two interfaces of a thin liquid film. The molecular migration by the mechanism of surface diffusion is found to be responsible for removing the material from the lumps. The particle radius was measured experimentally as a function of time. The data are fitted with the theoretical computations, and very good agreement is observed. In order to achieve the best fit we vary one parameter, whose value gives the maximum adsorption (Γ_b) along the rim where the lump is attached to the interface. Bigger clusters are more compressed, which leads to higher values of Γ_b. Protein multilayers form on the liquid surface.

Acknowledgements This work was financially supported by Kraft General Foods, Inc., and in part by the Bulgarian Ministry of Education, Science and Technologies.

References

1. Marinova KG, Gurkov TD, Velev OD, Ivanov IB, Campbell B, Borwankar RP (1997) Colloids and Surfaces A 123–124: 155–167

2. Clark DC (1995) In: Gaonkar AG (ed) Characterization of Food: Emerging Methods. Elsevier, Amsterdam, Chapter 2, pp 23–57

3. Clark DC, Coke M, Mackie AR, Pinder AC, Wilson DR (1990) J Colloid Interface Sci 138:207

T.D. Gurkov et al.
Protein lumps in thin liquid films

4. Coke M, Wilde PJ, Russell EJ, Clark DC (1990) J Colloid Interface Sci 138:489
5. Lalchev ZI, Wilde PJ, Clark DC (1994) J Colloid Interface Sci 167:80–86
6. Gurkov TD, Kralchevsky PA, Nagayama K (1996) Colloid Polym Sci 274: 227–238
7. Edwards DA, Brenner H, Wasan DT (1991) Interfacial Transport Processes and Rheology. Butterworth-Heinemann, Boston
8. Constantinides A (1987) Applied Numerical Methods with Personal Computers. McGraw-Hill, New York
9. Scheludko A, Exerowa D (1959) Kolloid-Zeitschrift 165:148–151
10. Gaonkar AG (1989) J Amer Oil Chem Soc 66:1090–1092
11. Brown JR, Shockley P (1982) In: Jost P, Griffith OH (eds) Lipid–Protein Interactions, Vol 1, Ch 2. Wiley, New York, pp 25–68
12. Graham DE, Phillips MC (1979) J Colloid Interface Sci 70:415–426
13. Peters Th Jr (1985) Adv Protein Chem 37:161–245

Progr Colloid Polym Sci (1998) 110:269–274
© Steinkopff Verlag 1998

U. Dahmen-Levison
G. Brezesinski
H. Möhwald

Enzymatic hydrolysis of monolayers: A polarization modulated-infrared reflection absorption spectroscopy study

U. Dahmen-Levison (✉)
G. Brezesinski · H. Möhwald
Max-Planck-Institut für Kolloid und
Grenzflächenforschung
Rudower Chaussee 5
D-12489 Berlin
Germany
E-mail: dahmen@mpikg.fta-berlin.de

Abstract Hydrolysis experiments have been performed on monolayers of L-dipalmitoylphosphatidylcholine (L-DPPC) and on monolayers of the ether-ester L-1-O-hexadecyl-2-stearoylphosphatidylcholine (L-HSPC). The investigation was carried out using phospholipase A_2 (PLA$_2$), derived from snake venom (*Crotalos Atrox*), which selectively hydrolyzes the sn-2 ester bond of L-phospholipids. Pressure-area isotherms and polarization-modulated infrared reflection absorption spectroscopy (PM-IRRAS), a versatile tool in the "in situ" investigation of interfacial reactions, were utilized to evaluate the dependence of the reaction rate on the lateral pressure. It has been determined that the reaction rate is enhanced in the liquid expanded/liquid condensed coexistence range, while at high pressures, the reaction rate comes almost to a halt. The influence of the reaction rate on the chemical structure of the phospholipid was also investigated. It was shown that the reaction rate is slower for ether–ester lipids than for diester lipids in the condensed phase. In an earlier study, we investigated the monolayer structures and the structural changes, which are being induced upon enzyme adsorption on D-enantiomeric monolayers. The results suggest that a better hydrolysis efficiency for condensed phases is related to a larger structural change upon adsorption.

Key words Phospholipase A_2 – monolayer – air-water interface – polarization modulated-infrared reflection absorption spectroscopy – FT-IR

Introduction

Phospholipase A_2 (EC 3.1.1.4, PLA$_2$) is a small, stereo-selective, calcium-dependent enzyme which hydrolyzes the sn-2 ester linkage of phosphatidylcholines. A characteristic feature of PLA$_2$ kinetics is the activation at the phospholipid concentration at which aggregates are formed (critical micellar concentration, CMC). Up to 10^4-fold higher enzyme activities are observed for substrate aggregates such as micelles, vesicles, membranes, or monolayers compared to the same substrate in a monomeric

dispersion [1–3]. One explanation which can account for this is called the *substrate theory* [4, 5], which states that enhanced activity can result from the fact that phospholipids within an aggregate are conformationally more favorable for enzymatic hydrolysis than those in the monomeric state. Alternatively, dehydration of interfacial substrate or a faciliated diffusion of interfacial phospholipids to the catalytic site of an interface-bound enzyme may be responsible for interfacial activation. Another explanation is the *enzyme theory* [6, 7], which postulates the occurrence of favorable, rate-enhancing conformational changes of the enzyme while bound to an interface. Both theories should not be applied exclusively.

The catalytic rate is dependent on the "quality" of the lipid-water interface. The "quality" is determined by the aggregate composition and physicochemical properties of the interface. One example of a well-defined interface is a Langmuir monolayer, where parameters like molecular density, lateral pressure, or ionic conditions can be varied independently. In this report, we describe hydrolysis experiments performed on Langmuir monolayers of the L-dipalmitoylphosphatidylcholine (L-DPPC) and of an ether–ester lipid, on L-1-O-hexadecyl 2-stearoylphosphatidylcholine (L-HSPC). The ether–ester-lipids are frequently present in biological membranes, e.g. plasmalogen-PC and platelet-activating factor.

The catalytic cleavage by PLA_2 can be monitored by fluorescence microscopy in the liquid-condensed/liquid-expanded coexistence range [8,9]. It has been shown that PLA_2 specifically recognizes condensed domains and that the reaction commences at the edges of the domains. The dependence of the reaction rate on the lateral pressure of the monolayers was evaluated, as well as the influence of the reaction rate on the chemical structure of the phospholipid (diester compared to ether-ester).

Experimental

Monolayer preparation

All reagents and solvents were of the highest purity available and used without further processing. L-dipalmitoylphosphatidylcholine (L-DPPC) was purchased from Sigma (Taufkirchen, Germany). L-1-O-Hexadecyl-2-stearoylphosphatidylcholin (L-HSPC) was synthesized by Jens Jakob (Institute of Pharmaceutical Chemistry, University of Halle, Germany) following the method of Bittman [10].

In all experiments, the subphase consisted of an aqueous buffer-solution containing 150 mM NaCl, 5 mM $CaCl_2$ and 10 mM Tris at a pH of 8.9. Water was purified with a Millipore desktop system, leading to a specific resistance of 18.2 MΩ cm. The film balance utilized in these investigations was built by R&K (Wiesbaden, Germany) and equipped with an Wilhelmy-type pressure measuring system.

The 1 mM phospholipid/chloroform solution (p.a. grade; Merck, Darmstadt, Germany) was spread and the film was compressed with a velocity of $\sim 15\,\text{Å}^2$/molecule $\times$ min after waiting 5 to 10 min.

Hydrolysis experiments

Phospholipase A_2 from *Crotalus atrox* venom (660 units/mg protein, Sigma, Taufkirchen, Germany) was dissolved in buffer and stored in aliquots at $-20\,°C$. The enzyme solution was injected into the subphase underneath the monolayer and below the barrier of the trough. For pressure-dependent investigations, an enzyme concentration of 0.3 ng protein/ml subphase was chosen. The amount of enzyme within the interface is at these low concentrations independent of the lateral pressure applied. The diffusion dependence of the experiments can be neglected, since all experiments were conducted consistently. The enzymatic solution was injected at an initial pressure of 40 mN/m (at which nearly no hydrolysis takes place) and the subphase carefully stirred. Next, the film was expanded, a pressure chosen and held constant. After 60 min the monolayer was compressed again to 40 mN/m and an IR-spectrum was measured.

Polarization modulated FT-IR spectroscopy

Our optical setup (Fig. 1a) consists of an IR-source, a Michelson interferometer from a Bruker IFS66 (Bruker, Karlsruhe, Germany) spectrometer, and an external reflection unit. At the exit of the interferometer, a mirror directs the IR-beam into the adjustable arm of the reflection unit. It passes both a polarizer (KRS5, Specac, Orpington, Great Britain) in p-position and a photoelastic modulator (ZnSe, Type II, Hinds), and is then focussed by a paraboloidal mirror onto the water surface. The trough is placed in a sealed box fitted with BaF_2-windows. The reflected intensity is focussed by a ZnSe lens onto a liquid-nitrogen-cooled MCT detector. The detected intensity is double modulated, first by the interferometer (300–1200 Hz) and then by the photoelastic modulator (73.9 kHz). The electronical setup (Fig. 1b) used is designed to obtain two signals $(R_p - R_s)$ and $(R_p + R_s)$, where R_p and R_s represent the reflectivities for p- and s polarization, respectively. Whereas one part $(R_p + R_s)$ is treated only by bandpass filtering (300–1200 Hz, Stanford Research System, Model SR 650), the other part $(R_p - R_s)$ is first bandpass filtered around 73.9 kHz, then demodulated (EG&G Lock-In amplifier, Model SR 650) and bandpass filtered (300–1200 Hz) again. The signals are multiplexed and connected with the standard electronics of the spectrometer. Here the resulting interferogram is amplified, digitalized (16 bit AD-board) and Fourier transformed with level 2 of zero filling and a triangle apodization function (OPUS-software). The dephasing of the photoelastic modulator is a function of the wave number. Maximum dephasing was set either to 2000 or to 1650 cm^{-1}. The IR spectra were collected using 200 scans at 4 cm^{-1} resolution with a scanning time of about 5 min. Only anisotropic absorptions contribute to the PM-IRRAS

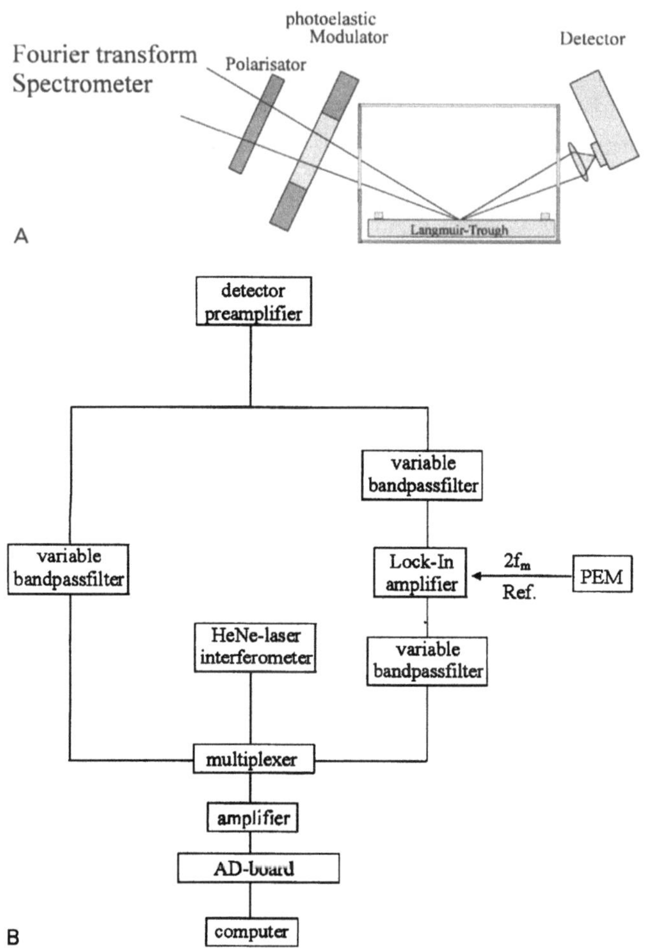

Fig. 1 (a) Optical setup of PM-IRRAS experiment, (b) electronical setup of PM-IRRAS experiment

signal S [11]

$$S = C \frac{R_p - R_s}{R_p + R_s} J_2(\phi), \tag{1}$$

where C is a constant of the electronical setup, and $J_2(\phi)$ the second-order Bessel function describing the wave number dependence of dephasing. Isotropic absorption by the water vapor or the subphase is not detected. Normalized difference spectra

$$\Delta S = \frac{(S_d - S_0)}{S_0} \tag{2}$$

are given, with S_0 the signal of the bare water surface and S_d the signal of the film covered surface. The angle of incidence was set to 74°. The spectra shown are raw spectra with no baseline correction or smoothing required. In all spectra, a broad negative band at around 1650 cm^{-1} appears. This band arises from the different reflectivities of covered and bare water surfaces, and is due to the wavelength dependence of the complex refractive index of water [12]. A restructuring of water molecules at the interface may also explain the very strong intensity of this band [13].

Results and discussion

Figure 2a gives the isotherm of a DPPC monolayer at 20°C. A main phase transition between liquid expanded and liquid condensed is observed at 4.7 mN/m. A collapse occurs at around 65 mN/m. The film is stable at a pressure of 40 mN/m, at which a PM-IRRAS spectra of the pure monolayer is shown (Fig. 2b). The signal-to-noise ratio is especially in the region between 1300 and 1800 cm^{-1} much better compared to IRRAS spectra. The water vapor bands are completely compensated.

The conformationally sensitive CH$_2$ antisymmetric and symmetric stretching vibrations at 2919.7 and 2851.2 cm^{-1} correlate with an increased order in the

Fig. 2 (a) Pressure-area isotherm of L-DPPC at 20 °C; (b) raw spectra of a DPPC monolayer at 40 mN/m

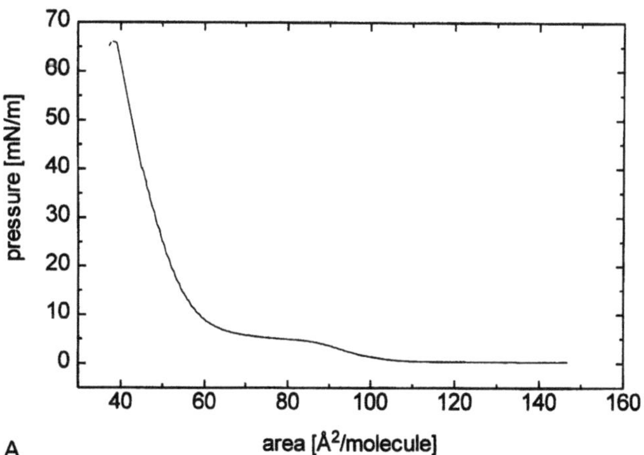

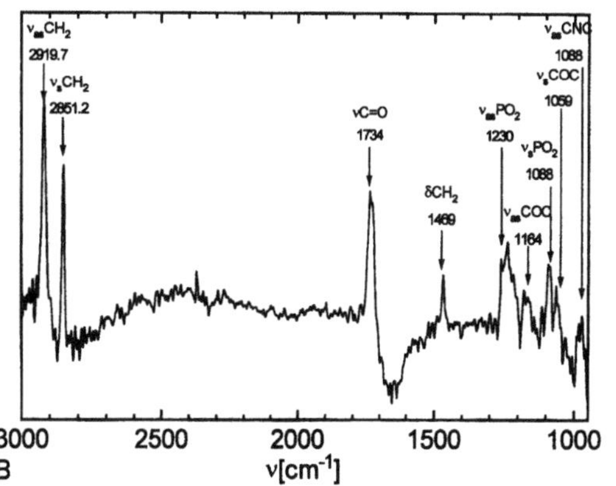

hydrocarbon chains. A continuous decrease in frequency with decreasing molecular area has been described [14]. The symmetric interfacial carbonyl ester CO stretching vibration appears at 1734 cm^{-1}. A CH$_2$ scissoring mode is found at 1469 cm^{-1}. Bands at 1230 and 1088 cm^{-1} are due to phosphate asymmetric and symmetric stretches, respectively. The phosphate stretches are each composed of at least two frequencies. The presence of multiple bands suggests a heterogeneous population of hydrogenated and not hydrogenated groups. For stretching vibrations, shifts to higher frequencies are indicative of increasing bond strength, which may result from decreased hydrogen bonding. The intensity of the bands are dependent on the lateral density of the molecules and on the orientation of the transition moments with respect to the plane of incidence. The ester band at 1734 cm^{-1} exhibits an almost linear relationship between the integrated band intensity and the area/molecule (data not shown).

In this report we have investigated the enzymatic hydrolysis of phosphatidylcholines. An increase in pressure is observed when an enzymatic solution is injected into the subphase at lower pressures. This is due to the penetration of PLA$_2$ into the film. After a certain reaction time the pressure decreases indicating that hydrolysis products must be partially soluble. The amount of PLA$_2$ within the lipid interface is pressure dependent. The desorption (see Fig. 3c) as well as the hydrolysis could be detected by PM-IRRAS. Spectra of DPPC monolayers before and after degradation are shown for 30, 10 and 1 mN/m (Fig. 3).

Bands at 1581 and 1540 cm^{-1} appear during hydrolysis and are due to an increasing amount of product in the monolayer. Both bands are assumed to be asymmetric stretching vibrations of the carboxylate. The carboxylate bands observed in presence of enzyme differ from those in spectra of pure fatty acid monolayers (or of mixtures DPPC/fatty acid/lysolipid) on the buffer subphase (see Fig. 4). The band at 1562 cm^{-1} represents a Ca^{2+}-associated carboxygroup and the weak band at 1540 cm^{-1} an unassociated carboxygroup. In an earlier analysis of spectra of Ca^{2+}/octadecanoate the appearance of a component at 1578 cm^{-1} in highly condensed monolayer states has been described [15].

The appearance of the band at 1581 cm^{-1} must be caused by an interaction of protein and fatty acid. It may be due to a strong association of protein and carboxy group (partially inhibited product release or a highly condensed state of the fatty acid induced by the protein). The strong intensity of the unassociated band at 1540 cm^{-1} may be due to the occurrence of enzyme domains, which are formed underneath phase separated areas containing fatty acids. The enzyme may act as a shield for Ca^{2+}-association (band at 1562 cm^{-1}). A broad band at

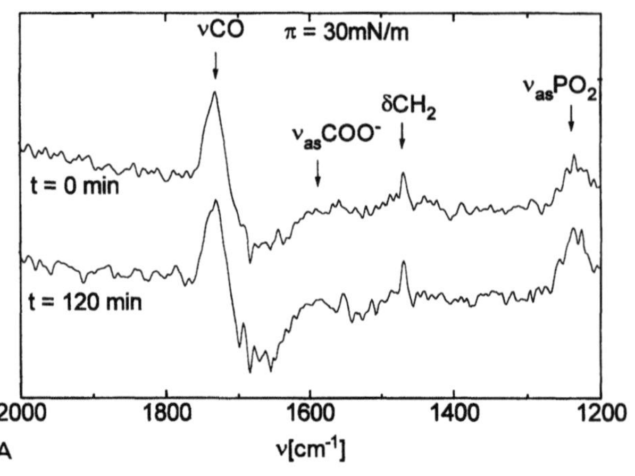

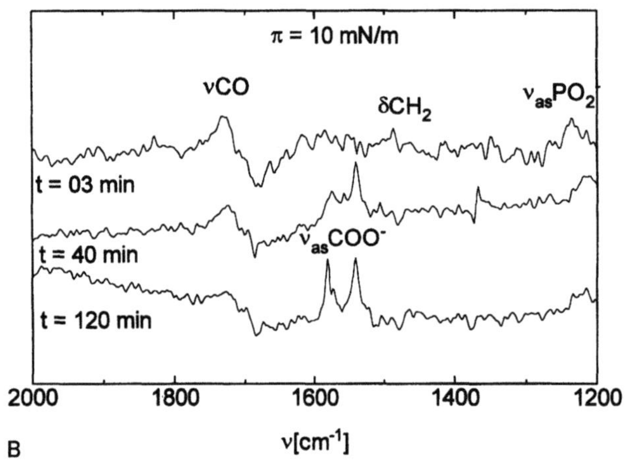

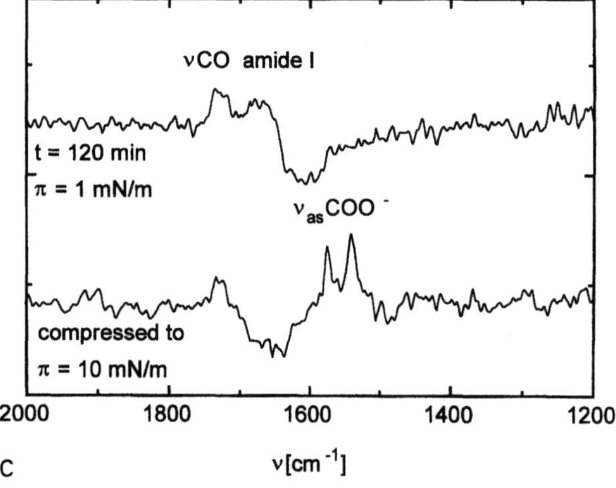

Fig. 3 Spectra of DPPC monolayer before and after injection of enzyme (6 ng/ml subphase) (a) at $\pi = 30$ mN/m after $t = 0$ and 120 min, (b) at $\pi = 10$ mN/m after $t = 3$, 40, and 120 min, (c) at $\pi = 1$ mN/m after $t = 120$ min, and after compression to 10 mN/m

1562 cm^{-1} was only observed when a spectra was measured directly after the film was being compressed and equilibrium (phase separation) was not yet reached.

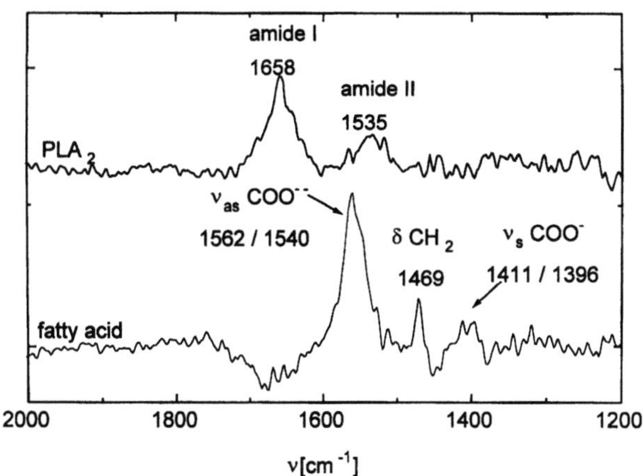

Fig. 4 Spectra of pure monolayers: PLA$_2$ (10 mN/m) and fatty acid (40 mN/m)

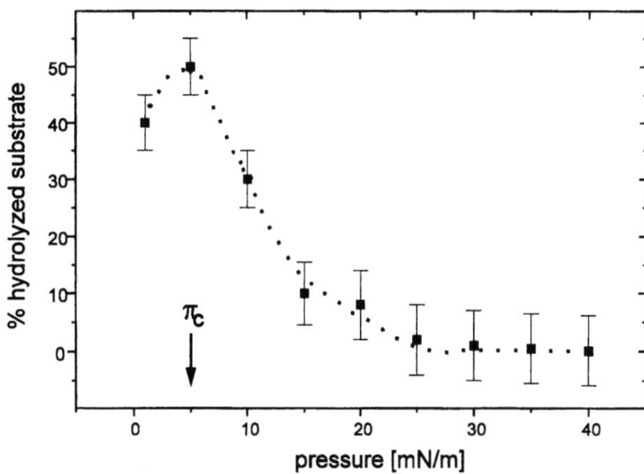

Fig. 5 Percentage of substrate hydrolyzed after 60 min of hydrolysis (0.3 ng/ml subphase) vs. lateral pressure at which the hydrolysis took place. The percentages were evaluated using the integrated intensity of the esterband at 1734 cm^{-1} measured at 40 mN/m and calibration curves obtained for defined ternary mixed monolayers

The spectral change occurs much faster at 10 mN/m than at 30 mN/m, where only little change is detected. At 1 mN/m, PLA$_2$ which is penetrated into the monolayer can be detected, whereas after compression the broad amide I band at around 1658 cm^{-1} is decreased and the enzyme desorped.

For comparison, spectra of a fatty acid monolayer and of pure PLA$_2$ at the air/water interface are shown (Fig. 4). The amide I and amide II bands, at 1658 and at 1535 cm^{-1}, respectively, confirm a mostly α helical structure of this very stable and rigid protein (7 disulfide bonds) [16]. The fatty acid spectrum shows an asymmetric carboxylate stretching vibration at 1562 cm^{-1}, with a shoulder at 1540 cm^{-1}, weak symmetric stretching vibrations at around 1411 and 1396 cm^{-1} and a CH$_2$ scissoring mode at 1469 cm^{-1}.

In order to evaluate the pressure dependence of the hydrolysis reaction, we also determined the amount of substrate being hydrolyzed after a certain reaction time (60 min) at a constant pressure. Due to catalytic cleavage, the number of ester groups are constantly being reduced. The integrated intensity of the ester band was utilized to determine the composition of the monolayer. Spectra of defined ternary mixed lipid monolayers were analyzed and calibration curves of integrated intensity of the ester band vs. monolayer composition were obtained for various pressures (not shown). This had become necessary due to the fact that ternary mixtures exhibit a nonideal behavior (phase separation) [17] and due to the partial solubility of lysolipid at high pressures. In order to work with optimum absolute intensities a relative high pressure of 40 mN/m was chosen for the measurements which were conducted after the enzyme was allowed to hydrolyze the monolayer for 60 min at various fixed pressures. The results (Fig. 5) show a maximum degradation, which in this case means

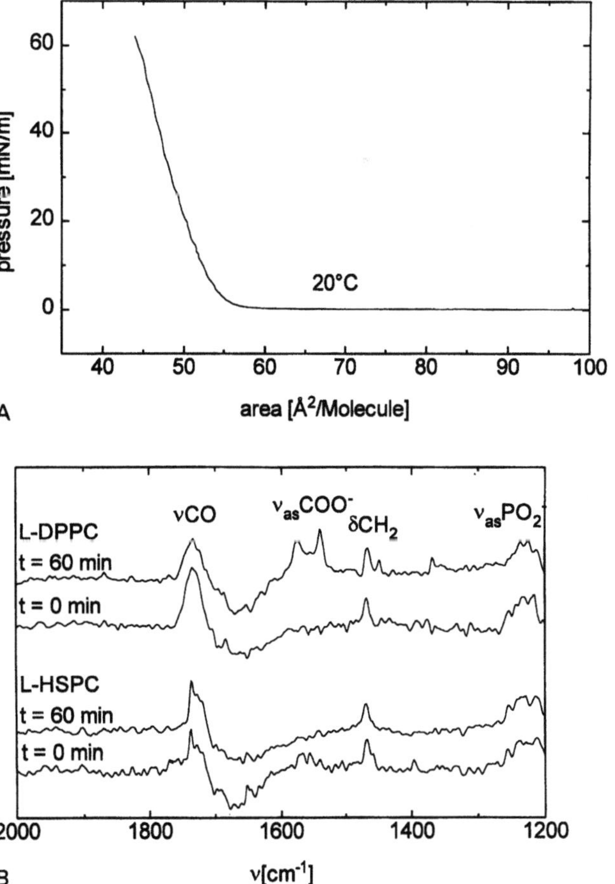

Fig. 6 (a) Pressure/area isotherm of L-HSPC at 20 °C, (b) spectra of monolayers before and after 60 min of hydrolysis (6 ng PLA$_2$/ml subphase) at 8 mN for L-DPPC and L-HSPC, spectra taken at 40 mN/m

maximum activity at a lateral pressure of 5 mN/m, where the main phase transition occurs. At higher pressures (in the condensed phase), the activity decreases and comes almost to a halt at pressures above 25 mN/m.

The pressure/area isotherm of L-HSPC at 20 °C (Fig. 6a) exhibits only a condensed phase. Compared to DPPC, HSPC lacks a carbonyl-group at the 1 position on the glycerol backbone and the chainlength in the 2-position is increased. Both modifications are known to increase the lateral packing density in monolayers [18].

The spectrum of L-HSPC is similar to the one of DPPC showing a symmetric carbonyl ester CO stretching vibration at $1737 \, \text{cm}^{-1}$ with a shoulder at $1727 \, \text{cm}^{-1}$, a CH_2 scissoring mode at $1469 \, \text{cm}^{-1}$, and a phosphate stretching vibration at $1225 \, \text{cm}^{-1}$. After a reaction time of 60 min at a lateral pressure of 8mN/m, a slight decrease in intensity is detected for the ester band and a very broad band arises in the area of the carboxylate band (Fig. 6b). We estimate, based on the calibration curves for DPPC, that a maximum of 30% HSPC was being hydrolyzed after 60 min. Applying the same reaction conditions to DPPC, the monolayer was degradated to about 70%.

Conclusion and outlook

PM-IRRAS is a versatile tool for "in situ" investigation of interfacial reactions. It was utilized to determine the dependence of the rate of the hydrolysis reaction on lateral pressure applied. It could be shown that the rate is largest in the coexistence region. This is in agreement with fluorescence microscopy studies, where it was shown that the enzyme accumulates at the interface expanded/condensed where the reaction commences.

Additionally we compared the rate of hydrolysis of a diester phospholipid and an ether–ester phospholipid. For both phospholipids the reaction was studied in the condensed phase at 8 mN/m. The reaction rate is slower for the ether ester–lipid than for the diester. In an earlier study, we investigated the monolayer structures and the structural changes, which are being induced upon enzyme adsorption on D-enantiomeric monolayers [19]. The results suggest that a better hydrolysis efficiency for condensed phases is related to a larger structural change upon adsorption.

Acknowledgments We thank J. Jacob, Universität Halle, Germany, for synthezising the ether–ester phosphatidylcholine. We are grateful to W. Goedel, MPI-KGF, Berlin, and B. Desbat, Université de Bordeaux, France, for advice about the experimental PM-IRRAS setup. Helpful discussions with A. Gericke, Universität Halle are gratefully acknowledged. This work was supported by the Deutsche Forschungsgemeinschaft (DFG).

References

1. Pieterson WA, Vidal JC, Volwerk JJ, de Haas GH (1974) Biochemistry 13:1455
2. Verger R, Mieras MCE, de Haas GH (1973) J Biol Chem 248:4023
3. Lichtenberg D, Romero G, Menashe M, Biltonen RL (1981) J Biol Chem 261:5334
4. Brockerhoff H (1968) Biochim Biophys Acta 159:296
5. Wells MA (1974) Biochemistry 13:2248
6. Entresanngler B, Desnuelle P (1974) Biochim Biophys Acta 159:437
7. Kilby PM, Primrose WU, Roberts GCK (1995) Biochem J 305:935
8. Grainger DW, Reichert A, Ringsdorf H, Salesse C (1989) FEBS Lett 252:73
9. Grainger DW, Reichert A, Ringsdorf H, Salesse C (1990) Biochim Biophys Acta 1023:365
10. Witzke NM, Bittman R (1986) J Lipd Res 27:344
11. Blaudez D, Buffeteau T, Cornut JC, Desbat B, Escafre N, Pezolet M, Turlet JM (1993) Appl Spectrosc 47:869
12. Handke M, Milosevic M, Harrick NJ (1991) Vib Spectrosc 1:251
13. Blaudez D, Buffeteau T, Cornut JC, Desbat B, Escafre N, Pezolet M, Turlet JM (1994) Thin Solid Films 242:146
14. Hunt RD, Mitchell ML, Dluhy RA (1989) J Mol Struct 214:93
15. Gericke A, Hühnerfuss H (1994) Thin Solid Films 245:74.
16. Arni RK, Ward RJ (1996) Toxicon 34:827
17. Maloney KM and Grainger DW (1993) Chem Phys Lipids 65:31
18. Brezesinski G, Dietrich A, Struth B, Böhm C, Bouwman WG, Kjaer K, Möhwald K (1995) Chem Phys Lipids 76:145
19. Dahmen-Levison U, Brezesinski G, Möhwald H, Thin Solid Films, in press.

Progr Colloid Polym Sci (1998) 110:275–279
© Steinkopff Verlag 1998

M.S. Romero-Cano
A. Martin-Rodriguez
G. Chauveteau
L. Nabzar
F.J. de las Nieves

Colloid stability of nonionic surfactant/latex complexes at high ionic strength

M.S. Romero-Cano · F.J. de las Nieves (✉)
Universidad de Almería
Departmento de Física Aplicada
La Cañada de San Urbano s/n
E-04120 Almería
Spain

A. Martin-Rodriguez
Biocolloid and Fluids Physics Group
Department of Applied Physics
Faculty of Sciences
University of Granada
E-18071 Granada
Spain

G. Chauveteau · L. Nabzar
Division Gisement
Institute Francais du Pétrole
1 & 4 Av. Bois Préau, BP 311
92506 Rueil Malmaison Cedex
France

Abstract An experimental study on the effect of the temperature on colloid steric stability of polystyrene particles covered with nonionic surfactants (ethoxylated fatty alcohols) is described. A sulfate latex, with relative low surface charge, was used as solid substrate for the adsorption of two surfactants having different hydrophilic tail lengths. After the adsorption of the surfactant the latex–surfactant complexes were very stable even at a concentration higher than 1 M of NaCl. The effect of fast-speed centrifugation and redispersion in ultrapure water on the stability of the complexes was first studied. The experimental results showed that after several centrifugation the stability of the complexes decreased. The second objective of this work was to find the thermodynamic condition for flocculation of non-centrifuged complexes prepared with different amounts of surfactant in a medium. The critical flocculation temperature (CFT) was found to rise with surfactant concentration.

Key words Polymer colloids – nonionic surfactant adsorption – steric stabilization

Introduction

Polymer colloids have a natural colloidal stability as a consequence of their charged surface groups produced during the synthesis process. This stability is due to electrostatic repulsion between electrical double layer as the DLVO theory explains [1–3]. In some cases an additional stabilization is required to improve the stability of dispersions and different kinds of polymers and surfactants are widely used. Nonionic surfactant adsorbed onto colloidal particles can also act as steric stabilizers [4]. The main advantage of this stabilization is that aqueous sterically stabilized dispersions are comparatively insensitive to the presence of electrolytes because the dimensions of nonionic chains vary relatively little with the electrolyte concentration. When a sterically stabilized dispersion is obtained, flocculation by salt addition is not possible because the steric contribution to the total potential, enough to stabilize the system, does not decrease with increasing electrolyte concentration. This stability is limited only by thermodynamic factors (Flory–Huggins solvency parameter) that control the solvency of the stabilizing moieties. The transformation from long-term stability to catastrophic flocculation for sterically stabilized dispersions occurs abruptly at the critical flocculation point (CFP). In our case we have used the temperature as thermodynamic variable of flocculation. Nevertheless, systems can be found where the steric contributions to the repulsive potential are of the same order as that of the electric one but not enough to stabilize the dispersion, and thus, although higher stability of the system is reached, it can be flocculated by adding electrolyte. The effectivity of these dispersions electrosterically stabilized is directly related to the hydrophilic moiety length, increasing as the hydrophilic tail is increased [5]. However, the characteristics of the hydrophobic moiety are determining factor in obtaining very stable dispersions. An adequate hydrophobic moiety can prevent, for example, the desorption of surfactant and therefore the dispersion destabilization.

In this work very stable latex dispersion were obtained by using nonionic surfactants with a relative short hydrophilic length (8–12 EO units) but an optimal hydrophobic moiety composed of a linear hydrocarbon chain of 16–18 carbons. The purpose of this work is to study the centrifugation effect and mainly the temperature effect on the colloidal stability of our complexes after adsorption of nonionic surfactant, when the free surfactant is in the solution.

Experimental

Chemicals

All chemicals in this study were of analytical grade and were used without further purification. The water in all experiments was ultrapure with specific electrical conductivity lower than $1\,\mu S/cm$ (ATAPA S.A. Spain).

Latex characterization

The latex used in this work was synthesized in our laboratories. Styrene (Merk) was previously distilled under low pressure (10 mm Hg and $40\,°C$). Negatively-charged polystyrene latex [PS-2] was prepared using the emulsifier-free method, with potassium persulfate as initiator, in a discontinuous reaction, according to Kotera et al. [6]. The surface charge density and the electron microscopy diameter were $-4.8 \pm 0.9\,\mu C/cm^2$ and $361 \pm 14\,nm$ with a polydispersity index of 1.0046.

Surfactant characterization

Genapol O-080 and Genapol O-120 (from Hoechst), were used without further purification. Surfactant concentration before and after adsorption was determined by turbidimetry using the Tanic acid method [7, 8] with a Spectronic Genesys 5 spectrophotometer (Milton Roy, USA). The molecular structure of Genapol O surfactant corresponds with a fatty alcohol polyglycol ether based on C16/C18 alcohol with 8 and 12 units of EO for Genapol O-080 and O-120, respectively. The cloud points for 1% solutions of these surfactants are $> 40\,°C$ for Genapol O-080 and $> 100\,°C$ for Genapol O-120 and their CMCs

are around $0.031\,g/l$. The last data were provided by Hoechst Corporation.

Complexes preparation

Complexes were prepared by surfactant adsorption in batteries for 24 h at $25.0 \pm 0.1\,°C$ with various added contents of surfactant on $0.25\,m^2$ of latex surface. The surfactant–latex mixtures were gently shaken during adsorption experiment.

Results and discussion

Adsorption

Adsorption isotherms in aqueous medium, in which the adsorbed amount (Γ) is plotted versus the total surfactant concentration (C) are shown in Fig. 1. The adsorption isotherms showed a well-defined plateau. The amount of surfactant adsorbed reached a saturation value at a concentration higher than the critical micelar concentration (CMC). The linear part of the adsorption isotherm corresponds to total adsorption of surfactant for surfactant concentration lower than CMC. This suggests the high affinity that these surfactants have for polystyrene surface. The isotherm plateau could be explained considering the total entropic change of the transition from surfactant in micelles to surfactant adsorbed as the main cause of free energy variation during adsorption process at surfactant concentration higher than CMC. The existence of a repulsive interaction energy between PEO chains can explain that the amount of surfactant adsorbed decreases as the number of ethylene oxide increases. This result is an argument in favor of a fairly flat orientation of the surfactant on the surface. It is likely that some EO segments are in contact with the surface [9–11].

Centrifugation effect

PS-2/Genapol O-080 complex was prepared under the same conditions that we have described previously with 0.1% of surfactant. At this concentration of surfactant the adsorption plateau was obtained as isotherms adsorption showed. The static colloidal stability of this complex in a medium $1\,M\ NaCl + 0.1\,M\ CaCl_2$ was studied measuring the hydrodynamic diameter by Photoncorrelation Spectroscopy (PCS) after 24 h. The PCS results were conclusive in showing a hydrodynamic diameter for this complex after 24 h of $359 \pm 3\,nm$. The hydrodynamic diameter of PS-2/Genapol O-080 complex in water was of

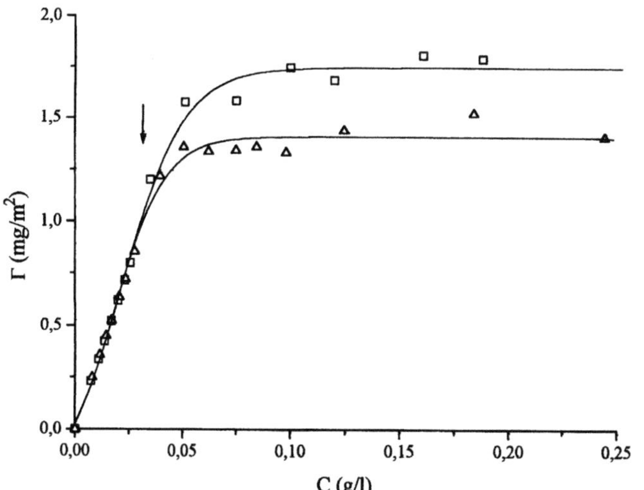

Fig. 1. Adsorption isotherm of Genapol O-080 (□) and Genapol O-120 (△) onto PS-2 latex

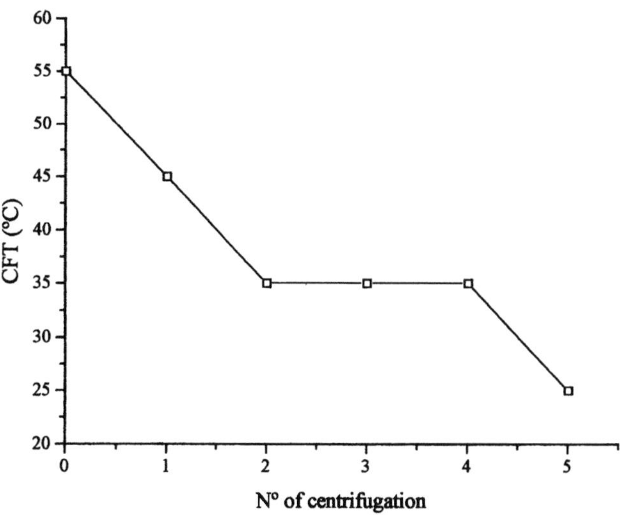

Fig. 2. Effect of centrifugation-redispersion process on colloidal stability of PS-2/Genapol O-080 (0.1%) in 1 M NaCl + 0.1 M CaCl$_2$ medium

359 ± 5 nm. Destabilization of this complex was obtained by heating. The critical flocculation temperature (CFT) was 55 °C. The fact that we had to heat the dispersions to obtain flocculation is due to enthalpic stabilization that PEO moieties produce [12]. In enthalpic stabilization, the enthalpy change on close approach of the particles promotes stabilization, whereas the corresponding entropy change promotes flocculation. Their relative contributions to the overall free energy change are such that the enthalpy contribution predominates.

In order to study the effect of fast-speed centrifugation (13 500 rpm during 40 min at 25 °C) on the colloidal stability of the complex in the saline medium referred to previously, an easy experiment was developed. Different aliquots of PS-2/Genapol O-080 complex were subjected to different centrifugations with redispersion in ultrapure water. After this treatment, the CFT in the saline medium was determined for the complex which has been subjected to different centrifugation and it was that after five centrifugations the complex flocculated at room temperature (25 °C). The CFT decreased as the number of centrifugation–redispersion increased (Fig. 2). This is a logical result considering the sequential desorption of surfactant during this drastic process. In the first and second centrifugation the free surfactant was almost practically removed. However, the removal of adsorbed surfactant was much more difficult as shown by the plateau found in the CFT plot. The last result is a consequence of the high affinity that the hydrophobic moiety of these surfactant presents for the polystyrene surface due to the great number of hydrophobic contacts that they can establish. Finally, the desorption could be almost complete after five centrifugations.

Surfactant concentration effect

PS-2/Genapol O-080 and PS-2/Genapol O-120 complexes were prepared at different surfactant concentrations: 0.005%, 0.010%, 0.020%, 0.025% and 0.1%. Flocculation in a 1 M NaCl + 0.1 M CaCl$_2$ medium was induced by heating. The complexes were kept at different temperatures for 24 h. Then the diameter was measured using PCS. Figures 3 and 4 show the colloidal stability behavior of complexes against the temperature. In ordinate the only values were 1 or 0, which means, stability or no stability. Thus, the sample with 0.005% of Genapol O-080 was unstable at room temperature, while the samples with 0.01%, 0.015–0.025%, and 0.1% of surfactant in solution were unstable at 35 °C, 45 °C and 55 °C, respectively. For the Genapol O-120 samples, the complexes with 0.005%, 0.01–0.015% and 0.02–0.1% of surfactant in solution were unstable at 45 °C, 55 °C and 65 °C, respectively. In all cases the complexes obtained with the larger surfactant (Genapol O-120) showed higher colloid stability. Napper's studies [12–14] showed that for dispersions with terminally anchored PEO chains the CFT was relatively insensitive to the molecular weight of the stabilizing moieties, and a great correlation was also found the θ-temperature of the stabilizing polymer and the CFT at high coverage. In our case the θ-temperature is approximately 90 °C [15], however, the CFT obtained at high coverage is of 55 °C for Genapol O-080 and 65 °C for Genapol O-120. This is in contrast, therefore, to Napper's studies. For physically adsorbed polymer the situation is much more complex. The CFT depends strongly on the degree of coverage [16, 17], and also is quite sensitive to the packing density

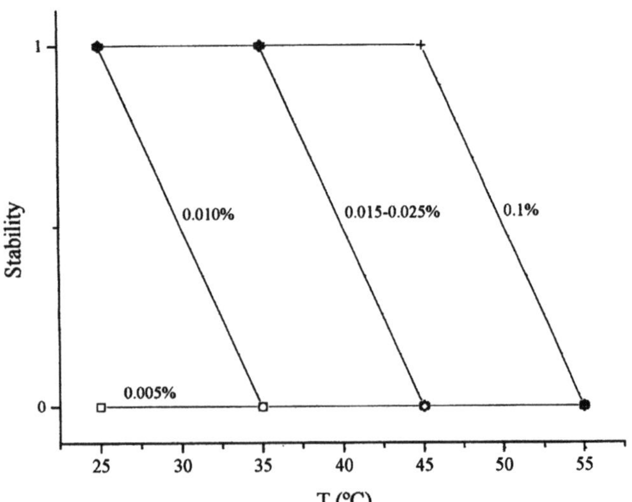

Fig. 3. Effect of temperature on colloidal stability of PS-2/Genapol O-080 with different percentage of surfactant in 1 M NaCl + 0.1 M CaCl$_2$ medium: 0.005% (□) 0.010% (O), 0.015% (△), 0.020% (∇), 0.025% (◇) and 0.1% (+)

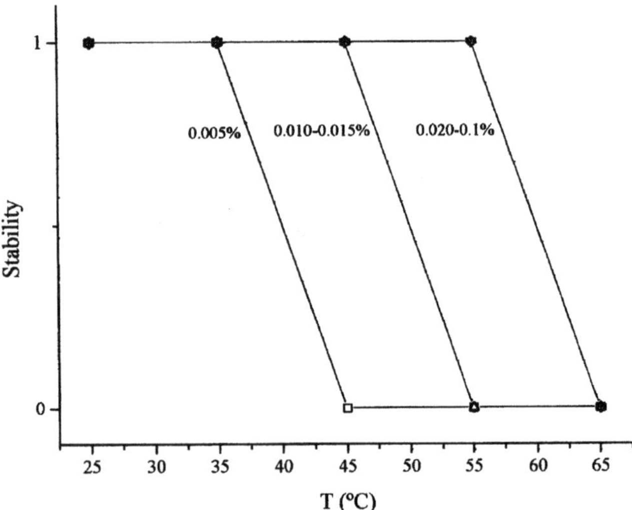

Fig. 4. Effect of temperature on colloidal stability of PS-2/Genapol O-120 with different percentage of surfactant in 1 M NaCl + 0.1 M CaCl$_2$ medium: 0.005% (□), 0.010% (O), 0.015% (△), 0.020% (∇), 0.025% (◇) and 0.1% (+)

of the stabilizing moieties for particles stabilized by relative low molecular weight chains [18]. The relative short chain tail or the possible fairly flat orientation of the surfactant on the surface may be the possible explanation by which this steric stabilized particles flocculated under better than θ-conditions. The dependence found between the hydrophilic length and the CFT could be explained if we consider that the large surfactant has large packing density of surfactant in the adlayer, as we have found with

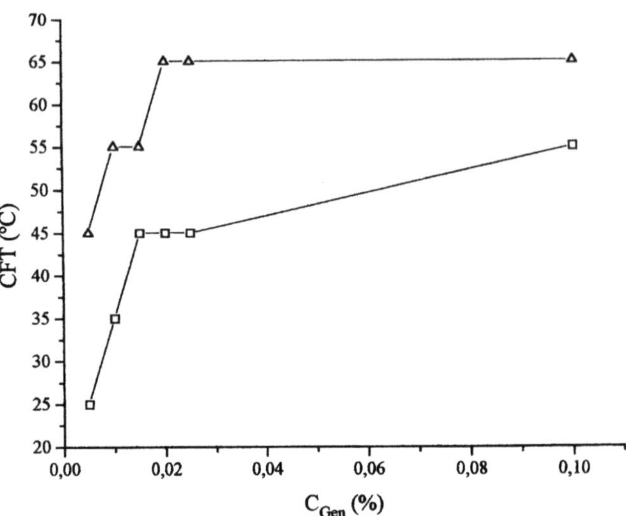

Fig. 5. Variation of CFT against surfactant concentration for PS-2/Genapol O-080 (□) and PS-2/Genapol O-120 (△) complexes

Triton X surfactants [19]. In a previous work [19] we found that the adlayer thickness of surfactant decreases as the hydrophilic length increases. As a consequence, the effective volume fraction of surfactant in the adlayer is increased causing the stabilization of the dispersion.

On the other hand, we have found that the CFT is increased with surfactant concentration until a plateau is found. If we consider (as the adsorption isotherm reveals) that the complexes studied have their surface saturated with surfactant, except for 0.005% complexes, the only difference between them is the free surfactant concentration. It is well-known that when the surface is not totally covered, the effectivity of steric stabilization is lower and the sterically stabilized dispersions flocculates in better solvent than in theta-solvent. The range where we found an increase in CFT corresponds with the adsorption isotherm region where the plateau is not still well defined (beginning of plateau), see Fig. 5. As a consequence, the colloidal stability of complexes from this region, related directly with the CFT, were lower. The plateau found for CFT corresponds obviously with the well-defined plateau of adsorption isotherms where the complexes have an optimal degree of coverage.

Conclusions

This study shows that Genapol O-080 and Genapol O-120 have a great affinity for polystyrene surface; therefore the expontaneous desorption is slightly probable. As a consequence of this high affinity caused by the linear hydrophobic moiety and the optimal hydrophilic chains length, these surfactants are good steric stabilizers as indicated by the results obtained in $1\,M$ $NaCl + 0.1\,M$ $CaCl_2$ aqueous medium. The temperature is a critical parameter that controls the colloidal stability of surfactant-latex complexes in a saline medium. This sterically stabilized particles flocculated under better than θ-conditions as a consequence of the relative short chain tail or the possible fairly flat orientation.

Acknowledgement The financial support provided by the "Institut Français du Pétrole (Paris) " (and permission for publishing these results) and by CICYT (under project MAT 96-1035-C03-03) is greatly appreciated.

References

1. Derjaguin BV, Landau LD (1941) Acta Physicochem URSS 14:633
2. Verwey EJW, Overbeek JThG (1946) Trans Faraday Soc 42B:117
3. Verwey EJW, Overbeek JThG (1948) Theory of Stability of Lypohobic Colloids. Elservier, Amsterdam
4. Hunter RJ (1995) Foundations of Colloid Science. Vol I The Universities Press Ltd, Belfast.
5. Romero-Cano MS, Rodriguez-Martin A, Chauveteau G, de las Nieves FJ (1998) J Colloid Interface Sci 198:273
6. Kotera A, Furusawa K, Tacite Y (1970) Kolloid ZZ Polym 239:677
7. Attia YA, Rubio J (1975) Br Polym J 7:135
8. Couture L, Van de Ven TGM (1991) Colloids Surfaces 54:245
9. Killmann E, Maier H, Baker JA (1988) Colloids Surfaces 31:51
10. Rupprecht H, Lieble H (1972) Kolloid ZZ Polym 250:719
11. Rubio J, Kitchener JA (1976) J Colloid Interface Sci 57:132
12. Napper DH (1970) J Colloid Interface Sci 32:106
13. Napper DH, Netschey A (1971) J Colloid Interface Sci 37:528
14. Napper DH (1977) J Colloid Interface Sci 58:390
15. Boucher CA, Hines PM (1976) J Polym Sci Polym Phys Ed 14:2241
16. Lambe R, Tadros ThF, Vincent B (1978) J Colloid Interface Sci 66:77
17. Dobbie JW, Evans R, Gibson DV, Smitham JB, Napper DH (1973) J Colloid Interface Sci 45:557
18. Cowell C, Vincent B (1982) J Colloid Interface Sci 87:518
19. Romero-Cano MS, Martin-Rodriguez A, Chaveteau G, de las Nieves FJ (1998) Colloids Surfaces A 140:345

Progr Colloid Polym Sci (1998) 110:280–284
© Steinkopff Verlag 1998

Study of DPPC/TC/water phase diagram by coupling of synchrotron SAXS and DSC: I equilibration kinetics

K. Andrieux
L. Forte
G. Keller
C. Grabielle-Madelmont
S. Lesieur
M. Paternostre
M. Ollivon
C. Bourgaux
P. Lesieur

K. Andrieux (✉) · L. Forte · G. Keller
C. Grabielle-Madelmont · S. Lesieur
M. Paternostre · M. Ollivon
Laboratoire de Physico-Chimie des
Systèmes Polyphasés
CNRS URA 1218
Université Paris-Sud
92296 Châtenay-Malabry
France

C. Bourgaux · P. Lesieur
Laboratoire pour l'Utilisation du
Rayonnement Electromagnétique
Bat. 209D
Université Paris-Sud
91405 Orsay
France

Abstract As a prerequisite for the determination of the phase diagram of the 1,2-dipalmitoylphosphatidyl-choline (DPPC)/sodium taurocholate (TC)/buffer system, equilibration kinetics of selected ternary mixtures were examined as a function of TC/DPPC molar ratio (r), storage time (4–30 days) and temperature (4 °C, 17 °C or 26 °C). For this purpose, aqueous suspensions containing mixtures of TC in a large range of concentrations (0–225 mM) and DPPC (50 mM) were analyzed by small angle X-Ray scattering (SAXS) and compared to differential scanning calorimetry (DSC) analysis obtained at a very low heating rate 0.08 °C/min). The results show that the molecular organization obtained strongly depends on r and storage conditions. The samples with a low molar ratio $r = 0.1$ were quickly equilibrated at 26 °C and not at 4 °C and 17 °C. At $0.2 \leq r \leq 0.3$, they were not at equilibrium within one month whatever experimental conditions. Lamellar phases were observed at $0.4 \leq r \leq 1.0$ after few days to one month depending on r and storage temperature. Samples corresponding to mixtures of lamellar and micellar phases observed at higher molar ratios (≥ 2.0) were slowly equilibrating at 4 °C.

Key words Bile salts – calorimetry – micelles – phospholipids – vesicles – X-ray diffraction

Introduction

Bile salts are with lecithins, cholestrol, bile pigments, inorganic ions and proteins, the main constituents of bile. They are polar amphipathic molecules which are essential by their solubilizing power, in the digestion of fat and fat-soluble vitamins, as well as for the absorption of other nutrients. Their amphiphilic character allows them to aggregate into micelles in aqueous solution [1]. In the presence of lipid suspensions, BS molecules partition between water and lipidic aggregates. Although, they display a very peculiar behavior, their surface-active properties are responsible for their classification as surfactants [1].

Surfactants are widely used as molecular tools in membrane studies and liposome-mediated drug delivery. In excess water, mixtures of lamellar phase-forming lipids, such as phospholipids, and micellar phase-forming surfactants, such as bile salts (BS), assemble into mixed amphiphilic structures, the nature of which depends on the relative amounts of phospholipid and surfactant [2–4]. The sequence of phases found for the solubilization of lipidic vesicles by surfactants is complex, not yet completely understood [4,5] and depends directly on the partition coefficient of the surfactant between aggregates and solution [6]. Depending on their molecular [BS]/[lipid] ratio, the self-aggregation of bile salts (BS) and phospholipids results either in vesicles at a low ratio or mixed micelles at a high ratio. A micelle–vesicle transition is induced by BS removal from mixed micelles, while the reverse process results from BS addition to vesicles [7].

The solubilization kinetics of egg phosphatidylcholine vesicles by sodium taurocholate (TC) have shown by light scattering techniques that the [TC]/[Lipid] molar ratio corresponding to the micelle formation is strongly dependent on the rate of TC addition [8].

Spink et al. [9] studied by differential scanning calorimetry (DSC) the transformations occurring in TC/1,2-dipalmitoylphosphatidylcholine (DPPC) mixtures as a function of temperature. Schurtenberger et al. [10] investigated by quasielastic light scattering the influence of temperature on the formation and the size of vesicles produced spontaneously by diluting a mixed micellar solution of dimyristoylphosphatidylcholine and glycocholate beyond the mixed micellar phase boundary. Phase diagrams involving DPPC and non-ionic surfactants were determined as a function of temperature by combinations of techniques including DSC, NMR and small-angle X-ray diffraction scattering (SAXS) [11, 12]. To our knowledge, except calorimetric studies [9, 13], no systematic characterization of the lamellar phases formed by phospholipids/BS mixtures has been published.

In this study, as a prerequisite for the determination of the phase diagram of DPPC/TC/water system, equilibration kinetics are examined by synchrotron SAXS as a function of TC/PC molar ratio (r), time and temperature and DSC.

Materials and methods

The 1,2-dipalmitoyl-sn-glycero-3-phosphocholine (DPPC) was purchased from Avanti Polar-Lipids Inc. (Alabaster, AL, USA) and sodium taurocholate (TC) from Sigma (St Louis, MO, USA). TC was purified as previously described [13].

Sample preparation

DPPC and TC were weighed and appropriate volumes of methanol and chloroform are added to solubilize TC and DPPC, respectively. Organic solvents were removed by a stream of dry nitrogen followed by overnight low vacuum exposition of the sample in a lyophilizer. Then, the dried films were suspended in 0.5 ml of buffer solution (145 mM NaCl, 10 mM HEPES, pH 7.4) by vortex mixing. The resulting suspensions were heated at 50 °C for 15 min. until clear homogeneous mixtures were achieved. Aliquots of these mixtures were introduced into thin glass capillaries (GLAS, Muller, Berlin, Germany) which were filled with argon gas, closed with a drop of melted paraffin to prevent water evaporation and stored at 4 °C, 17 °C and 26 °C for 4, 8, 15 and 30 days before the SAXS analysis.

Synchrotron small-angle X-ray diffraction/scattering (SAXS)

Capillaries were successively analyzed at 25 °C on the D24 ($\lambda = 1.489$ Å) and D22 ($\lambda = 1.377$ Å) lines of DCI synchrotron of LURE using one-dimensional position-sensitive proportional detectors (512 or 1024 channels) for detection as described previously [14].

Differential scanning calorimetry (DSC)

DSC recordings were performed in a microcalorimeter of flux type (ARION, sensitivity about 100 μV/mW at 30 °C) as previously described [13]. For each scan, the samples (about 400 μl), prepared as above and stored at room temperature, were first melted at about 45 °C, introduced in the calorimeter at room temperature and chilled to about 4 °C (1 h), then heated from 4 °C to 50 °C at a very low scan rate (0.08 °C/min.) to ensure as far as possible equilibrium conditions, and analyzed twice successively.

Results and discussion

The thermal and structural behavior of the DPPC/TC/water system was studied at constant DPPC concentration (50 mM) in the range $0 < [TC] \leq 225$ mM ($0 \leq r \leq 2.0$) by DSC and SAXS.

Figure 1 shows the recordings obtained at a very slow heating rate by DSC for DPPC/TC mixtures with $0 \leq r \leq 2.0$. The addition of TC decreases the onset temperature of the main transition and generates multiple transitions. This decrease is related to r; the more TC added, the larger the temperature depression. At low r ($r < 0.4$), only multiple endotherms are observed while at $r \geq 0.4$ exothermic transitions appear more and more clearly with increasing r. The very low heating rate used (0.08 °C/min.) generally insures scanning conditions allowing hydrated systems to reach an organization faster than the rate of disorganization imposed by heating itself [13, 15]. The succession of exothermic and endothermic transitions observed at $r \geq 0.4$ shows a slow process of crystallization. This last process is slower than the rate used for the sample chilling in the calorimeter and crystallization is delayed until the slow heating of DSC is applied to the sample.

Samples stored at different temperatures (4 °C, 17 °C and 26 °C) for periods varying from 4 to 30 days were examined at 25 °C by SAXS. This procedure allows to determine the conditioning level of the different mixtures relatively to equilibrium state as a function of storage conditions (time and temperature). Taking into account the

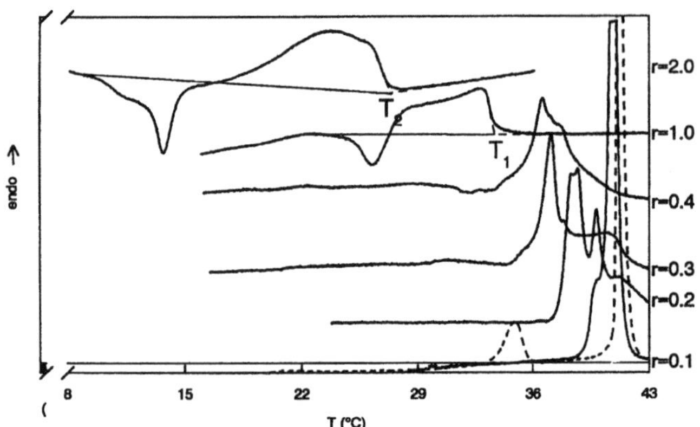

Fig. 1 Representative DSC recordings of the heating at 0.08 °C/min of DPPC/TC mixtures ([DPPC] = 50 mM) as a function of r in the range of $0.1 \leq r \leq 2.0$ (main transition of pure DPPC is also shown for comparison ----). $T_1 = 27.4$ °C, $T_2 = 33.7$ °C

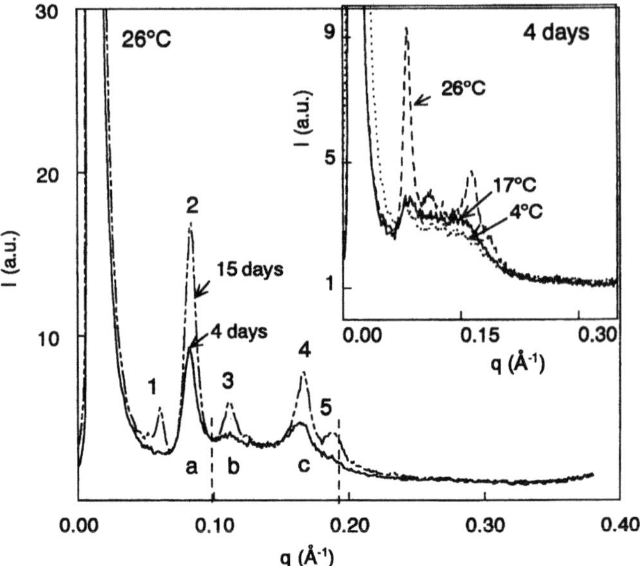

Fig. 2 Evolutions of the SAXS recordings (Intensity (I) vs q) for $r = 0.1$ ([DPPC] = 50 mM, [TC] = 5 mM) as a function of storage time at 26 °C (main figure) and temperature after 4 days (inset). Observed repeat distances: $1 = 102.2$ Å, $2 = 74.7$ Å, $3 = 56.1$ Å, $4 = 37.8$ Å, $5 = 33.5$ Å after 15 days and $a = 76.7$ Å, $b = 56.1$ Å, $c = 38.4$ Å after 4 days. (Positions of the d_{001} and d_{002} lines of pure hydrated DPPC are also shown ---- for comparison)

low cmc of TC (2–6 mM) [1, 16] relatively to the lipid concentration, the concentration of TC in the aqueous medium in equilibrium with the mixed phospholipid/surfactant aggregates can be neglected so that r can be assimilated to the TC/DPPC molar ratio in these aggregates [8].

The diffraction patterns observed in the different conditions will be examined below as a function of increasing r. These patterns evidence the formation of different structures either lamellar or not. The discussion of the molecular organizations is out of the scope of this paper and will not be considered here.

$r = 0.1$ – The inset of Fig. 2 shows the SAXS recordings obtained for three samples of the same mixture at $r = 0.1$ and stored for 4 days at temperatures ranging from 4 °C to 26 °C. The recordings corresponding to the samples stored at 4 °C and 17 °C show no distinct peaks while storage at 26 °C yields an organized structure. Low-temperature-stored samples seem not to be equilibrated. Upon prolonged storage at 26 °C, the molecular organization of the sample is improved as evidenced by the sharp diffraction peaks obtained (Fig. 2). The existence of several phases deduced from the repeat distances observed in the pattern recorded after stabilization agrees with the observation of overlapped endotherms by DSC (Fig. 1).

$r = 0.2$–0.3 – Samples corresponding to TC/DPPC ratios ranging from 0.2 to 0.3 are very viscous and very difficult to handle; it seems that the equilibration time required to obtain a pattern corresponding to an organized structure is longer than 30 days (the time limit investigated in this study), even if the temperature is maintained at 26 °C (data not shown). Longer storage of $r = 0.2$ sample led to the observation of a lamellar phase [13]. In this r range, the kinetic of equilibration is noticeably slow.

$r = 0.4$ – With increasing TC concentration (≥ 20 mM), the samples became easier to handle. Again,

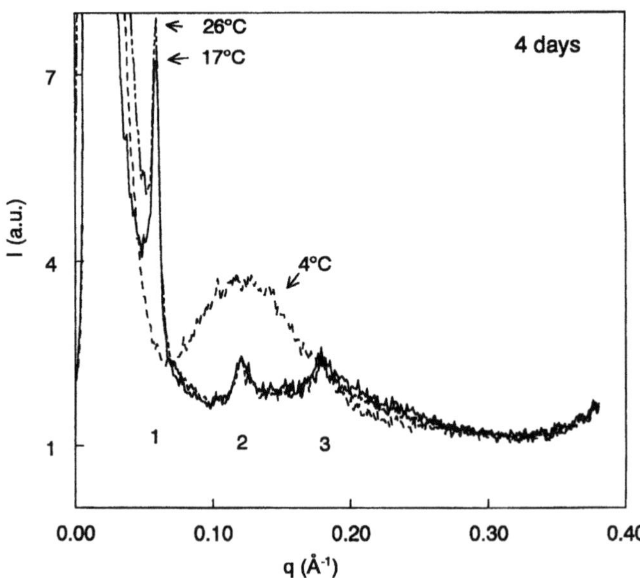

Fig. 3 Evolution of the SAXS recordings (Intensity vs q) for $r = 0.4$ ([DPPC] = 50 mM, [TC] = 20 mM) as a function of storage temperature after 4 days

no organized structure is observed when the sample is stored at 4 °C (Fig. 3), even if the storage is prolonged for 30 days (data not shown). Sharp peaks appear when the storage temperature is raised to 17 °C or 26 °C. This indicates that the mixtures self-equilibrate at upper temperature (≥ 17 °C) forming a lamellar phase with high periodicity

Table 1 Evolution of the repeat distances (d_{001}) of molecular structures formed vs. r after storage at 26 °C for 30 days

r	0.4	0.6	1
d_{001}	105.9	94.0	84.4
d_{002}	52.1	46.7	42.0
d_{003}	34.5	31.2	27.9

($d_{001} = 106$ Å) (Table 1). Then, we conclude again that storage temperature strongly influences the equilibration process. This is not surprising since the chain-melting temperature of these two samples observed by DSC is about 35.4 °C, i.e. 18.4 °C and 9.4 °C above the conditioning ones (17 °C and 26 °C) (Fig. 1).

$r = 0.6–1.0$ – The recordings corresponding to samples composed of TC/DPPC ratio = 0.6 and 1 and stored 30 days at 17 °C or 26 °C also exhibit three sharp peaks. The repeat distances and the intensities of these peaks change according to the composition of the sample as well as the conditioning temperature (Fig. 4 and Table 1). Unexpectedly, periodicities observed decrease with increasing concentration of TC. The evolution of intensities of the different orders and of the distances vs r might correspond to strong changes in the repartition function of electronic density along the direction perpendicular to the bilayers.

$r = 2.0$ – Storage at 4 °C results on defined peaks corresponding to a well organized structure, likely lamellar in spite of the absence of d_{001} line, together with smooth scattering broad peaks with a minimum at about $q = 0.1$ Å^{-1} showing the presence of some micelles (30 days) (Fig. 5). Again lamellar peak intensities are increasing vs storage time. The samples prepared only four days before the analysis, display even more intense peaks when storage temperature is raised (Fig. 5 inset). Sample equilibration seems to be improved by storage duration at 26 °C. However, Fig. 6 shows quite a reverse process. In fact, the storage and observation temperatures likely coincide with that of the chain melting of the DPPC/TC mixtures as found by DSC (Fig. 1). Indeed, there is a range of r for which the DSC transition temperatures induced by TC addition occur in the temperature domain chosen to equilibrate the samples.

$2.0 < r \leq 4.5$ – This behavior, due to the decrease of chain-melting temperature, persists at increasing TC ratios (data not shown).

The above results show that thermally active phospholipids in vesicles are not necessarily at equilibrium when formed at room temperature as observed previously by DSC measurements by Spink et al. [9]. The thermal and structural behaviors reported in this study shed some light on the thermal procedures required to obtain defined structures as well as the complexity of equilibration that

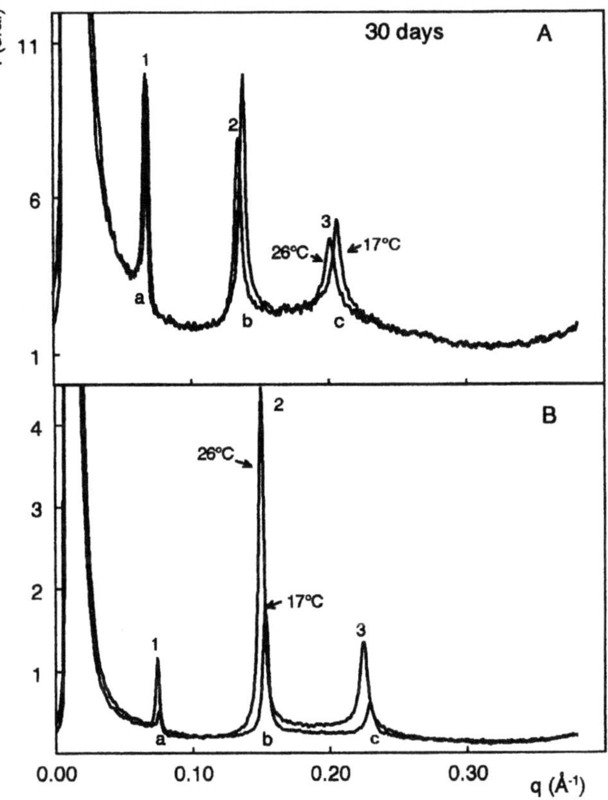

Fig. 4 Evolutions of the SAXS recordings (Intensity vs q) for $r = 0.6$ (A) and $r = 1.0$ (B) ([DPPC] = 50 mM, [TC] = 30 and 50 mM) as a function of storage temperature after 30 days

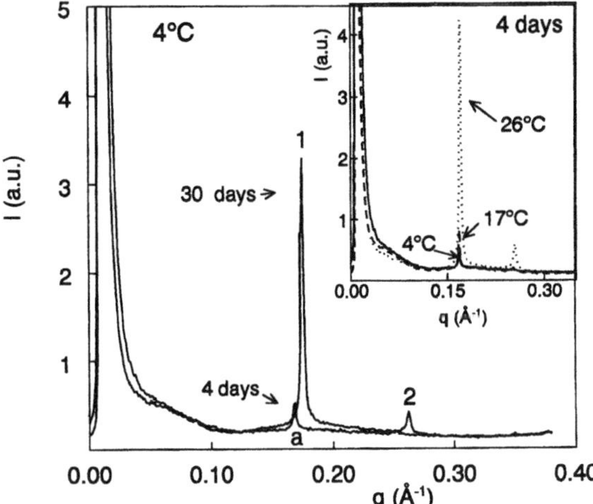

Fig. 5 Evolutions of the SAXS recordings (Intensity vs q) for $r = 2.0$ ([DPPC] = 50 mM, [TC] = 100 mM) as a function of time at 4 °C (main figure) and temperature at 4 days (inset). Observed repeat distances: 1 = 35.8 Å, 2 = 23.9 Å after 30 days and $a = 37.15$ Å after 4 days

might be related to a slow lateral diffusion process similar to that observed previously for nonionic lamellar forming phases [8, 17]. Further experiments are under way to complete our understanding of the various structures observed.

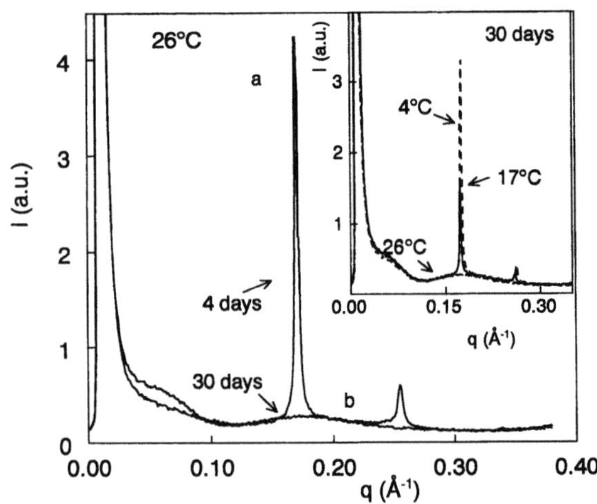

Fig. 6 Evolutions of the SAXS recordings (Intensity vs q) for $r = 2.0$ ([DPPC] $= 50$ mM, [TC] $= 100$ mM) as a function of time at $26\,^{\circ}$C (main figure) and temperature at 30 days (inset). Observed repeat distances: $a = 36.9$ Å, $b = 24.6$ Å after 4 days

Conclusion

Coupling DRXT and DSC allowed us to study the equilibration kinetics of mixtures composed of DPPC (50 mM) and TC in a large range of concentrations. The results show that the molecular organization obtained strongly depends on r and storage conditions. The samples with a low molar ratio $r = 0.1$ were quickly equilibrated at $26\,^{\circ}$C and not at $4\,^{\circ}$C and $17\,^{\circ}$C. At $0.2 \leq r \leq 0.3$, they were not at equilibrium within one month whatever the experimental conditions. Lamellar phases were observed at $0.4 \leq r \leq 1.0$ after few days to one month depending on r and storage temperature. At $r \geq 2.0$ and depending on storage and observation temperatures lamellar or micellar phases (or their mixtures) which were observed were slowly equilibrating at $4\,^{\circ}$C.

Acknowledgments We would like to thank the IFSBM (Institut de Formation Supérieure Biomédicale) for supporting K.A.

References

1. Cabral DJ, Small D (1990) In: Handbook of physiology, The Gastro-Intestinal System, III Ch. 31, Physical Chemistry of Bile, pp 621–662
2. Walter A, Vinson PK, Kaplun A, Talmon Y (1991) Biophys J. 60:1315
3. Lichtenberg D (1996) In: Barenholz Y and Lasic DD (eds) Handbook of Nonmedical Applications of Liposomes, Vol II, pp 199–218. CRC Press, Boca Raton, FL
4. Ollivon M, Eidelman O, Blumenthal R, Walter A (1988), Biochem 27(5):1695
5. Seras M, Edwards K, Almgren M, Carlson G, Ollivon M, Lesieur S (1996) Langmuir 12:330
6. Paternostre M, Meyer O, Grabielle-Madelmont C, Lesieur S, Ghanam M, Ollivon M (1995) Biophys J 69:2476
7. Lasch J (1995) Biochem Biophys Acta 1241:269
8. Ramaldes G, Fattal E, Puisieux F, Ollivon M (1996) Colloids and Surfaces 6:363
9. Spink CH, Lieto V, Mereand E, Pruden C (1991) Biochem 30:5104
10. Schurtenberger P, Bertani R, Kanzig W (1986) J Colloid Interface Sci 114:82
11. Dahim M (1995) PhD Thesis
12. Madler B, Klose G, Mops A, Richter W, Tschierske C, (1994) Chem Phys Lipids 71:1
13. Forte L, Andrieux K, Keller G, Grabielle-Madelmont C, Lesieur S, Paternostre M, Ollivon M, Bourgaux C, Lesieur P (1998) J Thermal Anal 51:773
14. Keller G, Lavigne F, Loisel C, Ollivon M, Bourgaux C (1996) In: Slade L and Levine H (eds) J Thermal Anal Special issue on Recent Advances in applications of thermal analysis to foods 47:1545
15. Kekicheff P, Grabielle-Madelmont C, Ollivon M (1989) J Colloid Interface Sci 131(1):112
16. Coello A, Meijide F, Nunez ER, Tato JV (1996) J Pharm Sci 85(1):9
17. Seras M, Gallay J, Vincent M, Ollivon M, Lesieur S (1994) J Colloid Interface Sci 167:159

Progr Colloid Polym Sci (1998) 110:285–290
© Steinkopff Verlag 1998

W. Schärtl
G. Lindenblatt
A. Strack
P. Dziezok
M. Schmidt

Spherical and rod-like colloids with polymer-brush surfaces

Dr. W. Schärtl (✉)
G. Lindenblatt · A. Strack
P. Dziezok · M. Schmidt
Institut für Physikalische Chemie
Universität Mainz
Welderweg 11
D-55099 Mainz
Germany
E-mail: wolfgang@hal2000.chemie.uni-mainz.de

Abstract In this paper, we describe a strategy to overcome incompatibility of colloidal particles and polymer coils as well as immiscibility of spherical and rod-shaped nanoparticles. Two new types of model colloids are presented, colloidal nanospheres with hairy surfaces (spherical brushes) and polymacromonomers to represent cylindrical brushes. The spherical brushes are synthesized from polyorganosiloxane-μ-gels of diameter 20 nm by grafting onto anionically prepared polystyrene macromonomers of molecular weight $M_w = 5000$ g/mol. On average, each sphere has a surface layer of 200 polymer chains. Compatibility of spherical nanoparticles with polymer coils was probed by turbidity of as-cast films as well as electron microscopy and atomic force microscopy. In contrast to their precursors without hairy surfaces, the spherical brushes are compatible with polymer coils of low molecular weight ($M = 4000$ g/mol) upto very high concentrations. The cylindrical brushes are synthesized by radical poylmerization of polystyrene–methylmethacrylate macromonomers of molecular weight $M_w = 4000$ g/mol. The product shows a high polydispersity in length of the PMMA-backbone. By applying continuous polymer fractionation (CPF) we were able to reduce length polydispersity of the cylindrical particles from above 4.0 to smaller than 1.5. Analogous to spherical brushes, these cylinders are also compatible with low molecular weight polymer coils. Further, we have prepared transparent films containing mixtures of spherical and cylindrical brushes in a matrix of low molecular weight polystyrene upto high particle concentrations.

Key words Colloidal nanosphere – spherical brushes – cylindrical brushes

Introduction

Colloidal additives play a very important role for enhancing the performance of polymeric materials. Some typical examples are fibres used for molecular reinforcement or carbon black in tire rubber. However, it is extremely difficult to disperse nanoparticles in polymer melts on a molecular level: depletion interactions usually lead to immiscibility of colloidal particles and polymer coils in highly concentrated solutions [1, 2]. A good review on the phase behaviour of colloidal spheres in general, discussing also the behaviour of mixtures of spherical colloids and polymer coils in solution, is given in [3]. It should be noted that, whereas colloid polymer mixtures in solution have been studied quite extensively, there are only a few papers on polymer melt composites. An example of miscibility enhancement of rigid rod and flexible macromolecules by ionic interactions to obtain homogeneous composite materials has recently been published by Eisenbach et al. [4].

Another strategy to obtain highly concentrated composites of nanoparticles and flexible polymer coils can be deduced from blockcopolymer micelles [5]. Blockcopolymer micelles form homogeneous mixtures with homopolymer coils up to weight fractions of 50%. This miscibility is caused by the fact that the micellar coronae, which also may be regarded as hairy surfaces of spherical colloidal particles, are identical in both chemical composition and structure to the polymer coils of the matrix. Therefore, one would expect that, in general, particles with brush surfaces should be compatible at least to polymer coils. Further, compatibility of particles with different topologies should also be enhanced by brush surfaces. The concept of compatibilizing stiff and flexible polymers by attaching flexible side chains to the stiff species has been used previously to prepare a mixture of polystyrene and a liquid crystalline graft polyester with polystyrene side chains [6], leading to a molecularly reinforced polymer composite.

On the other hand, making topologically incompatible colloidal particles like spheres and rods compatible by introducing identical polymer brush surfaces may also cause strongly enhanced miscibility. For common particles, Frenkel has shown in his simulations [7] that mixtures of spheres and needles, e.g., show demixing into two separate phases over a wide concentration range. Although such colloidal multicomponent systems are expected to exhibit a very interesting behavior concerning structure and dynamic, due to topological incompatibility only few experimental examples are found in literature [8–10]. Pecora [8, 9] as well as Miller [10] and coworkers have studied suspensions of spherical and rod-like particles in N,N-dimethylformamide (DMF), where miscibility of topologically incompatible particles might have been increased by strong dipole–dipole interactions of particles. However, the homogeneous mixtures described in [8–10] still have a comparatively low content of nanoparticles. Alternatively, using the concept of polymer brush surfaces, it should be possible to prepare highly concentrated sphere–rod mixtures.

Here, we present two new species of model colloidal particles with brush surfaces which are expected to show an increased miscibility with flexible polymers as well as with each other: spherical brushes are synthesized from polyorganosiloxane-μ-gels [11] by grafting onto of polystyrene macromonomers. Depending on the length of the grafted chains in relation to the polyorganosiloxane cores, one obtains either hairy spheres (short chains) or star polymers (long chains). Grafting reactions of polymer chains onto multifunctional cores as dendrimers [12] or silica particles [13] to prepare model star molecules or to enhance solubility of colloidal particles have been

described previously. Note, however, that the microgel cores in our case, in contrast to dendrimers, besides simpler chemical preparation provide also possibilities of later chemical modification like incorporation of dye labels [14] or noble metals [15]. Compared to conventional silica particles [13], our major advantage is the close match of particle density to the surrounding matrix. As rod-like model colloids with brush surfaces we choose polymacromonomers [16]. Usually, these highly branched macromolecules have a large polydispersity of their backbone. Here for the first time, to our knowledge, we present a method to obtain considerable amounts of rather monodisperse cylindrical brushes by continuous polymer fractionation [17] of the originally very polydisperse material.

Compatibility of the new nanoparticles in binary mixtures with polymer coils as well as in ternary systems has been investigated by microscopy (transmission electron microscopy (TEM) and atomic force microscopy (AFM)) techniques. Also, first small angle X-ray measurements showed that the spherical particles give sufficient scattering contrast to be distinguishable from the background matrix of cylindrical brushes or polystyrene coils, thereby allowing to probe the sample structure in the bulk whereas TEM only probes a sliced section, AFM the surface of the as-cast and hot-molded films. Since SAXS results are only very preliminary so far, they will be shown in a subsequent publication. As a more qualitative first criterion of particle miscibility we have also checked the transparency of the molded films.

Experimental

Spherical brushes: Figure 1 sketches the synthesis of highly crosslinked polyorganosiloxane-μ-gels with a surface of polystyrene (PS) chains: first, sperical microgels are formed by polycondensation of trimethoxymethylsilane. After the core has been formed, a functionalized shell consisting of Si-H-groups is introduced by addition of hydridotrimethoxysilane. These core–shell particles are endcapped using dihydridotetramethyldisiloxane, precipitated from the aqueous phase and dissolved in toluene. To obtain a high surface functionality, particles are once more endcapped with dihydridotetramethyldisiloxane in organic solution. PS macromonomers are synthesized by standard anionic polymerization of styrene, using para-bromomethylstyrene as a special termination agent to introduce a polymerizable group at the polystyrene chain end. This functionality may as well be used for polymerization reactions, leading to cylindrical brushes with PS side chains and PS backbone [16]. We use a hydrosilylation couplingreaction with platinum catalysis [18] to attach the PS chains onto the Si-H-functionalized polyorganosiloxane

Progr Colloid Polym Sci (1998) 110:285–290
© Steinkopff Verlag 1998

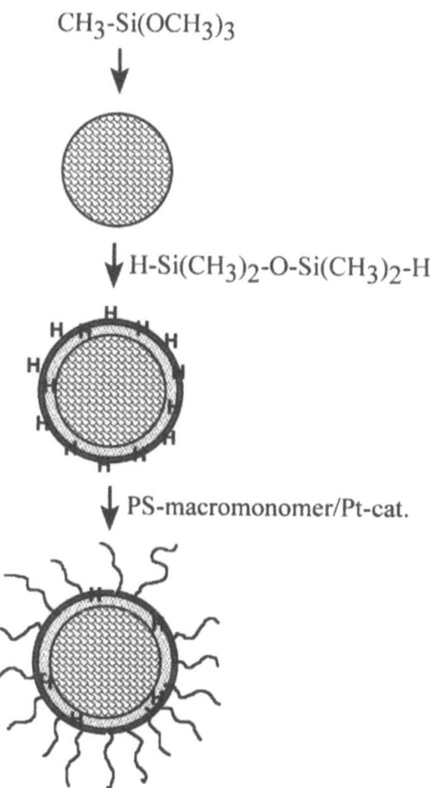

Fig. 1 Reaction scheme of synthesis of polyorganosiloxane microgels with surface grafted polystyrene macromonomer chains

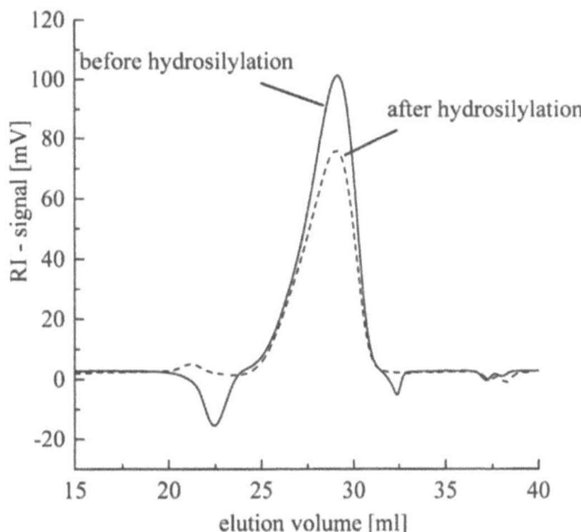

Fig. 2 GPC eluograms of polyorganosiloxane microgel cores, polystyrene macromonomer hairs before and after grafting reaction, detected by a refractive index detector. Peaks at 30 min correspond to non-grafted macromonomers (note the decrease of the peak area due to the grafting reaction). Peaks at 22 min correspond to the spherical particles. Whereas microgel cores show a negative peak area, spherical brushes give a positive peak due to a dramatic change in average refractive index. From these peak areas, the average number of macromonomer chains per spherical brush maybe calculated (see Table 1)

particles. Success of the coupling reaction is probed by gel permeation chromatography (GPC: PSS styragel columns, Waters differential refractometer detector), and dynamic light scattering using an Ar-laser and ALV5000 correlator.

Figure 2 shows the corresponding GPC elugrams. Note that the eluent peak of the spherical particles changes its sign from negative to positive due to the change in refractive-index-caused by the attached PS chains. Taking into account the molecular weight of each species and the refractive-indicex-increments of non-grafted polyorganosiloxane-μ-gels (-0.043) and polystyrene ($+0.110$) in toluene, one can determine the average number of polystyrene chains per microgel from the peak areas of the GPC eluogram. Table 1 gives a summary of particle properties like hydrodynamic radii (R_H), weight-average molecular masses (M_w) and refractive index increments (dn/dc). One

should note the considerable number of PS chains per particle, corresponding to a surface closely packed with polystyrene brush hairs.

Cylindrical brushes: Polymacromonomers as models for cylindrical brushes are synthesized by radical polymerization of PS-methylmethacrylate (MMA) macromonomers. The later have been prepared by standard anionic polymerization of styrene, using ethylenoxide and methacryloylchloride as a special termination agent to introduce a MMA functional group at the PS chain end. Details of this reaction have been described elsewhere [16]. These macromonomers form, after polymerization, cylindrical brushes with PMMA backbone and PS sidechains. Due to steric repulsion of the sidechains, the polymacromonomers exhibit a remarkable chain stiffness of the backbone. This can be seen from a Kuhn length of about 150 nm (depending on the molecular weight of the macromonomers) as determined by light scattering [19]. Due to the radical polymerization step, however, the

Table 1 Characterization of spherical brushes

	R_H	M_w	dn/dc (toluene)	No. of hairs
μ-gel cores	10.8 nm	2.7×10^6 g/mol	-0.043	—
PS macromonomers	—	5.0×10^3 g/mol	$+0.110$	—
Spherical brushes	16.6 nm	—	—	≈ 240

Table 2 Characterisation of cylindrical brushes

	M_w [g/mol]	M_w/M_n	P_w (aspect ration)	R_g [nm]	R_H [nm]
Macromonomer	3.79×10^3	1.05	1	—	—
Polymacromonomer	$3.36\ 10^6$	4.3	887 (12.0)	54.7	30.2
1. Sol fraction	0.49×10^6	2.1	129 (1.7)	—	—
1. Gel fraction	3.83×10^6	2.9	1008 (13.6)	—	—
2. Sol fraction	0.39×10^6	1.7	103 (1.4)	11.8	9.7
2. Gel fraction	3.91×10^6	2.4	1029 (13.9)	—	—
3. Sol fraction	0.51×10^6	1.5	134 (1.8)	—	—
3. Gel fraction	3.92×10^6	2.2	1032 (13.9)	56.1	32.7
4. Sol fraction	1.13×10^6	1.4	273 (3.7)	—	—
4. Gel fraction	6.24×10^6	1.5	1642 (22.2)	—	—
5. Gel fraction	6.40×10^6	1.5	1684 (2.7)	—	—

cylindrical brushes exhibit a large polydispersity in length while being uniform in diameter. So, the molecular weight typically ranges from about 400.000 g/mol (star-like molecules) upto 6.000.000 g/mol (long cylindrical brushes). To obtain monodisperse rods in considerable amounts (2–3 g), we have used continuous polymer fractionation (CPF) [17]. Successful application of this powerful fractionation technique on polymacromonomer samples has been proved by GPC. Table 2 gives an overview on molecular weights of macromonomer precursor, polydisperse rods and fractionated materials as determined by GPC and light scattering. For some samples (polydisperse polymacromonomer, 2. sol fraction and 3. gel fraction) the radius of gyration (R_g) and the hydrodynamic radius (R_H) have been determined by light scattering. In Table 2, P_w is the weight average polymerization degree of the PMMA backbone of the cylindrical brushes as calculated from the macromonomer molecular weight of 3800 g/mol. The aspect ratio length/diameter of the cylindrical brushes can be crudely estimated by the ratio $P_w/(2P_{macromonomer})$, assuming the cylinder diameter to be equal to the length of 2 stretched macromonomer chains ($2P_{macromonomer} = 74$).

Flexible polymer coils: The third component of our composite materials, the flexible polymer coil, is chosen either from commercially available PS standards (Polymer Standards Service) or from the PS chains used for grafting, part of which have been terminated with methanol instead of para-bromomethylstyrene.

Various mixtures have been prepared from these materials: all components are dissolved in chloroform (total polymer concentration 10 wt%). The solvent is slowly evaporated at room temperature (solvent casting) and the dry film is vacuum-annealed at a temperature above the glass transition of the polymer matrix ($T_{anneal} = 120\,°C$) for two days to remove all remaining solvent. After annealing, the samples are hot-pressed between two glass plates, separated by a teflon spacer of thickness 100 μm, at 150 °C. In the melt, we check transparency of the thin films. The cooled films are studied by AFM (film surface, "tapping

mode") and TEM (Philips EM 420 ST), using thin slices (2 μm thickness) for the latter. The results concerning enhanced miscibility, especially comparison of spherical

Fig. 3 Photographs of composite polymer melts showing the high transparency of spherical brush polymer mixtures. Sample description is observed through the film of 100 μm thickness. Whereas a film with even a small content of 1 wt% standard microgels is non-transparent, a film containing 30 wt% spherical brushes is transparent enough to allow for reading of the sample description

1 wt% standard microgels in polymer melt

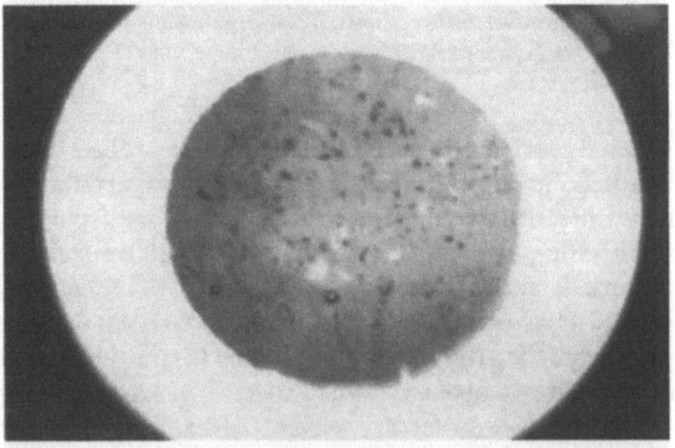

30 wt% spherical brushes in polymer melt

brushes and non-grafted polyorganosiloxane-μ-gels, are presented and discussed in the last section of this paper.

Results and discussion

Whereas a polymer film containing only 1 wt% of non-grafted spheres is totally non-transparent, films containing spherical brushes upto concentrations of 30 wt% spherical cores within the matrix of PS hairs and PS coils are still highly transparent (Fig. 3). Figure 4 shows a TEM micrograph of a mixture of 1 wt% spherical polyorgano-siloxane-μ-gels without grafted polymer surfaces and 99 wt% linear PS (M = 4000 g/mol), compared to a mixture of 20 wt% spherical brushes (concentration of polyorganosiloxane cores!) and 80 wt% PS. As expected, the particles without brush surface form large aggregates due

to the strong depletion demixing, thereby causing the above turbidity of the as-cast films. In contrast, the spherical brushes are dispersed on a molecular level within the polymer.

We therefore may conclude here that depletion demixing of polymer coils and colloidal particles indeed can be overcome using colloids with polymer brush surfaces. So far, we only have checked miscibility of polymer coils of molecular weight equal to or smaller than the brush hairs. In the near future, we will check more systematically how relative length of brush hairs and polymer coils influences compatibility.

Concerning ternary systems consisting of spheres, cylinders and flexible coils studied so far, optically homogeneous (transparent) films are obtained for concentrations upto 20 wt% short rods (4.sol, see Table 2) and 20 wt% spherical brushes (Fig. 5a), which is much higher

Fig. 4 Transmission electron micrographs of polymer colloid mixtures comparing standard polyorganosiloxane microgels (a) and spherical brushes (b). The white bar indicates a distance of 100 nm. See text for discussion

Fig. 5 Photographs of composite polymer melt showing the high transparency of spherical brush, cylindrical brush and coil polymer mixtures. Sample description is observed through the film of 100 μm thickness (see above). Films contain (a) 20 wt% short cylindrical brushes (4.sol, see Table 2), 20 wt% spherical brushes and 60 wt% polymer coils (M = 4000 g/mol), (b) 10 wt% long cylindrical brushes (5.gel, see Table 2), 10 wt% spherical brushes and 80 wt% polymer coils (M = 4000 g/mol).

(a) 3 wt% standard μ-gels in polymer melt

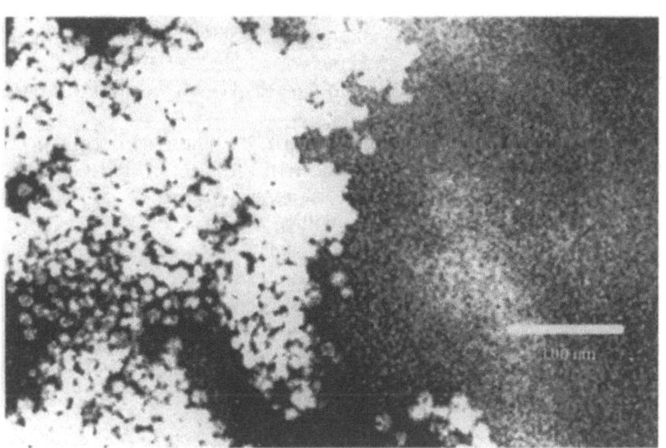

(a) 20 wt% spherical, 20 wt% short cylindrical brushes in polymer melt

(b) 20 wt% spherical brushes in polymer melt

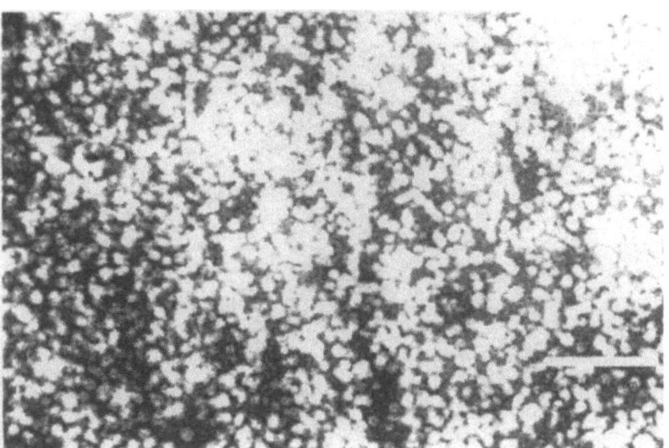

(b) 10 wt% spherical, 10 wt% long cylindrical brushes in polymer melt

than samples previously studied by Pecora or Miller [8–10]. In contrast, mixtures of spherical brushes and long rods (5.gel, see Table 2) are only transparent for concentrations up to 10 wt% spherical and 10 wt% cylindrical brushes (Fig. 5b).

Therefore, our studies show that preparation of highly concentrated mixtures of spherical and cylindrical particles based on application of identical brush surfaces seems to be possible. We should note here that the later results on ternary systems are still rather preliminary and a more careful check on the influence of parameters as cylinder length or chain molecular weight both of polymer coils and brush hairs has to be done. Whereas, as stated above, the length of the spherical brush hair could be adjusted from hairy to star-like particles, in case of the cylinders, however, we are limited in molecular weight of the macromonomers which can be used to synthesize poly-macromonomer cylinders: so far, for steric reasons it has not been possible to prepare cylinders of considerable length consisting of macromonomers of molecular weight larger than 10 000 g/mol. In the near future, we will present a more detailed study of the phase behavior of our new model particles.

Acknowledgement We would like to thank R. Würfel for taking the TEM pictures. Financial support by the Deutsche Forschungs-gemeinschaft (project D23-SFB264) is also gratefully acknowledged.

References

1. Asakura S, Oosawa F (1954) J Chem Phys 22:1255
2. Vrij A (1976) Pure Appl Chem 48:471
3. Poon WCK, Pusey PN (1995) In: Baus M et al (eds) Observation, Prediction and Simulation of Phase Transitions in Complex Fluids, Kluwer Academic Publishers, NL, pp 3–51
4. Eisenbach CD, Hofmann J, Fischer K (1994) Macromol Rapid Commun 15:117
5. Schaertl W, Tsutsumi K, Kimishima K, Hashimoto T (1996) Macromolecules 29:5297
6. Heitz T, Rohrbach P, Höcker H (1989) Makromol Chem 190:3295
7. Bolhuis P, Frenkel D (1994) J Chem Phys 101:9869
8. Tracy MA, Pecora R (1992) Macromolecules 25:337
9. Tracy MA, Garcia JL, Pecora R (1993) Macromolecules 26:1862
10. Gold D, Onyenemezu C, Miller WG (1996) Macromolecules 29:5710
11. Baumann F, Schmidt M, Deubzer B, Dauth J, Geck M (1990) Macromolecules 23:3796
12. Roovers J, Zhou LL, Toporowski PM, van der Zwan M, Iatrou H, Hadjichristidis N (1993) Macromolecules 26:4324
13. Philipse AP, Pathamanoharan C (1993) J Colloid Interface Sci 159:96
14. Schärtl W, Graf C, Schmidt M (1997) Progr Colloid Polym Sci 104:129
15. Roos C, Schmidt M (1997) in preparation
16. Tsukahara Y, Tsutsumi K, Yamashita Y, Shimada S (1990) Macromolecules 23:5201
17. Weinmann K, Wolf BA, Rätzsch MT, Tschersich L (1992) J Appl Polym Sci 45:1265
18. Lindenblatt G (1997) Diploma Thesis, Mainz
19. Wintermantel M, Gerle M, Fischer K, Schmidt M, Wataoka I, Urakawa H, Kajiwara K, Tsukahara Y (1996) Macromolecules 29:978

Progr Colloid Polym Sci (1998) 110:291–295
© Steinkopff Verlag 1998

A. Chakrabarti
R. Zajac

Monte Carlo study of layer formation and exchange kinetics in polymer adsorption

A. Chakrabarti (✉)
Department of Physics
Cardwell Hall
Kansas State University
Manhattan, KS 66506
USA

R. Zajac
Department of Arts, Sciences & Business
Kansas State University-Salina
Salina, KS 67401
USA

Abstract We have carried out extensive Monte Carlo simulations of the adsorption of both homo- and hetero-polymer chains from semidilute solution onto a solid/liquid interface. We investigate the specific conformations adopted by the adsorbed chains, the kinetics of layer growth and chain exchange, as well as the effects of impurities in the system. We consider the effects of impurities which are either on the surface or within the chains themselves. These impurities may be either repulsive or simply inactive. The deGennes self-similar grid layer, which we directly recover in the pure system, is found to be only slightly perturbed by even large amounts of impurities. In a pure system, we find that the chains comprise two broad groups: a tightly bound group and a loosely bound group. However, in impurity systems, the exchange of polymer with the solution is found to be accelerated, so that this distinction is greatly reduced. While we note close similarities between the static effects of surface impurities and of chain impurities, dramatic differences arise in the kinetic effects of these two cases. In particular, we note the appearance of a large group of chains which are conformationally "frozen" when repulsive impurities are incorporated into the chains, severely restricting their ability to slither along the surface. This yields considerably different behavior of the chains with respect to the layer's relaxation and exchange.

Key words Polymer adsorption – self-similarity – random copolymers

Introduction

When a solution of polymer chains is placed in contact with an adsorbing surface, the small preferential adsorption per segment adds up to a huge adsorption strength for the whole molecule, and an adsorbed layer is formed [1]. At equilibruim, this layer participates in a continual, balanced traffic of chains to and from the bulk solution. Since there are many adsorbed chains in a layer, it is possible for an adsorbed layer to contain much diversity in its chain conformations comprising various trains, tails and loops [2–5]. However, since all of these static properties arise from a process which is inherently dynamic, a thorough understanding of them can only be achieved by studying the migration of chains.

Studies have addressed the growth kinetics of adsorbed layer formation leading to equilibrium, as well as the washing of a layer with a solution of different chains [6]. Of particular interest is the case where the new chains are not very different from the original ones, because such experiments serve to probe the details of equilibrium chain exchange. Based on scattered clues from experiment, it has been suggested that the adsorbed chains might fall into two categories: a tightly bound and a more loosely bound category. Supposedly, chains arriving late at the surface would find it completely covered by the early ones, which would have had time to spread out and acquire relaxed conformations. Thus the late comers would not be able to attach to it thoroughly, and their adsorption would be short-lived.

Computer simulation offers us a unique perspective from which to study both static and dynamic properties of polymer adsorption with careful scrutiny. We have therefore applied a Monte Carlo simulation [7] to study adsorption of polymer systems. By following the flight of individual chains, we have obtained a detailed description of their structure and dynamics. Although such studies of pure systems are extremely useful, they are nevertheless an idealization. In reality, most cases of polymer adsorption are never quite homogeneous, containing chemical impurities which are also expected to have an effect on the layer structures. It is therefore of interest to examine what impact these impurities have on the structure and dynamics of the adsorbed layer, and how the case of impurities within the surface differs from that of impurities within the chains. In order to address these concerns, we have further expanded our simulation [8] to include systems containing chemical impurities either within the adsorbing surface, or within the chains themselves.

Model and simulation methods

In our study, the polymer chains are described by the standard Monte Carlo lattice model [9], wherein a molecule is represented by a number N $(= 50\,100\,200)$ of successive monomers which occupy adjacent sites on a 3D cubic lattice, without overlap. Vacant lattice sites correspond to solvent molecules. The no-overlap condition gives rise to an excluded volume interaction between monomers, characteristic of a good solvent. The chains evolve according to the augmented Verdier–Stockmayer moves, along with a slithering, serpentine motion of the whole molecule [9]. After a brief equilibration period, a contact interaction with the surface is "turned on". Thereafter, desorption of adsorbed monomers is disfavored, being accepted with a probability of $e^{-\chi/kT}$, where $\chi\,(= 1.8\,kT)$ is the enthalpy of adsorption of a segment over a solvent molecule. This causes an adsorbed layer to build up. In order to prevent the depletion of the bulk solution, we maintain the number of free chains constant by adding or deleting chains in the half-system furthest from the surface. In this way, the total number of chains in our system continually fluctuates. Throughout the growth and equilibrium of the layer, the physical quantities of interest are measured in great detail, and averaged over repeated experiments with different initial configurations. Typically, we average over 10 such "runs".

In the next set of simulations, we include the effects of impurities. For impurities within the chains, we choose at random a given number of monomers individually for each chain, which are then taken to be impurities (or B-type monomers). The remaining monomers are taken to be pure (A-type). While the A-type monomers behave as in the pure system, the B-type monomers are taken to be either (i) ambivalent to the surface (inactive), (ii) repulsive to the surface. Each case is considered separately. In the case of repulsive impurity monomers, the *adsorption* of the B-type monomer is disfavored, being accepted with a probability of $e^{-\chi/kT}$, where the repulsive interaction is taken to be as strong $(1.8kT)$ as the attractive interaction for A-type monomers. We have also considered cases where the B-type repulsive interaction is twice as strong as the attractive A-type interaction.

For impurities within the surface (studied separately), a given number of surface sites are chosen at random to be B-type. In this case, the chains are taken to be pure, and their interaction with an A- or B-type surface site follows the same procedure as for the case of a pure surface interacting with A- and B-type monomers, described above. For both cases, we have considered impurity concentrations of 30% and 50%, for both inactive impurities, and repulsive impurities.

Results

Growth of the layer in a pure system

Initially, the growth of the layer has been found to be governed by the diffusion of chains to the vacant surface, so that the number of adsorbed chains grows in proportion to $t^{1/2}$, and the bulk concentration ϕ_0. Beyond this point, diffusion is no longer the slowest process in the growth, and the growth occurs more slowly. The surface becomes crowded so that incoming chains must worm their way through previously adsorbed chains, seeking an unoccupied spot on the surface. We have examined the number $\theta(t)$ of occupied surface sites. The quantity $(1 - \theta(t))/\theta_{\text{eq}}$ was found to decay exponentially with the growth of the layer, at low concentrations ($\phi_0 = \phi^*$, the overlap concentration). This follows from considering the vacant surface sites as particles which decay off the surface as the monomers adsorb. The simple exponential behavior then implies that these "particles" decay independently at low concentrations. At high concentrations, or late times, however, this behavior becomes slower than a simple exponential, signaling a strong degree of coordination between the decays. This occurs as the chains spread out over late times. This spreading is the dominant contribution to the growth of $\theta(t)$ because the addition of a new chain, in comparison, fills very few surface sites at these densities.

We can examine the growing layer's structure in more detail by examining the conformations of the adsorbed chains. Figure 1 shows what *percentage of monomers* in these chains are distributed into trains, tails and loops, at different stages of growth. At early times, the chains have a very high percentage of their beads in trains, owing to the ease with which they can spread out over the bare surface. The distribution gradually becomes broader over time, while its maximum occurs at more and more intermediate values of the chain percentage.

Similarly, the distributions of chain length in tails and loops show that the chains adsorbed at early times have very few of their beads off the surface, since the loops and tails are easily "reeled" in to become trains. The broadening of these distributions suggests that the early-adsorbing chains are forcing the new chains to adopt looser, less bound conformations, with fewer surface contacts.

Static properties of the adsorbed layer

The structure of the adsorbed layer is shown in Fig. 2, for a system with a bulk concentration equal to the overlap concentration. A least-squares fit to the log–log plot of the density profile yields a decay exponent of -1.33 ± -0.04, in excellent agreement with de Gennes' scaling prediction [10], $\phi(z) \sim z^{-4/3}$, for a self-similar layer in contact with a semidilute solution near the overlap concentration. This scaling result is thus confirmed *directly* by our present study, without our having to make assumptions or inferences, as required with experimental methods. It is interesting that this result is matched by our profiles for both chains of lengths $N = 100$ and 200, and that these profiles coincide over a considerable range of z. This, despite the fact that the profiles correspond to equilibria with solutions of different concentration (ϕ^* being N-dependent). This follows from the fact that the blob size throughout the self-similar region is independent of chain length; increasing the length serves only to increase the range over which this self-similarity is valid. The fact that our simulation recovers this result gives us confidence that our finite chain lengths are sufficient to capture much of the essential physics.

Dynamic properties of the layer in equilibrium

A further step of our study involves instantaneously "labeling" each of the adsorbed chains. In the further evolution of the system, the unlabeled free chains progressively replace the labeled ones at the surface. In all the cases considered, the number of labeled chains, $Q'(t)$, is found to decay exponentially [11], with a decay rate proportional

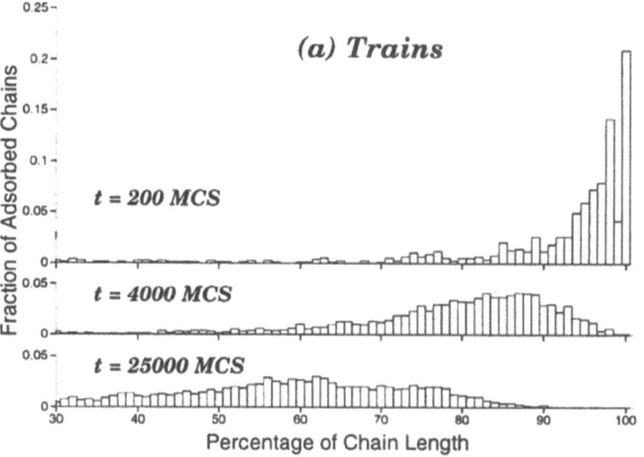

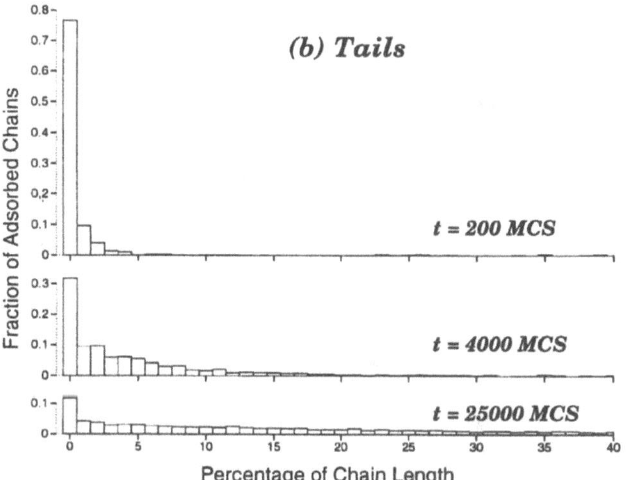

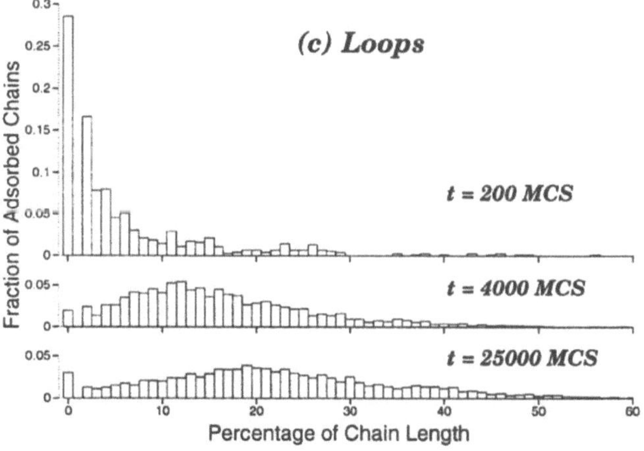

Fig. 1 Normalized distribution of monomers of adsorbed chains into (a) trains; (b) tails; (c) loops. This distribution does not show how many *actual* trains, tails and loops are formed, nor how long they are, but only what percentage of the chains they constitute. Note that the total number of chains represented in each distribution increases with time

(a)

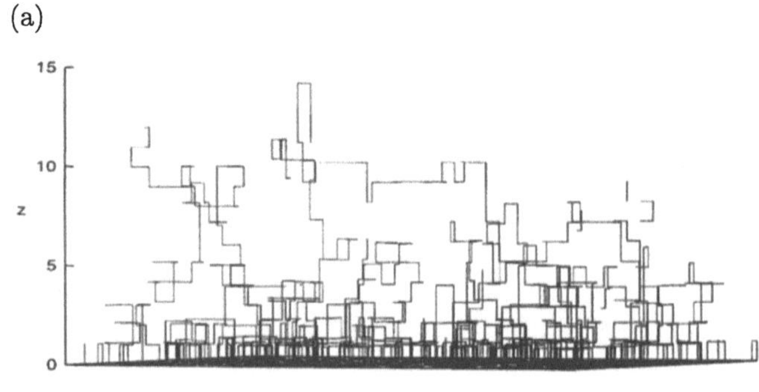

(b)

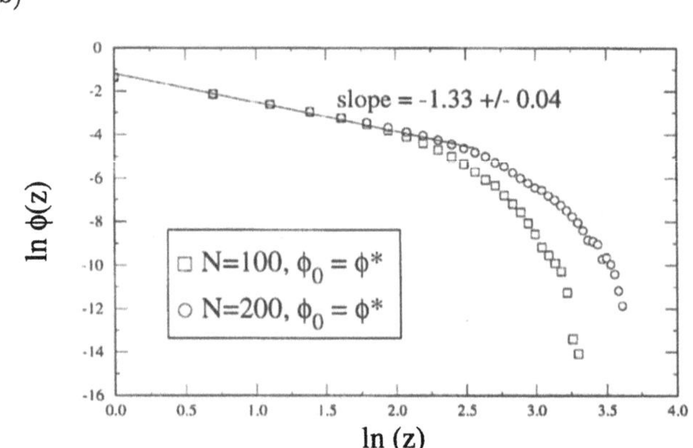

Fig. 2 (a) Illustration of a 45^2 area of the adsorbed layer at equilibrium. For this system ($N = 200$, $\phi_0 = \phi^*$) the layer corresponds to the de Gennes self-similar grid. Only adsorbed chains are shown. Note that the area comprises only 20% of the actual adsorbing surface used. (b) Logarithm of the corresponding density profile $\phi(z)$ of monomers of adsorbed chains for homopolymer layers in equilibrium with solutions at $\phi_0 = \phi^*$. Note that the simulation results for $N = 100$ and 200 coincide over some range, with a slope of $-4/3$

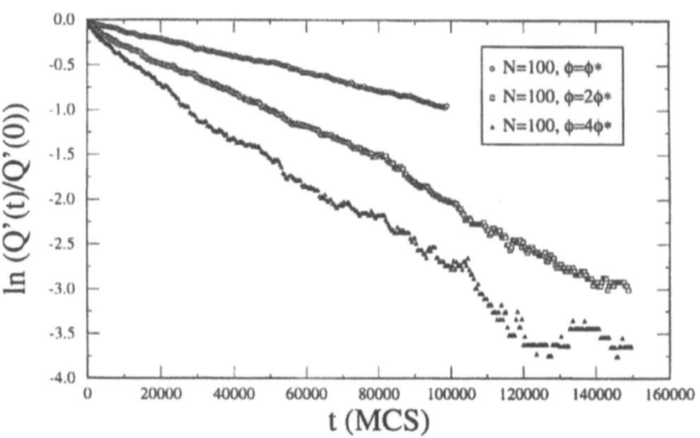

Fig. 3 Exponential decay of the number $Q'(t)$ of labeled adsorbed chains for systems of $N = 100$ at various solution concentrations. In all the cases considered the decay rate is proportional to the bulk concentration

to the bulk concentration (see Fig. 3). This is consistent with the picture of chain detachment occurring only with concomitant attachment of new chains from solution, as predicted by deGennes [10], and Semenov and Joanny [12]. This explains the slow desorption of a polymer layer seen when it is placed in contact with a pure solvent only.

Our simulational method allows us to probe this exchange in further detail. Specifically, we have found that in an equilibrium layer, the recently adsorbed chains have a distribution of surface residence times which decays sharply, while the chains adsorbed earlier on have a distribution with a broad peak at very long survival times. This disparity in chain behavior is further demonstrated by simulations in which the equilibrium layer is placed in contact with an equivalent solution of chemically different chains, in order to actively displace it. The distribution of surface residence times for the chains thus displaced is found to have two peaks, corresponding to contributions

from a short-lived, weakly bound population of chains, and from a long-lived, tightly bound one.

Effect of impurities on layer structure

When impurities are introduced into the surface, the initial growth of the layer is still essentially diffusive, except in the case of repulsive chain impurities, where the growth is slowed down almost immediately. This is because the "reeling in" of such chains is blocked by the impurities, nullifying the ability of early chains to take advantage of the initially bare surface. Because of this, the evolution of the conformational distributions is reversed in these systems, progressing from small-chain percentages in trains to gradually larger ones, as the chains relax over long times. Some chains, however, get kinetically trapped in this process, leading to a long-lived sub-population of chains with only about 20% of their length in trains. Nevertheless, the chains are moderately effective in adopting conformations which avoid unfavorable contacts, both for chain and surface impurities. As a result, the adsorbed layer is found to form a self-similar grid which is only slightly affected by the impurities, even in cases where they constitute 50% of the surface sites. For cases with *surface impurities*, the density profiles quickly regain the characteristic 4/3 profile away from the surface, except in the immediate proximal region. In the case of *chain impurities*, the same power-law decay is seen, except that the profile appears to be shifted outward by a small distance. Such a result has been predicted by Marques and Joanny [13] from their study of random copolymers. For adsorption from a semi-dilute solution, they conclude that the blob arrangement within the layer should resemble an Alexander-deGennes stretched-brush at small distances, and a self-similar grid

295

further out. Our results show good agreement with this prediction.

Effects of impurities on chain exchange

For pure systems, we compared the distributions of surface residence times for chains which adsorbed early to the surface, with that of chains which adsorbed late. A significant discrepancy was found: the distributions for early-adsorbing chains exhibited a peak at long times, whereas the distribution for the late-adsorbing chains showed only a peak at short times, followed by a monotonic decay. Thus, the late-adsorbing chains typically did not stay long at the surface, in comparison to the early chains. For impurity systems, however, this discrepancy is decreased when inactive impurity monomers are introduced into the chains; it is removed entirely when these impurities are made to be repulsive. Thus, the chain impurities effect on the "reeling in" process eliminates the segregation of adsorbed chains into tightly and loosely bound populations, based on early or late adsorption.

Conclusions

We have carried out simulations of polymer adsorption at a solid/liquid interface. Our results show that the simple model which we have implemented captures much of the essential physics of this many-body problem. It also allowed us to study details of chain conformation and dynamics in unprecedented detail. In particular, it has directly verified the formation of a self-similar grid layer, and the existence of different chain populations within the layer. These different groups have been shown to exhibit different properties with regard to exchange with the solution, and their sensitivity to impurities within the systems. The simulation also revealed that even large amounts of impurities do not seem to drastically perturb the layer's self-similarity.

Acknowledgements This work was supported by the National Science Foundation under Grant No. OSR-9255223 (NSF-EP-SCoR). Portions of this work were carried out on computational equipment purchased with the support of the National Science Foundation under Grant No. DMR-9413513.

References

1. Fleer G, Cohen-Stuart M, Scheutjens J, Cosgrove T, Vincent B (1993) Polymers at Interfaces. Chapman & Hall, London
2. Scheutjens JMHM, Fleer GJ (1979) J Phys Chem 83:1619–1635
3. Scheutjens JMHM, Fleer GJ (1980) J Phys Chem 84:178–190
4. Semenov AN, Joanny JF (1995) Europhys Lett 29:279–284
5. Semenov AN, Bonet-Avalos J, Johner A, Joanny JF (1996) Macromol 29:2179–2196
6. Granick S (1992) In: Sanchez I (ed) Physics of Polymer Surfaces and Interfaces, pp 227–244 Butterworth—Heinemann Manning Boston.
7. Zajac R, Chakrabarti, A (1996) J Chem Phys 104:2418–2437
8. Zajac R, Chakrabarti, A (1997) J Chem Phys 107:8637–8653
9. Kremer K, Binder K (1988) Comput Phys Rep 7:259–310
10. de Gennes PG (1987) Adv Colloid and Interface Sci 27:189–209
11. Frantz P, Granick S (1991) Phys Rev Lett 66:899–902
12. Semenov AN, Joanny JF (1995) J Phys II France 5:859–876
13. Marques CM, Joanny JF (1990) Macromol 23:268–276

Progr Colloid Polym Sci (1998) 110:296–299
© Steinkopff Verlag 1998

How much can you learn about thin adsorbed layers with optical techniques?

E.K. Mann
L. Heinrich
J.C. Voegel
P. Schaaf

E.K. Mann $^{\$}$ (✉)
Department of Physics
Kent State University
P.O. Box 5190
Kent, OH 44242-0001
USA
E-mail: mann@physics.kent.edu

$^{\$}$Federal Institute of Technology (ETH)
Zürich, Switzerland

L. Heinrich · P. Schaaf$^{\#}$
Institut Charles Sadron
Strasbourg
France

$^{\#}$Ecole Européenne de Chimie
Polymères et Matériaux
Strasbourg
France

J.C. Voegel
Institut National de la Santé et de la
Recherche Médicale
Unité 424
Strasbourg
France

Abstract The optical response of a dielectric interfacial film can be completely described, to second order in the film thickness, by three parameters called optical invariants. These allow the extraction of a maximum of information about the film while avoiding arbitrary models of that film. The information which can be obtained from experimental data on an isotropic film is, in order of decreasing precision: (i) the surface concentration of the adsorbed material, (ii) the average thickness of the layer through the first moment of the mass distribution in the layer, and (iii) a more complicated moment that gives an indication of the uniformity of the layer.

This method of analysis is applied to scanning angle reflectometry data on two kinds of films at a water/silica interface: (i) Protein films, and in particular, antibody/antigen complexes forming at the interface. This provides a model of a film that is essentially homogenous in the direction parallel to the surface, but which may have concentration gradients perpendicular to the surface. (ii) Films of polystyrene latex particles. Here the information gained is the size and number of particles adsorbed at the interface, with an indication of the uniformity of the particle distribution.

Key words Reflectometry – optical models – adsorbed layers – particle films

Optical methods for the analysis of adsorbed layers have many advantages: they can be used *in situ* during the adsorption process, they can be sufficiently rapid to follow adsorption kinetics, and they are readily available in a laboratory setting. Other techniques such as neutron reflectivity, with wavelength of the same order as the characteristic thickness of interfacial region, may be more sensitive to interfacial structure. However, using various tricks, such as working with polarized light near the Brewster angle, light can also be very sensitive to the material at the interface. In what follows, we will consider only the most classic of optical measurements, ellipsometry and reflectometry [1]. Further, we will confine ourselves to the case of dielectric layers, which lack the distinctive features in the frequency dependence of the dielectric constants which can make spectroscopic ellipsometry so rich in information.

The problem is to quantitatively interpret the optical signals to extract a maximum of reliable information about this interfacial region and the adsorbed layer. One would like to know the total quantity of adsorbed material, an average thickness of the layer, and any other information about the layer structure that one can get. Typically, one begins at the other end, by assuming a certain structure of the interfacial region, for example, a stratified layer with a particular optical density profile (most often a single uniform layer), and calculating the optical response of such a layer as a function of the parameters of the profile. This is compared to the experimental optical response, obtaining through a fitting procedure the profile parameters for which agreement is best.

The problem with this procedure is that typically one knows very little about the actual optical density profile: this is part of the information that one would like to gain. Furthermore, many different profiles give exactly the same optical response, within experimental accuracy. Using one particular profile to determine layer characteristics is equivalent to finding "optical averages" for these characteristics, where these averages are defined only indirectly through the particular profile used. For a true understanding of these values, one should reanalyze the data with a set of all physically plausible profiles.

Preferable would be a model-independent method of data analysis, yielding "optical averages" that are well defined, as integrals over mass distributions, for example. One would like to obtain a maximum of information from the data; one would also like to define the limits of such data, to avoid trying to extract information that is simply not there. Furthermore, it is desirable to have a method that is easy to use, for the purposes of data fitting.

Such a method exists. It consists of determining the *optical invariants* that describe the data. Two independent groups [2, 3] realized that the optical properties of any non-adsorbing thin layer could be described with a maximum of three independent film parameters, to second order in L/λ, the ratio between typical film thicknesses and the wavelength of light. The optical properties of the interfacial region can be described using a close analogy to the Gibb's interface used in thermodynamics [3]: The true system, with an interfacial region between two bulk phases, can be replaced conceptually with two bulk regions separated by an infinitely sharp interface. Any differences between the optical response of the idealized system and that of the real system can be corrected for by placing *surface excess* polarizations at the sharp boundary.

These surface excess polarizations have been calculated for a wide variety of systems [3, 4]. However, the surface excesses depend on the particular choice made for the position of the idealized interface within the true interfacial region. All observables, such as the optical response, must be independent of this choice. Only invariant combinations of the surface excess polarizations can enter into this response: these are the *optical invariants* referred to above. To second order in L/λ, the ellipticity coefficients derived from the ratio of the reflectivity coefficients for light polarized in and perpendicular to (p and s waves, respectively) the incident plane is a linear combination of one invariant (J_1) of first order in L/λ, and one second-order invariant (J_{23}) while the reflectivity is a linear combination of the square of J_1 and of two second-order invariants (J_{22} and J_{23}) [5].

These invariants can be readily extracted from experimental data [5], using standard LLSQ fitting schemes. The remaining difficulty is to interpret these invariants in physical terms that can be used in other contexts. In order to do this, it is necessary to resort to a physical model: an isotropic stratified layer may behave differently from a layer of spheres for example. However, the stratified layer model is a reasonable description of a wide variety of physical systems, as long as any inhomogeneity within each layer is on a sufficiently small scale.

The advantage of this method of analysis is that within the stratified layer model, it is not necessary to assume any particular dielectric constant profile. The major assumption is a relation between the optical density and the mass density. For dielectric materials, in which all dielectric constants are near 1, a linear relation is usually sufficient: for example, for proteins, the change in the dielectric constant ε of a solution with protein concentration c has been shown to follow the linear relation $d\varepsilon/dc \sim 0.5$ cm^3/g up to volume fractions of at least 0.4 [6]. With this assumption, and the further assumption that dielectric constants within the layer are not too different from that of water, the first-order invariant gives directly [5] the 0th-order moment of the mass distribution, where the mth moment is defined simply as $\mathcal{M}_n = \int dz\, z^n c(z)$: the zeroth moment is otherwise known as the total surface concentration. Similarly, the two second-order moments are, to lowest order in the mass density, proportional to $\mathcal{M}_1$, where the ratio $2\mathcal{M}_1/\mathcal{M}_0$ gives a well-defined average layer thickness.

Yet more information can be extracted from the three invariants, if they are known with sufficient accuracy and precision, by combining them to give a different invariant that is independent of both the average thickness and the total mass within the layer. This gives a parameter that is much more sensitive to other properties within the layer: its uniformity for example. We have recently introduced such a parameter [7], choosing it to give a null value if the film follows the model most commonly used for optical analysis, the uniform thin film. This parameter is thus a direct test of the uniformity of the film. It is related to a somewhat unusual moment of the mass within the layer:

$$F = 1 - \frac{\int\limits_0^\infty dz\, z\, c^2(z) + \int\limits_0^\infty dz_1 \int\limits_0^{z_1} dz_2\, c(z_1)\, c(z_2)}{\mathcal{M}_0^2} (1 + O(c)).$$

Clearly, the information on the profile available from a single parameter of this sort is limited, but it is the most that can be expected from reflectivity measurements of thin dielectric films: in itself, knowing the limits of a technique is useful information. However, even this information can be useful for setting limits on the film structure, consistent with the optical measurements. The parameter F yields a test for any two-parameter model, such as that of a uniform isotropic film or a collection of free spheres. Furthermore, in an isotropic stratified film, it cannot be negative unless the maximum density is at some distance from the surface, implying a depletion layer near the surface.

Such a negative value for uniformity parameter, $F = -0.3 \pm 0.05$, has been observed in the case of a multilayer system, built up by the successive adsorption of a protein and a polyclonal mixture of its antibody on a hydrophilic silica surface [5]. Preliminary AFM measurements [8] showed a smooth dense layer at the end of deposition, justifying the utilization of a stratified film layer to analyze the optical response. The optical technique used, scanning angle reflectometry [5], consists of measuring the reflectivity of p waves at a series of angles around the Brewster angle, where the reflectivity of light with that polarization would be zero in the absence of any interfacial structure.

A naive view, such as our own at the beginning of these studies, would expect that the IgG antibody would attach to the layer of antigens (also IgGs in this case) already on the surface, to form a double layer. In fact, the invariant method of analysis, in agreement with analyses assuming either a single or a double uniform layer, found that the overall thickness was equivalent to three layers, with as many as three or four antibodies to each antigen [5]. *A priori* this could indicate aggregation within a thick antibody layer, except that the antibody alone on the surface formed a single layer: any aggregation is due to the presence of the antigen.

This then is the information with which we can build up a model of the process, remembering that the F parameter is consistent with mass distributions as shown in Fig. 1a, but inconsistent with distributions like those shown in Fig. 1b. Furthermore we know that while a protein, once adsorbed, tends not to leave the surface on its own, it can be replaced by another protein in what is called exchange; this is particularly true on a substrate to which the protein is less tightly bound, as in hydrophilic silica. This exchange has been observed to take place on a time scale of hours for the system IgG/IgG on silica [9]. When, as here, one of the IgG molecules is the antibody of the other, and there is specific binding of the two, the replacement can still occur, but the antigen remains bound to the surface through the antibody. The antigen would then be free to bind to other (polyclonal) antibodies in the solution. Once this occurred, further rotation of the complex is liable to be sterically hindered at the surface. In this way, a roughly trilayer model can be built up as shown in Fig. 1c. Note that in this model, if there are as many as three or four antibodies to each antigen, it is the third,

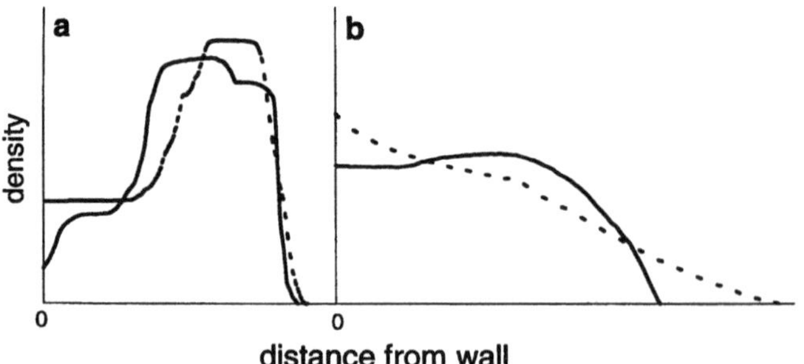

Fig. 1 (a) Density distributions giving $F = -0.3$. (b) Density distributions giving $F > 0$ (c) Structure of an antigen–antibody film, consistent with the measured thickness, density, and F values. Black molecules represent IgG antigens. White molecules represent IgG antibodies, where the dark regions indicate the areas that bind specifically with the antigens

outer layer that is likely to contain the most molecules, which is consistent with the negative value for F observed.

Use of the invariant method of analysis can thus give a maximum of information on the system, including, surprisingly, information about the density profile within the adsorbed layer. If used with a careful consideration of any systematic errors that can arise in the data, it can also contribute to confidence in the experimental results, avoiding such vague terms as "optical averages", and allowing models such as the above to be advanced.

However, so far we have considered only the case in which the film can be considered as stratified, uniform within each infinitesimal layer parallel to the surface. In many systems this is not the case. One extreme case, of spherical colloidal particles, again adsorbing at an interface, can be realized experimentally with polystyrene latex particles. Such experiments were performed with the scanning angle reflectometer, using, for example, latex particles with positively charged amidine surface groups adsorbing on the negatively charged silica surface [10]. This allows us to ask whether the invariant method as described above for the case of stratified layers can also be used for particulate ones.

In fact the invariant method, or for that matter the uniform or stratified layer model, gives reliable results for both the particle diameter and for the number of the particles on the surface, as long as the sphere diameter is less than about a tenth of the wavelength of the reflected light [11]: if we want no other information, it is not necessary to take into account the island nature of the films in this size range. However, since it was possible to get more information out of the stratified film case, using the F uniformity parameter, it should be possible here as well. It is, but only by taking into account the particular, particulate nature of the system, just as we used the stratified film model to interpret the F parameter in that case.

In the case of disordered particle layers, the natural parameters are the particle radius, number density, and distribution, as given by the radial distribution function. Indeed, Haarmans et al. [3] give relations for the polarization densities of a layer of spherical particles in terms of these parameters. These expressions can be used to derive expressions for the F parameter [12]. The results for two extremes are given in Fig. 2: one in which all particles remain so far from each other that multiple scattering between them can be ignored and one in which the distribution follows a simple step function, with a hard-core exclusion zone. Experimental results for particles with a diameter of 116 nm adsorbing from a dispersion with pure water are also given. The data are seen to fall close to the non-interacting case. Indeed, direct measurements [13] of the radial distribution function under similar, no-salt conditions reveal an exclusion zone more than twice that of the hard core, due to electrostatic interactions between particles which keep them at a distance one from another during the adsorption process.

Using these two examples, we see that a considerable amount of information about an isotropic adsorbed layer

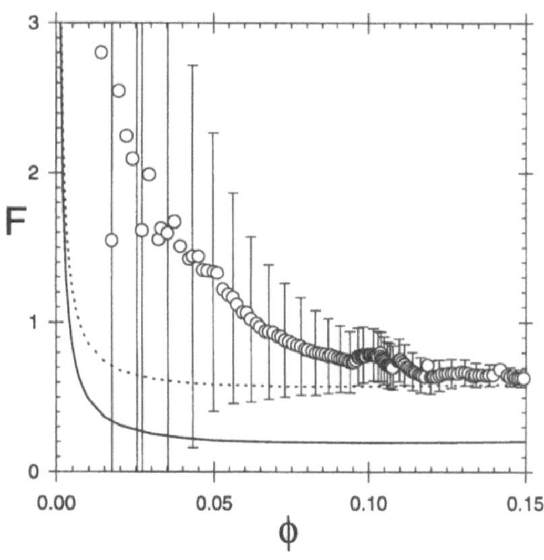

Fig. 2 The values of the uniformity ratio F deduced from scanning angle reflectometry data taken during the adsorption of positively charged polystyrene latex particles on silica. Experimental details are given in Ref. [10]. Circles refer to experimental data. Solid lines refer to values expected for a uniform distribution of hard spheres. Dashed line corresponds to ignoring multiple scattering between spheres

is available from optical data: using the method of optical invariants, both the quantity of adsorbed material and an average layer thickness can be deduced. Furthermore, with a minimum of other information (is the layer stratified, or does it consist of spheres), additional information about the layer structure can be deduced.

References

1. Azzam RMA, Bashara NM (1989) Ellipsometry and Polarized Light. North-Holland, Amsterdam, 1989
2. Lekner J (1987) Theory of Reflection. Martinus Nijhoff Publishers, Dordrecht
3. Bedeaux, D, Vlieger J (1973) Physica 67: 55–73
4. Blokhuis EM, Bedeaux, D (1990) Physica A 164: 515–548; Haarmans MT, Bedeaux D (1995) Thin Solid Films 258: 213–223
5. Heinrich L, Mann EK, Voegel JC, Schaaf P (1997) Langmuir 13:3177–3186
6. De Feijter JA, Benjamins J, Veer FA (1978) Biopolymers 17:1759–1768
7. Mann EK, Heinrich L, Voegel JC, Schaaf P (1996) J Chem Phys 105: 6082–6085
8. Heinrich L (1977) Doctoral Thesis, Université Louis Pasteur
9. Huetz Ph, Ball V, Voegel J-C, Schaaf, P (1995) Langmuir 11:3145–3152
10. Mann EK, van der Zeeuw EA, Koper GJM, Schaaf P, Bedeaux D (1995) J Phys Chem 99:790–797; Mann EK, Bollander A, Heinrich L, Koper GJM, Schaaf P (1996) J Opt Soc Am A 13: 1048–1056
11. Mann EK, Heinrich L, Schaaf P (1997) Langmuir 13:4906–4909
12. Mann EK, Heinrich L, Semmler M, Voegel J-C, Schaaf P (1998) J Chem Phys 108:7416–7425
13. Johnson CA, Lehnoff, A.M J Colloid Interface Sci (1996) 179:587–599

Progr Colloid Polym Sci (1998) 110:300–304
© Steinkopff Verlag 1998

M. Sikirić
S. Sarig
H. Füredi-Milhofer

The interaction of small and macromolecules with growing calcium hydrogenphosphate dihydrate crystals

M. Sikirić (✉)
Department of Animal Nutrition
Faculty of Agronomy
University of Zagreb
Svetošimunska 25
10000 Zagreb
Croatia

S. Sarig · H. Füredi-Milhofer
Casali Institute for Applied Chemistry
Graduate School of Applied Science
The Hebrew University of Jerusalem
Jerusalem
Israel

Abstract The influence of glutamate, aspartate, citrate and polyaspartate (MW = 5–15 kDa) on the growth morphology of calcium hydrogenphosphate dihydrate ($CaHPO_4 \cdot 2H_2O$, DCPD) was studied.

DCPD crystals were prepared under controlled conditions by fast mixing of the anionic and cationic reactant solutions and subsequent growth without further stirring in the course of 24 h at 37 °C. The initial conditions were: $c(CaCl_2) = c(Na_2HPO_4) = 0.021\ mol\,dm^{-3}$, $c(NaCl) = 0.3\ mol\,dm^{-3}$, pH 5.5. The respective additive was added to the ionic component prior to pH adjustment. The affected crystal faces were identified by light and scanning electron microscopy.

In the control system mostly large (approx. 200 μm × 100 μm) thin platelets with prominent (010) faces were obtained. Glutamate and aspartate had no significant effect, but both citrate and polyaspartic acid significantly inhibited DCPD crystallization and changed the morphology of the crystals by specifically affecting different crystal faces. Crystals grown in the presence of citrate ions appeared rod-like. When grown in the presence of polyaspartic acid crystals appeared much larger than in the control systems but the basic orientation was unchanged, indicating that polyaspartic acid adsorbs on (010) face.

Key words Calcium hydrogenphosphate dihydrate – aspartic acid – glutamic acid – citrate – polyaspartic acid – crystal morphology

Introduction

Interactions between inorganic crystals and organic molecules underlie crystallization processes in various fields, e.g. geology, biological and pathological mineralization, technology, scale formation in heating systems, etc. [1]. In biological mineralization organic macromolecules and/or molecular assemblies are utilized to control the size, shape and orientation of crystals [2]. The objective of our research was to understand the principles of these interactions because in many applications size and shape of the crystals are of utmost importance.

One way to study the influence of an additive is to study the morphology of crystals grown both in the presence and absence of an additive. In this way effects due to specific adsorption on certain crystal faces can be determined [3]. The habit of growing crystal is determined by the relative growth rates of its faces. The faster the growth rate in a direction perpendicular to a certain face, the smaller that face appears. If an effective growth inhibitor adsorbs on certain crystal faces, but not on the others, the affected faces will appear larger than in nonaffected crystal, resulting in change of crystal habit [3]. Thus, changes in crystal morphology can be used to deduce the mechanisms of crystal – additive interactions.

Calcium hydrogenphosphate dihydrate (DCPD, $CaHPO_4 \cdot 2H_2O$) is a frequent kidney stone constituent [4, 5]. It is used as additive to cattle feed and for different applications in cosmetics and pharmaceutical industries.

Glutamic and aspartic acid are the most abundant amino acids in the mucoprotein-like materials, which constitute the organic part of calcium stones [6]. Citrate is a urine constituent and it is used in the prevention of kidney stone formation [7]. Polyaspartic acid has partial β-sheet conformation and therefore presents an appropriate model molecule to study correlation between the crystal lattice and an acidic polymer of defined structure.

In this preliminary paper we report on the influence of the above-mentioned small and macromolecular inhibitors, i.e. glutamate, aspartate, citrate and polyaspartate on the growth morphology of DCPD crystals.

Materials and methods

Analytical grade chemicals and deionized water were used. Stock solutions of calcium chloride and sodium phosphate were prepared from $CaCl_2 \cdot 2H_2O$ and Na_2HPO_4, which were dried over night in a dessicator over silica gel. To the phosphate stock solution 0.01% sodium azide was added to avoid bacterial contamination. Anionic and cationic reactant solutions used for crystal preparation were diluted from the respective stock solutions. Commercially available sodium citrate (BDH Laboratory Supplies), aspartic acid, glutamic acid and polyaspartic acid (MW 5-15 kDa, Sigma) were used as additives.

Fig. 1 Scanning electron micrograph of DCPD crystals obtained (a) in control system: $c(CaCl_2) = c(Na_2HPO_4) = 0.021 \ mol \ dm^{-3}$ and $c(NaCl) = 0.3 \ mol \ dm^{-3}$, pH 5.5, 37 °C (without additives), (b) in the presence of citrate, $c_{cit} = 5 \times 10^{-5} \ mol \ dm^{-3}$, (c) in the presence of polyaspartic acid, $c_{pAsp} = 2 \times 10^5 \ mol \ AA \ dm^{-3}$

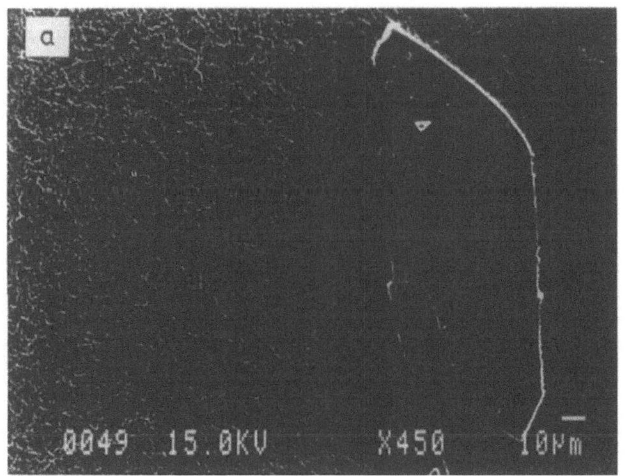

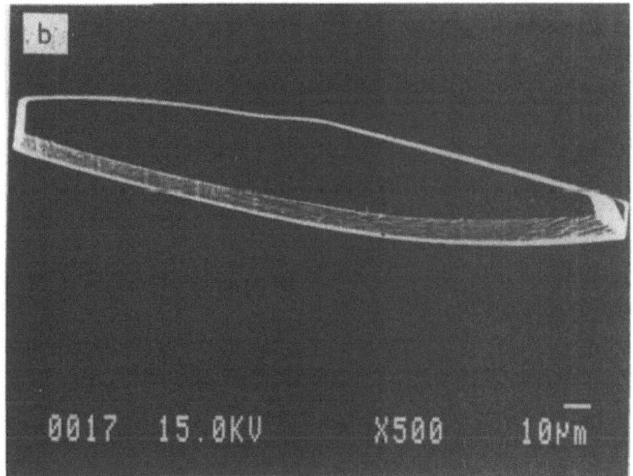

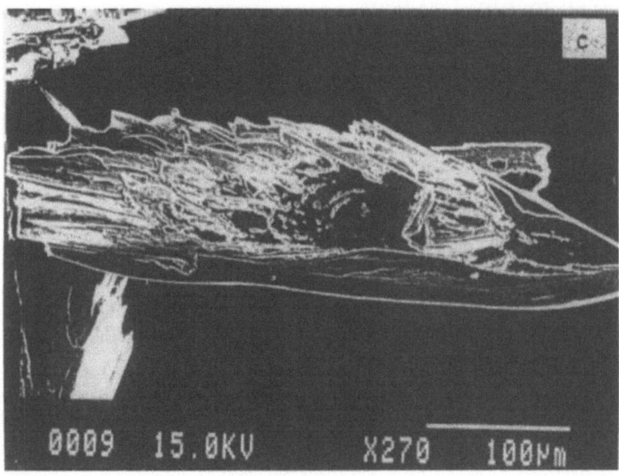

Cyrstallization experiments

Crystallization systems were prepared under controlled conditions by fast mixing of equal volumes of the anionic and cationic reactant solutions. Crystals were grown without further stirring at 37 °C. The initial conditions were: pH 5.5, $c(CaCl_2) = c(Na_2HPO_4) = 0.021$ mol dm^{-3} and $c(NaCl) = 0.3$ mol dm^{-3}. The respective additive was added to the anionic component prior to pH adjustment. In the control system calcium and phosphate concentrations were the same but no additive was added. After commencement of the reaction, changes in pH were monitored thus providing a qualitative estimate of the rate of crystallization [8]. In addition, samples were taken at given time intervals for observation by light and polarized microscopy. The time that elapsed before the detection of the first crystals was between 1 and 24 h depending on the nature and concentration of the additive. When the crystals were grown, the reaction was discontinued and the affected crystal faces were identified by scanning electron microscopy (JEOL JXA-8600 "Superprobe"). Samples for SEM were prepared by placing a drop of the suspension on a stub covered with carbon glue and removing the supernatant with filter paper. Crystals remaining on the stub were washed by placing a drop of deionized water and removing excess of water with filter paper. After drying samples in a dessicator over silica gel for several hours, the stubs were gold plated and used for observation.

Results and disscusion

The growth morphology of crystals that appeared in the control system was typical for DCPD which has been reported to grow in the form of thin platelets with prominent (010) and lateral (h01) faces [9, 10]. After 1 h of reaction time relatively large platelet-like crystals (approx. 200 × 100 μm and 5 μm thick) such as shown in Fig. 1(a) were obtained. Glutamic and aspartic acid showed no significant effect on the rate of growth and growth morphology of DCPD. In the investigated concentration range of 3×10^{-5}–5×10^{-4} mol dm^{-3}, crystals like those obtained in control system appeared after 1 h.

The influence of citrate ions was investigated in the concentration range 1×10^{-5}–5×10^{-4} mol dm^{-3}. Up to 1×10^{-4} mol dm^{-3} citrate ions affected the morphology but not the rate of growth of DCPD crystals, i.e. the pH vs. time curves were similar as in the control systems but crystals that appeared after 1 h were typically rod-like (Fig. 1(b)) instead of platelet-like as in the controls (Fig. 1(a)). When the concentration exceeded 1×10^{-4} mol dm^{-3}, both the morphology and the rate of

crystal growth were affected, i.e. crystallization could be observed only 24 h after sample preparation and the crystals were rod-like as at lower citrate concentrations. The observed change in crystal morphology is in accordance with previously reported observations [11].

The effect of polyaspartate was investigated in the concentration range of 0.2–20 × 10^{-6} mol AA dm^{-3} (calculated assuming an average MW of 10 kDa). At concentrations $5 < c < 7.5 \times 10^{-6}$ mol AA dm^{-3} very large crystals with the same basic orientation as in control systems were obtained within 24 h (Fig. 1(c)). Note that the prominent (010) face has a leaf-like appearance. At concentrations above 1.5×10^{-5} mol AA dm^{-3}, crystallization of DCPD was completely inhibited, i.e. even after several days no crystallization was apparent.

The above results can be explained by considering the ionic structure of the affected crystal faces and the molecular structure of the additives.

DCPD crystallizes in monoclinic system ($a = 5.812$ Å, $b = 15.810$ Å, $c = 6.239$ Å, $\beta = 116°$), space group Ia [12].

Fig. 2 Distribution curves of (a) glutamic acid and (b) aspartic acid species in the dependence of pH

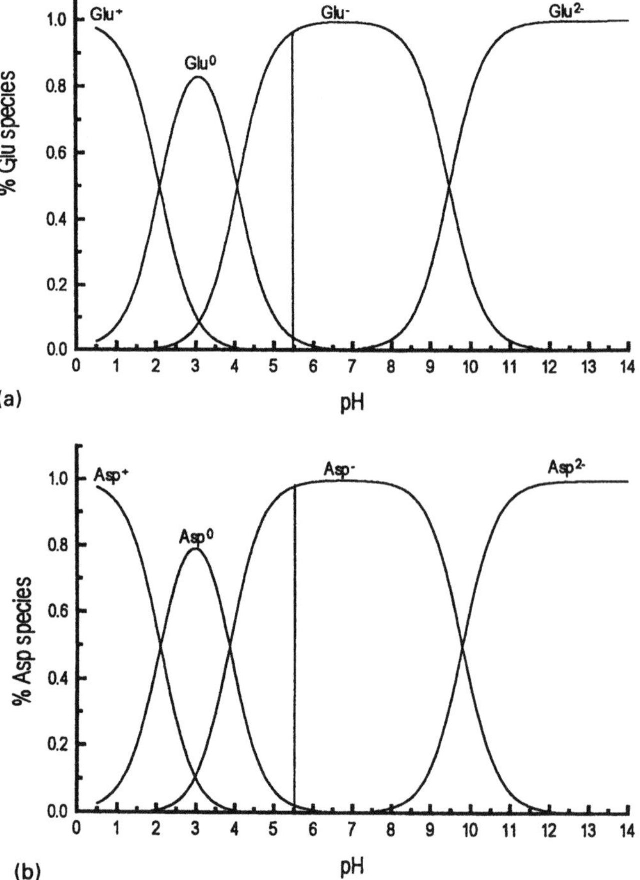

Progr Colloid Polym Sci (1998) 110:300–304
© Steinkopff Verlag 1998

The unit cell consists of alternating bilayers oriented parallel to the (010) plane, one of them consisting of calcium and hydrogenphosphate ions and the other one of water molecules. Typically, the growth morphology of DCPD crystals reflects the crystal structure so that (010) is the prominent crystal face [10,11]. It is a fair assumption that in aqueous solutions, bilayers of water molecules are for most of the real time exposed at the surface of the (010) face, while lateral faces have a mixed ionic character with intercalated water molecules [10]. Because of this kind of ionic structure DCPD represents a good model crystal for studying the several aspects of interactions between additives and crystals. It should be possible to assess:

(i) the importance of molecular size and structure of the additive, i.e. small or macromolecules, number of functional groups in the molecule and overall charge;

(ii) the importance of a structural fit between the organic molecule and the ionic structure of a particular crystal face;

(iii) the influence of the hydration layer exposed on the surface of the crystals.

The above described results can provide relevant information:

1. Glutamate and aspartate have virtually no effect under our experimental conditions. The distribution curves in Fig. 2 show that at the investigated pH both molecules behave as monocarboxylic acid, i.e. they exist as Glu^-

and Asp^- species. Clearly, one negative charge of these species is not sufficient for significant interaction with any of DCPD crystal faces. It has indeed been shown previously that even dicarboxylic acids have no significant effect on DCPD crystallization unless additional polar groups such as $-OH$ groups are present in the molecule [11].

2. Citrate has a profound effect on crystal morphology (Fig. 1(b)). Under the given experimental conditions the molecule exists as HL^{2-} (according to the distribution curves calculated for 25 °C in Ref. [11]); adding to the polarity of the molecule is the hydroxyl group. Assuming that the interaction with DCPD crystals is electrostatic the molecules would primarily recognize calcium ions exposed on the lateral crystal faces and therefore retard crystal growth in the direction perpendicular to these faces which explains the appearance of rod-like crystals. It seems that these electrostatic interactions are not strong enough to induce citrate to penetrate the hydration bilayer which is exposed on the (010) crystal face. For effective interaction on this crystal face some kind of specific recognition is necessary, which in this set of experiments is achieved only with

3. Polyaspartic acid: It has been shown on various systems that for effective interaction between specific macromolecular inhibitors and crystals of sparingly soluble salts a structural fit between the periodical spacings of the polar groups and the interionic distances on at least one of the crystal faces must exist [1, 3, 13]. In the case

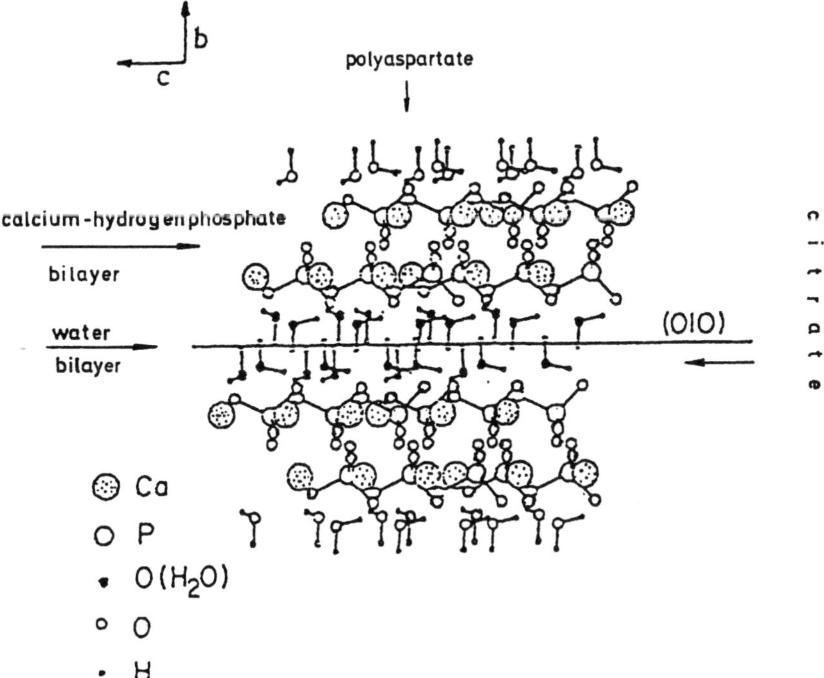

Fig. 3 Schematic representation of interaction between calcium hydrogenphosphate dihydrate crystals and citrate ions and polyaspartic acid (DCPD ionic structure after Ref. [10]).

of polyaspartic acid and DCPD such structural fit exists between distances of carboxylic groups in the polyaspartic β-sheet and distances of neighboring calcium ions from two adjacent layers which constitute one Ca–HPO$_4$ bilayer lying beneath hydrated layer parallel to the (010) plane. It thus seems that molecules of polyaspartic acid are intercalated which could explain the leaf-like appearance of the affected crystal plane (see Fig. 1(c)).

The above considerations are summarized in Fig. 3 which illustrates the mode of interaction of citrate and polyaspartate with DCPD crystals.

Conclusions

The above described studies show that

(i) For effective interaction of small organic molecules with DCPD crystals high charge density is required. Molecules with only one overall negative charge (aspartate and glutamate at pH < 5.5) do not inhibit crystal growth or influence the crystal growth morphology. Small molecules with several negatively charged groups, such as citrate ions do interact with the lateral faces of DCPD crystals slowing down crystallization and inducing changes in crystal morphology. The interactions are primarily electrostatic.

(ii) Highly charged macromolecules with partial β-sheet conformation such as polyaspartic acid specifically interact with the prominent (010) crystal plane although for most of the real time this plane is covered with a bilayer of structural water. The reasons for this specific interaction are high charge density and structural and stereochemical compatibility between polyaspartate carboxylates and lattice calcium ions exposed on the (010) crystal plane.

References

1. Füredi-Milhofer H, Sarig S (1996) Prog. Crystal Growth Charact Mater 32:45–74
2. Addadi L, Weiner S (1992) Angew Chemie Int Ed Engl 31:153–169
3. Addadi L, Weiner S (1985) Proc Natl Acad Sci (USA) 82:4110–4114
4. Daudon M, Donsimoni R, Hennequin C, Fellahi S, Le Moel G, Paris M, Troupel S, Lacour B (1995) Urol Res 23:319–326
5. Werness PG, Bergert JH, Smith LH (1981) J Crystal Growth 53:166–172
6. Garcia-Ramos JV, Carmona P (1982) J Crystal Growth 57:336–342
7. Tomson CRV (1995) British J Urol 76:419–424
8. Despotović R, Filipović N, Füredi-Milhofer H (1975) Calc Tissue Res 18:13–26
9. Abbona F, Christensson F, Franchini-Angela M, Lundager Madsen HE (1993) J Crystal Growth 131:331–346
10. Haninen D, Geiger B, Addadi L (1993) Langmuir 9:1058–1065
11. Brečević Lj, Sendijarević A, Füredi-Milhofer H (1984) Colloids and Surfaces 11:55–63
12. Curry NA, Jones DW (1971) J Chem Soc A 3725–3729
13. Füredi-Milhofer H, Moradian-Oldak J, Weiner S, Veis A, Mintz KP, Addadi L (1994) Connective Tissue Res. 30:251–26

Progr Colloid Polym Sci (1998) 110:305–306
© Steinkopff Verlag 1998

Progr Colloid Polym Sci (1998) 110:307–308
© Steinkopff Verlag 1998

SUBJECT INDEX